BIPOLAR DISORDERS

Medical Psychiatry

1. Handbook of Depression and Anxiety: A Biological Approach, *edited by Johan A. den Boer and J. M. Ad Sitsen*
2. Anticonvulsants in Mood Disorders, *edited by Russell T. Joffe and Joseph R. Calabrese*
3. Serotonin in Antipsychotic Treatment: Mechanisms and Clinical Practice, *edited by John M. Kane, H.-J. Möller, and Frans Awouters*
4. Handbook of Functional Gastrointestinal Disorders, *edited by Kevin W. Olden*
5. Clinical Management of Anxiety, *edited by Johan A. den Boer*
6. Obsessive-Compulsive Disorders: Diagnosis • Etiology • Treatment, *edited by Eric Hollander and Dan J. Stein*
7. Bipolar Disorder: Biological Models and Their Clinical Application, *edited by L. Trevor Young and Russell T. Joffe*
8. Dual Diagnosis and Treatment: Substance Abuse and Comorbid Medical and Psychiatric Disorders, *edited by Henry R. Kranzler and Bruce J. Rounsaville*
9. Geriatric Psychopharmacology, *edited by J. Craig Nelson*
10. Panic Disorder and Its Treatment, *edited by Jerrold F. Rosenbaum and Mark H. Pollack*
11. Comorbidity in Affective Disorders, *edited by Mauricio Tohen*

12. Practical Management of the Side Effects of Psychotropic Drugs, *edited by Richard Balon*
13. Psychiatric Treatment of the Medically Ill, *edited by Robert G. Robinson and William R. Yates*
14. Medical Management of the Violent Patient: Clinical Assessment and Therapy, *edited by Kenneth Tardiff*
15. Bipolar Disorders: Basic Mechanisms and Therapeutic Implications, *edited by Jair C. Soares and Samuel Gershon*
16. Schizophrenia: A New Guide for Clinicians, *edited by John G. Csernansky*
17. Polypharmacy in Psychiatry, *edited by S. Nassir Ghaemi*
18. Pharmacotherapy for Child and Adolescent Psychiatric Disorders: Second Edition, Revised and Expanded, *David R. Rosenberg, Pablo A. Davanzo, and Samuel Gershon*
19. Brain Imaging In Affective Disorders, *edited by Jair C. Soares*
20. Handbook of Medical Psychiatry, *edited by Jair C. Soares and Samuel Gershon*
21. Handbook of Depression and Anxiety: A Biological Approach, Second Edition, *edited by Siegfried Kasper, Johan A. den Boer, and J. M. Ad Sitsen*
22. Aggression: Psychiatric Assessment and Treatment, *edited by Emil Coccaro*
23. Depression in Later Life: A Multidisciplinary Psychiatric Approach, *edited by James Ellison and Sumer Verma*
24. Autism Spectrum Disorders, *edited by Eric Hollander*
25. Handbook of Chronic Depression: Diagnosis and Therapeutic Management, *edited by Jonathan E. Alpert and Maurizio Fava*
26. Clinical Handbook of Eating Disorders: An Integrated Approach, *edited by Timothy D. Brewerton*
27. Dual Diagnosis and Psychiatric Treatment: Substance Abuse and Comorbid Disorders: Second Edition, *edited by Henry R. Kranzler and Joyce A. Tinsley*
28. Atypical Antipsychotics: From Bench to Bedside, *edited by John G. Csernansky and John Lauriello*
29. Social Anxiety Disorder, *edited by Borwin Bandelow and Dan J. Stein*
30. Handbook of Sexual Dysfunction, *edited by Richard Balon and R. Taylor Segraves*
31. Borderline Personality Disorder, *edited by Mary C. Zanarini*
32. Handbook of Bipolar Disorder: Diagnosis and Therapeutic Approaches, *edited by Siegfried Kasper and Robert M. A. Hirschfeld*
33. Obesity and Mental Disorders, *edited by Susan L. McElroy, David B. Allison, and George A. Bray*
34. Depression: Treatment Strategies and Management, *edited by Thomas L. Schwartz and Timothy J. Petersen*
35. Bipolar Disorders: Basic Mechanisms and Therapeutic Implications, Second Edition, *edited by Jair C. Soares and Allan H. Young*

BIPOLAR DISORDERS

Basic Mechanisms and Therapeutic Implications

Second Edition

Edited by

Jair C. Soares
University of North Carolina School of Medicine
Chapel Hill, North Carolina, USA

Allan H. Young
University of British Columbia
Vancouver, British Columbia, Canada

informa
healthcare

New York London

Informa Healthcare USA, Inc.
52 Vanderbilt Avenue
New York, NY 10017

Printed in the United States of America on acid-free paper
10 9 8 7 6 5 4 3 2 1

International Standard Book Number-10: 0-8493-9897-5 (Hardcover)
International Standard Book Number-13: 978-0-8493-9897-1 (Hardcover)

Library of Congress Cataloging-in-Publication Data

Bipolar disorders : basic mechanisms and therapeutic implications—2nd ed./edited by Jair C. Soares, Allan H. Young.
p. ; cm. -- (Medical psychiatry ; 35)
Includes bibliographical references and index.
ISBN-13: 978-0-8493-9897-1 (hardcover : alk. paper)
ISBN-10: 0-8493-9897-5 (hardcover : alk. paper)
1. Manic-depressive illness. I. Soares, Jair C. II. Young, A. H. (Allan H.) III. Series.
[DNLM: 1. Bipolar Disorder--diagnosis. 2. Bipolar Disorder--therapy. W1 ME421SM v.35 2007/ WM 207 B61649 2007]

RC516.B5275 2007
616.89′5--dc22 2006103138

Visit the Informa Web site at
www.informa.com

and the Informa Healthcare Web site at
www.informahealthcare.com

Preface

Since the first edition of our book was published, increasing progress has been seen in the understanding of the basic mechanisms involved in the pathophysiology of bipolar mood disorders. When our first edition was prepared, there was comparatively little research being conducted on the mechanisms involved in this disorder. Recent years have seen an increasing interest in this field and an increasing amount of activity. Important research initiatives have begun to elucidate the pathophysiology of this disorder. These research initiatives are beginning to lead to breakthroughs in the understanding of the causation of bipolar mood disorders and the development of novel treatments. Some of these new advances have recently translated into newer treatments available for these disorders.

Of particular importance is the development of newer tools from neuropsychopharmacology, which have provided new ways to study various brain systems, including post-receptor and transcriptional mechanisms. Developments in neuroimaging have made possible the in vivo study of brain anatomy, neurotransmission, and metabolic processes. Important tools from genetics are becoming available and are being applied to further the understanding of mechanisms involved in bipolar disorders. Cognitive neuropsychology has also provided improved tools for the more refined study of brain functions in these disorders. These novel research avenues have provided new dimensions in exploring the biological mechanisms involved. New therapeutic developments have already become available in the past few years. These advances are expected to gradually continue to translate into new approaches for the treatment of bipolar disorder over the next few to several years.

The updated findings from this research have not been comprehensively summarized in a book focused specifically on the biological underpinnings of bipolar mood disorders. There are some excellent books available on the subject of bipolar disorders, but their focus is primarily on diagnostic issues, course of illness, and treatment. To fill this gap, we are proud to present the second edition of our book, *Bipolar Disorders: Basic Mechanisms and Therapeutic Implications*. This volume presents outstanding manuscripts by the leaders in the particular areas of biological research pertinent to bipolar disorders. Among other very important topics, we have included chapters on genetics, neuroimaging, neuropsychology, investigations of post-receptor and transcriptional abnormalities, potential interactions between biology and psychosocial factors, childhood onset and late-life bipolar disorder, and several other important topics. A chapter on the implications of these research areas for ongoing therapeutic developments in this field is also included. The potential therapeutic implications of new research, as in the first edition, are emphasized throughout the book.

We are very happy to have had the collaboration of some of the leading scientists in their respective fields of research, and believe this volume will be a valuable resource for researchers in this field and in related areas. It is presented as a complete and accessible reference to the most updated information on the biological basis and therapeutics of bipolar mood disorders. It should be useful as supplemental reading for graduate and postgraduate courses on the neurobiology of mental illness. Mental health practitioners will find it extremely useful as an updated source with the most recent research progress in this field. We hope you will share our excitement with these new developments.

Jair C. Soares
Allan H. Young

Contents

Preface .... *iii*
Contributors .... *vii*

1. Classification of Bipolar Disorders—Implications for Clinical Research *1*
Ralph W. Kupka and Willem A. Nolen

2. Prospects for the Development of Animal Models for the Study of Bipolar Disorder *19*
Haim Einat, Alona Shaldubina, Yuly Bersudsky, and R. H. Belmaker

3. Abnormalities in Catecholamines and the Pathophysiology of Bipolar Disorder *33*
Amir Garakani, Dennis S. Charney, and Amit Anand

4. Cholinergic-Muscarinic Dysfunction in Mood Disorders *67*
David S. Janowsky and David H. Overstreet

5. Serotonergic Dysfunction in Mood Disorders *89*
J. John Mann and Dianne Currier

6. Cell Membrane and Signal Transduction Pathways—Implications for the Pathophysiology of Bipolar Disorders *109*
Guang Chen and Husseini K. Manji

7. Searching for a Cellular Endophenotype for Bipolar Disorder *131*
Francine M. Benes

8. The Hypothalamic–Pituitary–Adrenal Axis in Bipolar Disorder *145*
David J. Bond and Allan H. Young

9. Brain Imaging Studies in Bipolar Disorder *161*
E. Serap Monkul, Paolo Brambilla, Fabiano G. Nery, John P. Hatch, and Jair C. Soares

10. Sleep and Biological Rhythms Abnormalities in the Pathophysiology of Bipolar Disorders *189*
Stephany Jones and Ruth M. Benca

11. Hypothesis of an Infectious Etiology in Bipolar Disorder *209*
Robert H. Yolken and E. Fuller Torrey

12. **EEGs and ERPs in Bipolar Disorders** *221*
R. Hamish McAllister-Williams

13. **Genetics of Bipolar Disorder** *233*
Nick Craddock

14. **Neurocognitive Findings in Bipolar Disorder** *251*
David C. Glahn and Carrie E. Bearden

15. **Biology vs. Environment: Stressors in the Pathophysiology of Bipolar Disorder** *275*
Morgen A. R. Kelly, Stefanie A. Hlastala, and Ellen Frank

16. **The Kindling/Sensitization Model: Implications for the Pathophysiology of Bipolar Disorder** *297*
Robert M. Post

17. **Biologic Factors in Different Bipolar Disorder Subtypes** *325*
Michael A. Cerullo and Stephen M. Strakowski

18. **Biological Factors in Bipolar Disorder in Childhood and Adolescence** *343*
Melissa A. Brotman, Daniel P. Dickstein,
Brendan A. Rich, and Ellen Leibenluft

19. **Biological Factors in Bipolar Disorders in Late Life** *361*
R. C. Young, J. M. deAsis, and G. S. Alexopoulos

20. **Perspectives for New Pharmacological Interventions** *377*
Charles L. Bowden

21. **Physical Comorbidity in Bipolar Disorder** *387*
Paul Mackin and Sylvia Ruttledge

22. **Toward a Pathophysiology of Bipolar Disorders** *401*
John F. Neumaier and David L. Dunner

Index 409

Contributors

G. S. Alexopoulos Institute for Geriatric Psychiatry, Weill Medical College of Cornell University, and New York Presbyterian Hospital, White Plains, New York, U.S.A.

Amit Anand Department of Psychiatry, Indiana University School of Medicine, Indianapolis, Indiana, U.S.A.

Carrie E. Bearden Semel Institute for Neuroscience and Human Behavior, University of California, Los Angeles, California, U.S.A.

R. H. Belmaker Division of Psychiatry, Faculty of Health Sciences, Ben-Gurion University of the Negev, Beer Sheba, Israel

Ruth M. Benca Department of Psychiatry, University of Wisconsin–Madison, Madison, Wisconsin, U.S.A.

Francine M. Benes Department of Psychiatry, Harvard Medical School, Boston, Massachusetts, U.S.A.

Yuly Bersudsky Division of Psychiatry, Faculty of Health Sciences, Ben-Gurion University of the Negev, Beer Sheba, Israel

David J. Bond Mood Disorders Centre of Excellence, University of British Columbia, Vancouver, British Columbia, Canada

Charles L. Bowden University of Texas Health Science Center at San Antonio, San Antonio, Texas, U.S.A.

Paolo Brambilla Department of Pathology and Experimental and Clinical Medicine, Section of Psychiatry, University of Udine, Udine, and Scientific Institute IRCCS E. Medea, Bosisio Parini, Italy

Melissa A. Brotman Mood and Anxiety Disorders Program, National Institute of Mental Health, Department of Health and Human Services, National Institutes of Health, Bethesda, Maryland, U.S.A.

Michael A. Cerullo Division of Bipolar Disorders Research, Department of Psychiatry, University of Cincinnati College of Medicine, Cincinnati, Ohio, U.S.A.

Dennis S. Charney Department of Psychiatry, Mount Sinai School of Medicine, New York, New York, U.S.A.

Guang Chen Laboratory of Molecular Pathophysiology, Mood and Anxiety Disorders Research Program, National Institute of Mental Health, Bethesda, Maryland, U.S.A.

Nick Craddock Department of Psychological Medicine, Wales School of Medicine, Cardiff University, Cardiff, U.K.

Dianne Currier Department of Psychiatry, Columbia University, New York, New York, U.S.A.

J. M. deAsis Institute for Geriatric Psychiatry, Weill Medical College of Cornell University, and New York Presbyterian Hospital, White Plains, New York, U.S.A.

Daniel P. Dickstein Mood and Anxiety Disorders Program, National Institute of Mental Health, Department of Health and Human Services, National Institutes of Health, Bethesda, Maryland, U.S.A.

David L. Dunner Department of Psychiatry and Behavioral Sciences, University of Washington, Seattle, Washington, U.S.A.

Haim Einat College of Pharmacy, University of Minnesota, Duluth, Minnesota, U.S.A.

Ellen Frank Western Psychiatric Institute and Clinic, University of Pittsburgh School of Medicine, Pittsburgh, Pennsylvania, U.S.A.

Amir Garakani Department of Psychiatry, Mount Sinai School of Medicine, New York, New York, U.S.A.

David C. Glahn Department of Psychiatry and Research Imaging Center, University of Texas Health Science Center at San Antonio, San Antonio, Texas, U.S.A.

John P. Hatch Mood Disorders Clinical Neurosciences Program, Department of Psychiatry, and Department of Orthodontics, University of Texas Health Science Center at San Antonio, San Antonio, Texas, U.S.A.

Stefanie A. Hlastala Children's Hospital and Regional Medical Center, Seattle, Washington, U.S.A.

David S. Janowsky Department of Psychiatry, University of North Carolina, Chapel Hill, North Carolina, U.S.A.

Stephany Jones Department of Psychiatry, University of Wisconsin–Madison, Madison, Wisconsin, U.S.A.

Morgen A. R. Kelly Western Psychiatric Institute and Clinic, University of Pittsburgh School of Medicine, Pittsburgh, Pennsylvania, U.S.A.

Ralph W. Kupka Bipolar Disorders Program, Altrecht Institute for Mental Health Care, Utrecht, The Netherlands

Ellen Leibenluft Mood and Anxiety Disorders Program, National Institute of Mental Health, Department of Health and Human Services, National Institutes of Health, Bethesda, Maryland, U.S.A.

Paul Mackin School of Neurology, Neurobiology, and Psychiatry, University of Newcastle upon Tyne, Newcastle upon Tyne, U.K.

Husseini K. Manji Laboratory of Molecular Pathophysiology, Mood and Anxiety Disorders Research Program, National Institute of Mental Health, Bethesda, Maryland, U.S.A.

J. John Mann Department of Psychiatry, Columbia University, New York, New York, U.S.A.

R. Hamish McAllister-Williams School of Neurology, Neurobiology, and Psychiatry, University of Newcastle upon Tyne, Newcastle upon Tyne, U.K.

E. Serap Monkul Mood Disorders Clinical Neurosciences Program, Department of Psychiatry, University of Texas Health Science Center at San Antonio, San Antonio, Texas, U.S.A., and Department of Psychiatry, Dokuz Eylül University School of Medicine, Izmir, Turkey

Fabiano G. Nery Mood Disorders Clinical Neurosciences Program, Department of Psychiatry, University of Texas Health Science Center at San Antonio, and Audie L. Murphy Division, South Texas Veterans Health Care System, San Antonio, Texas, U.S.A., and Institute of Psychiatry, University of São Paulo School of Medicine, São Paulo, Brazil

John F. Neumaier Department of Psychiatry and Behavioral Sciences, University of Washington, Seattle, Washington, U.S.A.

Willem A. Nolen Department of Psychiatry, University Medical Center Groningen, Groningen, The Netherlands

David H. Overstreet Center for Alcohol Studies, Department of Psychiatry, University of North Carolina, Chapel Hill, North Carolina, U.S.A.

Robert M. Post Mood and Anxiety Disorders Program, National Institute of Mental Health, Department of Health and Human Services, National Institutes of Health, Bethesda, Maryland, U.S.A.

Brendan A. Rich Mood and Anxiety Disorders Program, National Institute of Mental Health, Department of Health and Human Services, National Institutes of Health, Bethesda, Maryland, U.S.A.

Sylvia Ruttledge School of Neurology, Neurobiology, and Psychiatry, University of Newcastle upon Tyne, Newcastle upon Tyne, U.K.

Alona Shaldubina Division of Psychiatry, Faculty of Health Sciences, Ben-Gurion University of the Negev, Beer Sheba, Israel

Jair C. Soares Center of Excellence for Research and Treatment of Bipolar Disorders, Department of Psychiatry, University of North Carolina School of Medicine, Chapel Hill, North Carolina, U.S.A.

Stephen M. Strakowski Division of Bipolar Disorders Research, Department of Psychiatry, University of Cincinnati College of Medicine, Cincinnati, Ohio, U.S.A.

E. Fuller Torrey Stanley Medical Research Institute, Chevy Chase, Maryland, U.S.A.

Robert H. Yolken Stanley Division of Developmental Neurovirology, Johns Hopkins University School of Medicine, Baltimore, Maryland, U.S.A.

Allan H. Young Mood Disorders Centre of Excellence, University of British Columbia, Vancouver, British Columbia, Canada

R. C. Young Institute for Geriatric Psychiatry, Weill Medical College of Cornell University, and New York Presbyterian Hospital, White Plains, New York, U.S.A.

1 Classification of Bipolar Disorders—Implications for Clinical Research

Ralph W. Kupka
Bipolar Disorders Program, Altrecht Institute for Mental Health Care, Utrecht, The Netherlands

Willem A. Nolen
Department of Psychiatry, University Medical Center Groningen, Groningen, The Netherlands

INTRODUCTION

The classification of mood disorders has been a subject of scientific debate for more than 2500 years (1,2), and a precise delineation of these illnesses and its various clinical manifestations has yet to emerge. Many aspects of this discussion have recently been reviewed by Akiskal (2) and supplemented by commentaries from authoritative researchers. In this chapter we will give an overview of the current classification, the boundaries of bipolar disorder with other major psychiatric illnesses, the validity and reliability of diagnosis, and the implications for neuropsychiatric research. As all classifications that are based on clinical description rather than on etiology and pathophysiology are deemed to be temporary, we will begin with a brief historical overview and end with some areas that need further clarification.

A BRIEF HISTORY OF MOOD DISORDER

Mania and melancholia, already described by Hippocrates (460–377 BC), were linked as manifestations of one illness by Aretaeus of Cappadocia (c. 150 AD). His conception of mania being a consequence of melancholia was adopted throughout the Middle Ages and Renaissance by writers such as Vesalius, Burton, Willis, and Boerhaave (1). It was not until the 19th century that longitudinal clinical course was taken into account in the description of psychopathology. In 1854, Falret and Baillarger simultaneously but independently described a pattern of illness in which mania, depression, and a symptom-free interval appeared in more or less regular cycles over time (1). In addition to these severe forms of circular and periodic insanity, in 1882, Kahlbaum described cyclothymia and dysthymia with milder degrees of excitement and depression (1). It was Kraepelin who in 1899 brought all affective syndromes together under the name of manic-depressive insanity: circular and periodic forms, unipolar mania, recurrent depression, milder manifestations, and even subclinical forms that were considered part of personal predisposition or temperament (3). From his longitudinal descriptions of many cases he could not delineate clear boundaries between these clinical pictures and hypothesized that they were all manifestations of the same disease process. These observations from the pre-pharmacological era are of great importance for the understanding of the natural course of bipolar disorder, since most patients in later longitudinal studies have received both acute and prophylactic treatment, which may have modified the course of illness for the better or the worse (4).

In 1957, Leonhard (5) proposed the distinction between unipolar depression and bipolar illness, which was supported by Angst (6) and by Perris in 1966 (7), and by Winokur, Clayton, and Reich in 1969 (8). Subsequently, the unipolar–bipolar distinction was adopted by the American classification diagnostic and statistical manual of mental disorders-III (DSM-III) in 1980 (9), and the revised edition (DSM-III-R) in 1987 (10). At the same time the WHO classification International Classification of Diseases-9 (ICD) (1978) (11) still described all types of depressive and other forms of affective disorders under the category of manic-depressive psychosis. However, in the current ICD-10 (1992) (12), bipolar affective disorder is classified next to recurrent depressive disorder.

CURRENT CLASSIFICATION: DSM-IV AND ICD-10

DSM-IV (13) describes mood disorders in three parts: mood episodes, mood disorders, and specifiers, that is, characteristics of the most recent episode or the longitudinal course of recurrent illness. Mood episodes (manic, hypomanic, depressive, or mixed) are building blocks of mood disorders, and cannot be diagnosed as separate illnesses. Mood disorders are subdivided in depressive disorders and bipolar disorders, and two disorders based on etiology: somatic illness and substance abuse (Table 1). Specifiers describing the most recent episode refer to severity, whether or not it is in remission, and the presence of psychotic features or other clinical characteristics (Table 2). Longitudinal course specifiers refer to the degree of interepisodic recovery, seasonal pattern, and rapid cycling. These elements of classification aim at identifying relatively homogeneous subgroups, and are further enhanced by providing more or less detailed diagnostic criteria for every category described (see next section).

Compared to the DSM-IV, the text revision published in 2000 (DSM-IV-TR) (14) did not change the mood disorders section, nor the section on schizoaffective disorders. Therefore, we further refer to DSM-IV.

Unlike previous editions of DSM and ICD, DSM-IV and ICD-10 agree on major aspects of the classification of mood disorders, although there are several differences, as outlined in Table 1. One major difference is that ICD-10 allows for classifying a single manic/hypomanic or a single depressive episode next to bipolar affective disorder and recurrent depressive disorder, whereas in DSM-IV a single manic episode is always considered as being part of bipolar disorder, and likewise a single depressive episode is classified as major depressive disorder. Both classification systems agree that the relatively rare condition of repeated mania without a history of depressive episodes should be classified as bipolar disorder. A recent study from the Zurich group suggested that patients with a history of only mania(s) (M), or of mania(s) with only mild depression(s) (Md), have a lower morbidity risk in first degree relatives and a better prognosis than bipolar I patients with both mania(s) and full depression(s) (MD) (15).

DSM-IV gives a detailed definition of bipolar II disorder, characterized by one or more major depressive episodes accompanied by at least one hypomanic episode in the absence of a history of manic or mixed episodes (16,17). In contrast, in ICD-10, bipolar I and II disorders are not classified separately, but the latter is included in

TABLE 1 DSM-IV Classification of Mood Disorders with Corresponding ICD-10 Classification

DSM-IV/DSM-IV-TR[a]	Code[b]	ICD-10[c]	Code
Depressive disorders			
Major depressive disorder			
Single episode	296.2x	Depressive episode	F32.0–9
Recurrent	296.3x	Recurrent depressive disorder	F33.0–9
Dysthymic disorder	300.4	Dysthymia	F34.1
Depressive disorder NOS	311	Other depressive episode	F32.9
		Other recurrent depressive disorder	F33.9
Bipolar disorders			
Bipolar I disorder			
Single manic episode	296.0x	Manic (including hypomanic) episode	F30.x
Hypomanic[d]	296.40	Bipolar disorder, hypomanic[e]	F31.0
Manic[d]	296.4x	Bipolar disorder, manic[e]	F31.1–2
Mixed[d]	296.6x	Bipolar disorder, mixed[e]	F31.6
Depressed[d]	296.5x	Bipolar disorder, depressed[e]	F31.3–5
Unspecified[d]	296.7	Bipolar disorder, unspecified[e]	F31.9
Bipolar II disorder	286.89	Other bipolar affective disorder	F31.8
Hypomanic[d]			
Depressed[d]			
Cyclothymic disorder	303.13	Cyclothymia	F34.0
Bipolar disorder NOS	296.80	Other bipolar affective disorder	F31.8
Mood disorder due to			
General medical condition (depressive/manic/mixed)	293.83	Organic mood disorders (manic/bipolar/depressive/mixed)	F06.30–33
Substance induced mood disorder (depressive/manic/mixed)	292.84	Psychotic disorder due to	F1x.54
		Psychoactive	F1x.55
		Substance abuse (depressive/manic/mixed)	F1x.56
Mood disorder NOS	296.90	Other mood disorders	F38
		Unspecified mood disorders	F39
Schizoaffective disorder[f]	295.70	Schizoaffective disorders[f]	F25
Bipolar type		Manic type	F25.0
Depressive type		Depressive type	F25.1
		Mixed type	F25.2

[a]DSM-IV-TR is unchanged from DSM-IV with regard to mood disorders.
[b]For fifth digit see Table 2.
[c]Not all ICD-10 subcategories are listed.
[d]Most recent episode.
[e]Current episode.
[f]Included in the section: schizophrenia and other psychotic disorders.
Abbreviations: DSM, Diagnostic and Statistical Manual of Medical Disorders; ICD, International Classification of Diseases; NOS, not otherwise specified.

the category of "other bipolar affective disorders," without further description. However, in ICD-10 "bipolar affective disorder, current episode hypomanic" could also include those patients who have never experienced a full manic episode, that is, those with bipolar II disorder.

The problem of distinguishing hypomania from normal mood swings on the one hand and from mania on the other, which may occur in any classification, will be discussed in the next section.

TABLE 2 Diagnostic and Statistical Manual of Mental Disorders-IV Specifiers for Mood Disorders

Specifiers for most recent episode[a]	Code	Longitudinal course specifiers[a]
Severity		
Mild	xxx.x1	
Moderate	xxx.x2	
Severe	xxx.x3	
Severe with psychotic features	xxx.x4	
Mood-congruent psychotic features		
Mood-incongruent psychotic features		
Course		
In partial remission	xxx.x5	With/without interepisode recovery
In full remission	xxx.x6	Seasonal pattern[b]
Chronic[b]		Rapid cycling[c]
Postpartum onset		
Associated features		
Catatonic features		
Melancholic features[b]		
Atypical features[b]		

[a]DSM-IV-TR is unchanged from DSM-IV with regard to mood disorders.
[b]Specifier only for depressive episodes.
[c]Specifier only for bipolar I and II disorders.

MANIA AND HYPOMANIA

At the core of the manic syndrome is a persistently elevated, expansive and/or irritable mood during at least seven days (or any duration when hospitalized), accompanied by symptoms such as increased self-esteem, over-optimism or even grandiosity, pressure of speech, racing thoughts, distractibility, increased energy, increased sexual drive, overactivity or agitation, and loss of inhibitions leading to reckless involvement in pleasurable, hazardous, or embarrassing activities that may have serious marital, social, financial, or judicial consequences. Finally, symptoms cause marked impairment in social or occupational functioning. Error of judgment and lack of insight, which are often seen in manic patients, are not specifically included in DSM-IV diagnostic criteria. Even patients who did previously acknowledge having bipolar disorder may lose insight during a next manic episode. Highly characteristic of the manic syndrome is a decreased need for sleep combined with an increased feeling of energy, which is quite distinct from insomnia in depression or other psychiatric disorders where the diminished sleep coincides with feeling tired. In severe mania, psychotic symptoms may occur, either mood-congruent such as delusions of grandiosity, or mood-incongruent, such as persecutory delusions.

Diagnosing "classic" mania should not be too difficult a task for any clinician with some experience. However, during an initial interview, an intelligent patient may temporarily mislead the clinician by dissimulating the presence or the severity of his symptoms and by finding apparently meaningful explanations for his behaviors. In such cases, information from nurses and especially from close relatives should confirm the diagnosis of mania. The latter may be even truer in case of hypomania.

In DSM-IV, a hypomanic episode has essentially the same clinical features as a manic episode, but lasts a shorter period. In hypomania, there is an unequivocal change from normal functioning, which is observable by others, lasting at least four days. This separates hypomania from normal elevations of mood or the very mild mood elevations that occur in cyclothymic disorder. On the other hand, the symptoms are not severe enough to cause "marked impairment in social or occupational functioning" as in mania. In the ICD-10 clinical descriptions and diagnostic guidelines (12) hypomania does not lead to "severe or complete disruption of work or result in social rejection," although "considerable interference with work or social activity is consistent with a diagnosis of hypomania." Thus, ICD-10 hypomania includes conditions that would justify a diagnosis of (mild) mania according to DSM-IV criteria. Interestingly, the ICD-10 diagnostic criteria for research (18) are more in accordance with DSM-IV, stating that hypomania leads to "some interference with personal functioning in daily living." The problems that arise when defining the boundary between hypomania and (mild) mania are discussed by Goodwin, who points out that this boundary depends entirely upon the meaning of ill-defined qualifying words like "some," "considerable," "marked," "severe," or "complete" functional impairment (19). He also notes a tendency to avoid the somewhat pejorative diagnosis of "mania" in favor of "hypomania" in clinical settings.

The validity of a minimum duration of four days for a hypomanic episode has been tested by the Zurich group that found that patients with brief hypomania of one to three days duration did not significantly differ from those whose hypomanic episodes lasted at least four days (20).

In patients presenting with a current depressive episode, a history of prior hypomanic episodes is easily missed, especially since depression is the prevailing condition in bipolar II disorder (21). Thus, many of these patients will be misdiagnosed as suffering from (recurrent) unipolar depression. Revealing past hypomanic episodes may benefit from systematic inquiry in all aspects of the syndrome, especially the behavioral symptoms rather than elevation of mood. Patients and relatives will remember the short nights and the energetic overactivity more sharply than a period of cheerfulness or irritability.

In a recent survey among U.S. citizens using the Mood Disorders Questionnaire (22), which systematically checks (hypo)manic symptoms, 3.7% screened positive for bipolar I or II disorder. Of these, only 19.8% had previously received a diagnosis of bipolar disorder, 31.2% had received a diagnosis of unipolar depression, and 49.0% had received neither of these diagnoses (23).

Hypomanic or manic episodes may appear relatively late in the course of bipolar disorder, inevitably leading to an initial diagnosis of unipolar depression. The rate of spontaneous conversion from unipolar to bipolar mood disorder has been estimated at median 9.7%, with a reported maximum of up to 37.5% (24). Over the course of 11 years, hypomanic or manic episodes occurred in 12.5% of 559 prospectively followed unipolar patients (25). This phenomenon may in part explain the long delay between the occurrence of first mood symptoms and the diagnosis of bipolar disorder, which was on average 10 years in a survey among DMDA members (26) and also among clinical populations (27).

DEPRESSION

Depression is the main burden of bipolar illness. Longitudinal follow-up data from the Collaborative Depression Study showed that patients with bipolar I disorder, despite adequate treatment, on average reported depressive symptoms in 31.9%,

TABLE 3 Clues of Bipolarity in Patients with Major Depressive Episodes

Episode features	Longitudinal illness history
Rapid onset and end of episode	Family history of bipolar disorder
Brief episodes (<2 weeks)	First episode at younger age (<25)
Psychotic features	Frequent episodes
Atypical features	History of brief major depressive episodes
Psychomotor retardation and anergia	History of mania or hypomania
Hypersomnia	History of antidepressant-induced hypomania
Mood lability	History of postpartum depression
Nonresponse to antidepressants	Cyclothymic or hyperthymic temperament
Response to lithium	

manic/hypomanic symptoms in 8.9%, and cycling/mixed symptoms in 5.9% of weeks (28). In bipolar II disorder these percentages were even 50.3%, 1.3%, and 2.3% of weeks, respectively (28). The Stanley Foundation Bipolar Network reported similar outcomes in treated bipolar I patients using daily prospective life chart data: on average they were depressed 35.6% of the time, manic/hypomanic 12.6%, and ultradian cycling 3.3% (29). Another research group using prospective life chart data also found a depression-to-mania rating of 5.9 in bipolar I and 13.7 in bipolar II patients (30).

Although a depressive episode in the course of bipolar disorder in general has the same clinical features as unipolar depression, and both conditions have identical diagnostic criteria in DSM-IV and ICD-10, cross-sectional as well as longitudinal studies have revealed some features that may distinguish bipolar from unipolar depression (31–37). These features are summarized in Table 3.

In patients with depression, a family history of bipolar disorder may point to a bipolar course. The occurrence of brief episodes is also associated with bipolar depression. DSM-IV requires for a major depressive episode a minimum duration of at least two weeks. This may be appropriate in most cases of unipolar depression and many cases of bipolar disorder, but patients with bipolar disorder often also have brief episodes with rapid onset and remission, which may have the full range of severities (32). Prospective life chart data from the Stanley Foundation Bipolar Network showed that even patients with a nonrapid cycling course have on average twice as many brief depressive episodes, that is, of less than two weeks duration, than full-duration episodes (29). This was similar with regard to brief manic and hypomanic episodes.

MIXED STATES

In mixed states the complexity of bipolar disorder reaches its maximum. Pure depression and pure mania are the prototypical endpoints on a continuum of behavioral and emotional disturbances. Mixed states have originally been described by Kraepelin and his contemporary Weygandt as various admixtures of three dimensions: mood, thinking, and psychomotor activity (38). They distinguished six subtypes: depression with flight of ideas, excited depression, depressive-anxious mania, mania with thought poverty, inhibited mania, and manic stupor.

In DSM-IV, a patient with a mixed episode meets both criteria for a manic episode and a major depressive episode during at least one week, and may experience rapidly alternating moods of sadness, irritability, and dysphoria. It is thus essentially a type of manic episode that also contains full syndromal depression ("mixed mania"), which may be uncommon in clinical settings (39). This narrow definition excludes many patients with combinations of syndromal and subsyndromal symptoms of either polarity, for example, those with isolated depressive symptoms during a manic episode ("mixed mania") or with some manic symptoms during a major depressive episode ("mixed depression"). The prevalence of such more broadly defined mixed states in clinical practice is probably much higher than DSM-IV mixed episodes. It is estimated that mixed states occur in about 30% to 40% of acutely manic patients (32,40). Dysphoric mania, defined as a full manic syndrome with the simultaneous presence of some depressive symptoms, was present in 37% of manic patients in a clinical setting (41). In a recent study among 908 treated bipolar outpatients mixed hypomania, that is, the co-occurrence of depressive symptoms in patients with a hypomanic episode, was found in 57% of visits for hypomanic episodes (42). Mixed hypomania was equally prevalent in patients with bipolar I and II disorder, but more prevalent in women.

The concept and the terminology of mixed states are prone to confusion. ICD-10 takes a somewhat broader view than DSM-IV on mixed states, defined as either a mixture or a rapid alternation (usually within a few hours) of manic, hypomanic, and depressive symptoms; the minimum duration of mixed episodes is two weeks. The term "dysphoric mania," originally meant to indicate a mixed state, may be wrongly used to indicate "classic" mania with predominantly irritable rather than euphoric mood. Moreover, the distinction between these two conditions may be difficult and it is unclear whether this is clinically relevant since the evidence that mixed states should be regarded as separate clinical entities is controversial (40).

It is of clinical importance that mixed mania is less responsive to treatment, in particular with lithium, than "classic" euphoric mania (43).

ANTIDEPRESSANT-INDUCED MANIA AND HYPOMANIA

According to DSM-IV, patients with unipolar major depression who become hypomanic or manic when treated with antidepressants receive an additional diagnosis of substance-induced mood disorder with manic features. In these cases, it is assumed that the manic syndrome is a direct physiological effect of antidepressant medication, and such an episode should not lead to a diagnosis of bipolar disorder. This also applies for mood episodes induced by other medications, alcohol, or drugs of abuse. The concept that antidepressant-induced (hypo)mania in (unipolar) major depression is not indicative of bipolar disorder has been strongly debated, and evidence for an opposite point of view was recently reviewed by Chun and Dunner (24). They found that cases of treatment-induced hypomania in 121 studies of antidepressants in major depressive disorder patients were relatively rare, and within the rate of spontaneous conversion from unipolar to bipolar disorder. They conclude that these patients should be recognized as having bipolar disorder, and propose a revision of this category in DSM-V. Another argument for this position is that according to most guidelines, patients with a bipolar depression should not be treated with an antidepressant alone, but only in combination with antimanic agents, that is lithium, an anticonvulsant, or, according to some guidelines, an atypical antipsychotic. It appears logical to apply the same approach for depressed patients with a history of an antidepressant-induced hypomanic or manic episode, since when treated with antidepressant monotherapy there is a significant risk of another switch. Finally, it is also of interest that withdrawal of antidepressants can induce mania (44). It is likely that for this condition the same diagnostic considerations are relevant.

DIFFERENTIAL DIAGNOSIS: THE BOUNDARIES OF BIPOLAR DISORDER

External Boundaries

Cyclothymic disorder can be seen as the borderland between bipolar disorder and normal mood fluctuations. Patients with cyclothymia exhibit mood instability, that is (often numerous) periods with hypomanic symptoms that never meet criteria for a manic episode and with depressive symptoms that never meet criteria for major depressive episodes over the course of at least two years. A family history of mood disorders is more common in cyclothymic patients in whom the disorder proceeds to a full-blown bipolar disorder in the course of their life.

Borderline personality disorder is another disorder with affective instability. Therefore, it may be difficult to distinguish borderline personality disorder from cyclothymia and also from bipolar I or bipolar II disorder with rapid cycling (i.e., at least four episodes per year) and especially ultrarapid cycling (i.e., mood switches within a week of at least four episodes per month) or ultradian cycling (i.e., mood switches within a single day) (45). Moreover, both conditions may coexist. A controversial point of view is whether borderline personality disorder in fact is part of the bipolar spectrum (2,46–48).

Schizoaffective disorder, bipolar type, fills the gap between mania with mood-incongruent psychotic features and schizophrenia. A considerable number of severely ill patients admitted to psychiatric hospitals have manic or depressive episodes co-occurring with symptoms characteristic of schizophrenia, including mood-incongruent delusions or hallucinations. In addition, patients have delusions or hallucinations in the absence of prominent mood symptoms. The definition of schizoaffective disorders has varied over the years among classification systems and is still diffuse and uncertain (49). According to DSM-IV, such patients are classified as schizoaffective disorder when the co-occurrence of mood and psychotic symptoms as described occurs during the same period of illness, which may last for years. Nevertheless this definition is difficult to use in clinical practice, as shown by the small interrater reliability when diagnosing patients with these problems (50). It challenges the classic Kraepelinian dichotomy between manic-depressive illness and schizophrenia, and may be an indication of a considerable genetic overlap between these two groups of major psychiatric disorders (51).

Another diagnostic category that fits in the grey zone between bipolar disorder and schizophrenia but has been omitted in DSM-IV and ICD-10 is cycloid psychosis, as described by Leonhard (4). These conditions combine polymorphous psychotic symptoms with an episodic course and a generally favorable prognosis (52).

The boundary between recurrent unipolar depression and bipolar disorder has already been discussed, and will be further addressed in the section on the bipolar spectrum. If one accepts the unipolar–bipolar dichotomy then the question arises at which point these disorders split. This depends on the definition of hypomania and its lower limit towards normality, being the mildest manifestation of bipolarity. Current debate concentrates on the number of hypomanic symptoms needed, and the minimum duration of a hypomanic episode. DSM-IV requires four days, although at that time no data were present to determine whether three days would have been better (53). Recent studies by Angst et al. in Zurich (20) showed that hypomanic episodes of one to three days were of comparable clinical significance. Including brief hypomanic episodes into the definition of bipolar II disorder changes the rate of all major and minor forms of unipolar versus bipolar disorder (see later) (20).

Another area of potential diagnostic confusion lies in the distinction between agitated unipolar depression and bipolar mixed states, especially the "mixed depression" described earlier. There is some research evidence that agitated unipolar depression is part of the bipolar spectrum (54).

Internal Boundaries

Separating bipolar II from bipolar I disorder depends on the definition of hypomania and its upper limits towards mania (see earlier). In the longitudinal Collaborative Depression Study, patients with bipolar I disorder had more severe episodes whereas those with bipolar II disorder had a substantially more chronic course with significantly more major and minor depressive episodes and shorter inter-episodic well intervals (55). The authors conclude that these differences justify classification as two separate subtypes, although the overall clinical similarities of these subtypes suggest that they exist in a disease spectrum (55).

Rapid cycling, defined as the occurrence of at least four distinct mood episodes in one year, has been introduced as a course specifier of bipolar I and II disorder in DSM-IV (53,56). Rapid cycling is not mentioned at all in ICD-10. Taking a conventional categorical approach, a meta-analysis of 20 studies comparing rapid cyclers and nonrapid cyclers revealed some significant differences apart from episode frequency, in particular a slight overrepresentation of women and of bipolar II subtype among rapid cyclers (57). Associated with rapid cycling there was also a trend for depressive episode at onset of illness, a history of serious suicide attempts, a family history of affective disorder, and nonresponse to lithium prophylaxis. However, recent large studies (29,58,59) have shed doubt over the higher prevalence of rapid cycling among bipolar II patients. Moreover, if one takes a dimensional approach, differences occur gradually with increasing episode frequency and never reveal a cut-off point at four episodes per year or at any other episode frequency, suggesting that rapid cycling is not a distinct subtype but merely an extreme on a continuum of cycle frequency (29).

Finally, the boundary between mania and depression is blurred in patients with mixed states. There is also a potential overlap between mixed states with rapid mood shifts on the one hand, and ultra-rapid or ultradian cycling (45,47) on the other. The latter conditions are not defined in DSM-IV but can be classified as mixed episodes or alternatively in the residual category Bipolar Disorder Not Otherwise Specified. A summary of key criteria defining the major boundaries of bipolar disorder is given in Table 4.

THE BIPOLAR SPECTRUM CONCEPT

A broad definition of manic-depressive illness was originally proposed by Kraepelin (3), and reintroduced by Akiskal (2,60,61), Goodwin and Jamison (32), Cassano (63), and Angst (64), amongst others. This so-called bipolar spectrum includes syndromal and subsyndromal clinical conditions beyond the more narrowly defined DSM-IV and ICD-10 classifications of bipolar disorder. In recent years, different

TABLE 4 Key Boundaries of Bipolar Disorders in DSM-IV

Boundary	Criteria defining boundary
Normal mood fluctuations versus dys/hyper/cyclothymic temperament versus mood disorder	Occurrence and severity of depressive and hypomanic symptoms
Unipolar depression versus bipolar II disorder	Number and duration of hypomanic symptoms
Bipolar II versus bipolar I disorder	Impairment and duration of manic symptoms
Bipolar I versus schizoaffective disorder	Timing of psychotic symptoms
Schizoaffective disorder versus schizophrenia	Balance of psychotic and affective symptoms
Rapid cycling versus nonrapid cycling course	Occurrence of 4 distinct mood episodes/year

Source: From Ref. 73.

variants of a bipolar spectrum have been presented; an example as proposed by Akiskal is given in Table 5.

The bipolar spectrum concept puts emphasis on validators other than polarity (i.e., the occurrence of mania), such as a recurrent ("cycling") course of illness, positive family history of mood disorder, and unfavorable outcome of antidepressant treatment (62). In such a broad definition, recurrent brief depressions ("cycling within the depressive pole") that do not respond to or even get worse with antidepressants may be part of the spectrum. Lifetime prevalence rates of bipolar spectrum disorders range from 2.8% to 6.6% of the population (65).

One major rationale for a bipolar spectrum is that the application of narrow diagnostic criteria may result in misdiagnosis of many cases that would benefit from other treatments: mood stabilizing drugs rather than antidepressants.

A major objection towards the concept of a bipolar spectrum is that it shows an expansive trend, incorporating subsyndromal conditions and conditions that are only in part characterized by affective instability, such as borderline personality disorder (66). This may weaken the core concept of bipolar disorder, which appears to be a circumscript clinical entity suitable for genetic, biological, and treatment studies (66). Reanalyzing the data of the epidemiologic catchment area (ECA) study by taking into account all subsyndromal manic symptoms resulted in a lifetime prevalence rate of 5.1% on top of the 0.8% for manic episodes (bipolar I) and 0.5% for hypomanic episodes (bipolar II), yielding a total of 6.4% for bipolar spectrum disorders (67). The prevalence of bipolar (and unipolar) mood disorders

TABLE 5 A Bipolar Spectrum as Proposed by Akiskal

Bipolar $\frac{1}{2}$	Schizobipolar disorder
Bipolar I	Manic-depressive illness
Bipolar I $\frac{1}{2}$	Depression with protracted hypomania
Bipolar II	Depression with spontaneous hypomanic episodes
Bipolar II $\frac{1}{2}$	Depression superimposed on cyclothymic temperament
Bipolar III	Recurrent depression with antidepressant-induced hypomania
Bipolar III $\frac{1}{2}$	Recurrent depression with alcohol/substance-induced hypomania
Bipolar IV	Depression superimposed on hyperthymic temperament
Bipolar V	Recurrent depression ($\geq$5 episodes)

Source: From Ref. 61.

is further exploded when thresholds are set even lower, such as in the Zurich studies (20). Relaxing duration and severity criteria for hypomania resulted in an almost equal lifetime prevalence of 24.6% for the depressive spectrum (including major depression, dysthymia, minor depression, and brief recurrent depression) and 23.7% for "soft" bipolar spectrum (including bipolar I and broadly defined bipolar II, minor bipolar disorder, cyclothymia, and pure hypomania) (20). These authors state that 11% constitutes the spectrum of bipolar disorders proper, and another 13% "probably represent the softest expression of bipolarity intermediate between bipolar disorder and normality." If almost a quarter of the population is included, it is questionable whether such broad definitions are still meaningful indicators of psychopathology, given the lifetime prevalence of core bipolar I disorder of 0.5% in the same cohort, which is consistent with other epidemiological studies (19).

ASSESSMENT OF BIPOLAR DISORDER

Diagnostic assessment of bipolar disorders has several aspects: a lifetime diagnosis according to DSM-IV of ICD-10, a diagnosis of the current mood episode or a state of interepisodic remission, rating the severity of the current mood disturbance, and depicting the longitudinal course of the illness.

A diagnosis of bipolar disorder in clinical research settings is commonly obtained by applying one of the semi-structured interviews for axis I diagnoses by trained clinicians, such as the Structured Clinical Interview for DSM-IV axis I disorders, Research Version, Patient Edition (SCID-I/P) (68), the related Mini International Neuropsychiatric Interview (MINI) (69), or the Schedules for Clinical Assessment in Neuropsychiatry (SCAN) (70). In epidemiological studies, fully structured interviews are used, often by trained lay interviewers, such as the Composite International Diagnostic Interview (CIDI) (71) (REF WHO). All these instruments arrive at current or lifetime DSM-IV (SCID) or DSM-IV and ICD-10 (SCAN, CIDI) diagnoses. Since each diagnostic criterion of each mood episode as well as every psychotic symptom is rated separately, these instruments are suitable for assessing symptom profiles beyond the diagnostic boundaries described earlier. Semistructured interviews by trained clinicians may be somewhat more specific in diagnosing bipolar disorder than structured interviews by lay interviewers (72).

An self-rated instrument for screening for bipolar disorder is the Mood Disorders Questionnaire (MDQ) (22), which has been applied in a large U.S. population sample (23) and various European countries, and actually checks the previous presence of manic symptoms according to DSM-IV criteria. If rated positive it should be followed by a clinical interview.

A dimensional scale that was designed to serve as an adjunct to conventional categorical diagnosis is the Bipolar Affective Disorder Dimension Scale (BADDS) (73). It provides a description of some of the basic features of an individual's lifetime experience of psychopathology relevant to the bipolar spectrum. The scale is mainly based on the ICD-10 criteria and has four dimensions: (*i*) mania: the presence and severity of manic syndromes; (*ii*) depression: the presence and severity of depressive syndromes; (*iii*) psychosis: the presence of psychotic symptoms and the balance of mood and psychotic symptomatology; and (*iv*) incongruence: the mood congruence of psychotic symptoms and the temporal relationships between affective and psychotic symptomatology. Using multiple data sources, every subject is scored in the range 1–100 on each of the four dimensions. The instrument was developed within the context of family studies, and incorporates the boundaries between unipolar, bipolar, schizoaffective, and schizophrenic disorders (Table 4). It retains a measure of severity and also includes mild or subclinical cases, and avoids hierarchical loss of information inherent to classification systems. BADDS appears to be especially suitable to assess bipolar spectrum disorders next to or beyond the strict DSM-IV or ICD-10 categories.

There are various symptom rating scales for measurement of severity of mania and depression, such as the Young Mania Rating Scale (YMRS) (74), the Bech Rafaelsen Mania Scale (BR-MAS) (75), the Hamilton Rating Scale for Depression (HAMD) (76), the Montgomery Åsberg Depression Rating Scale (MADRS) (77), and the Inventory of Depressive Symptomatology (IDS) (78). The latter includes many "atypical" symptoms, which frequently occur in bipolar depression, and is thus particularly suitable for this condition.

A problem typical for the assessment of change in bipolar disorder is that improvement of a depressive episode may go hand in hand with emerging mania, and vice versa, and that rapid changes of polarity may add to the overall illness burden beyond depressive and manic symptoms per se. To address this problem, the Clinical Global Impressions scale, Bipolar Version (CGI-BP), was designed (79). This is a modification of the original CGI scale, providing global severity ratings of both depression and mania, as well as a global severity rating of overall bipolar disorder. A similar set of CGI scales was designed for the assessment of change in clinical trials (79).

These scales all make a cross-sectional assessment of the manic or depressive episode with a time span ranging from two days to two weeks. Given the waxing and waning course of bipolar disorder, repeated cross-sectional assessments will not reveal the frequently occurring mood episodes in between. For continuous longitudinal assessments, either retrospective or prospective, a graphic method such as the NIMH-Life Chart Methodology (NIMH-LCM) is more suitable (Fig. 1) (80,81). This instrument has been validated in recent years (82,83), and is available in a clinician-rated and a self-rated version. A similar instrument for longitudinal assessment is the prospectively self-rated computerized ChronoRecord (84).

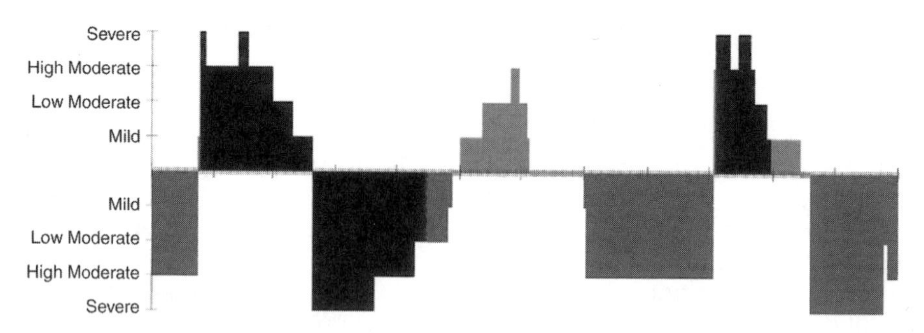

FIGURE 1 One-year prospective daily life chart showing prototypical bipolar I disorder with a continuous "rapid cycling" course in a man aged 67 with an illness history of over 50 years. Mania above and depression below baseline. *Source*: From Ref. 29.

DIAGNOSTIC VALIDITY AND IMPLICATIONS FOR NEUROPSYCHIATRIC RESEARCH

Although our current diagnostic categories have limited validity and are provisional until external validators are identified, the availability of clear diagnostic categories with its criteria has nevertheless greatly improved the reliability of diagnosis in clinical and research settings. In most neurobiological, genetic, and treatment studies of bipolar disorder, subjects will have received a diagnosis of (DSM-IV) bipolar disorder. Also, further distinctions are often made between subjects with bipolar I or bipolar II disorder and subjects with a rapid cycling course versus with slower cycle frequencies of less than four mood episodes per year (often indicated as nonrapid cyclers). As discussed earlier, the boundaries between these categories are arbitrary, and it is unlikely that these boundaries, which are entirely based on clinical descriptive criteria, exactly point out the potential boundaries between distinct disorders with a different etiologic or pathophysiological background (73,85).

Dimensional diagnoses tend to be impractical and difficult to communicate (85). However, the concurrent use of categorical and dimensional diagnostic instruments as proposed by Craddock et al. (73) may be particularly appropriate in bipolar spectrum disorders. The bipolar spectrum concept provides a compromise, since its prototypical categories lie along a unipolar–bipolar continuum and on continuums with other diagnoses such as psychotic disorders. It is also useful for clinical practice, alerting clinicians to search for clues for bipolar disorder in patients with mood disorders and treat them according to the best available evidence. However, for neurobiological research, a broadly defined bipolar spectrum as a whole may be over inclusive. Investigating relatively homogeneous subgroups delineated by relatively narrow diagnostic categories that are now widely accepted (major unipolar depression, bipolar I and bipolar II disorder) within the entire bipolar spectrum could help to identify biological markers and the genetics for core syndromes. Comparing shared and unshared neurobiological and genetic features may eventually lead to subgroups that are more firmly based on etiology and pathophysiology. This may also shed light on the true nature of overlapping areas between prototypical phenotypes, such as "pseudo-unipolar" and schizoaffective disorders.

The longitudinal evolution of bipolar disorder makes every diagnostic assessment in a given individual temporary, since apparent unipolar depression can turn out to be bipolar disorder, bipolar II disorder can progress to bipolar I disorder, and bipolar disorder can deteriorate to schizoaffective disorder. This may be particularly problematic in studies of unipolar depression. Patients who have only had depressive episodes at the time of their participation in such a study, but at a later point show features of bipolar disorder, may compromise the identification of potential differences between unipolar and bipolar disorder.

TOWARDS DSM-V

Despite the limitations and pitfalls of categorical diagnostic classifications (85), DSM-IV and ICD-10 have greatly improved the reliability of the different mood disorder diagnoses. For the sake of continuity in epidemiologic, neurobiologic, and genetic research, we should be reluctant to change diagnostic criteria for mood episodes and mood disorders as long as there is no compelling evidence to do so. Still, there are certain areas that are open for change.

First, the very restrictive definition of mixed episodes, in particular the requirement of full syndromal criteria for mania and depression, could be relaxed, as outlined earlier. Given all the possible variants of mixed states, it is obvious that they can occur not only in bipolar I but also in bipolar II disorder.

Second, there is now enough evidence to include antidepressant-induced (hypo)mania in the diagnostic category of bipolar disorder. Psychoactive substance- or treatment-induced (hypo)mania could be a separate subcategory (bipolar III disorder), or an episode specifier of bipolar I and II disorder.

Third, a better-operationalized and more consistent definition of the hypomania–mania boundary would improve the delineation of bipolar II versus bipolar I disorder, even if it remains uncertain whether these subtypes are fundamentally different.

Fourth, the concept of rapid cycling, which represents a dimension rather than a subtype, could be defined along a continuum as proposed either by counting episodes over a certain time period (e.g., a year) or at least by dividing it into more subcategories (nonrapid, rapid, ultra-rapid, and ultradian cycling) as suggested by Kramlinger and Post (45) which allows for the inclusion of briefer but still significant episodes (29).

Finally, a specifier for depressive episodes indicating bipolarity in up till then unipolar patients, including features as summarized in Table 3, could alert clinicians and researchers for latent bipolar disorder, without prematurely crossing the boundaries of the current diagnostic classifications.

CONCLUSION

Polarity and cyclicity have been described as core dimensions of manic-depressive illness (Fig. 1) (32), which, despite its heterogeneity, can still be regarded as one of the most consistently described disorders in psychiatry (2,12). From these dimensions, a bipolar spectrum can be constructed, and within this bipolar spectrum, various subtypes have been defined. In this chapter we have highlighted the main areas of uncertainty and controversy about the internal and external boundaries of this spectrum, which rely exclusively on clinical description and still lack external validators. It may well be that the most specific validators for bipolar disorder will be revealed by studying patients with the core bipolar I syndrome (66). Eventually, neurobiological and genetic studies must not only provide such validators for accurate and valid diagnoses, but above all direct towards a more targeted treatment for the individual patient with a variant of bipolar spectrum disorder.

REFERENCES

1. Jackson SW. Melancholia and Depression. From Hippocratic Times to Modern Times. New Haven: Yale University Press, 1986.
2. Akiskal H. Classification, diagnosis, and boundaries of bipolar disorders: a review. In: Maj M, Akiskal HS, Lopez-Ibor JJ, Sartorius N, eds. Bipolar Disorder. Evidence and Experience in Psychiatry. Vol. 6. Chichester: Wiley, 2002.
3. Kraepelin E. Psychiatrie. Ein Lehrbuch für Studirende und Aerzte. II. Band. Sechste Auflage. Leipzig: Verlag Barth, 1899.

4. Angst J, Sellaro R. Historical perspectives and natural history of bipolar disorder. Biol Psychiatry 2000; 48:445–457.
5. Leonhard K. Aufteilung der endogenen Psychosen. Berlin: Academie Verlag, 1957.
6. Angst J. Zur Ätiologie und Nosologie endogener depressiver Psychosen. Berlin: Springer Verlag, 1966.
7. Perris C. A study of bipolar (manic-depressive) and unipolar recurrent depressive psychoses. Acta Psychiatr Scand 1966; 42(suppl 194):68–82.
8. Winokur G, Clayton PJ, Reich T. Manic-Depressive Illness. St. Louis: Mosby, 1969.
9. American Psychiatric Association. Diagnostic and Statistical Manual of Mental Disorders. 3rd ed. (DSM-III). Washington DC: American Psychiatric Press, 1980.
10. American Psychiatric Association. Diagnostic and Statistical Manual of Mental Disorders. 3rd ed.—Revised (DSM-III-R). Washington DC: American Psychiatric Press, 1987.
11. World Health Organization. The ICD-9 Classification of Mental and Behavioural Disorders. Clinical Descriptions and Diagnostic Guidelines. Geneva: WHO, 1978.
12. World Health Organization. The ICD-10 Classification of Mental and Behavioural Disorders. Clinical Descriptions and Diagnostic Guidelines. Geneva: WHO, 1992.
13. American Psychiatric Association. Diagnostic and Statistical Manual of Mental Disorders. 4th ed. (DSM-IV). Washington DC: American Psychiatric Press, 1994.
14. American Psychiatric Association. Diagnostic and Statistical Manual of Mental Disorders. 4th ed.—Text Revision (DSM-IV-TR). Washington DC: American Psychiatric Press, 2000.
15. Angst J, Gerber-Werder R, Zuberbuhler HU, Gamma A. Is bipolar I disorder heterogeneous? Eur Arch Psychiatry Clin Neurosci 2004; 254:82–91.
16. Dunner DL, Tay LK. Diagnostic reliability of the history of hypomania in bipolar II patients and patients with major depression. Compr Psychiatry 1993; 34:303–307.
17. Berk M, Dodd S. Bipolar II disorder: a review. Bipolar Disord 2005; 7:11–21.
18. World Health Organization. The ICD-10 Classification of Mental and Behavioural Disorders. Diagnostic Criteria for Research. Geneva: WHO, 1993.
19. Goodwin G. Hypomania: what's in a name? Br J Psychiatry 2002; 181:94–95.
20. Angst J, Gamma A, Benazzi F, Ajdacic V, Eich D, Rossler W. Toward a re-definition of subthreshold bipolarity: epidemiology and proposed criteria for bipolar-II, minor bipolar disorders and hypomania. J Affect Disord 2003; 73:133–146.
21. Judd LJ, Akiskal HS, Schettler PJ, et al. A prospective investigation of the natural history of the long-term weekly symptomatic status of bipolar II disorder. Arch Gen Psychiatry 2003; 60:261–269.
22. Hirschfeld RM, Williams JB, Spitzer RL, et al. Development and validation of a screening instrument for bipolar spectrum disorder: the Mood Disorder Questionnaire. Am J Psychiatry 2000; 157:1873–1875.
23. Hirschfeld RM, Calabrese JR, Weissman MM, et al. Screening for bipolar disorder in the community. J Clin Psychiatry 2003; 64: 53–59.
24. Chun BJ, Dunner DL. A review of antidepressant-induced hypomania in major depression: suggestions for DSM-V. Bipolar Disord 2004; 6:32–42.
25. Akiskal HS, Maser JD, Zeller PJ, et al. Switching from "unipolar" to bipolar II. An 11-year prospective study of clinical and temperamental predictors in 559 patients. Arch Gen Psychiatry 1995; 52:114–123.
26. Lish JD, Dime-Meenan S, Whybrow PC, Price RA, Hirschfeld RM. The National Depressive and Manic-depressive Association (DMDA) survey of bipolar members. J Affect Disord 1994; 31:281–294.
27. Suppes T, Leverich GS, Keck PE, et al. The Stanley Foundation Bipolar Treatment Outcome Network. II. Demographics and illness characteristics of the first 261 patients. J Affect Disord 2001; 67:45–59.
28. Judd LJ, Akiskal HS, Schettler PJ, et al. The long-term natural history of the weekly symptomatic status of bipolar I disorder. Arch Gen Psychiatry 2002; 59: 530–537.
29. Kupka RW, Luckenbaugh DA, Post RM, et al. A comparative study of rapid and non-rapid cycling bipolar disorder using daily prospective mood ratings in 539 out-patients. Am J Psychiatry 2005; 162:1273–1280.

30. Joffe RT, MacQueen GM, Marriott M, Trevor Young L. A prospective, longitudinal study of percentage of time spent ill in patients with bipolar I or bipolar II disorders. Bipolar Disorders 2004; 6:62–66.
31. Beigel A, Murphy DL. Unipolar and bipolar affective illness. Differences in clinical characteristics accompanying depression. Arch Gen Psychiatry 1971; 24:215–220.
32. Goodwin FK, Jamison KR. Manic-Depressive Illness. New York: Oxford University Press, 1990.
33. Winokur G, Coryell W, Endicott J, Akiskal H. Further distinctions between manic-depressive illness (bipolar disorder) and primary depressive disorder (unipolar depression). Am J Psychiatry 1993; 150:1176–1181.
34. Winokur G, Coryell W, Keller M, Endicott J, Akiskal H. A prospective follow-up of patients with bipolar and primary unipolar affective disorder. Arch Gen Psychiatry 1993; 50:457–465.
35. Mitchell PB, Wilhelm K, Parker G, Austin MP, Rutgers P, Malhi GS. The clinical features of bipolar depression: a comparison with matched major depressive disorder patients. J Clin Psychiatry 2001; 62:212–216.
36. Ghaemi SN, Hsu DJ, Ko JY, Baldassano CF, Kontos NJ, Goodwin FK. Bipolar spectrum disorder: a pilot study. Psychopathology 2004; 37:222–226.
37. Bowden CL. A different depression: clinical distinctions between bipolar and unipolar depression. J Affect Disord 2005; 84:117–125.
38. Marnereos A. Origin and development of concepts of bipolar mixed states. J Affect Disord 2001; 67:229–240.
39. Swann AC. Depression, mania, and feeling bad: the role of dysphoria in mixed states. Bipolar Disord 2000; 2:325–327.
40. McElroy SL, Keck PE Jr, Pope HG Jr, Hudson JI, Faedda GL, Swann AC. Clinical and research implications of the diagnosis of dysphoric or mixed mania or hypomania. Am J Psychiatry 1992; 149:1633–1644.
41. Akiskal HS, Hantouche EG, Bourgeois ML, et al. Gender, temperament, and the clinical picture in dysphoric mixed mania: findings from a French national study (EPIMAN). J Affect Disord 1998; 50:175–186.
42. Suppes T, Mintz J, McElroy SL, et al. Mixed hypomania in 908 patients with bipolar disorder evaluated prospectively in the Stanley Foundation Bipolar Treatment Network: a sex-specific phenomenon. Arch Gen Psychiatry 2005; 62:1089–1096.
43. Swann AC, Bowden CL, Morris D, et al. Depression during mania. Treatment response to lithium or divalproex. Arch Gen Psychiatry 1997; 54:37–42.
44. Andrade C. Antidepressant-withdrawal mania: a critical review and synthesis of the literature. J Clin Psychiatry 2004; 65:987–993.
45. Kramlinger KG, Post RM. Ultra-rapid and ultradian cycling in bipolar affective illness. Br J Psychiatry 1996; 168:314–323.
46. Paris J. Borderline or bipolar? Distinguishing borderline personality disorder from bipolar spectrum disorders. Harv Rev Psychiatry 2004;12:140–145.
47. Mackinnon DF, Pies R. Affective instability as rapid cycling: theoretical and clinical implications for borderline personality and bipolar spectrum disorders. Bipolar Disord 2006; 8:1–14.
48. Akiskal HS, Yerevanian BI, Davis GC, King D, Lemmi H. The nosologic status of borderline personality: clinical and polysomnographic study. Am J Psychiatry 1985;142:192–198.
49. Marneros A, Rottig S, Wenzel A, Bloink R, Brieger O. Schizoaffective mixed states. In: Marneros A, Goodwin F, eds. Bipolar Disorders. Mixed States, Rapid Cycling and Atypical Forms. Cambridge: University Press, 2005:187–206.
50. Maj M, Pirozzi R, Formicola AM, Bartoli L, Bucci P. Reliability and validity of the DSM-IV diagnostic category of schizoaffective disorder: preliminary data. J Affect Disord 2000; 57:95–98.
51. Craddock N, Owen MJ. The beginning of the end for the Kraepelinian dichotomy. Br J Psychiatry 2005; 186:364–366.
52. Beckmann H, Pfuhlmann B. Re-examining the bipolar-schizophrenia dichotomy. In: Kasper S, Hirschfeld RMA, eds. Handbook of bipolar disorder. Diagnosis and therapeutic approaches. New York: Taylor and Francis, 2005:37–48.

53. Dunner DL. Bipolar Disorders in DSM-IV: Impact of rapid cycling as a course modifier. Neuropsychopharmacology 1998; 19:189–193.
54. Benazzi F, Koukopoulos A, Akiskal HS. Toward a validation of a new definition of agitated depression as a bipolar mixed state (mixed depression). Eur Psychiatry 2004; 19:85–90.
55. Judd LL, Akiskal HS, Schettler PJ, et al. The comparative clinical phenotype and long term longitudinal episode course of bipolar I and II: a clinical spectrum or distinct disorders? J Affect Disord 2003; 73:19–32.
56. Bauer MS, Calabrese J, Dunner DL, et al. Multisite data reanalysis of the validity of rapid cycling as a course modifier for bipolar disorder in DSM-IV. Am J Psychiatry 1994; 151:506–515.
57. Kupka RW, Luckenbaugh DA, Post RM, Leverich GS, Nolen WA. Rapid and non-rapid cycling bipolar disorders: a meta-analysis of clinical studies. J Clin Psychiatry 2003; 64:1483–1494.
58. Coryell W, Solomon D, Turvey C, et al. The long-term course of rapid-cycling bipolar disorder. Arch Gen Psychiatry 2003; 60:914–920.
59. Schneck CD, Miklowitz DJ, Calabrese JR, et al. Phenomenology of rapid cycling bipolar disorder: data from the first 500 participants in the Systematic Treatment Enhancement Program. Am J Psychiatry 2004; 161:1902–1908.
60. Akiskal HS. The bipolar spectrum: new concepts in classification and diagnosis. In: Grinspoon L, ed. Psychiatry Update 1983. Washington: American Psychiatric Press, 1983.
61. Akiskal HS. The bipolar spectrum: history, description, boundaries, and validity. In: Kasper S, Hirschfeld RMA, eds. Handbook of Bipolar Disorder. Diagnosis and Therapeutic Approaches. New York: Taylor and Francis, 2005:49–68.
62. Katzow JJ, Hsu DJ, Nassir Ghaemi S. The bipolar spectrum: a clinical perspective. Bipolar Disord 2003; 5:436–442.
63. Cassano GB, Rucci P, Frank E, et al. The mood spectrum in unipolar and bipolar disorder: arguments for a unitary approach. Am J Psychiatry 2004; 161:1264–1269.
64. Angst J, Cassano G. The mood spectrum: improving the diagnosis of bipolar disorder. Bipolar Disord 2005; 7(suppl 4):4–12.
65. Dunner DL. Clinical consequences of under-recognized bipolar spectrum disorder. Bipolar Disorders 2003; 5:456–463.
66. Baldessarini RJ. A plea for the integrity of the bipolar disorder concept. Bipolar disorders 2000; 2:3–7.
67. Judd LL, Akiskal HS. The prevalence and disability of bipolar spectrum disorders in the US population: re-analysis of the ECA database taking into account subthreshold cases. J Affect Disord 2003; 73:123–131.
68. First MB, Gibbon M, Spitzer RL, Williams JBW. Structured Clinical Interview for DSM-IV Axis I Disorders: SCID-I/P (Version 2.0). New York: Biometrics Research Department, 1996.
69. Sheehan DV, Lecrubier Y, Sheehan KH, et al. The Mini-International Neuropsychiatric Interview (M.I.N.I.): the development and validation of a structured diagnostic psychiatric interview for DSM-IV and ICD-10. J Clin Psychiatry 1998; 59(suppl 20):22–33.
70. Wing JK, Babor T, Brugha T, et al. SCAN. Schedules for Clinical Assessment in Neuropsychiatry. Arch Gen Psychiatry 1990; 47:589–593.
71. Robins LN, Wing J, Wittchen HU, et al. The Composite International Diagnostic Interview. An epidemiologic Instrument suitable for use in conjunction with different diagnostic systems and in different cultures. Arch Gen Psychiatry 1988; 45: 1069–1077.
72. Regeer EJ, ten Have M, Rosso ML, Hakkaart-van Roijen L, Vollebergh W, Nolen WA. Prevalence of bipolar disorder in the general population: a Reappraisal Study of the Netherlands Mental Health Survey and Incidence Study. Acta Psychiatr Scand 2004; 110:374–382.
73. Craddock N, Jones I, Kirov G, Jones L. The Bipolar Affective Disorder Dimension Scale (BADDS)—a dimensional scale for rating lifetime psychopathology in bipolar spectrum disorders. BMC Psychiatry 2004; 4:19.

74. Young RC, Biggs JT, Ziegler VE, Meyer DA. A rating scale for mania: reliability, validity and sensitivity. Br J Psychiatry 1978; 133:429–435.
75. Bech P, Rafaelsen OJ, Kramp P, Bolwig TG. The mania rating scale: scale construction and inter-observer agreement. Neuropharmacology 1978; 17:430–431.
76. Hamilton M. A rating scale for depression. J Neurol Neurosurg Psychiatry 1960; 23:56–62.
77. Montgomery SA, Asberg M. A new depression scale designed to be sensitive to change. Br J Psychiatry 1979; 134:382–389.
78. Rush AJ, Gullion CM, Basco MR, Jarrett RB, Trivedi MH. The Inventory of Depressive Symptomatology (IDS): psychometric properties. Psychol Med 1996; 26:477–486.
79. Spearing MK, Post RM, Leverich GS, Brandt D, Nolen W. Modification of the Clinical Global Impressions (CGI) Scale for use in bipolar illness (BP): the CGI-BP. Psychiatry Res 1997; 73:159–171.
80. Post RM, Roy-Byrne PP, Uhde TW. Graphic representation of the life course of illness in patients with affective disorder. American Journal of Psychiatry 1988; 145:844–848.
81. Leverich GS, Post RM. Life Charting of Affective Illness. CNS Spectrums 1998; 3:21–37.
82. Denicoff KD, Leverich GS, Nolen WA, et al. Validation of the prospective NIMH-Life-Chart Method (NIMH-LCM-p) for longitudinal assessment of bipolar illness. Psychol Med 2000; 30:1391–1397.
83. Meaden PM, Daniels RE, Zajecka J. Construct validity of life chart functioning scales for use in naturalistic studies of bipolar disorder. J Psychiatr Res 2000; 34:187–192.
84. Bauer M, Grof P, Gyulai L, Rasgon N, Glenn T, Whybrow PC. Using technology to improve longitudinal studies: self-reporting with ChronoRecord in bipolar disorder. Bipolar Disord 2004; 6:67–74.
85. Kendell R. The choice of diagnostic criteria for biological research. Arch Gen Psychiatry 1982; 32:1334–1339.

2 Prospects for the Development of Animal Models for the Study of Bipolar Disorder

Haim Einat
College of Pharmacy, University of Minnesota, Duluth, Minnesota, U.S.A.

Alona Shaldubina, Yuly Bersudsky, and R. H. Belmaker
Division of Psychiatry, Faculty of Health Sciences, Ben-Gurion University of the Negev, Beer Sheba, Israel

INTRODUCTION

Attitudes towards the study of lithium on behavior in animals and humans have often been influenced by preconceived notions of the nature of psychiatry and psychopharmacologic treatment. Many psychiatrists would like to see psychopharmacological agents as "magic bullets" in a sense similar to antibiotics. Antibiotics are not expected to have effects on organisms that are not infected with bacteria; effects that do occur are seen as side effects unrelated to the mode of action. Thus neuroleptic dopamine-blocking drugs are believed by many psychiatrists to have no effects on persons who are not psychotic; antidepressants are seen as distinguished from stimulants of abuse as not having mood-elevating effects in the absence of depression. The study of lithium's effects has often been carried out in a similar tradition. Moreover, lithium is seen as a difficult drug to give to normal volunteers for the period of three weeks or a month that would approximate the amount of time necessary for significant effect on a manic episode.

PSYCHOSTIMULANT-INDUCED MODELS

Animal studies have been a mainstay of the study of lithium effects on behavior other than clinical studies of psychophathology. These animal studies were reviewed in 1991 (1) and again in 2003 (2). The overriding majority of studies in this field has used the concept of pharmacologically induced mania and depression and has attempted to show lithium prevention. The usual agent for pharmacological induction of mania has been amphetamine, although more specific and direct dopamine agonists such as quinpirole have also been used (3). Using reserpine or tetrabenazine to induce depression has also been studied (4,5). A background concept has been the fact that low dose amphetamine has effects primarily on open field activity whereas higher dose amphetamine causes stereotypy (6,7). Dopamine blockers are well known to block both the low dose hyperactivity and the high dose stereotypy of amphetamine, and this nicely fits their usefulness in both mania and schizophrenia (8). Further support for the amphetamine-induced hyperactivity model comes from the association of psychostimulants with the onset of mania in susceptible individuals (9,10) and from some clinical studies that support an effect of lithium in preventing the behavioral effects of stimulants in people (11,12). To delineate the differences between the dissimilar responses to doses of amphetamine, attempts have been made to differentially look at the effect of lithium pretreatment on low dose amphetamine effects versus high dose amphetamine effects, the hypothesis being that lithium as an antimanic agent will prevent the effects of low-dose amphetamine but because it is devoid of true antipsychotic properties it will not be able to affect amphetamine-induced stereotypy. Often an underlying biochemical hypothesis was that low-dose amphetamine released mostly serotonin and noradrenaline whereas high dose amphetamine released more dopamine. All of these assumptions, hypotheses, and preconceptions are today viewed skeptically. They were heuristic as hypotheses that generated much good research; however, the yield of robust replicable data has been poor. Given the fact that modern studies of psychopharmacologic agents often require hundreds of patients to show statistically significant effects, it may not be surprising that animal studies with 10 to 15 rats in each group often come up with conflicting results. The heterogeneity of the amphetamine response is well known (13), and lithium response may also be heterogeneous in outbred rat strains (14,15). We shall discuss later the possibility of using this heterogeneity in responses to advance research.

The effect of lithium to block hyperactivity in rats has also been given new impetus by a paper from Caron's group (16). They injected dopamine transporter knockout mice with lithium (50, 100, or 200 mg/kg) a half hour after they were placed in an open field and found significant reduction in horizontal activity in mice injected with 100 and 200 mg lithium. This paradigm had previously been used by Nixon et al. (17) and positive results had also been found. However, in humans even very high doses of lithium do not have immediate effects in mania but clearly do have side effects such as nausea and muscle weakness. Therefore, it is difficult to evaluate these acute effects reported by Caron's group (16), especially in the light of the long history of contradictory results in this field. Several reports demonstrate inhibition of amphetamine-induced hyperactivity by acute and chronic lithium pretreatment in rats and mice (7,15,18–23). However, other publications show absence of lithium effect in this model (8,24–27). New data coming from Manji's laboratory (28) also indicate that lithium's effects on amphetamine hyperactivity may be related to genetic background. A comprehensive study of 12 strains of mice (three outbred and nine inbred strains) shows that acute lithium injection at 100 mg/kg attenuated amphetamine-induced hyperactivity in four strains, had no effect in four strains, and augmented the effects of amphetamine in one strain (other strains did not respond to amphetamine treatment at the dose used). Moreover, chronic oral lithium treatment at concentrations that had been previously demonstrated to result in therapeutic blood levels (four weeks, 2.4 g/kg in food), resulted in similar effects with acute administration in some strains but not in all (28). All in all, these results suggest that the effects of lithium on amphetamine hyperactivity may be strain (genetically) dependent and that there may be different mechanisms that are involved in the short term (acute) and long term (chronic) effects of lithium in this model. An additional level of complexity in evaluation of lithium's effects on the response to amphetamine is added by the fact that behavioral responses to psychostimulants are strongly affected by the testing environment and procedure. These variations are more prominent during chronic psychostimulant treatment but are also apparent after acute administration (29). For example, hyperactivity measures had been shown to increase more in large open field arena compared to smaller activity monitors (30) and an environment that is similar to the home cage had been demonstrated to hinder the development of ambulatory activity (31).

It is possible to give amphetamine or methylphenidate to humans on chronic lithium and this was done in many studies (11,14,21,32,33). While some have claimed marked effects of lithium to attenuate the amphetamine-induced response, others have found no effect at all (32). The numbers on subjects in these studies are smaller than the large numbers of patients required to show a lithium effect in a clinical situation and human heterogeneity may be the answer to the contradictory results. We await a clear paradigm that will give robust findings in the amphetamine hyperactivity model of mania in humans as in rodents. Perhaps the issue is blood lithium levels since these have varied greatly between studies and usually levels of 0.7 mM have been considered sufficiently similar to human treatment to be an acceptable model. It is unclear why behavioral effects are so difficult to demonstrate, whereas biochemical effects of lithium in normal subjects are marked and highly replicable (34).

Interestingly, the notion of endophenotypes that had recently been strongly emphasized in the research of bipolar disorder had renewed the interest of scientists in the strength of amphetamine-induced behaviors and underlying brain changes as important modeling tools.

Endophenotypes are quantifiable components in the genes-to-behaviors pathways, distinct from psychiatric symptoms that make genetic and biological studies of etiologies for disease categories more manageable (35). In the context of modeling, endophenotypes approach can be helpful as it reduces the complexity of symptoms and multifaceted behaviors, resulting in units of analysis that are simpler to model in animals (36).

One of the tentative endophenotypes that had been repeatedly suggested for bipolar disorder is dysregulation of dopaminergic function and hypersensitivity to psychostimulants. This possible endophenotype was suggested based on significant data in both human and animal studies. Impaired brain reward pathways, enhanced rewarding effects of psychostimulants in patients with affective illness, possible relationship between dopamine release in the ventral striatum, euphoric responses, and some evidence for genetic variance that may explain the individual differences in brain response to psychostimulants all suggest that behavioral changes observed after exposure to amphetamine may be useful as marker for bipolar disorder (37). However, this new line of study, exploring amphetamine responses not as a model of mania but as a model of an endophenotype of bipolar disorder, must include additional experimentation that will look beyond hyperactivity into different facets of amphetamine-induced behavior, the possible relationship between such behaviors, the effects of mood stabilizers, and the genetic predisposition related to individual variability in responses to amphetamine as well as to lithium effects on amphetamine-induced behavior. Some suggestions along these lines are detailed later in this chapter.

LITHIUM AS AN ANTIDEPRESSANT

Beyond the amphetamine-related models, a very exciting advance in this field has occurred recently, in a paper by O'Brien et al. (38) who used a specific regimen of lithium administration to mice: mice received 0.2% lithium chloride in food for a period of five days followed by 0.4% for 10 additional days and reported robust effects in the Porsolt forced swim test. Previous studies of lithium in Porsolt forced swim test were equivocal, although Bourin's group (39,40) showed that lithium could reliably potentiate the effects of other antidepressants. The study of Bourin et al. (39,40) used an acute lithium dose but it fit the preconceived notion that in the clinic lithium is an augmenter of antidepressant response and not a powerful antidepressant itself. However, the paradigm of O'Brien et al. (38) suggests a powerful effect of lithium in the Porsolt forced swim test. This is unlikely to be an artifact, since activity in the Porsolt forced swim test requires an increase in struggling behavior. Previous concerns about lithium artifacts have usually pointed out that lithium patients experience some sense of malaise and muscle weakness and nausea. These would be unlikely to cause the reported effects in the Porsolt forced swim test.

We (41) have been able to replicate the O'Brien et al. (38) finding and have shown that it is dependent on blood levels. Blood levels greater than 1 mM are necessary for the robust effect that O'Brien et al. (38) finds, whereas blood levels of 0.7 mM show no effect in the Porsolt forced swim test at all. Many studies of chronic lithium in the past were quite satisfied with levels of 0.7 mM on the average and even studies with higher blood levels had a significant portion of the animals with blood levels below 0.7 mM. This robust effect of lithium on the Porsolt forced swim test provides an opening for behavioral pharmacological analysis in the future. For instance, questions can be asked such as whether pretreatment with PCPA, a serotonin synthesis inhibitor, will prevent the effect of lithium in the Porsolt forced swim test or whether presynaptic $5HT_{1a/1b}$ or postsynaptic $5HT_2$ or β-adrenergic receptors agonists/antagonists will modulate this effect. A recent hypothesis of antidepressant action is induction of neurogenesis in the hippocampus. It could be an interesting question, whether TrkB (BDNF) receptor agonists/antagonists will affect the lithium's antidepressant effect in the Porsolt forced swim test. Interestingly, inhibition of the Erk-MAP kinase pathway was demonstrated to decrease immobility time in the forced swim test (42). However, the same treatment also increased activity in an open field and this effect was ameliorated by chronic lithium treatment. Hence, suggesting that the effects of Erk inhibition is less likely to be antidepressant-like and more likely to be promanic (42), a notion that is further supported by other studies on the behavioral effects of manipulating the Erk pathway (43,44). A key question would be whether other mood stabilizers such as valproate have a similar effect in the Porsolt swim test. Another key question will be whether the weight loss due to reduced appetite in chronically lithium-treated rats might cause increased activity in the Porsolt swim test. This needs to be done by "yoking" mice to others who are eating lithium and let them eat only the exact same amount a day as the lithium-treated animals eat. It is also possible to add a nontoxic bitter taste to the control food to reduce the food intake and to see if this affects Porsolt results. Our finding that lithium effects in the Porsolt swim test require a blood level greater than 1 mM is actually congruent with clinical reports that the antidepressant effects of lithium require higher blood levels than the prophylactic effect.

CURRENT PROBLEMS AND POSSIBLE SOLUTIONS

Although there were some advances in modeling bipolar disorders, it appears that the field had been quite limited for many years compared with model development for other psychiatric disorders. It is possible that, at least in part, the nature of the disease that includes oscillating between depression and mania episodes hindered scientists from making serious attempts to model it. Just a few attempts were done over the years to model the entire scope of bipolar disorder with manipulations such as sleep deprivation (45) or intermittent cocaine administration (46), but for a variety of reasons, these tentative models did not become a central tool to explore the biology of bipolar disorder or to screen new drugs for it (2,47).

Bioassays

Attempts had also been made over the years to develop models that are more a bioassay than a comprehensive behavioral model. An example of such incomplete model is the study of Bersudsky et al. (48) showing lithium inhibition of forskolin-induced hypoactivity. This study is based on the fact that lithium biochemically inhibits forskolin induced rises in cyclic AMP. The behavioral finding is therefore a bioassay of the chemical finding. However, to become a model it would need to be corroborated by a finding that forskolin induces hypoactivity or a depressive-like syndrome in humans. Another example is lithium augmentation of pilocarpine-induced seizures. This phenomenon had been repeatedly demonstrated and can be used to explore lithium-mimetic drugs. Furthermore, the increase in seizure susceptibility after lithium treatment was demonstrated to be dependant on inositol depletion as it is blocked by inositol administration (49) and augmented by inositol reuptake inhibition (50). Since the inositol depletion theory (51) is one of the leading hypotheses regarding the therapeutic effects of lithium, the use of pilocarpine-induced seizures as a rudimentary screening model can be justified. However, the behavioral phenomenon is unrelated to the features of the disease and therefore the utility of this paradigm as a real model is questionable.

Whereas the models mentioned above did contribute to the research efforts on bipolar disorder and its treatment, there is clearly a lack of better and more appropriate animal models for the disease. This deficiency is repeatedly emphasized as one of the major problems hindering bipolar disorder research (52). Some new approaches recently suggested in the literature are summarized below.

Modeling Facets of the Disease

One relatively simple approach stays within the realm of modeling based on face validity, that is, the similarity in behavior observed in the disease and in the model (53), and looking at components of the behavior rather than the entire disease (54). However, in contrast to present work that is based mainly on very few behavioral components of bipolar disorder (e.g., hyperactivity as a model for manic behavior), this approach suggests a more comprehensive battery of tests and models that will explore a broader range of the behavioral facets of the disease. Accordingly, it may be possible to develop separate models for facets of mania such as activity or restlessness; extreme irritability; reduced sleep; provocative, intrusive, or aggressive behavior; increased sexual drive; abuse of drugs; distractibility or reduced ability to concentrate; and poor judgment. Furthermore, many such models were already developed in the context of research of other disorders but they must be validated for bipolar disorder (54). If some of these models can be validated, it may be possible to develop a battery of models that will be appropriate for the screening of possible new mood stabilizers or to test new hypotheses regarding the underlying biology of the disorder. One example of such an initial validation attempt was recently demonstrated with a model for aggression (55). Since intrusive and aggressive behaviors are one of the facets of mania, this study tested the validity of a commonly used test for aggression, the resident intruder paradigm in mice, as a tentative model for this facet of mania. The results of the study demonstrate that chronic administration of lithium or valproate, at therapeutically relevant doses, ameliorates the aggressive behavior in the resident intruder paradigm without affecting other aspects of social behavior. Accordingly, this study suggests that the paradigm has predictive validity and can be used as part of a battery of models for the study of new mood stabilizers (54). Additional new models emphasized aggression in a competition for food task (56) and irritability measured as resistance to capture (57,58).

Interestingly, even within this relatively simple approach, a number of candidate manipulations were identified that had been previously demonstrated to result in a number of behavioral changes that are similar to facets of mania. Therefore, if all these specific models will be validated, the resulting battery will represent a group of bipolar-like behaviors (54). For example, psychostimulant administration does not result only in hyperactivity (as discussed above) but was also reported to induce reduced sleep, distractibility, risk taking behavior, and increased responses to reward—all facets of mania (54).

From Molecules to Behavior

Although the approach described above may be conducive to further research, modeling methods that concentrate on face validity of one component of the disorder have been repeatedly criticized (59). Recent developments in basic studies of the etiology of bipolar disorder coming from research using modern techniques of brain imaging and novel methods of molecular biology may now assist in the search for more comprehensive models that can be based more on construct validity than on face validity, that is, models that will be developed based on a possible mechanism rather than on behavioral similarities. One possible strategy that can be employed was recently alluded to in a paper from the Soares group (59). These authors suggest that genetic models can now be developed for the disease and show that appropriate and relevant genes can be identified by comparing genetic changes in available animal models to changes in patients. For example, Machado-Vieira et al. (59) summarize findings regarding the genes encoding GRK proteins and show they are related to defects in dopamine transmission, to behavioral sensitization to psychostimulants in animals, and to a more severe form of bipolar disorder in patients. Furthermore, postmortem studies have shown changes in GRK genes in the prefrontal cortex of patients who suffered from severe mood disorders (59). Altogether, the authors suggest that modifications of the GRK genes (using transgenic techniques in mice for example) may result in a better model for the disease that will be hypothesis-driven.

Other genes and intracellular pathways had been implicated during the last decade in bipolar disorder and indeed some of these ideas can be used to create hypothesis-driven models with strong construct validity.

Manipulations of many of these tentative genes proteins and intracellular pathways in animals do indeed result in behaviors that resemble bipolar disorder and were recently summarized in a review paper (44). Data regarding a variety of manipulations, pharmacological and genetic, were summarized, and the conclusion of the authors was that there is strong evidence for the involvement of PKC, GSK3, and the Erk-pathway in bipolar-like changes with some evidence supporting additional mechanisms including AMPA receptors, inositol, glucocorticoid receptors, and Bcl-2 (44). Much of the behavioral information presented in that review was collected while studying other disorders or the functions of normal brain behavior and studies in the context of bipolar disorder are now needed to further explore the role of these molecules and pathways in bipolar-like behavior in animals. Yet, the data support the notion that bipolar-like behavioral changes correspond with manipulations of bipolar-related molecules and this now may be a reasonable approach to develop more specific, hypothesis-driven models for the disease.

Individual Variability

Further exploration of the relationship between specific genes, molecules, pathways, and behavioral models, may be enhanced by looking at individual variability of responses. The issue of individual differences in behavioral modeling has been grossly neglected for many practical reasons, but this neglect may represent one of our major failures. It is apparent that the etiology of bipolar disorder (as of other psychiatric and nonpsychiatric diseases) is based on an interaction between the underlying genetics and the environmental effects on the biology where susceptible individuals that are exposed to environmental precipitating factors will express the disease phenotype (60–62). However, in most of our attempts to model bipolar disorder we expose a group of "normal" animals to a specific manipulation (whether it is a lesion, a drug, or an environmental stimulus), and we expect them to become "sick" and allow us to explore possible new therapies or the underlying biology of the "sickness." Alternatively, with the developments in transgenic technology, we manipulate a mouse gene that is implicated in bipolar disorder or its treatment and expect the entire population of mutant mice to behave differently than the wild type controls. These approaches to modeling can be helpful when there is an expectancy that a single gene mutation may be responsible for a major part of a disease or its treatment, but this is probably not the case for bipolar disorder, and accordingly, this approach may be limited to demonstrating involvement of specific genes in the disease.

Any scientist who has been studying behavior knows the wide range of individual variability within groups. Usually we try to overcome this variability by increasing group size, but further attention to individual responses may in fact be conducive to our research. If indeed subgroups of animals within a group exposed to a specific manipulation can be identified as responders versus nonresponders (higher vs. lower behavioral change), it will enable us to (*i*) use the responders as a better model for the disease and (*ii*) explore the biological differences between the subgroups. Some work using such methods has been done in the context of other psychiatric disorders with interesting results demonstrating a relationship between the extent of a behavioral response and biological changes (63–66). For example, Cohen and her colleagues (67–69) exposed outbred rats to a traumatic experience (inescapable cat odor) and tested them for anxiety-like measures immediately and 10 days after the exposure. Whereas all rats showed anxiety-like responses immediately after exposure, only about 30% of animals remained anxious at the later testing. Interestingly, the animals that had a long-term effect on behavior also had long lasting changes in physiological measures (heart rate variability) and biochemical measures (higher plasma corticosterone and ACTH levels, increased sympathetic activity, diminished vagal tone, and increased sympathovagal balance) suggesting that these animals may be an excellent model for post-traumatic stress disorder (66). It may now be interesting to explore the underlying genetic differences that may account for the differential responding in these subgroups of outbred rats. A number of attempts to look at individual variability have also been done in the context of depression and bipolar disorder (65,70,71), but for most of it, scientists who identified differential responses tried to amplify them by breeding the different subgroups to create different lines of responders versus nonresponders (72,73). This approach may clarify some of the genetics by making an extreme "caricature" of the initial strain. However, the process of inbreeding may also mask the variability of the normal population since other biological changes that evolve during inbreeding may overshadow the specific differences that were responsible for the initial differential responses.

Research that emphasizes diversity in responding presents two main problems. First, a technical issue: if we want to identify subgroups in a general population, we must start with a much larger number of animals. In light of constraints such as money, space, and constant ethical concerns about animal research, this may not always be easy. The second problem is more conceptual. Looking at individual variability demands that we first identify subgroups within a population and then test them in the context of our study. For example, if one hypothesizes that animals that show a higher response to psychostimulants may model the susceptibility of manic patients to these drugs and wants to test the effects of a new mood stabilizer in this model, the first stage would be to treat a large group of animals with a psychostimulant, identify the high- versus low-responding subgroups, then treat with the new mood stabilizer, and see if indeed it has an effect in the susceptible group but not in the resilient animals (as we may expect from a good mood stabilizer). However, in testing for the effects of the new drug, the behavior is not only influenced by the new treatment or the initial differences between the subgroups, but also by the experience the animals had during the screening procedure. Yet, if we can identify screening procedures that are minimally intrusive or invasive, further attention to individual variability may open many new avenues for our research and may result in significantly better models.

Modeling Endophenotypes

An additional approach that in a way combines many of the tentative methods described above is modeling endophenotypes of disease. As mentioned earlier, endophenotypes are quantifiable components in the genes-to-behaviors pathways, distinct from psychiatric symptoms. Endophenotypes are heritable; they are associated with illness in the population; they are state-independent (manifest in an individual whether or not illness is active) and may need to be elicited by a challenge (36). In the context of animal models, it is important to emphasize that endophenotypes are not synonymous with symptoms. As such, an animal model of an endophenotype of bipolar disorder may not have face validity for any facet of the disease but will have strong construct validity for the endophenotype [for in-depth discussion of the validity of models in psychiatry see (2,47,74)]. Animal models based on the endophenotypes approach may not be ideal for drug screening purposes but may have great importance in the attempts to explore genes and validate neurobiological mechanisms in model organisms (36).

Current theories regarding tentative endophenotypes for bipolar disorder based on genetic and biological studies of patients and families include attention deficits, circadian rhythm instability, irregularities in motivation and reward, brain structural changes, increased sensitivity to stress and psychostimulants, and limbic-hypothalamo-pituitary-adrenocortical (LHPA) axis malfunction (36,37). Animal models that will represent any of these tentative endophenotypes will be helpful to decipher the biological basis of these specific endophenotypes and the relationship between a specific endophenotype and susceptibility to the disease. Models for endophenotypes can be developed genetically, by modulation of specific genes that are implicated in the endophenotype, but can also be developed based on individual variability within groups of "normal" animals as described above. For example, it was recently demonstrated that the aggression displayed in the resident-intruder test is ameliorated by chronic mood stabilizers treatment (55). This behavior had been previously shown to be LHPA-axis dependent (56,75) and it would be interesting to see now if variability in this behavior may be related to other behavioral measures related to the LHPA axis and to any specific genetic features (55).

CONCLUSIONS

There are several directions in which this field can go heuristically:

1. Development of an entirely novel model. For instance, dogs are more difficult to study than rats, involving more expense and greater ethical concerns. However, male dogs exposed to the scent of vaginal secretions of a female dog in heat become hyperactive, aggressive, hypersexual, and will not sleep or eat for days while under the influence of this scent. Since hypersexuality and hyperactivity are clearly parts of mania and since a new love affair is a frequent stimulus for the onset of a manic episode, this model could have face validity. The effects of lithium and other mood stabilizers on this model could be an important direction. The biochemical effects of the pheromones of female canines in heat on the brain of the male dog might also elicit important information.
   Recent papers (16,38) suggest that the classic field of study of lithium effects on amphetamine hyperactivity or on the forced swim test might actually have been a correct direction and that the contradictory results might have been due to inadequate lithium dosing. A major effort is now underway to resolve whether this is the case. If so, studies of other mood stabilizers and the biochemical effects of higher dose lithium in these models could lead to rapid new information.
2. Validation of additional facets of the disease may provide researchers with a larger and broader arsenal of tools to explore the different components of mania and depressive behavior, especially in the context of drug and mutant animals screening (54).
3. Further attention to individual variability in behavioral response may be critical for the development of more clinically relevant models and can forward the understanding of genetic differences that may account for behavioral diversity. Individual variability can also be of great importance in the exploration of models for the endophenotypes of disease.
4. The notion of endophenotypes in bipolar disorder, suggesting an intermediate level of exploration between symptoms and disease that may be genetically and biologically relevant, poses a set of new challenges in modeling (36). Each of the

tentative endophenotypes of bipolar disorder includes a set of biological and behavioral components that may be possible to model using either genetic techniques or other manipulations. Appropriate models for endophenotypes will have a major impact on further research into this rejuvenated hypothesis.

Animal models for bipolar disorder have been used for many decades with significant success in studies related to both the development of new drugs and the exploration of the biological basis of the disorder. Yet, it is now clear that with the recent major developments in molecular and genetic methodologies and brain imaging techniques, the available models cannot adequately respond to the new challenges (52,76). It is now the time for behavioral scientists to make a major effort, possibly combining all the approaches discussed above, to detect, create, and validate new models that may provide better and more adequate tools to further the research of bipolar disorder.

REFERENCES

1. Kofman O, Belmaker RH. Animal models of mania and bipolar affective disorders, In: Soubrie P, ed. Anxiety, Depression and Mania. New York: Karger, 1991:103–121.
2. Einat H, Manji HK, Belmaker RH. New approaches to modeling bipolar disorder. Psychopharmacol Bull 2003; 37(1):47–63.
3. Shaldubina A, Einat H, Szechtman H, et al. Preliminary evaluation of oral anticonvulsant treatment in the quinpirole model of bipolar disorder. J Neural Transm 2002; 109(3):433–440.
4. Lerer B, Ebstein RP, Felix A, et al. Lithium amelioration of reserpine-induced hypoactivity in rats. Int Pharmacopsychiatry 1980; 15(6):338–343.
5. Einat H, Karbovski H, Korik J, et al. Inositol reduces depressive-like behaviors in two different animal models of depression. Psychopharmacology (Berl) 1999; 144(2): 158–162.
6. Belmaker RH, Lerer B, Klein E, et al. The use of behavioral methods in the search for compounds with lithium-like activity. In: Levy A, Spiegelstein MY, eds. Behavioral Models and the Analysis of Drug Action. Amsterdam: Elsevier, 1982:343–356.
7. Borison RL, Sabelli HC, Maple PJ, et al. Lithium prevention of amphetamine-induced "manic" excitement and of reserpine-induced "depression" in mice: possible role of 2-phenylethylamine. Psychopharmacology (Berl) 1978; 59(3):259–262.
8. Ebstein RP, Eliashar S, Belmaker RH, et al. Chronic lithium treatment and dopamine-mediated behavior. Biol Psychiatry 1980; 15(3):459–467.
9. Anand A, Verhoeff P, Seneca N, et al. Brain SPECT imaging of amphetamine-induced dopamine release in euthymic bipolar disorder patients. Am J Psychiatry 2000; 157(7):1108–1114.
10. Murphy DL, Brodie HK, Goodwin FK, et al. Regular induction of hypomania by L-dopa in "bipolar" manic-depressive patients. Nature 1971; 229(5280):135–136.
11. Huey LY, Janowsky DS, Judd LL, et al. Effects of lithium carbonate on methylphenidate-induced mood, behavior, and cognitive processes. Psychopharmacology (Berl) 1981; 73(2):161–164.
12. Van Kammen DP, Murphy DL. Attenuation of the euphoriant and activating effects of d- and l-amphetamine by lithium carbonate treatment. Psychopharmacologia 1975; 44(3):215–224.
13. Tecce JJ, Cole JO. Amphetamine effects in man: paradoxical drowsiness and lowered electrical brain activity (CNV). Science 1974; 185(149):451–453.
14. Angrist B, Gershon S. Variable attenuation of amphetamine effects by lithium. Am J Psychiatry 1979; 136(6):806–810.
15. Gould TJ, Keith RA, Bhat RV. Differential sensitivity to lithium's reversal of amphetamine-induced open-field activity in two inbred strains of mice. Behav Brain Res 2001; 118(1):95–105.

16. Beaulieu JM, Sotnikova TD, Yao WD, et al. Lithium antagonizes dopamine-dependent behaviors mediated by an AKT/glycogen synthase kinase 3 signaling cascade. Proc Natl Acad Sci USA 2004; 101(14):5099–5104. Epub 2004 Mar 24.
17. Nixon MK, Hascoet M, Bourin M, et al. Additive effects of lithium and antidepressants in the forced swimming test: further evidence for involvement of the serotoninergic system. Psychopharmacology (Berl) 1994; 115(1–2):59–64.
18. Berggren U. Effects of chronic lithium treatment on brain monoamine metabolism and amphetamine-induced locomotor stimulation in rats. J Neural Transm 1985; 64(3–4): 239–250.
19. Berggren U, Tallstedt L, Ahlenius S, et al. The effect of lithium on amphetamine-induced locomotor stimulation. Psychopharmacology (Berl) 1978; 59(1):41–45.
20. Berggren U, Engel J, Liljequist S. The effect of lithium on the locomotor stimulation induced by dependence-producing drugs. J Neural Transm 1981; 50(2–4):157–164.
21. Flemenbaum A. Does lithium block the effects of amphetamine? A report of three cases. Am J Psychiatry 1974; 131(7):820–821.
22. Hamburger-Bar R, Robert M, Newman M, et al. Interstrain correlation between behavioural effects of lithium and effects on cortical cyclic AMP. Pharmacol Biochem Behav 1986; 24(1):9–13.
23. Lerer B, Globus M, Brik E, et al. Effect of treatment and withdrawal from chronic lithium in rats on stimulant-induced responses. Neuropsychobiology 1984; 11(1):28–32.
24. Arriaga F, Dugovic C, Wauquier A. Effects of lithium on dopamine behavioural supersensitivity induced by rapid eye movement sleep deprivation. Neuropsychobiology 1988; 20(1):23–27.
25. Cappeliez P, Moore E. Effects of lithium on an amphetamine animal model of bipolar disorder. Prog Neuropsychopharmacol Biol Psychiatry 1990; 14(3):347–358.
26. Fessler RG, Sturgeon RD, London SF, et al. Effects of lithium on behaviour induced by phencyclidine and amphetamine in rats. Psychopharmacology (Berl) 1982; 78(4): 373–376.
27. Pittman KJ, Jakubovic A, Fibiger HC. The effects of chronic lithium on behavioral and biochemical indices of dopamine receptor supersensitivity in the rat. Psychopharmacology (Berl) 1984; 82(4):371–377.
28. Gould TD, O'Donnell KC, Picchini AM, Manji HK. Strain differences in lithium attenuation of D-amphetamine-induced hyperlocomotion: a mouse model for the genetics of clinical response to lithium. Neuropschopharmacology 2006.
29. Einat H, Einat D, Allan M, et al. Associational and nonassociational mechanisms in locomotor sensitization to the dopamine agonist quinpirole. Psychopharmacology (Berl) 1996; 127(2):95–101.
30. Decker S, Grider G, Cobb M, et al. Open field is more sensitive than automated activity monitor in documenting ouabain-induced hyperlocomotion in the development of an animal model for bipolar illness. Prog Neuropsychopharmacol Biol Psychiatry 2000; 24(3):455–462.
31. Einat H, Szechtman H. Environmental modulation of both locomotor response and locomotor sensitization to the dopamine agonist quinpirole. Behav Pharmacol 1993; 4(4):399–403.
32. Silverstone PH, Pukhovsky A, Rotzinger S. Lithium does not attenuate the effects of D-amphetamine in healthy volunteers. Psychiatry Res 1998; 79(3):219–226.
33. Wald D, Ebstein RP, Belmaker RH. Haloperidol and lithium blocking of the mood response to intravenous methylphenidate. Psychopharmacology (Berl) 1978; 57(1):83–87.
34. Ebstein R, Belmaker R, Grunhaus L, et al. Lithium inhibition of adrenaline-stimulated adenylate cyclase in humans. Nature 1976; 259(5542):411–413.
35. Gottesman II, Gould TD. The endophenotype concept in psychiatry: etymology and strategic intentions. Am J Psychiatry 2003; 160(4):636–645.
36. Gould TD, Gottesman II. Psychiatric endophenotypes and the development of valid animal models. Genes Brain and Behavior 2006; 5(2):113–119.
37. Hasler G, Drevets WC, Gould TD, Gottesman II, Manji HK. Toward constructing an endophenotype strategy for bipolar disorder. Biological Psychiatry 2006; 60(2):93–105.

38. O'Brien WT, Harper AD, Jove F, et al. Glycogen synthase kinase-3beta haploinsufficiency mimics the behavioral and molecular effects of lithium. J Neurosci 2004; 24(30): 6791–6798.
39. Bourin M, Hascoet M, Colombel MC, et al. Differential effects of clonidine, lithium and quinine in the forced swimming test in mice for antidepressants: possible roles of serotoninergic systems. Eur Neuropsychopharmacol 1996; 6(3):231–236.
40. Redrobe JP, Bourin M. Evidence of the activity of lithium on 5-HT1B receptors in the mouse forced swimming test: comparison with carbamazepine and sodium valproate. Psychopharmacology (Berl) 1999; 141(4):370–377.
41. Bersudsky Y, Shaldubina A, Belmaker RH. Lithium's effect in forced-swim test is blood level dependent but not dependent on weight loss. Behav pharmacol 2007; 18(1):77–80.
42. Einat H, Yuan P, Gould TD, et al. The role of the extracellular signal-regulated kinase signaling pathway in mood modulation. J Neurosci 2003; 23(19):7311–7316.
43. Einat H, Manji HK, Gould TD, et al. Possible involvement of the ERK signaling cascade in bipolar disorder: behavioral leads from the study of mutant mice. Drug News Perspect 2003; 16(7):453–463.
44. Einat H, Manji HK. Cellular plasticity cascades: gene to behavior pathways in animal models of bipolar disorder. Biological Psychiatry 2006; 59(12):1160–1171.
45. Gessa GL, Pani L, Fadda P, et al. Sleep deprivation in the rat: an animal model of mania. Eur Neuropsychopharmacol 1995; 5(suppl):89–93.
46. Antelman SM, Caggiula AR, Kucinski BJ, et al. The effects of lithium on a potential cycling model of bipolar disorder. Prog Neuropsychopharmacol Biol Psychiatry 1998; 22(3):495–510.
47. Einat H, Kofman O, Belmaker RH. Animal models of bipolar disorder: from a single episode to progressive cycling models. In: Myslobodsky M, Weiner I, eds. Contemporary Issues in Modeling Psychopharmacology. Boston: Kluwer Academic Publishers, 2000:165–180.
48. Bersudsky Y, Patishi Y, Bitsch Jensen J, et al. The effect of acute and chronic lithium on forskolin-induced reduction of rat activity. J Neural Transm 1997; 104(8–9):943–952.
49. Kofman O, Sherman WR, Katz V, et al. Restoration of brain myo-inositol levels in rats increases latency to lithium-pilocarpine seizures. Psychopharmacology (Berl) 1993; 110(1–2):229–234.
50. Einat H, Kofman O, Itkin O, et al. Augmentation of lithium's behavioral effect by inositol uptake inhibitors. J Neural Transm 1998; 105(1):31–38.
51. Berridge MJ, Irvine RF. Inositol phosphates and cell signalling. Nature 1989; 341(6239): 197–205.
52. Nestler EJ, Gould E, Manji H, et al. Preclinical models: status of basic research in depression. Biol Psychiatry 2002; 52(6):503–528.
53. Willner P. The validity of animal models of depression. Psychopharmacology (Berl) 1984; 83(1):1–16.
54. Einat H. Modelling facets of mania — new directions related to the notion of endophenotypes. J Psychopharmacol 2006; 9:9.
55. Einat. Establishment of a battery of simple models for facets of bipolar disorder: a practical approach to achieve increased validity, better screening and possible insights into endophenotypes of disease. Behavior Genetics 2007; 37(1):244–255.
56. Malatynska E and Knapp RJ. Dominant-submissive behavior as models of mania and depression. Neurosci Biobehav Rev 2005; 29(4–5):715–737.
57. Kalynchuk LE, Pinel JP, Treit D, et al. Changes in emotional behavior produced by long-term amygdala kindling in rats. Biol Psychiatry 1997; 41(4):438–451.
58. Kalynchuk LE, Pinel JP, Treit D, et al. Persistence of the interictal emotionality produced by long-term amygdala kindling in rats. Neuroscience 1998; 85(4): 1311–1319.
59. Machado-Vieira R, Kapczinski F, Soares JC. Perspectives for the development of animal models of bipolar disorder. Prog Neuropsychopharmacol Biol Psychiatry 2004; 28(2):209–224.
60. Shih RA, Belmonte PL, Zandi PP. A review of the evidence from family, twin and adoption studies for a genetic contribution to adult psychiatric disorders. Int Rev Psychiatry 2004; 16(4):260–283.

61. Kieseppa T, Partonen T, Haukka J, et al. High concordance of bipolar I disorder in a nationwide sample of twins. Am J Psychiatry 2004; 161(10):1814–1821.
62. Craddock N, Jones I. Molecular genetics of bipolar disorder. Br J Psychiatry 2001; 178(suppl 41):S128–S133.
63. Isgor C, Slomianka L, Watson SJ. Hippocampal mossy fibre terminal field size is differentially affected in a rat model of risk-taking behaviour. Behav Brain Res 2004; 153(1):7–14.
64. Kabbaj M, Devine DP, Savage VR, et al. Neurobiological correlates of individual differences in novelty-seeking behavior in the rat: differential expression of stress-related molecules. J Neurosci 2000; 20(18):6983–6988.
65. Taghzouti K, Lamarque S, Kharouby M, et al. Interindividual differences in active and passive behaviors in the forced-swimming test: implications for animal models of psychopathology. Biol Psychiatry 1999; 45(6):750–758.
66. Cohen H, Zohar J. An animal model of posttraumatic stress disorder: the use of cut-off behavioral criteria. Ann NY Acad Sci 2004; 1032:167–178.
67. Cohen H, Zohar J, Matar M. The relevance of differential response to trauma in an animal model of posttraumatic stress disorder. Biol Psychiatry 2003; 53(6):463–473.
68. Cohen H, Zohar J, Matar MA, et al. Setting apart the affected: the use of behavioral criteria in animal models of post traumatic stress disorder. Neuropsychopharmacology 2004; 29(11):1962–1970.
69. Cohen H, Zohar J, Matar MA, et al. Unsupervised fuzzy clustering analysis supports behavioral cutoff criteria in an animal model of posttraumatic stress disorder. Biol Psychiatry 2005; 58(8):640–650. Epub 2005 Jun 22.
70. Nielsen CK, Arnt J, Sanchez C. Intracranial self-stimulation and sucrose intake differ as hedonic measures following chronic mild stress: interstrain and interindividual differences. Behav Brain Res 2000; 107(1–2):21–33.
71. Dietz DM, Tapocik J, Gaval-Cruz M, et al. Dopamine transporter, but not tyrosine hydroxylase, may be implicated in determining individual differences in behavioral sensitization to amphetamine. Physiol Behav 2005; 86(3):347–355. Epub 2005 Aug 29.
72. Overstreet DH, Rezvani AH, Janowsky DS. Genetic animal models of depression and ethanol preference provide support for cholinergic and serotonergic involvement in depression and alcoholism. Biol Psychiatry 1992; 31(9):919–936.
73. Kamens HM, Burkhart-Kasch S, McKinnon CS, et al. Sensitivity to psychostimulants in mice bred for high and low stimulation to methamphetamine. Genes Brain Behav 2005; 4(2):110–125.
74. Willner P, ed. Behavioral models in psychopharmacology: theoretical, industrial and clinical perspectives. In: Behavioral Models in Psychopharmacology. Cambridge: Cambridge University Press, 1991:3–19.
75. Ebner K, Wotjak CT, Landgraf R, et al. Neuroendocrine and behavioral response to social confrontation: residents versus intruders, active versus passive coping styles. Horm Behav 2005; 47(1):14–21.
76. Tecott LH, Nestler EJ. Neurobehavioral assessment in the information age. Nat Neurosci 2004; 7(5):462–466.

3 Abnormalities in Catecholamines and the Pathophysiology of Bipolar Disorder

Amir Garakani and Dennis S. Charney
Department of Psychiatry, Mount Sinai School of Medicine, New York, New York, U.S.A.

Amit Anand
Department of Psychiatry, Indiana University School of Medicine, Indianapolis, Indiana, U.S.A.

INTRODUCTION

The catecholamine (CA) hypothesis of bipolar disorder (BD)—a deficiency of CA in depression and excess in mania—was proposed nearly three decades ago. CA abnormalities remain the most replicated finding in the pathophysiology of BD. However, the role of CA abnormalities in the pathophysiology of BD still remains unclear. For example, it is unclear whether changes in CAs seen in manic and depressed states are secondary to the mood state or primary, and it remains to be clarified whether abnormalities in the CA system are presynaptic or postsynaptic. Rapid advances in the field of neuroscience in the last three decades have increased our knowledge of the role of CAs in the working of the nervous system and provided new tools to explore CA abnormalities. Clinical research in CA abnormalities in BD has evolved from measurement of changes in CAs in bodily fluids and peripheral tissue to neuroendocrine challenge studies to molecular analysis of postmortem tissue and direct visualization of CA system with brain imaging methods such as single photon emission computed tomography (SPECT) and positron emission tomography (PET).

Preclinical and clinical literature on the role of CAs in depression and psychiatric illnesses and mode of action of psychotropic drugs is fairly extensive. In this review, the main focus is on studies that have specifically investigated the role of CAs in BD. There are only a few preclinical studies regarding pathophysiology of BD because of a lack of suitable animal models for bipolar illness. However, there is an extensive preclinical literature regarding pathophysiology of depression using animal models of depression. In this review, findings from depression research are reviewed where they are relevant to understanding of pathophysiology of BD.

This chapter first reviews the role of CAs in physiology of mood and mood regulation. Next, studies that have investigated CA abnormalities in BD, using different methodological paradigms, are reviewed. The interaction of CAs with other neuromodulators is discussed, and a model for the role of CAs in mood regulation is presented. Finally, methodological difficulties in conducting research in the pathophysiology of BD and future directions of research in this area are discussed.

NEUROCHEMISTRY AND NEUROPHYSIOLOGY OF CATECHOLAMINE

Neurochemistry and Neurophysiology of the Dopaminergic System

Dopaminergic cell bodies located in the ventral mesencephalon form most dopamine (DA) cell bodies and project widely throughout the central nervous system (CNS). These cell bodies give rise to the nigrostriatal, mesocortical, and mesolimbic DA projections. A separate set of dopaminergic cell bodies projecting to the hypothalamus and pituitary arise from a different brain region—the arcuate nucleus—and are referred to as the tuberoinfundibular (TIDA) and tuberohypophysial neurons (2). The arcuate nucleus receives input from cortical regions and is involved in production of hormones such as growth hormone (GH) and prolactin in response to different mood states.

Dopamine receptors have traditionally been divided into D1 and D2 types based on the presence and absence of a positive coupling between receptors and adenylate cyclase activity. D1 receptors mediate the dopamine-stimulated increase in adenylate cyclase activity. D2 receptors are thought to mediate effects that are independent of D1-mediated effects and also to exert an opposing influence on adenylate cyclase activity (3). Recently, a number of subtypes of these receptors have been discovered that are of particular importance to the study of psychiatric disorders. D3 receptors, a subtype of D2 receptors, and D5 receptors, a subtype of D1 receptors, are present in high levels in the limbic brain structures. The D4 receptor, a subtype of D2 receptor that has high levels in the frontal cortex, midbrain, amygdala, medulla, and lower levels in the basal ganglia, has been implicated in the action of clozapine, which reverses both the negative and positive symptoms of schizophrenia (4). Self et al. (5) have described opposite modulation of reward behavior by D1 and D2 receptors agonists. D1 receptor agonists decrease the reinforcement of reward-seeking behavior, whereas D2 receptor agonists increase reinforcement of reward-seeking behavior (5). It can be postulated that the state of anhedonia seen in depression could be a manifestation of either increase in activity of the D1 receptors and/or a decrease in activity at the D2 receptor site and opposite changes could lead to mania.

Role of Dopamine in Reward Mechanisms

Dopamine has been implicated in the neurochemical mechanisms involved in reward behavior. Schultz (6) has reported that DA neurons in the ventral tegmental area (VTA) and substantia nigra are preferentially activated in response to a novel rewarding stimulus and encode for reward predictability. Wainer (7), in a series of studies, reported that anhedonia seen in an animal model of depression is related to dopamine receptor subsensitivity that is reversed by a variety of antidepressant drugs. Furthermore, the reward potential of a number of addicting drugs such as cocaine and opiates seem to be mediated via the mesolimbic dopamine pathways involving the nucleus accumbens (3). Alteration in reward-related behavior has been thought to be central to the pathophysiology of BD (8,9). Mesocortical and mesolimbic dopaminergic pathways have been shown to be involved in reward-related behavior. Manic behavior is frequently associated with reward-seeking behavior and depression with withdrawal and inability to derive pleasure from a rewarding stimulus (anhedonia). Therefore, it is likely that abnormalities of mesocorticolimbic dopamine system may be present in BD.

NEUROCHEMISTRY OF THE NORADRENERGIC SYSTEM

The major noradrenergic (NA) nucleus in the brain is the locus coeruleus (LC), which is located on the floor of the fourth ventricle in the rostra/pons (10). NA neurons give rise to diffuse axonal projections and innervate virtually all areas of the brain and spinal cord. The mammalian brain also contains smaller collections of additional NA neurons and adrenergic neurons that are located in discrete regions of the pons and medulla. These neurons show more restricted patterns of axonal projections. The NA cell bodies exert influence on the brain and the body. Therefore, they are involved in both modulating brain function and producing the body's response to emotions.

The NA cell bodies projecting to other brain regions seem to exert a modulatory effect on the target site. Not all NA-containing nerve terminals in the cortex make synaptic contact with the local cortical neurons; rather, some of these neurons release NA in a manner similar to that through which hormones are secreted and thus have generalized effects on the CNS regions (11). The LC is also very sensitive to both external environmental stimuli and also changes in the body's internal homeostasis. The LC output is involved in flight-and-fight responses and regulates level of arousal, the responses of the sympathetic nervous system including pulse rate and blood pressure, and the signaling of the danger signal of the organism.

LC neurons receive a number of inputs that provide information about the state of the body's external and internal environment. These inputs include other neurotransmitter systems, for example, the serotonin [5-hydroxytryptamine (5-HT)], opioid, gamma aminobutyric acid (GABA), acetylcholine (ACh), dopamine, and glutamate systems. A number of peptides influence the firing rate of the LC neurons, most notable being the corticotropin-releasing hormone (CRH). Finally, the NA system itself provides negative feedback to the LC neurons (10,12). The synthetic pathway for CA involves a series of enzymatic reactions. Tyrosine hydroxylase is the rate-limiting enzyme for the synthesis of both norepinephrine (NE) and DA. Dopamine 3-hydroxylase, which converts DA to NE, is present only in NA neurons. In adrenergic neurons, the enzyme phenylethanolamine-N methyl transferase converts NE to epinephrine.

Adrenergic receptors have been classified as being either α- or β-adrenergic subtype. Each of these subtypes has two secondary subtypes, (α_1 and α_2; β_1 and β_2). Each of these receptors has been cloned (4). Variant forms of these receptors may exist with different regional distributions and functional properties. Activation of the β-adrenergic receptors leads to physiological responses by stimulating adenylate cyclase via coupling with G_s protein (13). Activation of the α_1-adrenergic receptors leads to physiological responses through activation of G_χ proteins (13). Activation of α_2-adrenergic receptors leads to physiological responses via coupling with G_i and or G_o proteins, which leads to activation of specific K^+ channel and/or inhibition of adenylate cyclase (13). There are also α_2-adrenoreceptors present on 5HT neuron terminals in the hippocampus, and electrophysiological studies suggest that they exert a tonic inhibitory influence on the firing of 5HT neurons (14).

Role of Norepinephrine in Reward Mechanisms

As noted earlier, abnormalities in reward mechanisms are likely to be present in BD. DA function has been implicated in maintenance of reinforcing properties of a rewarding stimulus. The role of NE in reward mechanisms is less clearly understood. However, considering its role in attention and arousal, NE may be involved in the initial phase of learning by increasing attention on the rewarding stimulus (15), Increased LC firing has been reported with exposure to a rewarding stimulus (15). Furthermore, LC firing has been shown to modulate the firing of DA neurons (16). In mania, increased NE neurotransmission may be responsible for increased attention to rewarding stimuli, and in depression a decrease in NE may lead to lack of attention or interest in rewarding stimuli (8,9).

CATECHOLAMINE AND MOOD

Mood and Mood Regulation

Our understanding of what constitutes emotion or mood has changed with changing sociocultural views about emotions and with scientific progress in the investigation of neural substrate of emotions. A century ago, William James (17) postulated that emotions arose from bodily reactions to stimuli, such as changes in the autonomic and motor system led to changes in heart rate, blood pressure, increased or decreased bodily secretions, and changes in motor activity, which was then perceived as an emotion. Selye's (18) discovery of the stress response highlighted the role of hormones, particularly stress hormones such as cortisol, in the regulation of emotion. Until recently, the neural substrate of mood and emotion was thought to be confined to the older part of the brain, the so-called emotional brain or the limbic circuit (19,20). Papez (20) described a circuit in the brain, which he called the emotional brain, and postulated that changes in this primitive part of the brain were responsible for changes in emotions. Recent advances in neuroimaging and neurophysiology have brought to our attention the role of cognition and neocortical function in the formulation and regulation of mood. Damasio (21) postulated that emotions, instead of just a function of primitive aspects of the brain, are more accurately conceptualized as arising from interactive effects of the functioning of neocortex, limbic system, basal ganglia, brain-stem autonomic nuclei, and bodily responses such as changes in blood pressure and pulse rate.

Reciprocal links between the corpus striatum and the cerebral cortex have been shown to be involved in the production of movements and more recently have been implicated in the production of normal thought processes (22,23). The basal ganglia plays a central role in regulation of the motor and cognitive circuits. Recently, it has been proposed that a parallel medial prefrontal cortex–striatum–palladium–thalamic circuit is responsible for mood regulation (24–29). Such a hypothetical mood-regulating circuit (MRC) (30) is depicted in Figure 1. Within this circuit the principal neurotransmitters for fast conductance (<1 msec) (31) are glutamate (e.g., cortical–subcortical connections) (32) and GABA (e.g., striatopallidial and pallidothalamic) (29,32). The double inhibition mediated through GABA between the striatopallidial and pallidothalamic links can confer an oscillating property to the activity of this circuit (29). The different states that this circuit oscillates through can be conceptualized as varying mood states. The fast-conductance feedback loops within this circuit can stabilize mood within certain limits and prevent extreme changes in mood. However, feedback loop circuits of this type have the disadvantage of uncontrollable oscillations when any part of the circuit is damaged (33). Similarly, abnormalities in external modulators could change the oscillatory properties of the circuit.

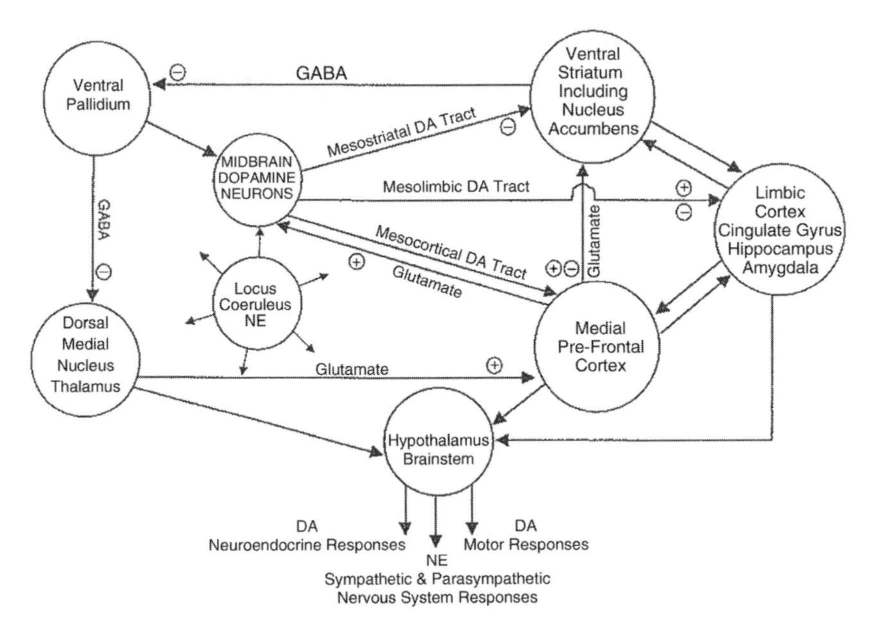

FIGURE 1 Relationship of catecholamine tracts originating from the midbrain to the MRS. *Abbreviations*: DA, dopamine; GABA, gamma aminobatyric acid; NE, norepinephrine.

Catecholamines as Modulators of the Mood-Regulating Circuit

A number of modulatory neurotransmitters modulate the fast-conductance neurotransmission mediated via glutamate and GABA (30,34) by acting on signal transduction factors and change in neuronal excitability. The striatum plays an integral role in modulation of cortical outputs in the motor system and is postulated to do the same for thought and emotions (22,23,29). The principal modulatory neurotransmitter within the striatum is dopamine, which may act directly via G protein-coupled receptors or indirectly via control of neuropeptide expression (35). Mesostriatal dopamine projections from the midbrain provide the dopaminergic modulation of the MRC in response to external rewards (6).

Diffuse projections from the LC also modulate the activity of the MRC. NE neurons project more diffusely to the brain. NE has been shown to decrease the signal-to-noise ratio for firing of neurons (7). Adrenergic input can modulate neuronal excitability through its actions on such signal transduction mechanisms as the cAMP system or G proteins (34,36).

Additional modulation is provided by other neurotransmitters such as 5HT, ACh, neuropeptides, and hormones. The activity of the MRC may also be regulated by intracellular factors such as internal variations in signal transduction factors and genomic factors. The central role of CAs in mood regulation is illustrated by mood changes seen in a number of neuropsychiatric illnesses involving CA dysfunction. Parkinson's disease, which involves degeneration of DA neurons, is frequently associated with depression. Patients with both BD and Parkinson's disease have been noted to have increased motor symptoms when depressed and decreased symptoms during manic states, suggesting increased DA neurotransmission during mania and decreased transmission in depression. Folstein et al. noted that

Huntington's disease, an autosomal dominant illness associated with DA neuronal atrophy in the caudate and putamen region, is frequently accompanied by depression (40% comorbidity) and mania (10% comorbidity). Familial calcification of basal ganglia (Fahr's disease) and striatal infarcts are also accompanied by depression and mania (37). Together, these observations suggest that dopamine abnormalities that are central to the motor abnormalities seen in these neuropsychiatric illnesses may also lead to dysregulation of the MRC. For example, Swerdlow and Koob (29) have drawn a parallel between dyskinesias seen in movement disorder associated with DA abnormalities and mood dysregulation seen in BD.

The kindling and sensitization model proposed by Post and Weiss (38) draws an analogy between seemingly increased incidence of mood episodes seen during the natural course of BD and the increasing ability of a repeatedly administered electrical or chemical stimulus to induce a seizure or changes in motor behavior. There are some indications that CAs may be involved in this phenomenon because kindled seizures are abolished by monoamine depletion with reserpine (39) and administration of haloperidol. As discussed earlier, the neuromodulatory properties of CAs can alter excitability of neurons; therefore, repeated release of CAs may contribute to the kindling phenomena by altering signal transduction mechanisms. Phasically applied DA has been shown to potentiate corticostriatal neurotransmission and alter striatal neuronal plasticity (6,40). Recently, Antelman et al. (1) reported a cycling animal model for BD. Repeated exposures to cocaine led to cyclicity of amphetamine-induced efflux of dopamine from slices of rat nucleus accumbens that was prevented by lithium treatment.

Catecholamines as Translators of Activity of MRC into the Body's Emotional Response

Changes in activity of the MRC are translated into bodily changes by the output of the MRC directed at the hypothalamus, leading to neuroendocrinological changes (e.g., through TIDA DA projections), actions on the parasympathetic and sympathetic (mediated through NE) nuclei in the brainstem, and actions on the motor circuit (DA mediated). Activity of the sympathetic nervous system is mediated by production of NE and epinephrine from the adrenal gland and nerve endings. Increase in CM and their metabolites in serum, urine, and cerebrospinal fluid (CSF) frequently accompany increased arousal and activity. Therefore, besides regulating mood, CAs are also the product of any changes in mood arising from dysfunction of the MRC. Biochemical changes in the CA system, particularly in the periphery, are therefore dependent on the mood state and are not necessarily indicative of etiology of the change in mood state. A review of abnormalities of CAs in BD is best done keeping in mind these two separate components of the CA system.

ABNORMALITIES OF THE CATECHOLAMINE SYSTEM IN BIPOLAR DISORDER

"This hypothesis, which has been designated the 'catecholamine hypothesis of affective disorders,' proposes that some, if not all, depressions are associated with an absolute or relative deficiency of catecholamines, particularly norepinephrine, at functionally important receptor sites in the brain. Elation conversely may be associated with an excess of such amines" (41).

The biogenic amine, or catecholamine, hypothesis was first described in the literature, concurrently by Schildkraut (41) and Bunney and Davis (42). It was noted that depletion in levels of norepinephrine and its metabolites, which could be induced by antihypertensive agents such as reserpine, caused a depressive state. Likewise, monoamine oxidase inhibitors, such as the antituberculosis agent iproniazid, which blocked the enzyme responsible for the deactivation of norepinephrine, were shown to have antidepressant properties. Imipramine, initially developed for the treatment of schizophrenia, was also found to have antidepressant properties, by blocking the reuptake of norepinephrine.

Shortly after, researchers focused on the contribution of other neurotransmitters on the pathophysiology of BD. Mania was conceptualized as a state in which there is an excess of the catecholamines while in depression there is a decrease in CAs (43)—the so-called second generation CA hypothesis. Another hypothesis incorporated the role of serotonin in CA dysfunctions—the permissive hypothesis that stated that both mania and depression may be associated with decreased serotonin function. Both Coppen (44) and Lapin (45) proposed an indoleamine hypothesis, stating that decreased levels of serotonin contribute to depressive illness. Prange and others (46) expounded on this idea by proposing the permissive hypothesis of biogenic amines, which states that catecholamine activity is mediated by a deficiency of serotonin transmission. Thus, in the context of decreased serotonin, an increase in CA produces a manic state, while decrease in CA produces a depressive state.

Abnormalities in Serum, Urine, and CSF Levels of Catecholamines and Their Tetabolites

The original CA hypothesis of depression (41,42) postulated that depression was caused by a decrease in the amount of CA production and mania was due to a compensatory increase in CA production. Consequently, much of the earlier studies of the pathophysiology of BD were directed toward measuring CAs and their metabolites in bodily fluids. The results of these studies shed only limited light on the pathophysiology of BD because many of these studies did not give consistent results and because it is difficult to determine the relative contribution of peripheral versus central origin of plasma and urinary CA metabolites homovanillic acid (HVA) and 3-methoxy-4-hydroxyphenylglycol (MHPG) levels. CSF HVA levels are also considerably influenced by plasma levels and the brain (47), and many studies did not make a correction for plasma levels while reporting CSF levels of CAs and their metabolites. Furthermore, most CSF dopamine metabolites are derived from the nigrostriatal pathways and do not necessarily reflect the function of the mesocorticolimbic (MCL) system (7). Finally, it is difficult to completely tease out state-related confounds. As peripheral MHPG, HVA, NE, and epinephrine levels are considerably influenced by motor activity and degree of arousal, changes in these measures in mania and depression are more likely to be a state-related rather than a trait abnormality.

Abnormalities in Norepinephrine and Its Metabolites

No consistent relationship has been found in levels of MHPG in the CSF, serum, or urine in studies of depressed patients (48). Schatzberg et al. (49) reported that bipolar depressed patients had significantly lower urinary MHPG levels than unipolar nonendogenous depressed patients. Similar results were obtained for plasma NE and MHPG levels. Some recent studies have supported this finding, whereas others have not. Negative findings were reported from the Depression Collaborative studies (50) and in a Swedish study (51). According to Schatzberg and Schildkraut (52), the negative findings of these studies could be due to the inclusion of more bipolar II patients than bipolar I patients. Other studies have used more complex measures of CA levels in an effort to find a relationship with the state of depression or to differentiate between different types of depression. Ratio of NE and epinephrine to their metabolites, total body CA turnover, ratio of NE to NE plus metabolites, and epinephrine to epinephrine plus metabolites and discriminant functional analysis of 24-hour urinary CAs and metabolites [depression (D)-type scores] have been studied. These studies showed that unipolar depressed subjects have a higher excretion of CA than bipolar depressed subjects and control subjects (53). These measures may be useful in differentiating unipolar from bipolar depression; however, the significance of these findings for the pathophysiology of BD is not clear.

Dopamine Abnormalities

A number of studies of turnover of dopamine (analogous to studies of NE and its metabolites) and its metabolites have been carried out. CSF levels of the major dopamine metabolite HVA have been consistently found to be decreased in depression associated with psychomotor retardation but not in agitated depression and are increased in mania (54). Therefore, reduced CSF HVA is thought to be related more to the symptom of psychomotor retardation than to depressed mood (7). A decrease in prolactin levels in seasonal affective disorder has been reported and was seen in both unipolar and bipolar patients throughout the year irrespective of depressive symptoms (9,55). The decreased prolactin level has been proposed as a trait abnormality that could be secondary to increased DA receptor sensitivity secondary to decreases in DA levels in bipolar depression.

CHALLENGE STUDIES

Depletion Studies

Effects of acute depletion of CA using alphamethylparatyrosine (AMPT), a tyrosine hydroxylase inhibitor, have been studied in patients with depression and in healthy subjects. In healthy subjects, chronic administration of AMPT does not cause depressive symptoms (56). In bipolar depressed patients, AMPT has been noted to increase depression and in manic subjects it can decrease the severity of mania (57). Berman et al. (58) reported an increase in depressive symptoms after AMPT-induced CA depletion in euthymic subjects with a remote history of unipolar depression.

Anand et al. (59) reported on the effects of CA depletion in euthymic bipolar subjects stable on lithium therapy. AMPT administration in these subjects did not lead to relapse of depressive symptoms. However, 36 to 48 hours after depletion was completed and the subjects were recovering from the depletion with return of plasma MA and MHPG levels back to baseline, subjects had a transient relapse of hypomanic symptoms. This relapse of hypomanic symptoms was not accompanied by increase in plasma HVA and MHPG levels from baseline. It was hypothesized that the relapse of hypomanic symptoms could have been due to either an increase in receptor sensitivity or increased central production or release of CAs that was not detected in plasma.

Stimulation Studies

CA-releasing agents such as amphetamine have been used to increase CA levels in the brain. Though several studies have shown increased arousal and activation after amphetamine injection in bipolar subjects and even nonbipolar subjects, sustained mania has not been consistently recorded (60,61,62). The psychostimulant effect of amphetamines is blocked by dopamine antagonists (63,64). McTavish et al. (65) used a methamphetamine challenge in acutely manic and healthy control subjects who four hours earlier had received a tyrosine-free amino acid mixture, in a double-blind crossover with a control mixture. The tyrosine depletion caused a lowering of the physiological and psychological responses to methamphetamine.

The observation that of all pharmacological challenges, L-dopa is most likely to induce mania lends support to the CA hypothesis for mania. Similarly, direct DA agonists such as bromocriptine and piribedil can relieve bipolar depression and even precipitate mania (66), and the antidepressant response to piribedil has been associated with low pretreatment levels of HVA in the CSF (67). Cocaine, a potent inhibitor of dopamine transporter (DAT), can cause mania-like symptoms in healthy control subjects and precipitate mania in bipolar subjects. There has been one report of precipitation of manic symptoms with administration of yohimbine, an α_2-adrenergic presynaptic autoreceptor antagonist (68). There are also two case reports of secondary mania caused by caffeine (69,70).

Neuroendocrine Challenge Studies of Catecholamine Receptor Function in Bipolar Disorder

Neuroendocrine challenge studies measure plasma levels of hormones after a challenge with a drug that acts on a particular receptor. Most of these studies have been done in depression, but no consistent abnormalities have emerged. Similarly, in BD no consistent results have been found in DA systems using this method. The neuroendocrine challenge paradigm has a number of limitations: it is difficult to find agents that are specific to only a particular receptor subtype because in many cases the challenge paradigm cannot differentiate between pre- and postsynaptic effects of the challenging agent (71), and the evidence for receptor sensitivity is indirect and is influenced by a number of intervening variables. Some of these limitations can be circumvented by using agents that are more specific and by combining challenge studies with brain imaging studies in which the changes in brain events can be measured directly.

Dopamine Receptor Abnormalities

Meltzer et al. (72) and Hirschowitz et al. (73) did not find an increased GH response to apomorphine (a postsynaptic D2 receptor stimulator) in manic subjects. McPherson et al. (74) also failed to show a GH or prolactin change in response to apomorphine in bipolar or unipolar depression. Linkowski et al. (75) and Nurnberger et al. (64) were unable to differentiate manic subjects from other diagnostic groups on hormonal responses to L-dopa and amphetamine.

Neuroendocrine Receptor Abnormalities

The GH response to clonidine that has been shown to be mediated through postsynaptic α_2 receptors (76) has been used in a number of studies. A blunted GH response to clonidine was shown to be present in depression (77), and a number of studies have replicated this finding. However, Ansseau et al. (78) did not find any differences between manic and depressed subjects on GH response to clonidine. Moreover, GH response to the dopamine antagonist apomorphine (79), amphetamines blockade of NA and DA uptake (80), and the serotonin agonist *m*-Chlorophenylpiperazine (*m*CPP) (81) has also been found to be reduced. GH release is also mediated by somatomedins and somatostatin, which have been reported to be altered in depressed patients (61,82). Therefore, it is possible that an abnormality intrinsic to GH release may be present in affective disorders. More studies in drug-free bipolar depressed subjects are needed to elucidate the cause of the blunted response in depressed subjects (76).

In summary, no consistent change in a particular CA receptor type is seen in BD. Methodological issues, particularly the lack of ideal animal models of BD and the limitations of the neuroendocrine challenge paradigm, make it difficult to draw firm conclusions from studies done so far. Brain imaging studies offer some promise in the measurement of the DA and the 5T system. However, radioligands for the adrenergic system have not been used often in clinical studies.

MECHANISM OF ACTION OF PHARMACOLOGICAL AGENTS USED IN THE TREATMENT OF BIPOLAR DISORDER

The biochemical effects of medication used in the treatment of BD have been used to elucidate the role of CAs in the pathophysiology of BD.

Medication Effects on the Dopamine System

A number of effective antimanic agents act on the dopamine system. Neuroleptics such as haloperidol, which are D2 receptor antagonists, are one of the most effective antimanic agents, lending support to the hypothesis of increased DA neurotransmission in mania. Lithium and sodium valproate have both been shown to decrease postsynaptic D2 receptor sensitivity. Waldmeier (83) reviewed the mechanism of action of drugs useful in the treatment of BD and concluded that downregulation of DA neurotransmission seemed to be a common property of medications useful in the treatment of BD.

Willner (7) recently proposed the hypothesis that even selective serotonin reuptake inhibitors (SSRIs) and tricyclic antidepressants improve depression by increasing the sensitization of D2/D3 receptors in the MCL dopamine system. The MCL dopamine system is primarily responsible for reward reinforcement and experience of pleasure. Using the paradigm of locomotor response to amphetamine or apomorphine as a measure of DA receptor function, Willner (7) showed that most antidepressants increase the psychomotor stimulant response to dopamine agonists. This response is primarily mediated through the MCL DA system, and behavioral stereotypes mediated through the mesostriatal system and neuroendocrine responses mediated through the TIDA DA system are not affected. The effects of chronically administered tricyclic antidepressants are reversed by administration of the DA antagonist sulpiride in the nucleus accumbens but not in the dorsal striatum (84).

Medication Effects on the Neuroendocrine System

Özerdem et al. (85) demonstrated lithium administration (after 5 days and 4 weeks) in healthy volunteers produces increased plasma NE and increased response to idazoxan, an α_2-adrenoreceptor antagonist. Antimanic effects of lithium have been shown to be accompanied by decreases in plasma and urinary MHPG levels (86). Clonidine, an α_2-agonist that decreases presynaptic NE release, has been reported to have some antimanic activity (87,88). Nearly all antidepressants have been implicated in precipitation of mania when used for the treatment of bipolar depression. However, tricyclic antidepressants that predominantly decrease NE reuptake have been identified as most likely to precipitate mania or increase cycling (89). Paradoxically, bupropion, a dopamine reuptake inhibitor, is thought to be an antidepressant least likely to precipitate mania.

Catecholamine Studies in Peripheral Tissue

Due to the inaccessibility of the brain cells themselves, investigators have studied biochemistry in blood cells for general abnormalities that may be common to both blood cells and to neurons. In this regard, platelet studies have been conducted because platelets have a number of important similarities with neurons. However, the relationship of biochemical abnormalities in peripheral tissue to that in the CNS is not clear.

Enzyme Studies

Enzymes responsible for production and degradation of CA are also present in platelets. Platelet dopamine β-hydroxylase (DBH), the enzyme that converts DA and NE, has been found to be lower in bipolar depressed subjects compared with unipolar subjects (90–93). The lower levels of DBH in bipolar subjects compared with unipolar subjects have been found most in subjects with a family history of affective illness (94). Ikeda et al. (95) reported higher DBH levels in the manic phase compared with the depressed phase. The value of platelet DBH levels for the diagnosis of BD is limited by a large variation between individuals that precludes accurate statistical evaluation.

Monoamine oxidase (MAO) deaminates both DA and NE to their inactive metabolites. MAO is of two types: MAO-A is found predominantly in neurons and metabolizes both DA and NE, whereas MAO-B is found in platelets and metabolizes DA. Therefore, the interpretation of changes in platelets for changes in neurons is questionable. MAO levels have been found to be lower than controls in bipolar I depressed subjects, but this difference is not found in bipolar II depressed subjects and unipolar depressed subjects (96).

Catechol-o-methyltransferase (COMT) is an extraneuronal enzyme that degrades NE. It is also found in red blood cell membranes and is under genetic control. In one study, this enzyme was found to be lower in bipolar depressed subjects compared with unipolar subjects (97), but this finding has not been replicated (98). Two interesting studies of velo-cardio-facial syndrome (22q11.2 deletion syndrome or DiGeorge syndrome) show that patients have a deletion of an allele that results in low activity of COMT, and causes a rapid cycling BD (99). Graf and colleagues (100) treated three of four patients with metyrosine, which was given to reduce CA levels, and did in fact lower the baseline levels of CSF HVA, and improve their aggression and behavioral symptoms. A larger study of COMT is needed to clarify these findings.

Receptor Studies

Platelets and leukocytes have been most frequently used for peripheral tissue receptor studies. Decreased prostaglandin E stimulation of platelet α_2

receptor-mediated cAMP production has been reported in unipolar depression (101). Garcia-Sevilla et al. (102) investigated the α_2-adrenergic receptor-mediated inhibition of platelet adenylate cyclase and induction of platelet aggregation. They found a hypersensitivity to α_2-adrenergic receptor agonists in drug-free depressed patients that was decreased after long-term antidepressant treatment (102). Platelet α_2 receptor number measurement has revealed an increase in some studies, but no change has been found in other studies (103). Results of studies of peripheral α_2-adrenergic receptors have also been contradictory (104). α_2-Adrenergic receptor studies for bipolar depression patients need to be conducted.

β-Adrenergic receptors have been studied on human leukocytes. Though no studies have investigated the bipolar group separately, a decreased cAMP response to β-adrenergic receptors seems to be predominantly found in unipolar depressed subjects (105). Wright et al. (106) reported decreased β-adrenergic receptors in bipolar subjects and their ill relatives, a finding that has not been replicated (107,108).

Signal Transduction Mechanisms

Manji and Lenox (109) reported an increase in leukocyte membranes of immunolabeling of the 45-kDa form of $G_{\alpha,s}$, in bipolar affective disorder group considered as a whole (lithium-treated or untreated) compared with control subjects. It has been shown that lithium competitively inhibits the phosphatidylinositol second messenger signal transduction pathway (110–112). Manji et al. (113) reported effects of lithium and sodium valproate on the protein kinase C (third messenger) pathway, and the protein kinase C inhibitor tamoxifen has been shown to have some antimanic efficacy. Both lithium and sodium valproate also seem to have genomic effects (113,114). Freidman et al. (115) reported altered protein kinase C activity during the manic phase of BD. Recently, Kaya and colleagues found that in bipolar patients, lithium, thought to inhibit inositol monophosphatase (IMPase), does in fact increase erythrocyte IMPase activity with prolonged use (116).

POSTMORTEM STUDIES IN BIPOLAR DISORDER

Postmortem studies can directly assess brain neurotransmitters and their receptor systems. Though numerous postmortem studies have investigated the role of DA receptors in schizophrenia, few or none have examined DA receptors in autopsy specimens from patients with BD. A more detailed discussion of postmortem studies can be found in Chapter 8. Young et al. (117–120) in a series of studies reported abnormalities in CA and signal transduction mechanisms in BD. Young et al. (117) reported an elevation of $G_{s\alpha}$ subunit in prefrontal cortex in BD subjects; increased forskolin-stimulated cyclic AMP production in prefrontal, occipital, and temporal cortex; but no increase in GTP-induced CAMP production (118). Young et al. (119) did not find any differences between bipolar subjects and control subjects on basal levels of NE or DA but did find an increased MHPG/NE ratio in BD subjects. Young et al. (120) did not report any differences in β receptor binding in any part of the brain in BD subjects and did not find any change in $G_{s\alpha}$ subunit. Rahman et al. (121) reported a decrease in cytosolic cAMP-dependent phosphokinase levels that could be secondary to increased AMP signaling in BD subjects. Freidman and Wang (122) reported that in bipolar brain membrane there is enhanced receptor–G protein coupling and an increase in the trimeric state of the G proteins and concluded that these changes may contribute to produce exaggerated transmembrane signaling and to the alterations in affect that characterize bipolar affective disorder. Wang and Freidman (123) also reported an increase in phosphokinase C activity in postmortem brains of bipolar subjects. Recently, Dean and others (124) measured zolpidem-insensitive and zolpidem-sensitive (3H) flumazenil binding in the hippocampus of postmortem patients with schizophrenia and BD. They found an increase in zolpidem-insensitive binding, suggesting an increase in GABA receptors containing the $\alpha5$ subunit, which may be related to cognitive symptoms seen in BD.

Postmortem studies point to a possible increased CA turnover and abnormal postsynaptic signal transduction mechanisms. However, postmortem studies have several limitations: dynamic changes in CA cannot be measured, the affective state of the subject at the time of death cannot be controlled, frequently the medication status of the subjects is also not known, and cause of death can affect results. Obtaining a more homogenous sample of postmortem brains in regard to cause of death, documentation of medication status before death, and clinical diagnosis at the time of death can circumvent these difficulties. Setting up brain banks and consortiums (e.g., the Stanley Foundation Brain Consortium) can help researchers with access to a larger number of brain samples.

BRAIN IMAGING OF CATECHOLAMINE SYSTEM IN BIPOLAR DISORDER

Recent advances in brain imaging have opened new ways to directly measure neurotransmitter receptor function in vivo. Compared with the number of studies done in schizophrenia and depression, fewer brain-imaging studies have been conducted in BD. Brain imaging studies are reviewed in detail in Chapter 9. A short discussion is presented here of brain imaging studies that have direct relevance to the role of CAs in BD.

Structural Brain Imaging Studies

These studies in BD have suggested the following abnormalities may be present: increased rates of subcortical white matter and periventricular hyperintensities in elderly and nonelderly patients (125,126), increased third and lateral ventricular measures (127,128), and smaller cerebellar measures (129,130). There is also equivocal evidence of temporal lobe abnormalities (131). Several studies support the finding of enlargement of the amygdala (132,133). Some studies have shown an increased hippocampus volume (134) and increased caudate volumes in bipolar subjects (126), a finding different from unipolar depression in which a decreased hippocampal volume and decreased size of the caudate have been reported (135,136). Drevets et al. (28) reported a structural defect in the subgenual prefrontal cortex in subjects with familial depression (unipolar and bipolar depressed), a finding that has been replicated (137).

Blood Flow Studies

Cerebral blood flow studies have provided evidence for functional abnormalities in structures in the MRC in BD. Work by Al Mousawi et al. (138) reported lower blood flow in the frontal lobe compared with the occipital lobe in BD. Gyulai et al. (139)

reported asymmetries in anterior temporal lobes in both bipolar mania and depression. Migliorelli et al. (140) reported reduced blood flow in right basotemporal cortex, whereas O'Connell et al. (141) reported increased activity in the temporal lobes in manic subjects. Drevets et al. (28) reported decreased blood flow in the prefrontal subgenual cortex in depressed BD subjects and increased flow in mania.

Together the structural and functional studies of BD tend to point to abnormalities in the prefrontal-striatal-thalamic-hippocampal area that may be abnormal in BD. As discussed earlier, CAs, and in particular DA, are principal modulators of this circuit.

Neurochemical Studies

PET, SPECT, and magnetic resonance spectroscopy studies explore the neurochemical abnormalities that may underlie the structural and blood flow abnormalities in BD. Again, compared with a number of studies done in schizophrenia and depression, very few neurochemical studies have been done in BD. PET studies of DA receptors have shown an elevation of striatal D2 receptor numbers in psychotic mania but not in nonpsychotic mania (142–145). In these studies, psychotic manic subjects were more similar to schizophrenic subjects and non-psychotic manic subjects were more similar to healthy control subjects on measures of D2 receptor binding. In another study that did not make a distinction between unipolar and bipolar depressed subjects, D'haenen and Bossuyt (146) reported bilateral increase in D2 receptors in the basal ganglia. It is difficult to reconcile the findings of these studies unless increased D2 receptor numbers is a trait marker for BP; however, euthymic bipolar subjects were not studied using these methods.

Anand et al. (147) using single photon emission tomography investigated the presysnaptic dopamine release in response to an amphetamine challenge in euthymic bipolar patients. Baseline postsynaptic receptor concentration as measured using the D2 specific radioligand iodobenzamide [(^{123}I)IBZM] was not different from healthy subjects. Moreover, presynaptic dopamine released in response to an amphetamine challenge measured indirectly by the degrees of displacement of IBZM from baseline was also not different in euthymic bipolar patients and healthy subjects. However, BD patients had a greater euphoric response to amphetamine challenge than healthy subjects. Together, these findings were consistent with postsynaptic intracellular signal transduction abnormalities in the CA neurotransmission in BD.

A more recent PET study compared [(18)F]6-fluro-L-dopa uptake in manic patients and healthy controls before and after treatment with divalproex (148). Before treatment, there is no difference between the two groups, but post-treatment, [(18)F]DOPA rate constants were significantly reduced compared to controls, indicating that the bipolar patients had lower presynaptic dopamine activity.

Genetic Studies

BD is a psychiatric illness with a strong genetic basis. However, the exact genetic abnormality is still not known. Chapter 13 details the role of genetic factors in BD, and here a brief review of genetic studies pertaining to CA abnormalities is presented. Potential phenotypic abnormalities of the CA system from CA synthesis to CA signal transduction have been reviewed above. A number of investigators have tried to find a genetic association for these phenotypic abnormalities. However, no consistent results have emerged.

A number of studies have investigated the association between the tyrosine hydroxylase (TH, the rate-limiting enzyme in CA synthesis) gene and BD. Though some investigators have found a possible linkage between the M gene and BD (149–150), most studies have found no such linkage (151–154). Turecki et al. (153) conducted a meta-analysis of studies and concluded that there was no overall association between BD and the TH gene. No linkage has been found between BD and the DBH or dopa decaxboxylase gene. Some studies show there is evidence of no genetic or allelic association between BD and the COMT gene (155), although one study found a link between low-activity COMT and a female bipolar I proband (156), and Kirov et al. (157) showed that the low activity allele is common in rapid cycling patients. Lim et al. (158) and Rubinsztein et al. (159) reported a possible allelic association between BD and the MAO-A enzyme, whereas Parsian and Todd (160) did not find allelic association of BD with either the MAO-A or MAO-B enzyme. Some investigators reported a possible linkage between BD and the DAT gene (161,162), whereas others have been unable to find this association (152,163). No association between any of the DA type 2, 3, or 4 receptor genes and BD has been found (152,163,164), though Parsian and Todd (160) reported a weak association. There has also been no link found between the norepinephrine transporter (NET) and BD (165). Furthermore, ongoing research has suggested a correlation between BD and the GABA-A receptor α subunit genes (GABRA), such as GABRA-1 (166), GABRA-3 (167), GABRA-5 (168), although some studies contradict these findings (169,170). Finally, Ram et al. (171) did not find genetic linkage between BD and the gene coding for the $G_{s,\alpha}$ subunit protein.

In summary, no convincing linkage between a gene encoding for a particular aspect of CA function and BD has been found. However, with rapid progress in genetic molecular biology, more evidence may become available.

MODULATION OF CATECHOLAMINE NEUROTRANSMISSION AND IMPLICATIONS FOR THE PATHOPHYSIOLOGY OF BIPOLAR DISORDER

Modulation by Other Neurotransmitters

As discussed earlier, mood regulation can be conceptualized as stable oscillatory activity of the MRC maintained by fast feedback mechanisms within the circuit and external modulation by other neuromodulators such as DA, NE, 5HT, ACh, and neuropeptides. Fluctuations of mood seen in BD can be conceptualized as breakdown of these feedback and modulatory mechanisms. CAs modulate the MRC system directly by their effects on MRC fast conductance pathways (mediated via GABA and glutamate) and in turn are modulated by other neurotransmitters themselves (e.g., 5HT, ACh, neuropeptides, etc.). Therefore, abnormal modulation of the CA system by other modulators can lead to abnormalities in MRC. The role of each of these modulators is described in more detail in other chapters of this book and here a brief summary of interactions with the CA system is presented.

NE Modulation of DA Neurotransmission

There is a close relationship between NE and DA neurotransmission, NE being involved in the attention and arousal of the organism and DA involved in the motivational/reward aspects of a task (6,172). There has been a more detailed investigation of the effect of NE on the DA system and less so of the effect of DA on the NE system. It has been suggested that, teleologically speaking, because DA is the precursor of NE, it seems reasonable that the latter would have evolved as the more developed system, modulating its antecedent, the more primitive one (54). However, considering the complementary role of these neurotransmitters, both have significant effect on the other's function.

Earlier investigations of NE modulation of DA function revealed contradictory findings. Antelman and Caggiula (173) reported that NE inhibits DA neurotransmission, whereas Archer et al. (174) suggested a facilitatory role of NE on DA-mediated function. More recent evidence points to an excitatory role of NE on the mesocortical DA system and inhibitory effect of noradrenergic system on the prefrontal cortical DI receptor-mediated neurotransmission that is accompanied by an increase in subcortical DA neurotransmission (175). Depletion of both NE and DA results in greater blunting of GH response to clonidine in rats than depletion alone (176). Therefore, overall the evidence points to a facilitatory role of NE on the DA system within the MRC.

5HT Modulation of Catecholamine Neurotransmission

5HT neurons in the raphe nucleus in the brainstem have direct projections to the striatum, frontal cortex, and the limbic system. Lesions of the 5HT system increase low-affinity β-adrenergic receptor density (177), and 5HT has been shown to have an inhibitory effect on the NE system through the presynaptic heteroreceptors (14). However, other investigators have reported a synergistic role of 5HT and NE neurotransmissions. Stimulation of postsynaptic $5HT_2$ receptor has been reported to increase the response of midbrain NE neurons to sensory stimuli (178). A number of different types of 5HT and DA receptors have been discovered, and therefore the interactions between the two systems are likely to be complex. However, most studies indicate that the 5HT system has an inhibitory effect on the DA neurotransmission (179–182). The so-called permissive hypothesis of the role of 5HT in BD postulates that both mania and depression are due to low levels of 5HT. However, increase in 5HT, for example by treatment with SSRIs, has been shown to induce mania. Lithium, which has been shown to facilitate 5HT neurotransmission (183), has both antimanic and antidepressant properties. The mechanism of the differential effect of 5HT on mood is not known at present and needs to be further investigated. CAs also have a modulatory effect on the 5HT system, but this has been less well investigated. For example, it has been shown that NA denervation prevents tricyclic antidepressants from causing sensitization of forebrain neurons to 5HT in laboratory animals (184).

Excitatory Amino Acids Modulation of Catecholamine Neurotransmission

Excitatory amino acids such as glutamate and aspartate act through the N-methyl-D-aspartate (NMDA) and non-NMDA receptors. They influence monoamine transmission, including dopamine and NE, and are in turn influenced by these catecholamines (185). Chronic, but not acute, administration of noncompetitive NMDA antagonists is associated with decreased density of β-adrenergic receptors in mouse cortex (186). Chronic desipramine binding has been shown to increase total NMDA receptor binding (187).

Most neurotransmission is thought to involve the glutamate system, CAs modulate glutamate neurotransmission, and DA has a predominantly inhibitory modulatory effect on glutamate transmission (4,35). However, the glutamate system itself has been shown to have a modulatory effect on the CA system.

NMDA receptors such as ketamine, phencyclidine, and MK-801 are known to cause increased subcortical DA neurotransmission (188). This effect is frequently associated with increased motor activity, mood elevations, and cognitive dysfunction—symptoms similar to those seen in mania. Furthermore, medications such as lamotrigine, which decrease glutamate release, have been shown to have a mood-elevating effect (189,190). Cortical glutamate projections to the subcortical regions have a net inhibitory effect on the subcortical system activity. One mechanism of this inhibitory effect has been postulated by Grace et al. (191), who proposed that the inhibition occurs by increased tonic DA release that then has an inhibitory effect on the subcortical system neurons (including DA neurons) by a negative feedback mechanism. Inhibition of this cortical glutamate projection to the subcortical MRC could lead to increased firing of DA neurons and elevation in mood and increased inhibition could lead to a depressed state. Anand et al. (189) reported that decreasing glutamate neurotransmission by administration of the glutamate release inhibitor lamotrigine leads to increased mood-elevating effects of the NMDA receptor antagonist ketamine in healthy subjects.

In recent studies, Dager and colleagues (192) found that unmedicated patients with BD, as compared to healthy controls, had increased levels of Glx (glutamate + glutamine + GABA) and lactate in the gray matter, as measured on proton echo-planar spectroscopic imaging (PEPSI). This may represent a shift in redox states, from oxidation to glycolysis. Their group also demonstrated that longitudinal treatment of BD patients, versus controls, with lithium caused a decrease in Glx concentrations, but not in lactate levels (193).

GABA Modulation of Catecholamine Neurotransmission

GABA is the most prevalent inhibitory neurotransmitter. However, GABA neurotransmission can be facilitatory in a circuit with two sequential GABA linkages. This is the case in the MRC as depicted in Figure 1 in the linkages of the ventral striatum to the palladium and the connection of the palladium to the thalamus. Therefore, GABA neurotransmission could either facilitate or inhibit neurotransmission in the MRC. Serum GABA levels have been found to be low in both depression and mania and euthymic BD. Moreover, many of the mood-stabilizing agents increase GABA neurotransmission in vitro and also have been shown to increase CSF and plasma GABA levels (194–196).

Neuropeptide Modulation of Catecholamine Neurotransmission

Neuropeptides such as somatostatin, CRH, substance P, and neuropeptide Y can alter neurotransmission function by direct action on noradrenergic or DA neurons. CRH has been shown to acutely increase LC firing rate. However, effects of chronically elevated CRH as seen in depression (197) on norepinephrine neurons have not been delineated. Chronic desipramine treatment attenuates the stress-induced activation of LC neurons mediated by CRH neurotransmission (195). Desipramine treatment has been shown to reduce CSF–CRH concentration (198). Neuropeptide Y (NPY) is another peptide that is colocalized with NE (199). Treatment with norepinephrine uptake inhibitors such as desipramine results in decreased NPY receptor density that could possibly be a clue to increased NPY levels (200). Somatostatin, a tetradecapeptide, is rich in the hypothalamus, amygdala, and nucleus accumbens. It is found to be involved in NE and DA neurotransmission (201). Chronic desipramine dosing in rats results in increased binding to somatostatin receptors in the nucleus accumbens (202). In the striatum, DA and various neuropeptides interact together to influence neurotransmission via GABA and glutamate pathways (32). Depressed patients show decreased CSF concentrations of somatostatin (201,203), a nonspecific finding because it is also decreased in a variety of other neuropsychiatric illnesses (201). However, Berrettini et al. (204) did not find any abnormality in a number of neuropeptide levels in the CSF of euthymic bipolar subjects compared with healthy control subjects.

ACh Modulation of Dopamine Neurotransmission

ACh is a major neurotransmitter with a predominant modulatory action on synaptic neurotransmission. Its affect on the DA system has been well studied in relation to the motor system where ACh has been shown to have an inhibitory effect on DA-mediated function (32). Symptoms of Parkinson's disease are thought to arise from an imbalance between the ACh and the DA system. Similarly, a decreased cholinergic tone is thought to remove the inhibitory effect on the CA system, leading to mania, and an increased cholinergic tone could lead to depression. The hypothesis that ACh could be involved in modulation of the CA system and lead to changes in mood is supported by the observation that physostigmine (a central cholinesterase inhibitor/cholinergic agonist) can lead to a switch from mania to depression (205). The ability of lithium to prevent supersensitivity of peripheral and central ACh receptors (as induced by denervation of atropine) (206) may contribute to its mood-stabilizing properties. Thus, ACh by its action on the CA system may play a critical role in mood stabilization, an area that has been understudied and merits further investigation.

Hormonal Modulation of Catecholamines

Hormonal changes frequently accompany emotional changes. CAs themselves directly regulate the production of a number of hormones such as GH and prolactin, and these hormones have an inhibitory feedback effect on CA neurotransmission at the level of the hypothalamus. Other hormones can also regulate CA neurotransmission and thereby modulate mood. Thyroid hormone has a significant effect on the adrenergic system, and alterations in thyroid function have been implicated in rapid cycling BD and refractory RD. Whybrow and Prange (207) proposed that the ability of thyroid hormone to increase P-adrenergic receptor sensitivity to norepinephrine may underlie its ability to modulate mood. Steroids can modulate CA function. Cortisol can lead to increased production of CM in the periphery and the brain.

Modulation of CA Neurotransmission by Intracellular Factors

In addition to changes at the circuit and synaptic levels, CA neurotransmission may be altered by changes in intracellular factors such as signal transduction, neurotrophic factors, and genetic factors. These effects may alter information processing or neurotransmitter release by the target neurons in a fashion that alters neural circuit behavior, and eventually depressive symptoms are alleviated (208). Phasic intracellular changes in the neurons of the MRC can lead to changes in the activity of these circuits that can then manifest as changes in mood. These intracellular changes are described in detail in other sections of this book and here a brief review of their role in CA neurotransmission is presented.

Wachtel (209) postulated that a dysregulation of neuronal second messenger function is involved in depression. This hypothesis suggests that, in depression, there is an imbalance of the major second messenger systems in the CNS resulting from diminished adenylate cyclase pathway activity and increased phospholipase C pathway activity, and in mania, the reverse occurs. Since it was discovered that lithium may work by inhibiting signal transduction mechanisms such as the G proteins and phosphatidylinositol pathways, extensive research has been done in this area. Lachman and Papolos (210,211) presented a hypothesis of cyclical changes in G proteins that could lead to cyclical changes in mood. More recently, investigators have started to look at protein kinase abnormalities (113) that could change CA receptor sensitivity such as DA receptor sensitivity in BD.

Other intracellular factors have been recently discovered that may have a role in the pathophysiology of BD and the mechanism of action of antidepressants and mood stabilizers. In preclinical studies, brain-derived neurotrophic factor (BDNF) and its receptor trkB have been shown to be increased with electroconvulsive therapy and BDNF mRNA is increased with chronic administration of several different classes of antidepressant drugs but not with acute administration of these drugs and not by administration of nonantidepressant psychotropic drugs (212). Local infusion of BDNF in the brain has been shown to have antidepressant effects in two behavioral models of depression, the forced-swim and learned helplessness paradigms (213,214). Neurotrophins such as BDNF and neurotrophin-3 (NT-3) may therefore be targets of long-term antidepressant drugs. Their putative antidepressant effects may be a result of the ability of neurotrophins to increase monoaminergic neurotransmission and to increase the survival of monoamine neurons. In this regard, NT-3 has been shown to be protective to norepinephrine (NE) neurons (215). Glial cell line-derived neurotrophic factor (GDNF) has been shown to increase sprouting of adult midbrain DA neurons (216).

Finally, a new area of study is clock genes, genes that encode for biological rhythms through production of certain proteins at a certain cycle length (217,218). Keeping in mind the close relationship of BD to biological rhythms, further study of the relationship of these genes to CA activity may provide clues to the pathophysiology of BD.

Bipolar and Unipolar Depression

A distinction between bipolar and unipolar depression (depression without history of hypomania or mania) is often made, and efforts have been made to investigate biological differences between unipolar and bipolar depression. At the same time, many studies of neurobiology of depression have not reported results separately for bipolar and unipolar depression. Though a number of research studies have looked at this issue, it is not clear whether a distinction between the two groups can be made in terms of CA abnormalities. Many studies have reported a decreased excretion of NE and its metabolites in bipolar depression compared with unipolar depression (52). Other findings are less well-replicated [e.g., lower DBH activity (91,92) and reduced platelet MAO activity (96) in bipolar depression compared with unipolar depression]. Clinical experience indicates that some depressed subjects have never had hypomania or mania and do not have a family history of BD. Therefore, investigation of the unipolar versus bipolar group has the potentia1 for uncovering the nature of abnormality that makes the bipolar group vulnerable to manic episodes. This area needs further study.

Bipolar Disorder and Psychotic Illness

BD has a close relationship with psychotic disorder in that both bipolar depression and mania can be associated with psychotic illness and BD can coexist with schizophrenia-like symptoms (schizoaffective disorder). A common link with CA abnormality may exist, as CA abnormalities (particularly DA abnormalities) are present in both disorders. Patients with delusional depression show an increase in peripheral HVA excretion as opposed to nondelusional patients who show decreased HVA secretion (219). Hyperactivity of the CA system is also suggested by an increase in peripheral MHPG excretion in patients with delusional depression (220). Agren (51) reported increased CSF HVA in psychotic depression and that increased CSF HVA levels correlate with the degree of psychosis. Other studies have not reported an increase in plasma HVA in psychotic depression (221). DBH, the enzyme that converts DA to NE, has been found to be lower in psychotic depression than in nonpsychotic depression. Hyperactivity of the DA system is one of the prominent hypotheses for positive psychotic symptoms (222) as it is for mania. Therefore, questions regarding the specificity of CA abnormality for BD are raised (e.g., why are acutely psychotic subjects not manic or how can acutely psychotic subjects be often depressed) (223). These contradictions can be reconciled by conceptualizing that the effect of CA abnormalities on different parallel cortical–subcortical circuits may lead to different illnesses (8,26,29) depending to what extent the mood, motor, or cognitive circuit is affected. For example, the same CA abnormality leading to dysregulation of the cortical–subcortical circuit may lead to mainly movement disorder if it affects predominantly the motor circuit, thought disorder if it affects the cognitive circuit, and mood disorder if it predominantly affects the mood circuit. Keeping in mind the close proximity and distributed nature of these circuits (27), CA abnormality would be expected to lead to a dysregulation in all three aspects of behavior but lead to a different clinical picture depending on the extent of involvement of the motor-, cognitive-, or mood-regulating part of the circuit (25,29). This is the case for most neuropsychiatric disorders that usually present as a combination of motor, mood, and cognitive abnormalities. In BD, the predominant dysfunction is mood; however, involvement of the motor circuit in the form of increased or decreased motor activity is frequently seen, and thought disorder in the form of flight of ideas and frank psychosis is not uncommon.

PERSPECTIVE

A review of the role of CAs in the pathophysiology of BD presents a complex picture (Table 1). On one hand, CA abnormalities seem to be definitely present during manic and depressed states but on the other the nature of the abnormality remains elusive. The increase in CAs during mania and decrease in depression in serum, urine, and CSF seems to be largely derived from changes in CAs in the periphery, and therefore is likely to be, to a large extent, secondary to the manic or depressed states themselves. If state-related CA changes are discounted, then what kind of CA abnormality is likely to lie underneath in mania or depression? The fact that drugs that increase CAs (e.g., amphetamine, cocaine, tricyclics, and yohimbine) can induce mania suggests that an increase in CAs could underlie the pathophysiology of mania. However, investigators have questioned whether the transient euphoric state induced by CA-increasing drugs is a good model of mania that involves a sustained change in mood over a period of time (43,61).

TABLE 1 Role of Catecholamines in the Etiology of Depression and Mechanism of Action of Antidepressants

Site of Action	Findings	Replicability
Synthesis	AMPT inhibition of TH leads to improvement in mania and worsening of bipolar depression	+
	Withdrawal from AMPT leads to improvement in mood in healthy subjects and emergence of hypomanic symptoms in euthymic bipolar disorder subjects.	++
Storage	Reserpine can improve mania.	++
Turnover	↓ MHPG in serum and urine in bipolar depression	++
	↓ HVA in bipolar depression associated with psychomotor retardation	++
	Lithium decreases CA turnover	++
	Clonidine, an $\alpha 2$ receptor agonist, decreases manic symptoms	+
Autoreceptor function	Yohimbine, an $\alpha 2$ autoreceptor antagonist, can induce mania	+
	D2 autoreceptor antagonists have antidepressant properties	+
	D2 autoreceptor stimulation in the striatum is associated with decreased firing of DA neurons	+
Postsynaptic receptor function	GH response to clonidine blunted in both mania and depression	++
	D2 receptor increased in psychotic mania	+
	D2 receptor unchanged in euthymic bipolar disorder	+
	D2 receptor down regulation may be a common mode of action in antimanic drugs	++
Reuptake inhibition	CA reuptake inhibitors are effective in bipolar depression	+++
	Amphetamine and cocaine inhibit CA uptake mechanisms and can lead to mania-like states	+++
Second and third messengers	Lithium mechanism of action linked with inhibition of PIP	++
	Antiphosphokinase drugs such as tamoxifen may be useful in mania	+
	G protein abnormalities seen in euthymic bipolar disorder	++
Neurotrophic factors	Neurotrophic factor 3 increases NE transmission and increases survival of NE neurons	+
	Transplantation of NE neurons in the frontal cortex reverses depression in animal models	+
Modulatory factors	NMDA receptor stimulation by excitatory amino acids causes increased subcortical CA transmission	++
	NMDA antagonism leads to mood elevation	+
	Glutamate release inhibition can lead to mood elevation	+
	Somatomedins and neuropeptide Y influence NE transmission	+
	5HT has inhibitory effects on the CA system through heteroreceptors	++
	Thyroid hormone increases CA receptor sensitivity	++
	Steroids can lead to increase DA release	+

Abbreviations: +, one or few studies; ++several studies; +++, highly replicated by several research groups; +/−, mixed or inconsistent results; 5-HT, 5-hydroxytryptamine (serotonin); AMPT, alphamethylparatyrosine; CA, catecholamine, DA, dopamine; GH, growth hormone; HVA, homovanillic acid; MHPG, methoxyhydroxyphenylglycol; NE, norephinephrine; NMDA, *N*-methyl-D-aspartate; PIP, phosphatidylinositolphosphate; TH, tyrosine hydroxylase.

If not a presynaptic increase in CA production, then an abnormality in postsynaptic receptor sensitivity is suggested by findings such as decreased prolactin levels in bipolar depressed subjects (55) and by emergence of hypomania during recovery from CA depletion with AMPT (43,59). However, extensive research using the neuroendocrinological challenge paradigm has been unable to uncover a specific abnormality in receptor sensitivity in bipolar depression or mania (76). Nonetheless, change in postsynaptic receptor sensitivity, either due to change in the receptors themselves or due to changes in second or third messenger systems, is suggested by the effects of medications useful in the treatment of BD (e.g., neuroleptics, antidepressants, lithium, and valproate). Current understanding of mode of action of these medications, the action of which is usually delayed and prolonged, suggests effects on signal transduction mechanisms, genetic effects, and changes in neurotrophic factors. However, more research needs to be done in this area to ascertain whether systems implicated in mode of action of medications are also found to be abnormal in BD. Brain imaging studies promise state-of-the-art technology to study brain structure and function in vivo. However, there is a paucity of brain imaging studies of the CA system in BD. The studies that have been conducted point to possible changes in postsynaptic CA receptors in BD (142,146).

The role of CA abnormalities is best understood in the context of the role of CAs in emotions and mood. In this regard, variations of the cortico–striatopallidial–thalamic circuit (depending on which areas of the brain are included) can be conceptualized as circuits involved in motor, thought, and mood production (8,25,26,29). Just as abnormalities of regulatory factors in the motor circuit can lead to motor abnormalities such as dyskinesias, tremors, or rigidity, abnormalities in cognitive circuit could lead to disruption of thought processes (22), and abnormalities of the MRC could lead to disruption in mood regulation, leading to fluctuations of mood between mania and depression (29). Therefore, BD can be conceptualized as an abnormality of regulation of activity of the MRC. As reviewed above, regulation of activity of the MRC is a function of a number of different but interacting systems. Intracellular signal transduction mechanisms and genetic mechanisms, some of which may be regulated by clock genes, regulate intrinsic rhythmicity and sensitivity of neurons. Within the circuits themselves, tight feedback control mediated by fast conductance via GABA and excitatory amino acids (EAAs) maintains the oscillating activity of the system within a certain range (33). Further regulatory control is provided by extrinsic neuromodulators such as CAs, 5HT, ACh, neuropeptides, and hormones. Among these, the CAs are likely to have a primary role as is suggested by their prominent role in reward mechanisms (6), widespread effects on the CNS, and as predominant neuromodulators in the striatum (32), which has an important function in regulating cortical output (224). Furthermore, both DA and NE have been shown to have a prominent role in arousal mechanisms, initiation and maintenance of motor activity, diurnal rhythms, and sleep and cognitive functions that are frequently abnormal in BD. Some investigators have suggested that the effect of other neuromodulators (e.g., 5HT) on mood may ultimately be through effect on the CA system, particularly the DA system (7). Therefore, the CA system may act as a bridge between the MRC and other external neuromodulators and therefore play a central role in mood regulation.

To further understand the factors involved in pathophysiology of BD, new models of interactions of mood-regulating factors need to be developed. Some of the models proposed for the role of CA in BD have been discussed above:

Bunney and Garland's (43) model of phasic changes in postsynaptic receptor sensitivity, Post and Weiss's (38) kindling hypothesis, Antelman et al.'s (1) sensitization-induced cycling model, the permissive hypothesis regarding 5HT modulation of DA function, Grace et al.'s (191) model of cortical–subcortical interaction, Lachman and Papolos' (210) model of cyclic changes in signal transduction mechanism, and the more recent but still developing model of modification of behavior by clock genes (217). New models need to be developed that can integrate the role of all known neuromodulators. Such models will probably emerge with greater understanding of mood regulation both at the molecular and the circuit levels.

The results of the studies reviewed earlier suggest an increase of CAs in mania and a decrease in depression (though there is an increase of CAs in bipolar depression associated with psychotic features). One major difficulty in ascertaining the role of CA in BD is that CAs, besides modulating the activity of the MRC, are also part of the output of the MRC. There is a close relationship between CAs and neurophysiological mechanisms involved in arousal, neuroendocrine response, sympathetic and parasympathetic nervous system activity, and motor activity and reward-related behavior. Therefore, it is difficult, if not impossible, to tease out whether CA abnormalities are secondary to the depressed or manic state or whether they have a more primary role in the induction of these states. It follows that a study of the manic or depressed state is unlikely to reveal the central abnormality in BD, particularly regarding CA abnormalities. Strategies that may be more successful are investigation of euthymic or well state, investigation of early or prodromal stages of BD before a full-blown manic or depressed episode have occurred, and investigation of well relatives of BD subjects. Studies of variables that are state independent (e.g., genetic studies that are not affected by these state-related constraints) need to be more intensively investigated.

Brain imaging studies have the greatest potential to unravel the role of CAs in the pathophysiology of BD. Future studies that are able to measure different aspects of the CA system at the same time (i.e., CA release, uptake, pre- and postsynaptic receptors) will be able to provide a more dynamic picture of CA abnormalities in BD. Measurement of different neurotransmitters and investigation of the interactions of other neurotransmitters with the CA system will provide a better understanding of abnormalities of modulation of the MRC in BD. Moreover, these brain imaging studies need to be conducted in all three states of the illness, preferably in the same subjects. Finally, the mechanism of action of medications useful to treatment of BD on the CA system needs to be further investigated.

Beside the methodological problems enumerated above, other difficulties in conducting research into the pathophysiology of BD are multiple: constantly changing phases of the illness (i.e., depression, mania, mixed states, and euthymia), confounding effects of medication or substance abuse, and difficulty in recruitment of subjects. Nevertheless, study of CA abnormalities in BD remains an exciting area of investigation. Rapid developments in our ability to directly observe brain events and study molecular and intracellular processes promise to reveal CA abnormalities that may be central to the pathophysiology of BD.

REFERENCES

1. Antelman SM, Caggiula AR, Kucinski BJ, et al. The effects of lithium on a potential cycling model of bipolar disorder. Prog Neuropsychopharmacol Biol Psychiatry 1998; 22:495–510.

2. Moore H, Todd CL, Grace AA. Striatal extracellular dopamine levels in rats with haloperidol-induced depolarization block of substantia nigra dopamine neurons. J Neurosci 1998; 18:5068–5077.
3. Hyman SE, Nestler E. Molecular Basis of Psychiatry. Washington, DC: American Psychiatric Press, Inc., 1993.
4. Cooper JR, Bloom FE, Roth RH. The Biochemical Basis of Neuropharmacology. New York: Oxford University Press, 1996.
5. Self DW, Barnhart WJ, Lehman DA, et al. Opposite modulation of cocaine-seeking behavior by Dl- and D2-like dopamine receptor agonists. Science 1996; 271:1586–1589.
6. Schultz W. Dopamine neurons and their roles in reward mechanisms. Curr Opin Neurobiol 1997; 7:191–197.
7. Willner P. Dopaminergic mechanisms in depression and mania. In: Bloom FE, Kupfer DJ, eds. Psychopharmacology: the Fourth Generation of Progress. New York: Raven Press, 1995:921–931.
8. Carroll BJ. Psychopathology and neurobiology of manic-depressive disorders. In: Carroll BJ, Barret JE, eds. Psychopathology and the Brain. New York: Raven Press, 1991:265–285.
9. Depue RA, Iacono W. Neurobehavioral aspects of bipolar disorder. Annu Rev Psychol 1989; 40:457–492.
10. Redmond DE. Studies of the Nucleus Locus Coeruleus in Monkeys and Hypothesis for Neuropsychopharmacology. New York: Raven Press, 1987.
11. Woodward DJ, Moises HC, Waterhouse BD, et al. Modulating actions of norepinephrine in the central nervous system. Federation Proc 1979; 38:2109–2116.
12. Charney DS, Southwick SM, Delgado PL, et al. Current Status of the Receptor Sensitivity Hypothesis of Antidepressant Action: Implications for the Treatment of Severe Depression. Basel: Marcel Dekker, 1990.
13. Bylund DB. Subtypes of c2-adrenoreceptors: pharmacological and molecular biological evidence converge. Trends Pharmacol Sci 1988; 9:356–361.
14. Mongeau R, Blier P, de Montigny C. In vivo electrophysological evidence for tonic activation by endogenous noradrenaline of a2-adrenoreceptors on 5-hydroxytryptamine terminals in the rat hippocampus. Naunyn Schrniedberg's Arch Pharmacol 1993; 347:266–272.
15. Sara SJ. Learning by neurones: role of attention, reinforcement and behaviour. Comp Rend Acad Sci 1998; 321:193–198.
16. Grenhoff J, Nisell M, Ferre S, et al. Noradrenergic modulation of midbrain dopamine cell firing elicited by stimulation of the locus coeruleus in the rat. J Neural Transm 1993; 93:11–25.
17. James W. The Principles of Psychology Vols.1 and 2. New York: Dover Publications, 1890 and 1950.
18. Selye H. Selye's Guide to Stress Research. Vol. 1. New York: Van Nostrand Reinhold, 1980.
19. MacLean PD. Some psychiatric implications of physiological studies on frontotemporal portion of limbic system (visceral brain). Electroencephaologr Clin Neurophysiol 1952; 4:407–418.
20. Papez JW. A proposed mechanism of emotion. Arch Neurol Psychiatry 1937; 38: 725–743.
21. Damasio AR. Descarte's Error: Emotion, Reason, and the Human Brain. New York: Grosset/Putnam, 1994.
22. Graybiel AM. The basal ganglia and cognitive pattern generators [see comments]. Schizophrenia Bull 1997; 23:459–469.
23. Nauta WJH. Reciprocal links of the corpus striatum with the cerebral cortex: a common substrate for movement and thought. In: Mueller E, ed. Neurology and Psychiatry: A Meeting of Minds. Basel: Karger, 1989:43–63.
24. Alexander GE, Crutcher MD. Functional architecture of basal ganglia circuits: neural substrates of parallel processing [see comments], Trends Neurosci 1990; 13:266–271.
25. Alexander GE, Crutcher MD, DeLong MR. Basal ganglia-thalamocortical circuits: parallel substrates for motor, oculomotor, "prefrontal" and "limbic" functions. Progr Brain Res 1990; 85:119–146.

26. Cummings JL. Anatomic and behavioral aspects of frontal-subcortical circuits. Ann Acad Sci 1995; 769:1–13.
27. Damasio AR. Towards a neuropathology of emotion and mood. Nature 1997; 386:769–770.
28. Drevets WC, Price JL, Simpson JR, et al. Subgenual prefrontal cortex abnormalities in mood disorders. Nature 1997; 386:824–827.
29. Swerdlow NR, Koob GP. Dopamine, schizophrenia, mania and depression: toward a unified hypothesis of cortico-striato-pallido-thalamic function. Behav Brain Sci 1987; 10:197–245.
30. Anand A, Li Y, Wang Y, et al. Activity and connectivity of mood regulating circuit in depression: a functional magnetic resonance study. Biol Psychiatry 2005; 15:1079–1088.
31. McCormick D. Membrane properties and neurotransmitter action. In: Shepherd GM, ed. Synaptic Organization of the Brain. New York: Oxford University Press, 1998:37–78.
32. Graybiel AM. Neurotransmitters and neuromodulators in the basal ganglia. Trends Neurosci 1990; 13:244–254.
33. Shepherd GM. Neurobiology. New York: Oxford University Press, 1994.
34. Lopez HS, Brown AM. Neuromodulation. Curr Opin Neurobiol 1992; 2:317–322.
35. Graybiel AM. The basal ganglia and the initiation of movement. Rev Neurol 1990; 146:570–574.
36. Dunwidde TV, Taylor M, Heginbotham LP, et al. Long-term increases in excitability in the CAI region of the rat hippocampus induced by beta-adrenergic stimulation: possible mediation by cAMP. J Neurosci 1992; 12:506–507.
37. Berthier ML, Kulisevsky J, Gironell MD, et al. Poststroke bipolar affective disorder. J Neuropsychiatry Clin Neurosci 1996; 8:160–167.
38. Post RM, Weiss SR. A speculative model of affective illness cyclicity based on patterns of drug tolerance observed in amygdala-kindled seizures. Mol Neurobiol 1996; 13:33–60.
39. Wikinson DM, Halpern LM. The role of biogenic amines in amygdalar kindling. I. Local amygdalar after discharge. J Pharmacol Exp Ther 1979; 211:151–158.
40. Wickens JR, Begg Al. Dopamine reverses the depression of rat corticostriatal synapses which normally follows high-frequency stimulation of cortex in vitro. Neuroscience 1996; 70:1–5.
41. Schildkraut J. The catecholamine hypothesis of affective disorder: a review of supporting evidence. Am J Psychiatry 1965; 122:509–522.
42. Bunney WEJ, Davis JM. Norepinephrine in depressive reactions. Arch Gen Psychiatry 1965; 13:483–494.
43. Bunney WE, Garland BL. A second generation catecholamine hypothesis. Pharmacopsychiatry 1982; 15:111–115.
44. Coppen AJ. Depressed states and indolealkylamines. Adv Pharmacol 1968; 6 (Pt B):283–91.
45. Lapin IP, Oxenkrug GF. Intensification of the central serotoninergic processes as a possible determinant of the thymoleptic effect. Lancet 1969; 1(7586):132–136.
46. Prange AJ Jr, Wilson IC, Lynn CW, et al. L-tryptophan in mania: Contribution to a permissive hypothesis of affective disorders. Arch Gen Psychiatry 1974; 30(1):56–62.
47. Kopin IJ, Gordon EK, Jimerson, DC, et al. Relation between plasma and cerebrospinal fluid levels of 3-methoxy-4-hydroxyphenylglycol. Science 1983; 219:73–75.
48. Charney DS, Heninger GR, Sternberg DE, et al. Receptor sensitivity and the mechanism of action of antidepressant treatment. Arch Gen Psychiatry 1981; 38:1160–1180.
49. Schatzberg AF, Orsulak PJ, Rosenbaum AU, et al. Toward a biochemical classification of depressive disorders. V. Heterogeneity of unipolar depressions. Am J Psychiatry 1982; 139:471–475.
50. Koslow SH, Maas Jam, Bowden CL, et al. CSF and urinary biogenic amines and metabolites in depression and mania. Arch Gen Psychiatry 1983; 40:999–1010.
51. Agren H. Depressive symptom patterns and urinary MHPG excretion. Psychiatry Res 1982; 6:185–196.

52. Schatzberg AF, Schildkraut AF. Recent studies on norepinephrine systems in mood disorders. In: Bloom FE, Kupfer DJ, eds. Psychopharmacology: The Fourth Generation of Progress. New York: Raven Press, 1995:911–920.
53. Schatzberg AF, Samson JA, Bloomingdale KL, et al. Toward a biochemical classification of depressive disorders. X. Urinary catecholamines, their metabolites, and D-type scores in sub-groups of depressive disorders. Arch Gen Psychiatry 1989; 46:260–268.
54. Goodwin FK, Jamison KR. Manic Depressive Illness. New York: Oxford University Press, 1990.
55. Depue RA, Arbisi F, Krauss S, et al. Seasonal independence of low prolactin concentration and highspontaneous eye blink rates in unipolar and bipolar II seasonal affective disorder. Am J Psychiatry 1990; 47:356–364.
56. Engelman K, Horowiz D, Jequier E. Biochemical and pharmacological effects of alpha-methyltyrosine in man. J Clin Invest 1968; 47:577–594.
57. Brodie HKH, Murphy DL, Goodwin FK, et al. Catecholamines and mania: the effect of alpha-methyl-paratyrosine on manic behavior and catecholamine metabolism. Clin Pharmacol Ther 1971; 12:218–224.
58. Berman RM, Narasimhan M, Miller HL, et al. Transient depressive relapse induced by catecholamine depletion potential phenotypic vulnerability marker? Arch Gen Psychiatry 1999; 56:395–403.
59. Anand A, Darnell A, Miller HL, et al. Effect of catecholamine depletion on lithium-induced long-term remission of bipolar disorder. Biol Psychiatry 1999; 45:972–978.
60. Anand A, Verhoeff P, Seneca N, et al. Brain SPECT imaging of amphetamine-induced dopamine release in euthymic bipolar disorder patients. Am J Psychiatry 2000; 157(7):1108–1114.
61. Gerner RM, Post RM, Bunney WE. A dopaminergic mechanism in mania. Am J Psychiatry 1976; 113:1177–1180.
62. Martinez D, Slifstein M, Broft A, et al. Imaging human mesolimbic dopamine transmission with positron emission tomography. Part II: amphetamine-induced dopamine release in the functional subdivisions of the striatum. J Cereb Blood Flow Metab 2003; 23(3):285–300.
63. Jacobson D, Silverstone T. Dextroamphetamine-induced arousal in human subjects as a model of mania. Psychol Med 1986; 16:323–329.
64. Nurnberger JIJ, Gershon ES, Simmons S, et al. Behavioral, biochemical and neuroendocrine responses to amphetamine in normal twins and "well-state" bipolar patients. Psychoneuroendocrinology 1982; 7:163–176.
65. McTavish SF, McPherson MH, Harmer CJ, et al. Antidopaminergic effects of dietary tyrosine depletion in healthy subjects and patients with manic illness. Br J Psychiatry 2001; 179:356–360.
66. Silverstone T. Response to bromocriptine distinguishes bipolar from unipolar depression [letter]. Lancet 1984; 1:903–904.
67. Post RM, Lake CR, Jimerson DC, et al. Effects of a dopamine agonist piribedil in depressed patients: relationship of pre-treatment homovanillic acid to antidepressant response. Arch Gen Psychiatry 1978; 35:609–615.
68. Price LH, Charney DS, Heninger GR. Three cases of manic symptoms following yohimbine administration. Am J Psychiatry 1984; 141:1267–1268.
69. Machado-Vieira R, Viale CI, Kapczinski F. Mania associated with an energy drink: the possible role of caffeine, taurine, and inositol. Can J Psychiatry 2001; 46(5):454–455.
70. Ogawa N, Ueki H. Secondary mania caused by caffeine. Gen Hosp Psychiatry 2003; 25(2):138–139.
71. Anand A, Charney DS. Catecholamines in depression. In: van Praag RM, Honig A, eds. Depression: Neurobiological, Psychopathological and Therapeutic Advances. London, UK: John Wiley & Sons Ltd., 1997:147–178.
72. Meltzer H Y, Kolakowska T, Fang VS, et al. Growth hormone and prolactin response to apomorphine in schizophrenia and major affective disorders. Relation to duration of illness and affective symptoms. Arch Gen Psychiatry 1984; 41:512–519.
73. Hirschowitz J, Zemlan FP, Hitzemann RJ, et al. Growth hormone response to apomorphine and diagnosis: a comparison of three diagnostic systems. Biol Psychiatry 1986; 21:445–454.

74. McPherson H, Walsh A, Silverstone T. Growth hormone and prolactin response to apomorphine in bipolar and unipolar depression. J Affect Disord 2003; 76(1–3):121–125.
75. Linkowski P, Brauman H, Mendlewicz J. Prolactin and growth hormone response to levodopa in affective illness. Neuropsychobiology 1983; 9:108–112.
76. Matussek N. Catecholamines and mood: neuroendocrine aspects. Curr Top Neuroendocrinol 1988; 8:145–182.
77. Charney DS, Heninger GR, Sternberg DE, et al. Adrenergic receptor sensitivity in depression: effects of clonidine in depressed patients and healthy subjects. Arch Gen Psychiatry 1982; 39:290–294.
78. Ansseau M, von Erenckell R, Cerfontaine JL, et al. Neuroendocrine evaluation of catecholaminergic neurotransmission in mania. Psychiatry Res 1987; 22:193–206.
79. Ansseau M, Cerfontaine R, et al. Blunted response of growth hormone to clonidine and apomorphine in endogenous depression. Br J Psychiatry 1988; 153:65–171.
80. Langer G, Heinze O, Reim B, et al. Growth hormone response to d-amphetamine in normal controls and depressive patients. Neurosci Lett 1975; 1:185–189.
81. Anand A, Charney DS, McDougle CJ, et al. Neuroendocrine and behavioral responses to m-chlorophenylpiperazine (MCPP) in depression. Am J Psychiatry 1994; 151:1626–1630.
82. Rubinow DR, Gold PW, Post RM, et al. CSF Somatostatin in affective illness. Arch Gen Psychiatry 1983; 40:409–412.
83. Waldmeier PC. Is there a common denominator for the anti-manic effects of lithium and anticonvulsants? Pharmacopsychiatry 1987; 20:37–47.
84. Cervo L, Samanin R. Repeated treatment with imipramine and amitryptline reduces the immobility of rats in the swimming test by enhancing dopamine mechanisms in the nucleus accumbens. J Pharmac Pharmacol 1988; 40:155–156.
85. Özerdem A, Schmidt ME, Manji HK, et al. Chronic lithium administration enhances noradrenergic responses to intravenous administration of the alpha2 antagonist idazoxan in healthy volunteers. J Clin Psychopharmacol 2004; 24(2):150–154.
86. Swann AC, Koslow SH, Katz MM, et al. Lithium carbonate treatment of mania. Arch Gen Psychiatry 1987; 44:345–354.
87. Giannini AJ, Pascarazi GA, Loiselle RH, et al. Comparison of clonidine and lithium in the treatment of mania. Am J Psychiatry 1986; 143:1608–1609.
88. Hardy M, Lecrubier 17, Widlocher D. Efficacy of clonidine in 24 patients with acute mania. Am J Psychiatry 1986; 143:1450–1453.
89. Altshuler LL, Post RM, Leverich GS, et al. Antidepressant-induced mania and cycle acceleration: a controversy revisited [see comments]. Am J Psychiatry 1995; 152:1130–1138.
90. Kjellman BF, Karlberg BE, Morel LE. Serum dopamine-beta-hydroxylase activity in patients with major depressive disorders. Acta Psychiatr Scand 1986; 73:266–270.
91. Levitt N, Dunner DL, Mendlewicz J, et al. Plasma dopamine hydroxylase activity in bipolar disorder. Psychopharmacologia 1976; 46:205–210.
92. Rihmer Z, Bagdy G, Arato M. Serum dopamine-beta-hydroxylase activity and family history of patients with bipolar manic-depressive illness. Acta Psychiatr Scand 1983; 68:140–141.
93. Strandman E, Wetterberg L, Perris C, et al. Serum dopamine-beta-hydroxylase in affective disorders. Neuropsychobiology 1978; 4:248–255.
94. Puzynski S, Rode A, Zaluska M. Studies on biogenic amine metabolizing enzymes (DBH, COMT, MAO) and pathogenesis of affective illness. I. Plasma dopamine-beta-hydroxylase activity in endogenous depression. Acta Psychiatr Scand 1983; 67:89–95.
95. Ikeda Y, Ijima M, Nomura S. Serum dopamine-beta-hydroxylase in manic-depressive psychosis. Br J Psychiatry 1982; 140:209–210.
96. Murphy DL, Coursey RD, Haenel T, et al. Platelet monoamine oxidase as a biological marker in the affective disorders and alcoholism. In: Usdin E, Hanin 1, eds. Biological Markers in Psychiatry and Neurology. Oxford: Pergamon Press, 1982: 123–134.
97. Dunner DL, Levitt M, Kumbaraci T, et al. Erythrocyte catechol-O-methyltransferase activity in primary affective disorder. Biol Psychiatry 1977; 12:237–244.

98. Maj M, Grazia A, Arena F, et al. Plasma cortisol, catecholamine and cyclic AMP levels, response to dexamethasone suppression test and platelet MAO activity in manic-depressive patients. Neuropsychobiology 1984; 11:168–173.
99. Papolos DF, Veit S, Faedda GL, et al. Ultra-ultra rapid cycling bipolar disorder is associated with the low activity catecholamine-O-methyltransferase allele. Mol Psychiatry 1998; 3(4):346–349.
100. Graf WD, Unis AS, Yates CM, et al. Catecholamines in patients with 22q11.2 deletion syndrome and the low-activity COMT polymorphism. Neurology 2001; 57(3):410–416.
101. Siever LJ, Uhde TW, Insel TR, et al. Growth hormone response to clonidine unchanged by chronic clorgyline treatment. Psychiatry Res 1982; 7:139–144.
102. Garcia-Sevilla JA, Padro D, Giralt T, et al. a2-Adrenoreceptor mediated inhibition of platelet adenylate cyclase and induction of aggregation in major depression: effect of long-term cyclic antidepressant drug treatment. Arch Gen Psychiatry 1990; 47:125–132.
103. Piletz JE, Schubert DSP, Halaris A. Evaluation of studies on platelet $\alpha 2$ adrenoreceptors in depressive illness. Life Sci 1986; 39:1589–1616.
104. Kindler SL. Norepinephrine and depression: a reappraisal. In: Pohl RG, ed. The Biologic Basis of Psychiatric Treatment. Progress in Basic and Clinical Pharmacology. Vol. 3. Basel: Karger, 1990:120–141.
105. Buckholtz NS, Davies AO, Rudorfer MV, et al. Lymphocyte beta adrenergic receptor function versus catecholamines in depression. Biol Psychiatry 1988; 24:451–457.
106. Wright AF, Crichton DN, Loudon JB, et al. Beta-adrenoreceptor binding defects in cell lines from families with manic-depressive disorder. Ann Hum Genet 1984; 48:201–204.
107. Berrettini WH, Bardakjian J, Barnett AL Jr, et al Beta-adrenoceptor function in human adult skin fibroblasts: a study of manic-depressive illness. Ciba Foundation Symp 1986; 123:30–41.
108. Berrettini WH, Cappellari CB, Numberger JI Jr, et al. Beta-adrenergic receptors on lymphoblasts. A study of manic-depressive illness. Neuropsychobiology 1987; 17:15–18.
109. Manji HK, Lenox RH. Long-term action of lithium: a role for transcriptional and posttranscriptional factors regulated by protein kinase C. Synapse 1994; 16:11–28.
110. Baraban JM, Worley PF, Snyder SH. Second messenger and psychoactive drug action: focus on the phosphoinositide system and lithium. Arch J Psychiatry 1989; 146:1251–1260.
111. Berridge MJ. The Albert Lasker Medical Awards. Inositol trisphosphate, calcium, lithium, and cell signaling. JAMA 1989; 262:1834–1841.
112. Lenox RH, McNamara RK, Papke RL, et al. Neurobiology of lithium: an update. J Clin Psychiatry 1998; 59(suppl 6):37–47.
113. Manji HK, Bebchuk JM, Moore GJ, et al. Modulation of CNS signal transduction pathways and gene expression by mood-stabilizing agents: therapeutic implications. J Clin Psychiatry 1999; 60(suppl 2):27–39, discussion 40–41, 113–116.
114. Chen G, Huang LD, Jiang YM, et al. The mood-stabilizing agent valproate inhibits the activity of glycogen synthase kinase-3. J Neurochem 1999; 72:1327–1330.
115. Freidman E, Hoau YW, Levinson D, et al. Altered platelet protein kinase C activity in bipolar affective disorder-manic episode. Biol Psychiatry 1993; 33:520–525.
116. Kaya N, Resmi H, Özerdem A, et al. Increased inositol-monophosphatase activity by lithium treatment in bipolar patients. Prog Neuropsychopharmacol Biol Psychiatry 2004; 28(3):521–527.
117. Young LT, Li PP, Kish SJ, et al. Postmortem cerebral cortex Gs alpha-subunit levels are elevated in bipolar affective disorder. Brain Res 1991; 553:323–326.
118. Young LT, Li PP, Kish SJ, et al. Cerebral cortex Gs alpha protein levels and forskolin-stimulated cyclic AMP formation are increased in bipolar affective disorder. J Neurochem 1993; 61:890–898.
119. Young LT, Warsh JJ, Kish SJ, et al. Reduced brain 5-HT and elevated NE turnover and metabolites in bipolar affective disorder. Biol Psychiatry 1994; 35:121–127.
120. Young LT, Li PP, Kish SJ, et al. Cerebral cortex beta-adrenoceptor binding in bipolar affective disorder. J Affect Discord 1994; 30:89–92.

121. Rahman S, Li PP, Young LT, et al. Reduced 3[M]cyclic AMP binding in postmortem brain from subjects with bipolar affective disorder. J Neurochem 1997; 68:297–304.
122. Friedman E, Wang HY. Receptor-mediated activation of G proteins is increased in postmortem brains of bipolar affective disorder subjects. J Neurochem 1996; 67(3):1145–1152.
123. Wang HY, Friedman E. Enhanced protein kinase C activity and translocation in bipolar affective disorder brains. Biol Psychiatry 1996; 40:568–575.
124. Dean B, Scarr E, McLeod M. Changes in hippocampal GABAA receptor subunit composition in bipolar 1 disorder. Brain Res Mol 2005; 138(2):145–155.
125. Altschuler LL, Conrad A, Hauser P, et al. T2 hyperintensities in bipolar disorder: magnetic resonance imaging comparison and literature meta-analysis. Am J Psychiatry 1995; 152:1139–1144.
126. Aylward EH, Roberts-Twille JV, Barta PE, et al. Basal ganglia volumes and white matter intensities in patients with bipolar disorder. Am J Psychiatry 1994; 151:687–693.
127. Strakowski SM, Wilson DR, Tohen M, et al. Structural brain abnormalities in first-episode mania. Biol Psychiatry 1993; 33:204–206.
128. Strakowski SM, DelBello MP, Zimmerman ME, et al. Ventricular and periventricular structural volumes in first- versus multiple-episode bipolar disorder. Am J Psychiatry 2002; 159:1841–1847.
129. DelBello MP, Strakowski SM, Zimmerman ME, et al. MRI analysis of the cerebellum in bipolar disorder: a pilot study. Neuropsychopharmacology 1999; 21(1):63–68.
130. Mills NP, Delbello MP, Adler CM, et al. MRI analysis of cerebellar vermal abnormalities in bipolar disorder. Am J Psychiatry 2005; 162(8):1530–1532.
131. Soares JC, Mann JJ. The anatomy of mood disorders—review of structural neuroimaging studies. Biol Psychiatry 1997; 41:86–106.
132. Altshuler LL, Bartzokis G, Grieder T, et al. An MRI study of temporal lobe structures in men with bipolar disorder and schizophrenia. Biol Psychiatry 2000; 48:147–162.
133. Brambilla P, Harenski K, Nicoletti M, et al. MRI investigation of temporal lobe structures in bipolar patients. J Psychiatr Res 2003; 37:287–295.
134. Kemmerer M, Nasarallah HA, Sharma S, et al. Increased hippocampal volume in bipolar disorder [abstract]. Biol Psychiatry 1994; 35:626.
135. Krishnan KR, Doraiswamy PM, Figiel GS, et al. Hippocampal abnormalities in depression. J Neuropsychiatry Clin Neurosci 1991; 3:387–391.
136. Krishnan KR, McDonald WM, Escalona PR, et al. Magnetic resonance of the caudate nuclei in depression. Preliminary observations. Arch Gen Psychiatry 1992; 49:553–557.
137. Hirayasu Y, Shenton ME, Salisbury DF, et al. Subgenual cingulate cortex volume in first-episode psychosis. Am J Psychiatry 1999; 156:1091 – 1093.
138. Al-Mousawi AH, Evans N, Ebmeier KP, et al. Limbic dysfunction in schizophrenia and mania. A study using 18F-labelled fluorodeoxyglucose and positron emission tomography. Br J Psychiatry 1996; 169:509–516.
139. Gyulai L, Alavi A, Broieh K, et al. I-123 iofetamine single-photon computed emission tomography in rapid cycling bipolar disorder: a clinical study. Biol Psychiatry 1997; 41:152–161.
140. Migliorelli R, Starkstein SE, Teson A, et al. SPECT findings in patients with primary mania. J Neuropsychiatry Clin Neurosci 1993; 5:379–383.
141. O'Connell RA, Van Heertum RL, Luck D, et al. Single-photon emission computed tomography of the brain in acute mania and schizophrenia. J Neuroimaging 1995; 5:101–104.
142. Pearlson GD, Wong DF, Tune LE, et al. In vivo D2 dopamine receptor density in psychotic and nonpsychotic patients with bipolar disorder. Arch Gen Psychiatry 1995; 52:471–477.
143. Wong WF, Pearlson GD, Tune LE, et al. Quantification of neuroreceptors in the living human brain. IV. Effect of aging and elevations of D2-like receptors in schizophrenia and bipolar illness. J Cereb Blood Flow Metab 1997; 17:331–342.
144. Gjedde A, Wong DF. Quantification of neuroreceptors in living human brain. V. Endogenous neurotransmitter inhibition of haloperidol binding in psychosis. J Cereb Blood Flow Metab 2001; 21(8):982–994.

145. Yatham LN, Liddle PF, Shiah IS, et al. PET study of [(18)F]6-fluoro-L-dopa uptake in neuroleptic- and mood-stabilizer-naive first-episode nonpsychotic mania: effects of treatment with divalproex sodium. Am J Psychiatry 2002; 159(5):768–774.
146. D'haenen H, Bossuyt A. Dopamine D2 receptors in the brain measured with SPECT. Biol Psychiatry 1994; 35:128–132.
147. Anand A, Verhoeff P, Seneca N, et al. Brain SPECT imaging of amphetamine-induced dopamine release in euthymic bipolar disorder patients. Am J Psychiatry 2000; 157(7):1108–1114.
148. Yatham LN, Liddle PF, Lam RW, et al. PET study of the effects of valproate on dopamine D(2) receptors in neuroleptic- and mood-stabilizer-naive patients with nonpsychotic mania. Am J Psychiatry 2002; 159(10):1718–1723.
149. Perez de Castro I, Santos J, Torres P, et al. A weak association between TH and DRD2 genes and bipolar. J Med Genet 1995; 32:131–134.
150. Smyth C, Kalsi G, Brynjolfsson J, et al. Further tests for linkage of bipolar affective. Am J Psychiatry 1996; 153:271–274.
151. Kawada Y, Hattori M, Fukuda R, et al. No evidence of linkage or association between tyrosine hydroxylase gene. J Affect Dis 1995; 34:89–94.
152. Souery D, Lipp O, Mahieu B, et al. Association study of bipolar disorder with candidate genes involved in catecholamine neurotransmission: DRD2, DRD3, DAT1, and TH genes. Am J Med Genet 1996; 67:551–555.
153. Turecki G, Rouleau GA, Mari J, et al. Lack of association between bipolar disorder and tyrosine hydroxylase: a meta-analysis. Am J Med Genet 1997; 74(4):348–352.
154. Souery D, Lipp O, Rivelli SK, et al. Tyrosine hydroxylase polymorphism and phenotypic heterogeneity in bipolar affective disorder: a multicenter association study. Am J Med Genet 1999; 88(5):527–532.
155. Kunugi H, Vallada HP, Hoda F, et al. No evidence for an association of affective disorders, Biol Psychiatry 1997; 42:282–285.
156. Mynett-Johnson LA, Murphy VE, Claffey E, et al. Preliminary evidence of an association between bipolar disorder in females and the catechol-O-methyltransferase gene. Psychiatr Genet 1998; 8(4):221–225.
157. Kirov G, Murphy KC, Arranz MJ, et al. Low activity allele of catechol-O-methyltransferase gene associated with rapid cycling bipolar disorder. Mol Psychiatry 1998; 3(4):342–345.
158. Lim LC, Powell J, Sham P, et al. Evidence for a genetic association between alleles of monoamine oxidase A gene and bipolar affective disorder. Am J Med Genet 1995; 60(4):325–331.
159. Rubinsztein DC, Leggo J, Goodburn S, et al. Genetic association between monoamine oxidase A and bipolar disorder. Hum Mol Genet 1996; 5:779–782.
160. Parsian A, Todd RD. Genetic association between monoamine oxidase and bipolar disorder. Am J Med Genet 1997; 74(5):475–479.
161. Homer JP, Flodman PL, Spence MA. Bipolar disorder: dominant or recessive. Genetic Epidemiol 1997; 14:647–651.
162. Waldman ID, Robinson BF, Feigon SA. Linkage disequilibrium between the dopamine transporter gene (DAT1). Genet Epidemiol 1997; 14:699–704.
163. Heiden A, Schussler P, Itzlinger U, et al. Association studies of candidate genes in bipolar disorders. Neuropsychobiology 2000; (42 suppl 1):18–21.
164. Stober G, Jatzke S, Hells A, et al. Insertion/deletion variant (-141C Ins/Del) in the 5′ regulatory region. J Neural Transm 1998; 105:101–109.
165. Leszczynska-Rodziewicz A, Czerski PM, Kapelski P, et al. A polymorphism of the norepinephrine transporter gene in bipolar disorder and schizophrenia: lack of association. Neuropsychobiology 2002; 45(4):182–185.
166. Horiuchi Y, Nakayama J, Ishiguro H, et al. Possible association between a haplotype of the GABA-A receptor alpha 1 subunit gene (GABRA1) and mood disorders. Biol Psychiatry 2004; 55(1):40–45.
167. Massat I, Souery D, Del-Favero J, et al. Excess of allele1 for alpha3 subunit GABA receptor gene (GABRA3) in bipolar patients: a multicentric association study. Mol Psychiatry 2002; 7(2):201–207.

168. Otani K, Ujike H, Tanaka Y, et al. The GABA type A receptor alpha5 subunit gene is associated with bipolar I disorder. Neurosci Lett 2005; 381(1–2):108–113.
169. Massat I, Souery D, Del-Favero J, et al. Lack of association between GABRA3 and unipolar affective disorder: a multicentre study. Int J Neuropsychopharmacol 2001; 4(3):273–278.
170. Ambrosio AM, Kennedy JL, Macciardi F, et al. A linkage study between the GABAA beta2 and GABAA gamma2 subunit genes and major psychoses. CNS Spectr 2005; 10(1):57–61.
171. Ram A, Guedj F, Cravchik A, et al. No abnormality in the gene for the G protein stimulatory alpha subunit in. Arch Gen Psychiatry 1997; 54:44–48.
172. Garvey MJ, Noyes R Jr, Cook B, et al. Preliminary confirmation of the proposed link between reward-dependence traits and norepinephrine. Psychiatry Res 1996; 65:61–64.
173. Antelman SM, Caggiula AR. Norepinephrine-dopamine interactions and behavior. Science 1977; 195:646–653.
174. Archer T, Fredriksson A, Jonsson G, et al. Central noradrenaline depletion antagonizes aspects of d-amphetamine-induced hyperactivity in the rat. Psychopharmacology 1986; 88:141–146.
175. Tassin J. Norepinephrine-dopamine interactions in the pre-frontal cortex and the ventral tegmental area; relevance to mental diseases. Adv Pharmacol 1998; 42:712–716.
176. Soderpalm B, Andersson L, Carlsson M, et al. Serotonergic influence on the growth hormone response to clonidine in the rat. J Neural Transm 1987; 69:105–114.
177. Gillespie DD, Manier DH, Sanders-Bush E, et al. The serotonergic/noradrenergic link in brain. II. Role of serotonin in the regulation of beta adrenoreceptors in the low agonist affinity conformation. J Pharmacol Exp Ther 1988; 244:154–459.
178. Chiang C, Aston-Jones G. A 5-hydroxytryptamine-2 agonist augments GABA and excitatatory amino acid inputs to nor-adrenergic locus coeruleus neurons. Neuroscience 1993; 54:409–420.
179. Azmitia EC, Segal M. An autoradiographic analysis of the differential ascending projections of the dorsal and median raphe nuclei in the rat. J Comp Neurol 1978; 179:641–668.
180. Gershon SC, Baldessarini R. Motor effects of serotonin in the central nervous system. Life Sci 1980; 27:1435–1451.
181. Prisco S, Pagannone S, Esposito E. Serotonin-dopamine interaction in the rat ventral tegmental area: an electrophysiological study in vivo. J Pharmacol Exp Ther 1994; 271:83–90.
182. Spoont MR. Modulatory role of serotonin in neural information processing: implications for human psychopathology. Psychol Bull 1992; 112:330–350.
183. deMontigny C, Tan AT, Caille G. Short-term lithium enhances 5-HT neurotransmission in rats administered chronic anti-depressant treatments. Paper presented at the Society for Neuroscience Annual Meeting, Los Angeles, CA, Oct 18–23, 1981.
184. Gravel P, de Montigny C. Noradrenergic denervation prevents sensitization of rat forebrain neurons to serotonin by tricyclic antidepressant treatment. Synapse 1987; 1:233–239.
185. Javitt DC, Zukin SR. Recent advances in the phencyclidine model of schizophrenia (see comments). Am J Psychiatry 1991; 148:1301–1308.
186. Paul IA, Trullas R, Skolnick P, et al. Down-regulation of cortical b-adrenoceptors by chronic treatment with functional NMDA antagonists. Psychopharmacology 1992; 106:285–287.
187. Kitamura Y, Zhao X-H, Takei M, et al. Effects of antidepressants on the glutamatergic system in mouse brain. Neurochem Int 1991; 19:247–253.
188. Olney JW, Farber NB. Glutamate receptor dysfunction and schizophrenia [see comments]. Arch Gen Psychiatry 1995; 52:998–1007.
189. Anand A, Charney DS, Cappiello A, et al. Lamotrigine reduces the psychotomimetic—but not the mood elevating—effects of ketamine in humans. Paper presented at the American College of Neuropharmacology Meeting, Hawaii, 1997.
190. Calabrese JR, Rapport DJ, Shelton MD, et al. Clinical studies on the use of lamotrigine in bipolar disorder. Neuropsychobiology 1998; 38:185–191.

191. Grace AA, Moore H, O'Donnell P. The modulation of corticoaccumbens transmission by limbic afferents and dopamine: a model for the pathophysiology of schizophrenia. Adv Pharmacol 1998; 42:721–724.
192. Dager SR, Friedman SD, Parow A, et al. Brain metabolic alterations in medication-free patients with bipolar disorder. Arch Gen Psychiatry 2004; 61(5):450–458.
193. Friedman SD, Dager SR, Parow A, et al. Lithium and valproic acid treatment effects on brain chemistry in bipolar disorder. Biol Psychiatry 2004; 56(5):340–348.
194. Shiah S-I, Yatham LN. GABA function in mood disorders; an update and critical review. Life Sci 1998; 63:1289–1303.
195. Valentino RJ, Curtis AL. Antidepressant interactions with corticotropin-releasing factor in the noradrenergic nucleus locus coeruleus. Psychopharmacol Bull 1991; 27:263–269.
196. Swann AC, Petty F, Bowden CL, et al. Mania: gender, transmitter function, and response to treatment. Psychiatry Res 1999; 88(1):55–61.
197. Nemeroff CB, Owens MJ, Bissette G, et al. Reduced corticotropin releasing factor binding sites in the frontal cortex of suicide victims. Science 1984; 226:1342–1344.
198. Veith RC, Lewis N, Langohr JI, et al. Effect of desipramine on cerebrospinal fluid concentrations of corticotropin-releasing factor in human subjects. Psychiatry Res 1993; 46:1–8.
199. Heilig M, Widerlov B. Neuropeptide Y: an overview of central distribution, functional aspects, and possible involvement in neuropsychiatric illnesses. Acta Psychiatr Scand 1990; 82:95–114.
200. Widdowson PS, Ordway GA, Halaris AE. Reduced neuropeptide Y concentrations in suicide brain. J Neurochem 1991; 59:73–80.
201. Rubinow DR, Davis CL, Post RM. Somatostatin in neuropsychiatric disorders. Prog Neuropsychopharrnacol Biol Psychiatry 1988; 12:S137–S155.
202. Gheorvassaki EG, Thermos K, Liapakis G, et al. Effects of acute and chronic desipramine treatment on somatostatin receptors in brain. Psychopharrnacology 1992; 108:363–366.
203. Gerner RH, Yamada T. Altered neuropeptide concentrations in cerebrospinal fluid of psychiatric patients. Brain Res 1982; 238:298–302.
204. Berrettini WH, Nurnberger JI Jr, Zerbe RL, et al. CSF neuropeptides in euthymic bipolar patients and controls. Br J Psychiatry 1987; 150:208–212.
205. Janowsky DS, Yousef MK, Davis JM, et al. Parasympathetic suppression of manic symptoms by physostigmine. Ach Gen Psychiatry 1973; 28:542–547.
206. Belmaker RH, Zohar J, Levy A. Unidirectionality of lithium stabilization of adrenergic and cholinergic receptors. In: Emric H, Aldenhoff JB, Lux HD, eds. Basic Mechanisms in the Action of Lithium. Amsterdam: Excerpta Medica, 1982:146–153.
207. Whybrow PC, Prange AJ. A hypothesis of thyroid-catecholamine-receptor interaction. Arch Gen Psychiatry 1981; 38:106–113.
208. Hyman SE, Nestler EJ. Initiation and adaptation: a paradigm for understanding psychotropic drug action. Am Psychiatry 1996; 153:151–162.
209. Wachtel H. Dysbalance of neuronal second messenger function in the etiology of affective disorders: apathophysiological concept hypothesizing defects beyond first messenger receptors. Neurotransm 1989; 75:21–29.
210. Lachman HM, Papolos DF. Abnormal signal transduction: a hypothetical model for bipolar affective disorder. Life Sci 1989; 45:1413–1426.
211. Lachman HM, Papolos DF. A molecular model for bipolar affective disorder. Med Hypoth 1995; 45:255–264.
212. Nibuya M, Morinobu S, Duman RS. Regulation of BDNF and trkB mRNA in rat brain by chronic electroconvulsive seizure and antidepressant drug treatments. J Neurosci 1995; 15:7539–7547.
213. Siuciak JA, Anthony Altar CA, Wiegand SJ, et al. Antinociceptive effect of brain derived neurtrophic factors and neurotrophin-3. Brain Res 1994; 633:326–330.
214. Siuciak JA, Lewis D, Wiegand SJ, et al. Brain derived neurotrophic factor (BDNF) produces an anti-depressant like effect in two animal models of depression. Paper presented at the Society of Neuroscience Annual Meeting, Miami Beach, FL, Nov 13–18, 1994.

215. Arenas E, Persson H. Neurotrophin-3 prevents the death of adult central noradrenergic neurons in vivo. Nature 1994; 367:368–371.
216. Gash DM, Zhang Z, Gerhardt G. Neuroprotective and neurorestorative properties of GDNF. Ann Neurol 1998; 44(3 suppl 1):S121–S125.
217. Bunney WE, Wu JC, Gillin C, et al. Clock genes: circadian abnormalities and therapeutic strategies in depression. Paper presented at the American College of Neuropsychopharmacology Annual Meeting, San Juan, Puerto Rico, 1998.
218. Hastings M. The brain, circadian rhythms, and clock genes. Br Med J 1998; 317:1704–1707.
219. Lykouras L, Markianos M, Hatzimanolis J, et al. Biogenic amine metabolites in delusional depression and melancholia subtypes of major depression. Prog Neuropsychopharmacol Biol Psychiatry 1994; 18(8):1261–1271.
220. Lykouras L, Markianos M, Hatzimanolis J, et al. Association of biogenic amine metabolites with symptomatology in delusional and non-delusional depression. Prog Neuropsychopharmacol Biol Psychiatry 1995; 19(5):877–887.
221. Devanand DP, Bowers MBJ, Hoffman FJJ, et al. Elevated plasma homovanillic acid in depressed female patients with melancholia and psychosis. Psychiatry Res 1985; 42:923–926.
222. Crow TJ. Positive and negative schizophrenic symptoms and the role of dopamine. Br J Psychiatry 1980; 137:383–386.
223. Fibiger HC. The dopamine hypothesis of schizophrenia and mood disorders: contraindications and speculations. In: Winner P, Scheel-Kruger J, eds. The Mesolimbic Dopamine System: From Motivation to Action. London: John Wiley and Sons Ltd., 1991:615–636.
224. Grace AA. Cortical regulation of subcortical dopamine systems and its possible relevance to schizophrenia. J Neural Transm 1993; 91:111–134.

4 Cholinergic-Muscarinic Dysfunction in Mood Disorders

David S. Janowsky
Department of Psychiatry, University of North Carolina, Chapel Hill, North Carolina, U.S.A.

David H. Overstreet
Center for Alcohol Studies, Department of Psychiatry, University of North Carolina, Chapel Hill, North Carolina, U.S.A.

HISTORICAL OVERVIEW AND HIGHLIGHTS

For more than a century, acetylcholine has been postulated to be a factor in the regulation and etiology of affect. In 1889, Willoughby (1) reported a case in which pilocarpine, now known to be a muscarinic cholinergic agonist, was used to alleviate acute mania. Subsequently, in the late 1940s, 1950s, and early 1960s, a number of authors observed the anergic, inhibitory, anxiety-enhancing and mood-depressing effects of centrally acting cholinesterase inhibitors, compounds that inhibit the breakdown of acetylcholine. The mood altering effects of these compounds, used as insecticides in the agriculture industry and as nerve agents by the military, were described naturalistically and as tested in experimental settings. The observations by Grob et al. (2), Gershon et al. (3,4), Bowers et al. (5), and Rowntree et al. (6) led to a series of reports suggesting that increases in central acetylcholine led to depression, anxiety, and anergia.

In the early 1970s, based on the above studies and on animal data reported by Domino and Olds (7), Stark and Boyd (8), and Carlton (9), Janowsky et al. (10) developed an adrenergic-cholinergic balance hypothesis of manic depression. This hypothesis proposed that depression represents an overabundance of central acetylcholine, relative to central adrenergic neurochemicals, and that mania represents the converse. Part of the work of Janowsky et al. (10,11) involved infusing the short acting reversible central cholinesterase inhibitor physostigmine on one occasion and the noncentrally acting cholinesterase inhibitor, neostigmine on another. Comparing behavioral effects, Janowsky and colleagues used this paradigm in manics, depressives, schizophrenics, and normals and observed decreased manic and increased depressive symptoms only in their physostigmine treated subjects (11,12). This work was replicated during the 1970s and early 1980s by a variety of investigators. In the late 1970s Sitaram et al. (13) observed that shortening of the cholinergic-sensitive sleep parameter, rapid eye movement (REM) latency, by cholinomimetic drugs was exaggerated in affective disorder patients, suggesting cholinergic supersensitivity in these patients, a finding subsequently replicated in a number of studies.

In the late 1970s and early 1980s, Davis and Davis (14) and Risch et al. (15) began a series of experiments in which they evaluated the effect of cholinergic influences on stress-sensitive neurohormones including ACTH, beta-endorphin, cortisol, epinephrine, vasopressin, and prolactin in depressed patients, normals, and patient controls. These authors discovered that the above neurohormones were increased by cholinergic stimulation and antagonized by centrally acting antimuscarinic agents such as atropine and scopolamine. Risch, Janowsky, and colleagues (15,16) also observed that exaggerated increases in serum ACTH and beta-endorphin levels occurred in depressed patients following cholinomimetic administration, compared to the changes that occurred in patient controls who did not have an affective disorder.

In the 1990s, peripheral cholinergic supersensitivity in affective disorder patients was observed following administration of the muscarinic agonist pilocarpine, given to induce pupillary constriction (17), and following administration of non-centrally acting cholinesterase inhibitor, pyridostigmine, given to stimulate growth hormone release. In addition, in the mid- and late 1990s, Charles et al. (18), Renshaw et al. (19) and others, using proton magnet resonance spectroscopy techniques, found that the acetylcholine precursor choline and its related metabolites were increased in the brains of affective disorder patients. Furthermore, the acetylcholine precursors lecithin and choline and the cholinesterase inhibitor donepezil were reported to be useful in the treatment of mania and related bipolar conditions (20,21). Finally, in the 2000s, the latest findings supporting a role for acetylcholine in the etiology and phenomenology of the affective disorders has involved discovering muscarinic receptor gene variants linked to major depression.

BEHAVIORAL FINDINGS

Cholinomimetic Effects on Manic Symptoms

As noted above, some of the most direct and graphic evidence of a role for acetylcholine in the etiology and phenomenology of affective disorders is derived from studies of manic patients who have been administered centrally acting cholinesterase inhibitors or directly acting muscarinic cholinergic agonists. As noted above, the first report of this phenomenon was published by Willoughby (1), who reported alleviation of mania by pilocarpine in 1889. Later, in the 1950s, Rowntree et al. (6) gave DFP, an irreversible centrally acting cholinesterase inhibitor, to a group of nine bipolar disorder patients and ten normal controls. Normal subjects developed depression, apathy, lassitude, irritability, and slowness of thoughts and some became depressed. These phenomena occurred before the onset of peripheral cholinergic symptoms such as nausea, cramping, and diarrhea, suggesting a central mechanism. Two of the bipolar patients who were tested while in remission showed mental changes like those observed in the normals, showing anergia and nausea without serious affective symptoms. Two hypomanic patients' symptoms improved with DFP administration, and these patients continued to be euthymic after DFP administration had ceased. One hypomanic patient became less manic, becoming slightly depressed after each of two courses of DFP, and this patient became manic again after DFP withdrawal. One nearly remitted hypomanic patient became floridly manic once DFP had been withdrawn, and one depressed bipolar patient showed a considerable increase in depression.

Beginning in 1972, Janowsky et al. (10,11) noted that the centrally active cholinesterase inhibitor physostigmine caused a short lived and very obvious reduction in hypomanic and/or manic symptoms in eight bipolar patients with manic or mixed manic and depressive symptoms. Saline placebo and the non-centrally acting cholinesterase inhibitor neostigmine, which does not enter the brain, produced no changes in mood or behavior. Furthermore, physostigmine's antimanic effects were reversed by the centrally acting antimuscarinic drug, atropine, suggesting that the antimanic effects of physostigmine were caused by a central muscarinic mechanism. After the patients received physostigmine, the average Beigel-Murphy mania "Elation-Grandiosity" subscale score was reduced by 78% and the "manic intensity" scale score was reduced by 48% (10,11). Physostigmine caused a decrease in specific manic symptoms including "is talking," "is active," "jumps from one subject to another," and "looks happy and cheerful." "Grandiosity" was significantly decreased in the three patients in whom it was present during the baseline period. "Irritability" was decreased in three patients and was increased in five. Following physostigmine administration, depression, as measured by the Bunney-Hamburg Depression Scale, showed an overall two-fold increase, with five of the eight manic patients studied developing a depressed mood. Physostigmine's effects lasted for a period of 20 to 90 minutes and were observed to begin a few minutes after infusion occurred. The total amount of physostigmine given varied from 0.25 mg to 3.0 mg. Physostigmine did not cause sedation as such, and the patients were not obtunded. They showed no slurred speech or ataxia and did not fall asleep. In addition to antimanic and mood depressing effects, nausea and vomiting were also a common concomitant of physostigmine administration (11,12,22), as had been noted occur in earlier studies (5,6).

Other studies subsequently replicated the antimanic effects of physostigmine. In 1973, Modestin et al. (23,24) reported a lessening of manic symptoms following the infusion of physostigmine in two of four manic patients. This effect did not occur when neostigmine was administered. Davis et al. (25) reported that physostigmine caused significant antimanic effects, especially in patients who were not hostile and/or irritable. In addition, Carroll et al. (26) studied a manic patient who had a corticosteroid-induced mania and noted that physostigmine caused a decrease in euphoria and mobility. Similarly, Krieg and Berger (27) reported data suggesting that the relatively specific muscarinic (M1) cholinergic agonist RS86 had significant antimanic effects.

However, several authors have wondered whether centrally active cholinomimetic agents were only effecting the affective and motoric components of mania, and not effecting the cognitive aspects of mania. Thus, Carroll et al. (26) and Shopsin et al. (28) brought up the question of whether cholinomimetics actually effect what they considered "core" aspects of mania, such as manic grandiosity and expansive thinking.

Manic Symptom Rebound Following Cholinomimetic Administration

There is some evidence that a late occurring effect of physostigmine administration is the enhanced activation of manic symptoms and an increase in its animal analogue, hyperactivity. Fibiger et al. (29) demonstrated in rats that increased central cholinergic activity, caused by the administration of physostigmine, led first to motor inhibition, and later to an increase over baseline in locomotion. This hyperactivity was presumed to be due to compensatory increases in adrenergic neurotransmitter activity. Hyperactivity became apparent as the cholinergic behavioral inhibition induced by the physostigmine wore off. The hyperactivity was exaggerated if a centrally acting antimuscarinic drug (i.e., scopolamine) was given at the beginning of the hyperactivity phase. The rebound hyperactivity was completely prevented if the centrally active antimuscarinic drug was given prior to initial physostigmine administration (29).

In parallel with the above preclinical study, an infrequent occurring exaggeration of baseline manic symptoms following cholinesterase inhibitor administration has also been found to occur in bipolar patients (10,11,23,24). Rowntree et al. (6) observed that one of the manics to whom he had administered DFP subsequently became more manic than at baseline. Later, Shopsin et al. (28) studied three highly manic patients given physostigmine up to 6 mg intravenously and observed rebounding. All three initially experienced varying degrees of sedation, drowsiness, a desire but inability to sleep, some dysthymia, and mild slurring of speech. A reduction in spontaneous speech and activity were apparent during this time. During this initial phase, the patients' flight of ideas, rambling speech, tangentiality, irritability, and cheerfulness were also attenuated. In this study, all three patients spontaneously stated, that after receiving physostigmine, they were "talked out" and did not want to be bothered. No vomiting occurred. Apathy and anergia were apparent, yet no patient actually became depressed. The most striking feature of this study was the late appearance in two patients of a "rebound," taking place approximately two hours after the physostigmine infusion. A marked exacerbation of the manic state over baseline levels occurred, lasting three to four hours, with a return to baseline at approximately six hours following the last physostigmine injection.

Mood Effects of Centrally Active Anticholinergic Drugs

Whereas cholinomimetic agents appear to cause depression, there is evidence that anticholinergic and anti-Parkinsonian medications have mood-elevating properties. Jellinec et al. (30) and Smith (31) summarized data showing that anti-Parkinsonian drugs, given to patients with Parkinsonian symptoms, caused positive feelings and a reversal of depressed mood. Also, schizophrenics who used or abused anti-Parkinsonian drugs have reported experiencing euphoria, being "buzzed or high," having a reduction in anxiety, having a sense of well being, and feeling more sociable and more confident, cheerful, and energetic. Coid and Strang (32) reported a case in which the anticholinergic agent procyclidine appeared to cause mania in a bipolar patient. Furthermore, there are several reports indicating that high doses of atropine and other centrally acting anticholinergics, such as ditran and scopolamine, can cause euphoria and alleviate depression (33,34,35,36). Similarly, Kasper et al. (33) observed antidepressant effects with the anticholinergic anti-Parkinsonian drug biperiden, and this most often occurred in patients with endogenous depression who had a non-suppressing dexamethasone suppression test.

In a promising recent report, Furey and Drevets (37), using a placebo-controlled double-blind crossover design in currently depressed patients, reported that scopolamine four mcg/kg caused a significant and relatively dramatic reduction in depression, as measured by the MADRAS depression scale and the Montgomery Asberg Depression Rating Scale. Scopolamine relieved depression both when compared to baseline and as compared to placebo. The dose utilized was relatively high compared to previous trials evaluating the effects of scopolamine in alleviating depression.

In contrast, some studies have been less promising with respect to the ability of centrally acting anticholinergic drugs to alleviate depression. Fritze et al. (38,39) added centrally acting anticholinergic agents to a treatment regime consisting of standard antidepressant drugs and did not show increased efficacy. Similarly, Gillin et al. (40) were unable to demonstrate that treatment of depressed patients with biperiden led to alleviation of depression.

Marijuana-Physostigmine Interactions

Marijuana often induces a sense of well-being, euphoria, hilarity, increased verbalizations, and flight of ideas not unlike some of the symptoms of hypomania (41). El-Yousef et al. (42) reported that small doses of physostigmine antagonized the intoxicating effects of marijuana in two normal volunteers. In addition, both volunteers became very severely depressed following physostigmine administration. Thus, marijuana was able to induce wittiness, creativeness, and hilarity in the subjects who smoked it; following physostigmine administration, a lethargic, drained, sad, extremely depressed state manifested by utterances of hopelessness, uselessness, and worthlessness, sobbing and crying, and extreme psychomotor retardation occurred (42). This state was much more extreme than that noted after physostigmine was given alone. This observation was inadvertently replicated by Davis et al. (43) in normal volunteers who had covertly smoked marijuana before receiving physostigmine. Thus, it appears that marijuana augments and amplifies the effects of physostigmine.

The above clinical observations concerning a marijuana-physostigmine interaction were paralleled in a preclinical study performed by Rosenblatt et al. (44). This study demonstrated that the active ingredient in marijuana, Δ-9 tetrahydrocannabinol, significantly increased physostigmine-induced lethality in rats. This effect was prevented by the centrally and peripherally acting anticholinergic agents, atropine and methylscopolamine respectively. Subsequently, cholinergic behavioral effects were also found by Duncan and Dagirmanjian to be augmented in rats by Δ-9 tetrahydrocannabinol (45).

Cholinomimetic-Catecholaminergic Interactions

A pharmacological model of naturally occurring adrenergic-cholinergic balance is found in the interactions and reciprocal effects of psychostimulants, which increase dopaminergic/noradrenergic activity, and cholinomimetics, which increase acetylcholine activity. Psychostimulant-induced increases in locomotor activity, self-stimulation, and gnawing behavior in rats, which have been considered to be animal models of mania, are rapidly antagonized by physostigmine, but not by neostigmine (7,8,10,46). Conversely, physostigmine's inhibitory effects in rats can be reversed by methylphenidate (46).

In a study performed in the early 1970s by Janowsky et al. (47), manic and schizophrenic patients were given intravenous physostigmine first, followed by methylphenidate and vice versa. Physostigmine alone decreased average mania ratings of talkativeness, happiness, activity, flight of ideas, and the overall Manic Intensity Scale scores and Elation/Grandiosity scores on the Bunney-Hamburg Mania Rating Scale. It also decreased activation and increased inhibition on the Janowsky-Davis Activation-Inhibition Scale (47). When methylphenidate alone was administered, six of the eight manics who received it rapidly and significantly increased their talkativeness, activity, flight of ideas, manic intensity, and Janowsky-Davis Activation Scale scores. The increase in the Janowsky-Davis Activation Scale scores which methylphenidate induced was partially reversed by physostigmine, but not neostigmine. Conversely, physostigmine-induced increases in the Behavioral Inhibition Scale scores were partially reversed by methylphenidate (46).

There is evidence that central cholinergic and catecholaminergic mechanisms not only balance each other, but are interactive. One human study by Ostrow et al. (48) demonstrated that physostigmine caused a rapid and dramatic drop in the urinary norepinephrine metabolite, serum 3-methyoxy-4-hydroxphenylglcol (MHPG) in a manic patient, presumably reflecting a drop in CNS noradrenergic activity. This phenomenon was associated with the induction of a tearful depressed state and improvement in the patients' manic symptoms (48).

Also, a negative correlation was noted between amphetamine-induced behavioral excitation and the ability of the muscarinic agonist arecoline, given on another occasion to decrease REM latency, an acetylcholine-sensitive sleep parameter (49). Likewise, Siever et al. (50) demonstrated that those individuals who showed the most extreme physostigmine and arecoline-induced anergy and negative affect had a blunted growth hormone response to the noradrenergic agonist clonidine, a sign of decreased noradrenergic responsiveness. Similarly, Schittecatte et al. (51) demonstrated that human depressives are subsensitive to the REM sleep-suppressing effects of the noradrenergic agonist, clonidine. However, it is not clear whether this subsensivity to clonidine reflects subsensitivity of the α-noradrenergic system as such, or represents the consequences of cholinergic overactivity.

As with the above described behavioral and phenomenological studies, there is a growing body of preclinical evidence suggesting that such monoamines as dopamine, norepinephrine, and serotonin on the one hand, and acetylcholine on the other, are reciprocally interactive. For example, Hasey and Hanin (52) showed that the immobility-promoting effects of physostigmine could be modified by manipulating the β-noradrenergic system. Ikarashi et al. (53) found that dopamine 2 (D2) receptor stimulation in striatum led to a decrease in striatal acetylcholine release, suggesting a decrease in acetylcholine availability. Downs et al. (54) demonstrated that brain dopamine depletion caused an exaggerated ACTH response following physostigmine administration in rats, suggesting that the dopamine depletion led to unantagonized acetylcholine activity. Similarly, imipramine, a noradrenergic antidepressant, has been found to decrease acetylcholinesterase activity in the hippocampus by Camarini and Benedito (55), who suggest that this decreased acetylcholinesterase activity is a reflection of decreased acetylcholine release.

With respect to serotonin, although the selective serotonin reuptake inhibitors do not appear to directly block muscarinic receptors, Saito et al. (56) demonstrated that acetylcholine release appears decreased by inhibitory serotonin (5HT) 1B hetero-receptors found on cholinergic nerve terminals, and Crespi et al. (57) found that 5HT3 receptor agonists decrease acetylcholine release by effecting 5HT3 heteroreceptors. Consistent with the above, the 5HT1A agonist 8-OH-DPAT turned off cholinergic REM—on neurons, which normally activate REM sleep. Inconsistent with the above antagonistic effects, however, 8-OH-DPAT also enhanced acetylcholine release from rat hippocampus and cerebral cortex (58), and the 5HT1A agonist MKC-242 increased extracellular acetylcholine activity (59).

Depressive Effects of Cholinomimetic Agents

Along with the antimanic effects of cholinomimetic agents, some of the most convincing evidence that acetycholine is involved in the regulation of affect is the observation that centrally active cholinomimetic drugs can rapidly induce a depressed mood. Cholinomimetic insecticides, as reported by Gershon et al. (3,4) and experimental nerve agents, as reported by Bowers et al. (5) and Rowntree et al. (6), cause depression in normals. A significant proportion of the manic patients who receive centrally acting cholinomimetics develop depressive symptoms, and this appears true for some normal volunteers and depressed patients as well (10,11,12). For example, Janowsky et al. (10,11) reported that six of the eight manics and two depressives studied showed increased depressed mood following physostigmine infusion (12). In addition, five of six schizoaffective patients (four excited, two depressed types) showed depressed mood and sadness after physostigmine infusion. Similarly, Davis et al. (25) and Modestin et al. (23,24) reported an increase in depression, following physostigmine administration in some of the manic patients they studied, as did Risch et al. (15) when giving arecoline to depressed patients.

Furthermore, Risch et al. (60) found a statistically significant mean increase in self and observer rated negative affect, including depression, in normals receiving intravenous physostigmine. Risch et al. (60) also found that normal volunteers given the directly acting muscarinic cholinergic agonist arecoline developed depression and other forms of negative affect including hostility and anxiety. Likewise, Mohs et al. (61) reported severe depression occurring in some Alzheimer's patients receiving the cholinergic agonist oxotremorine.

Consistent with the above information, acetylcholine precursors including deanol, choline, and lecithin have been reported to cause depression. Thus, Tamminga et al. (62) observed that a depressed mood was precipitated in some schizophrenic patients treated with choline. Casey (63) observed that depressed mood occurred in a subset of deanol-treated patients who had tardive dyskinesia. Similarly, Bajada observed that depressed mood was a side effect of choline and lecithin treatments employed to try to reverse the memory deficits of Alzheimer's disease (64).

Increasing central cholinergic activity also induces an anergic-inhibitory syndrome which appears very similar to the psychomotor retardation component of endogenous depression. This "inhibitory syndrome" has been operationalized in the Janowsky-Davis Activation-Inhibition Scale (11,12) and has been observed by a number of authors who have administered or observed the effects of centrally acting cholinomimetic drugs (6,11,15,25). The "Inhibition" part of the Janowsky-Davis scale consists of a composite of items rating for having lethargy, having slow thoughts, wanting to say nothing, being withdrawn, being apathetic, lacking energy, being drained, being hypoactive, lacking thoughts, being motor retarded, and being emotionally withdrawn.

Although a consistent effect of the administration of centrally acting cholinomimetic drugs is the induction of depressed mood and behavioral inhibition, a separate question is whether or not affective disorder patients show a differential sensitivity to these agents. There is a growing body of evidence that while non-affective disorder patients show the psychomotor retarding and inhibitory effects after receiving centrally acting cholinomimetics, they less often show the depressive effects of these agents when compared to patients with an affective disorder. Thus, Janowsky et al. (11) showed that of eight schizophrenics without an affective component to their illness, only one showed increased depressive symptoms following physostigmine infusion. Oppenheimer et al. (65) found no increases in depressed mood in his normal subject cohort when they were given physostigmine, although behavioral inhibition did occur. Similarly, Silva et al. (unpublished data) showed no increase in depressed mood after giving physostigmine to her carefully screened normal controls, although behavioral inhibition and nausea occurred in most of her subjects. Conversely, as described above, most studies in which cholinomimetic agents are given to affective disorder patients report depressive symptoms in the majority of subjects (11,12,65). Thus, in the studies by Janowsky et al. (11,12), approximately 25% of normals and nonaffective disorder patients were found to have increases in negative mood following physostigmine administration, in contrast to approximately 75% of the affective disorder patients studied. In addition, Edelstein et al. (66) reported that schizophrenic patients who responded to physostigmine with a clearing of psychotic symptoms were significantly more likely to respond with symptomatic improvement when given lithium, presumably because their illness represented a variant of affective disorder. Furthermore, Steinberg et al. (67) has found that increases in negative affect after physostigmine administration occurred selectively in those borderline personality disorder patients who had pre-existing affectively unstable personalities. Patients with personality disorders who were affectively stable (i.e., borderline patients who were primarily impulsive) did not show negative affect after physostigmine infusion.

In addition to evidence demonstrating that affective disorder patients are more likely to become depressed while receiving physostigmine and other centrally active cholinomimetic drugs than are controls or nonaffective disorder patients, affective disorder patients may also be relatively more sensitive to the general behavioral effects of centrally acting cholinomimetic agents. Rater and patient evaluated increases in the Janowsky-Davis Inhibition Scale Score and on the self-rated anxiety, hostility, and confusion subscales of the Profile of Mood States Scale showed significantly greater increases in the depressed patients than in non-affective disorder patients or normals after arecoline (15,16) or physostigmine infusion (16).

Whether or not behavioral supersensitivity to cholinomimetic drugs is a state or trait marker of affective disorders is uncertain. Oppenheimer et al. (65) found that most of the euthymic lithium-treated bipolar patients he studied developed a depressed mood after receiving physostigmine. Similarly, Casey (63) noted that tardive dyskinesia patients having a significant past history of affective disorder were more likely to show increased affective symptoms when administered the probable acetylcholine precursor, deanol, than were those without an affective disorder history. However, in contrast, Nurnberger (49,68,69) observed no difference in behavior or mood sensitivity when euthymic affective disorder patients and normals were compared after receiving arecoline. Thus, whether behavioral supersensitivity to centrally active cholinomimetic drugs in affective disorder patients is a state- or trait-linked phenomenon is uncertain, although much evidence favors it being a trait.

In spite of evidence suggesting that selective behavioral supersensitivity to central cholinomimetic agents in affective disorder patients exists, it is alternatively possible that cholinomimetic agents are actually affecting those underlying personality characteristics which are risk factors for mood disorders. Thus, as noted above, Steinberg et al. (67) found that increases in negative affect after physostigmine administration occurred selectively in those personality disorder patients with pre-existing affectively unstable personalities, as compared to those who were affectively stable or had primarily impulsive traits. This differential effect was relatively neurotransmitter-specific, since affectively unstable patients reacting to physostigmine with negative affect did not show mood changes following noradrenergic, serotonergic, or placebo challenges. In work complementary to the above study, Fritze et al. (38,39) noted that behavioral sensitivity to physostigmine (i.e., increased inhibition) correlated with baseline irritability and emotional lability, and with habitually passive stress coping strategies. These authors proposed that cholinergic sensitivity may be predominantly related to stress supersensitivity and coping profiles, rather than to specific affective disorder diagnoses.

BIOLOGICAL FINDINGS

Choline in Bipolar Patients

While there is considerable evidence suggesting that centrally acting muscarinic agonists and cholinesterase inhibitors can effectively decrease manic symptoms and/or precipitate depression, nonbehavioral biological markers also exist, suggesting a role for acetylcholine in the affective disorders. One potential marker of cholinergic changes in the affective disorders is erythrocyte choline activity, and one function of choline is to be a precursor of acetylcholine. Slight elevations in erythrocyte choline have been noted in patients with bipolar disorders by Bidzinski et al. (70), and have also been observed in unipolar depressives and schizophrenics. Furthermore, Stoll et al. (71) found that relatively increased levels of red blood cell choline existed in a subgroup of manic patients, and it was these manic patients who had relatively more symptoms at admission and a poor outcome at discharge. In addition, bipolar patients having relatively low levels of red blood cell choline had a history of having four times as many prior episodes of mania compared to episodes of depression. In contrast, patients with high erythrocyte choline levels had a history of similar numbers of manic and depressive episodes (71).

SPECTROSCOPIC STUDIES IN AFFECTIVE DISORDER PATIENTS

In vivo proton magnetic resonance spectroscopy provides a means for more directly assessing human brain choline activity in vivo, and possibly for indirectly assessing central acetylcholine function, since choline is a major precursor of acetylcholine, as well as many other compounds. Charles et al. (18) observed that there is a state-dependent increase in choline in the brains of patients with major depression when compared to controls. This increase in choline was noted to revert to normal after successful antidepressant treatment of the depression. More recently, Renshaw et al. (19) studied the basal ganglia of depressed and control subjects, and noted an alteration in the metabolism of cytosolic choline compounds in the depressives, particularly those who subsequently were responsive to fluoxetine. In addition, Hankura et al. (72) found that depressed bipolar disorder patients had higher absolute subcortical choline-containing compounds than did normals. Thus, it would appear that depression is associated with increased central choline activity, a probable marker of increased central acetylcholine activity.

Cholinomimetic Induced Changes in REM Sleep in Affective Disorder Patients

Depression is generally associated with characteristic sleep changes. Among these is a decrease in the time until rapid eye movement (REM) sleep occurs (REM latency), increased REM duration, and increased REM density (13). Significantly, as with naturally occurring depression, centrally acting cholinergic agonists such as arecoline, pilocarpine, the muscarinic 1 (M1) agonist, RS86, and cholinesterase inhibitors such as physostigmine have been shown to cause a shortening of REM latency and an increase in REM density. Conversely, centrally acting anticholinergic, dopaminergic, noradrenergic, and serotonergic agents cause an increase in REM latency and a decrease in REM density and REM duration (73).

Significantly, in the vast majority of studies performed to date, after central cholinomimetic drug administration, REM latency is supershortened and REM density relatively increased in affective disorder patients. Thus, in a groundbreaking study utilizing affective disorder patients, Sitaram et al. (13) found that following arecoline infusion, mean REM latency was significantly more shortened in euthymic bipolar patients and in a unipolar patient, compared to the shortening that occurred in normal volunteers. Sitaram et al. (13) also found similar results in six bipolar and two unipolar euthymic patients who had been kept off all medications for at least four months, thus, suggesting a trait phenomenon. Gillin et al. (74) also demonstrated cholinergically supershortened REM latency in a group of predominantly nonbipolar symptomatic depressives after infusion of arecoline, compared to controls. Similarly, Berger et al. (75) demonstrated a super-shortening of REM latency in nonbipolar depressives when compared to normals and to eating disorder patients following administration of the cholinergic-muscarinic agonist RS86. Berger also found that cholinomimetic-induced arousal and awakening from sleep occurred more frequently in acutely symptomatic affective disorder patients than in normals (75).

Gann et al. (76) investigated sleep EEG profiles following administration of RS86 to patients with major depression and anxiety disorders and to normal controls. RS86 caused supershortening of REM latency and an increase in REM density and REM duration in patients with major depression. Patients with anxiety disorders having secondary depression did not show enhanced REM abnormalities following RS86 administration, and anxiety disorder patients showed actually decreased REM density compared to controls. Similar results with respect to REM sleep responses to RS86 were noted by Riemann et al. in 1994 (77). Likewise, Dube and coworkers (78) showed that the REM sleep response to cholinergic stimulation with arecoline was significantly more pronounced in primary depressives than in patients with manic disorders, or those with mixed anxious/depressive symptoms, and Dahl et al. (79) noted similar results in children.

Using a converse approach, Poland et al. (80) demonstrated that the anticholinergic agent scopolamine caused a differential effect on REM density, reducing REM activity in a way consistent with a cholinergic abnormality in depression. Rao et al. (81) observed that scopolamine 1.5 mcg/kg IM, administered using a randomized double-blind crossover design, suppressed REM sleep and nocturnal cortisol levels. Importantly, and not supportive of cholinergic supersensitivity, both in the depressed patient group and in the controls, scopolamine suppressed REM sleep and cortisol levels equally. Likewise, Gillin et al. (82) noted that depressives, withdrawn from chronically administered scopolamine, did not show expected exaggerated cholinergic rebound effects, as measured by sleep EEGs. Similarly, the muscarinic receptor blocker biperiden was not capable of reversing the relapse back into depression following napping, which occurred in patients whose depression had been alleviated by sleep deprivation (83).

Studies demonstrating cholinergic supersensitivity of REM variables in symptomatic affective disorder patients are remarkably consistent in their results. What is less certain is whether or not this REM linked supersensitivity is a trait or a state phenomenon. The work of Sitaram et al. (13) and Nurnberger et al. (68,69) suggest that the changes are a trait phenomena. Remitted bipolar patients, previously untreated by drugs or off medications for months, were shown to have exaggerated REM latency shortening after receiving arecoline (68,69). Furthermore, Schreiber et al. (84) observed exaggerated shortening of REM latency and increased spontaneous sleep onset REM periods following RS86 administration in healthy nondepressed first degree relatives of nonbipolar patients with a DSM III diagnosis of major depression. However, in contrast, Berger et al. (75,85) noted exaggerated REM latency shortening following administration of RS86 only in actively depressed major depressive disorder patients and not in remitted ones. Similarly, Lauriello et al. (86) did not find an overall supersensitive REM latency shortening response to pilocarpine in mildly depressed patients, although these authors did note a greater cholinomimetic induced shortening of REM latency in their most highly symptomatic depressed patients, again suggesting a state phenomena.

The presumed hypersensitivity of REM sleep parameters to cholinomimetic agents in affective disorder patients appears to have a genetic component. A significant concordance of REM sleep parameter changes in monozygotic twins to whom arecoline was administered (69) was noted. Also, as observed in the work of Sitaram et al. (87,88), affectively ill members of the same families showed exaggerated shortening of REM latency after an arecoline infusion. Furthermore, as described above, the work of Schreiber et al. (84) suggests a genetic relationship to RS86-induced REM shortening. In an early data analysis, it was shown that those nondepressed first degree relatives who initially showed the greatest degree of REM shortening following RS86 administration were eventually found more likely to become clinically depressed upon later follow-up (Holsboer et al., personal communication).

Supersensitive Pupillary Responses to Pilocarpine

Sokolski and DeMet (17) have recently reported that the pupillary constriction response to pilocarpine is exaggerated in patients with major depression. They suggest that this supersensitivity is trait dependent. These authors note that the pupillary response to pilocarpine is probably mediated by muscarinic 3 (M3) receptors, possibly exerting their influence through G protein-phosphoinositol mechanisms. Similarly, Sokolski and DeMet have found that lithium and valproate acid-induced improvements in manic patients were correlated with increases in pupillary sensitivity to pilocarpine (89). Conversely, these authors also found that manic patients showed decreasing pupillary sensitivity to pilocarpine as the intensity of mania increased (90). These authors also demonstrated that individuals with more severe mania required higher concentrations of the muscarinic agonist pilocarpine to elicit a 50% reduction in pupil size (90). Thus the more symptomatic the manic patient was, the less sensitive he/she was to a cholinergic agonist, a finding consistent with an adrenergic-cholinergic balance hypothesis of mania and depression.

Cardiovascular Effects of Cholinomimetic Drugs

There is evidence that patients having major depressive disorder have increased mean urinary epinephrine excretion, and to a lesser extent, norepinephrine excretion. Depressed patients also have elevated pulse rates and blood pressure levels (91,92). Physostigmine, administered to normal and affective disorder patients causes profound increases in serum epinephrine levels, and slight increases in serum norepinephrine levels (91). Interestingly, the release of epinephrine is blunted, rather than exaggerated, in affective disorder patients (94). Furthermore, physostigmine and arecoline have both been shown to increase pulse rates and blood pressure in subjects pretreated with peripherally acting anticholinergic drugs, an effect which occurs to a similar extent in affective disorder patients and controls (90,91). These changes parallel preclinical observations in animals (93), and provide one more parallel between the phenomenology of naturally occurring depression and central cholinomimetic drug effects.

Growth Hormone Supersensitivity

Acetylcholine causes the release of growth hormone from the pituitary (95,96). Thus, pilocarpine, acetylcholine, and physostigmine all increase growth hormone release in vivo in rats and in vitro in rat pituitaries, and this increase is prevented and/or reversed by administration of centrally and noncentrally acting anticholinergic drugs (97). With respect to growth hormone release by cholinomimetics in humans, Janowsky et al. (96) found no increase in serum growth hormone levels following physostigmine infusion. However, their subjects had been pretreated with methscopolamine or propantheline (probanthene), both peripherally acting anticholinergic agents that block peripheral cholinergic effects including nocturnal hormone secretion (97). O'Keane et al. (98) reported growth hormone release following administration of the peripherally acting cholinomimetic agent pyridostigmine to depressed patients who had not been treated with a peripheral anticholinergic drug. These depressed patients showed exaggerated release of growth hormone as compared to controls, a finding also noted in manic patients (99). This exaggerated growth hormone response was most predominant in males with high baseline cortisol levels (99). Possibly suggesting some nonspecificity to the growth hormone response, Lucey et al. (100) reported exaggerated pyridostigmine induced growth hormone release in obsessive-compulsive disorder patients and O'Keane et al. (101) noted an enhanced growth hormone response to pyridostigmine in schizophrenics. However, Cooney et al. (102) noted that patients with schizophrenia and those with panic disorder who had low depression scores did not differ from a control group with respect to pyridostigmine-induced growth hormone release. Rubin et al. (103) found that low dose physostigmine (8 mcg/kg, IV) caused exaggerated growth hormone levels in depressed women, as compared with depressed men and controls. In a related study, Coplan et al. (104) in 2000 reviewed evidence that early sleep is associated with an increased secretion of growth hormone, and that this increase is due to muscarinic inhibition of somatostatin, a growth hormone suppressant. In a decade long follow-up study of depressed subjects and controls Coplan et al. (104) noted that initially "normal" subjects who subsequently developed depression/dysthymia or suicidality over the next decade on average had a more rapid increase in nocturnal growth hormone secretion and greater growth hormone secretion over the next four hours of sleep, respectively (104). This was most likely due to continuingly increased muscarinic activity.

Hypothalamic-Pituitary-Adrenal Axis Supersensitivity

A major characteristic of clinical depression is the activation of the hypothalamic-pituitary-adrenal (HPA) axis, and the associated finding that some depressed patients fail to have suppression of cortisol secretion after the administration of dexamethasone (105). Cholinomimetic drugs can release corticotropin (ACTH) releasing factor (CRF) and elevate serum ACTH and cortisol levels in animals and in humans (97). Physostigmine has also been shown by Doerr and Berger (106) to reverse dexamethasone-induced suppression of cortisol in normals and in depressives. Significantly, physostigmine-induced serum ACTH increases (but not cortisol increases) are exaggerated in affective disorder patients (15), again suggesting cholinergic supersensitivity. Thus, it appears that cholinomimetic-induced increases in the HPA axis occur, and that these parallel hormonal phenomena noted in endogenous depression, such as increased cortisol secretion, cortisol resistance to suppression by dexamethasone, and elevated ACTH levels.

Beta-endorphin secretion, also regulated by CRF, like ACTH appears naturalistically elevated in depressives and, like ACTH and cortisol, serum β-endorphin levels are significantly increased by physostigmine and other cholinomimetics (15,16). Furthermore, affective disorder patients have been shown to have significantly greater increases in β-endorphin levels after physostigmine infusion, when compared to normal controls and to nonaffective disorder patients (16), suggesting cholinergic supersensitivity.

A controversy has existed as to the interpretation of the above neuroendocrine results. Davis and Davis (14) observed that serum prolactin, cortisol, and growth hormone levels did not increase after physostigmine infusion unless other unpleasant symptoms occurred, such as dizziness, nausea, vomiting. They postulated that cholinomimetic-induced increases in HPA axis hormone levels and in other hormones may be due to a nonspecific stress effect, such as feeling nauseated or vomiting, rather than to direct cholinergic mediation of the release of hormones. Janowsky et al. (107) reviewed evidence suggesting that motion sickness, which includes nausea, dizziness, and vomiting, almost certainly involves a central cholinergic mechanism, and motion sickness is a potent stimulator of growth hormone, prolactin, and cortisol secretion.

However, there is much evidence available to indicate that the increase in HPA axis and other stress sensitive hormones occurring after cholinomimetic infusion is not due to nonspecific stress as such. Hasey and Hanin (52) demonstrated that centrally acting physostigmine caused significantly greater increases in cortisol release in rats than did noncentrally acting neostigmine. This finding occurred even though the peripheral toxicity of both drugs was recorded to be severe and equal to one another. Risch et al. (15,108,109) have observed that in arecoline-treated subjects in whom serum beta-endorphin, ACTH, and cortisol levels significantly increased, a sizable proportion of the subjects could not tell when active drug and when placebo had been administered.

Furthermore, Janowsky et al. (97) has noted that physostigmine's anergic effects precede its nauseating effects, and Raskind et al. (110) noted that increases in serum ACTH, epinephrine, and cortisol occurred following administration of physostigmine in aged controls and in Alzheimer's patients, whether or not nausea had occurred. Steinberg et al. (67) noted no correlations between the mood response to physostigmine and changes in cortisol, prolactin, growth hormone, or nausea. In addition, Janowsky and Risch (97) reported increases in serum prolactin and cortisol in physostigmine-treated patients and normals that concurrently manifested no nausea, emesis, or dizziness. Most recently, Rubin et al. (111) observed that very low doses of physostigmine, causing elevation of ACTH levels, caused only minimal or no subjective distress or nausea (111). Rubin et al. (112) furthermore noted that a heightened sensitivity to low doses of physostigmine occurred in female depressives not associated with increased side effects, with this group showing increased ACTH and cortisol levels when compared to male depressives and controls. Thus, it is likely that the stress-like effects of centrally acting cholinomimetic agents occurs via a direct mechanism, rather than working by causing a nonspecific stress.

Acetycholine as a Regulator of Stress

As implied above, it is possible that acetylcholine as such actually has a major role in moderating the body's various stress responses. Stress, being multidimensional, includes gastrointestinal, cardiovascular, behavioral, analgesic, immunological, endocrinological, and psychopathological changes. Consistent with the stress-activating effects of central acetylcholine, centrally active cholinomimetic drugs cause many, even most, of the same effects in man as do naturally occurring stressors. This includes development of negative affect, including depression; irritability; and anxiety; increases in stress sensitive neuroendocrines including ACTH, cortisol, beta-endorphin, growth hormone, prolactin, epinephrine, and possibly norepinephrine; increases in blood pressure and pulse rate; and increases in analgesia and serum glucose levels (35,36,93). Furthermore, information from preclinical studies suggests that many of the manifestations of stress may be mediated by acetylcholine, acting alone and interacting with other depression-relevant neurotransmitters such as norepinephrine, dopamine, serotonin, and gamma aminobutyric acid (GABA) (97).

Conversely, stress as such can cause significant changes in central acetylcholine activity (97). Gilad et al. (113) demonstrated that stress causes an increase in central acetylcholine release and a compensatory downregulation of muscarinic receptors. Gilad et al. (113) also demonstrated that acetylcholine release is differentially exaggerated in stress sensitive rats. Other investigators have noted that hypothalamic acetylcholine turnover increases after continuing stress, and that central acetylcholine receptor sites are increased during uncontrollable stress (114).

A more recent study by Mizuno and Kimura found that hippocampal acetylcholine release, as well as cortisol release, is increased following stress in young but not aged rats (115). Mark et al. (116), using microdialysis techniques, have demonstrated that inescapable stress selectively enhances acetylcholine release in rat hippocampus and prefrontal cortex, a phenomenon that they found increased further when the stress was lifted. Consistent with the above results, Day et al. (118,119) observed that prenatally stressed rats, when they became adults, showed a greater release of hippocampal acetylcholine when exposed to a mild stress, or after being given corticotropin-releasing factor. In addition, Kaufer et al. (117) has observed that stress and cholinesterase inhibitors alter the expression of genes that ultimately alter acetylcholine receptor function.

CENTRAL MUSCARINIC REGULATION OF CHOLINOMIMETIC EFFECTS

It would appear that cholinomimetic-induced changes in mood and behavior, increases in cortisol, ACTH, prolactin, beta-endorphin, and epinephrine as well as increases in blood pressure and pulse are due to a central rather than a peripheral muscarinic effect. Janowsky et al. (95) and Modestin et al. (23,24) noted that in contrast to physostigmine, the peripherally acting cholinesterase inhibitor neostigmine did not exert behavioral effects. Janowsky et al. (96) also noted that increases in serum ACTH, cortisol, prolactin, and serum epinephrine levels, as well as increases in blood pressure and pulse rate, nausea, and negative affect caused by physostigmine did not occur following neostigmine administration, suggesting a central mechanism for physostigmine's effects. In addition, physostigmine-induced effects, as described above, can be blocked by administration of the centrally acting anticholinergic drug, scopolamine, but not by the noncentrally acting anticholinergic drugs, methscopolamine and propantheline, suggesting a central muscarinic mechanism. Conversely, it would appear that some aspects of peripheral cholinergic supersensivity also exist in affective disorder patients. Exaggerated release of growth hormone and increased pupillary sensitivity to pilocarpine, both peripheral manifestations of cholinergic supersensitivity, exist in affective disorder patients and can be blocked by peripherally acting anticholinergic agents.

MUSCARINIC RECEPTOR GENE AND BINDING ALTERATIONS IN MAJOR DEPRESSION

Since the year 2000, several studies have suggested that aspects of the muscarinic cholinergic 2 receptor (CHRM2) gene are selectively associated with major depression. In 2002, Comings et al. (120) observed that there was a significant increase in the frequency of 11 homozygotes of the CHRM2 receptor gene in 126 women with major depression, as compared to 304 women without major depression. This finding did not occur in men with major depression. Subsequently, Wang et al. in 2004 (121) reported that variation in the CHRM2 gene predisposed to alcohol dependence and major depressive syndrome. These authors assert that their results provide strong evidence that variance within or close to the CHRM2 gene locus influenced the risk for Major Depressive Syndrome and alcohol dependence. Subsequently, Luo et al. (122) concluded that variation in the CHRM2 gene differentially predisposed to affective disorders, alcohol dependence, and drug dependence.

However, with respect to muscarinic binding, the evidence for an alteration in binding in mood disorder patients is essentially negative. For example, there is little evidence of alterations in binding to M2 and M4 receptors in anterior cingulate cortex between major depressives, bipolars, schizophrenics, and controls (123). Similarly, Katerina et al. (124), again using quantitative autoradiography to measure [(3)H] pirenzepine binding to M1 and M4 receptors found no difference in any laminae of the anterior cingulate cortex between bipolar and major depressive patients and controls, although a trend toward decreased binding in major depressives compared to controls was found. A significant effect in those who had suicided was also noted (124).

THERAPEUTIC IMPLICATIONS

Application of the adrenergic (monoaminergic)-cholinergic balance hypothesis of affective disorders to the treatment of depression and mania have sporadically been attempted. As described above, centrally acting anticholinergic drugs such as biperiden only equivocally have antidepressant efficacy, although results with relatively high doses of scopolamine are promising (37). The treatment of mania with centrally acting cholinomimetic agents has been more consistently rewarding. The choline precursor lecithin was used by Cohen in the early 1980s to treat mania, with promising results (125). Stoll et al. (20) reported that choline augmentation of lithium therapy in rapidly cycling bipolar disorder patients caused a substantial reduction in mania in five, and a marked reduction of all symptoms in four patients studied. Related to the above, Leiva (126) reported that phosphotidal choline was effective in the treatment of mania. More recently, Burt et al. (21) observed that the cholinesterase inhibitor donepezil (Aricept) 5.0 mg each day was useful in alleviating manic symptoms in six of eleven treatment resistant manic patients. Thus, the use of cholinomimetic agents to treat mania appears to have therapeutic potential, although the widespread use of cholinomimetic agents to treat mania has not occurred.

PERSPECTIVES

As reviewed above, there is considerable physiological and phenomenological data indicating that muscarinic cholinergic mechanisms play an important part in the etiology and modulation of affective disorders. However, it is very possible that pharmacologically or naturally induced changes in acetylcholine can cause relevant perturbations in downstream neurochemical modulators and neurotransmitters (i.e., serotonin, dopamine, norepinephrine, GABA, etc.) or in second messengers (127,128), or the converse, since most neurotransmitters and neuromodulators considered important in causing affective changes interact with acetylcholine, and all these neurochemicals exert important regulatory influence on downstream phenomena such as second messengers and G proteins. Evaluation of these complex interactions will likely yield promising results with respect to understanding the pathophysiology of affective disorders.

Significantly, exploration of the role of acetylcholine as it relates to bipolar disorders and other mood disorders such as major depressive disorder has remained a relatively under-explored area, in spite of much evidence supporting an adrenergic-cholinergic balance hypothesis of mood disorders. Other neurotransmitters and neurochemicals such as GABA, serotonin, norepinephrine, NMDA, and dopamine continue to be more popular targets of psychobiological and psychopharmacological research. However, applying 21st century technology such as advances in molecular genetics or brain imaging to the understanding of the relationship between acetylcholine and the affective disorders will help clarify the role of this neurotransmitter. Studying the effects on the central cholinergic nervous system of conventional and newer antidepressant and mood stabilizer medications (129,130,131), and utilizing genetically determined animal models of depression such as the hypercholinergic Flinders sensitive line rats (132,133), should also yield especially promising leads. Such techniques have much potential for supporting the possibility that acetylcholine is directly or indirectly involved in the etiology and the expression of affective disorders, acting alone or through other relevant neurotransmitters and/or second messengers.

REFERENCES

1. Willoughby EF. Pilocarpine in treating mania. The Lancet 1889; 1:1030.
2. Grob, A, Harvey AM, Langworthy OR, et al. The administration of diisopropylfluorophosphonate (DFP) to man. Bull of Johns Hopkins Hospital 1947; LXXXI: 257–266.

3. Gershon S, Shaw FH. Psychiatric sequelae of chronic exposure to organophosphorous insecticides. Lancet 1961; 1:1371–1374.
4. Gershon S, Angrist B. Effects of alterations of cholinergic function on behavior. In Cole JO, Freedman AM, Friedhoff AJ, eds. Psychopathology and Psychopharmacology. Baltimore: Johns Hopkins University Press, 1972:15–36.
5. Bowers MB, Goodman E, Sim VM. Some behavioral changes in man following anticholinesterase administration. J Nerv Ment Dis 1964; 138:383–389.
6. Rowntree DW, Neven S, Wilson A. The effect of diisopropylfluorophosphonate in schizophrenia and manic depressive psychosis. J Neurol Neurosurg Psychiatry 1950; 13:47–62.
7. Domino EF, Olds ME. Cholinergic inhibition of self-stimulation behavior. J Pharmacol Exp Ther 1968; 164:202–211.
8. Stark P, Boyd ES. Effects of cholinergic drugs on hypothalamic self-stimulation in dogs. Am J Physiol 1963; 205:745–748.
9. Carlton Pl. Cholinergic mechanisms in the control of behavior by the brain. Psychol Rev 1963; 70:16–39.
10. Janowsky DS, El-Yousef MK, Davis JM, et al. A cholinergic-adrenergic hypothesis of mania and depression. Lancet 1972; 2:632–635.
11. Janowsky DS, El-Yousef MK, Davis JM, et al. Parasympathetic suppression of manic symptoms by physostigmine. Arch Gen Psychiatry 1973; 28:542–547.
12. Janowsky DS, El-Yousef MK, Davis JM. Acetylcholine and depression. Psychosom Med 1974; 36:248–257.
13. Sitaram N, Nurnberger J, Gershon ES, et al. Cholinergic regulation of mood and REM sleep: A potential model and marker for vulnerability to depression. Am J Psychiatry 1982; 139:571–576.
14. Davis BM, Davis KL. Cholinergic mechanisms and anterior pituitary hormone secretion. Biol Psychiatry 1980; 15:303–310.
15. Risch SS, Kalin NH, Janowsky DS. Cholinergic challenge in affective illness: behavioral and neuroendocrine correlates. J Clin Psychopharmacol 1981; 1:186–192.
16. Janowsky DS, Risch SC, Judd LL, et al. Cholinergic supersensitivity in affect disorder patients: behavioral and neuroendocrine observations. Psychopharmacol Bull 1981; 17:129–132.
17. Sokolski KA, DeMet EM. Increased pupillary sensivity to pilocarpine in depression. Prog Neuropsychopharmacol Biol Psychiatry 1996; 20:253–262.
18. Charles HC, Lazeyras F, Krishnan KR, et al. Brain choline in depression: in vivo detection of potential pharmacodynamic effects of antidepressant therapy using hydrogen localized spectroscopy. Prog Neuro-Psychopharmacol Biol Psychiatry 1993; 18:1121–1127.
19. Renshaw PF, Lafer B, Babb SM, et al. Basal ganglia choline levels in depression and response to fluoxetine treatment: An in vivo proton magnetic resonance spectroscopy study. Biol Psychiatry 1997; 41(8):837–843.
20. Stoll AL, Sachs GS, Cohen BM, et al. Choline in the treatment of rapid-cycling bipolar disorder: clinical and neurochemical findings in lithium-treated patients. Biol Psychiatry 1996; 40(5):382–388.
21. Burt T, Sachs GS, Demopulos C. Donepezil in the treatment of bipolar disorder. Biol Psychiatry 1999; 45(8):959–964.
22. Janowsky DS, Risch SC. Role of acetylcholine mechanisms in the affective disorders. In: Meltzer HY, ed. Psychopharmacology. The third generation of progress. New York: Raven Press, 1987:527–534.
23. Modestin JJ, Hunger J, Schwartz RB. Uber die depressogene wirkung von physostigmine. Arch Psychiatrie Nervenkr 1973a; 218:67–77.
24. Modestin JJ, Schwartz RB, Hunger J. Zur frage der beeinflussung schizophrener symptome physostigmine. Pharmacopsychiatria 1973b; 3:300–304.
25. Davis KL, Berger PA, Hollister LE, et al. Physostigmine in man. Arch Gen Psychiatry 1979; 35:119–122.
26. Carroll BJ, Frazer A, Schless A, et al. Cholinergic reversal of manic symptoms. Lancet 1973; 1:427–428.

27. Krieg JC, Berger M. Treatment of mania with the cholinomimetic agent RS-86. Br J Psychiatry 1986; 1(48):613–615.
28. Shopsin B, Janowsky DS, Davis JM, et al. Rebound phenomena in manic patients following physostigmine. Neuropsychobiology 1975; 1:180–187.
29. Fibiger HD, Lynch GS, Cooper HP. A biphasic action of central cholinergic stimulation on behavioral arousal in the rat. Psychopharmacologia 1971; 20:366–382.
30. Jellinec T, Gardos G, Cole J. Adverse effects of antiparkinsonian drug withdrawal. Am J Psychiatry 1981; 138(12):1567–1571.
31. Smith JA. Abuse of antiparkinsonian drugs: a review of the literature. J Clin Psychiatry 1980; 41:35–354.
32. Coid B, Strang M. Mania secondary to procyclidine ("Kemadrin") abuse. Br J Psychiatry 1982; 141:81–84.
33. Kasper S, Moises HW, Beckman H. The anticholinergic biperiden in depressive disorders. Pharmacopsychiatry 1981; 14:195.
34. English DC. Reintegration of affect and psychic emergence with ditran. J Neuropsychiat 1962; 3:304–310.
35. Meduna LJ, Abood LG. Studies of a new drug (ditran) in depressive states. J Neuropsychiat 1959; 11:20–23.
36. Safer DJ, Allen RP. The central effects of scopolamine in man. Biol Psychiat 1971; 3:437–455.
37. Furey ML, Drevets WC. The old drug scopolamine offers new promise as a potent antidepressant agent: a randomized, placebo-controlled clinical trial. Neuropsychopharmacology 2005; 30(1):S170.
38. Fritze J, Lanczik M, Sofic E, et al. Cholinergic neurotransmission seems not to be involved in depression but possibly in personality. J Psychiatry Neurosci 1995; 20(1): 39–48.
39. Fritze J. The adrenergic-cholinergic imbalance hypothesis of depression: a review and a perspective. Rev Neurosci 1993; 4(1):63–93.
40. Gillin JC, Lauriello J, Kelsoe JR, et al. No antidepressant effect of biperiden compared with placebo in depression: a double-blind 6-week clinical trial. Psychiatry Res 1995; 58(2):99–105.
41. Hollister L. Marijuana in man: three years later. Nature 1970; 227:968.
42. El-Yousef MK, Janowsky DS, Davis JM, et al. Induction of severe depression in marijuana intoxicated individuals. Br J Addict 1973; 68:321–325.
43. Davis KL, Hollister LE, Overall J, et al. Effects on Cognition and effect in normal subjects. Psychopharmacology (Berl.) 1976; 51:23–27,1.
44. Rosenblatt JE, Janowsky DS, Davis JM, et al. The augmentation of physostigmine toxicity in the rat by Δ9-tetrahydrocannabinol. Res Com Chem Path Pharm 1972; 3:478–482.
45. Duncan E, Dagirmanjian R. Tetrahydrocannabinol sensitization of the rat brain to direct cholinergic stimulation. Psychopharmacology 1979; 60:237–240.
46. Janowsky DS, El-Yousef MK, Davis JM. Cholinergic antagonism of methylphenidate-induced stereotyped behavior. Psychopharmacologia 1972; 27:295–303.
47. Janowsky DS, El-Yousef MK, Davis JM. Antagonistic effects of physostigmine and methylphenidate in man. Am J Psychiatry 1973; 130:1370–1376.
48. Ostrow D, Halaris A, Dysken M, et al. State dependence of noradrenergic activity in a rapidly cycling bipolar patient. J Clin Psychiatry 1984; 45(7):306–309.
49. Nurnberger JL, Berrettini W, Mendelson WB, et al. Measuring cholinergic sensitivity: I. Arecoline effects in bipolar patients. Biol Psychiatry 1989; 25:610–617.
50. Siever LJ, Risch SC, Murphy DL. Central cholinergic-adrenergic balance in the regulation of affective state. Psychiatry Res 1981; 5:108–109.
51. Schittecatte M, Charles G, Machowsky R, et al. Reduced clonidine rapid eye movement suppression in patients with primary major affective illness. Arch Gen Psychiatry 1992; 49:637–642.
52. Hasey G, Hanin I. The cholinergic-adrenergic hypothesis of depression reexamined using clonidine, metoprolol, and physostigmine in an animal model. Biol Psychiatry 1991; 29:127–138.

53. Ikarashi Y, Takahashi A, Ishimaru H, et al. Suppression of cholinergic activity via the dopamine D2 receptor in the rat striatum. Neurochem Int 1997; 30(2):191–197.
54. Downs NS, Britton KT, Gibbs DM, et al. Supersensitive endocrine response to physostigmine in dopamine-depleted rats: a model of depression? Biol Psychiatry 1986; 21(8–9):775–786.
55. Camarini R, Benedito MA. Chronic imipramine treatment-induced changes in acetylcholinesterase (EC 3.1.1.7) activity in discrete rat brain regions. Braz J Med Biol Res 1997; 30(8):955–960.
56. Saito H, Matsumoto M, Togashi H, et al. Functional interaction between serotonin and other neuronal systems: focus on in vivo microdialysis studies. Jpn J Pharmacol 1996; 70(3):203–205.
57. Crespi D, Gobbi M, Mennini T. 5-HT3 serotonin hetero-receptors inhibit [3H] acetylcholine release in rat cortical synaptosomes. Pharmacol Res 1997; 35(4):351–354.
58. Fujii T, Yoshizawa M, Nakai K, et al. Demonstration of the facilitatory role of 8-OH-DPAT on cholinergic transmission in the rat hippocampus using in vivo microanalysis. Brain Res 1997; 761(2):244–249.
59. Sombonnthum P, Matsuda T, Asano S, et al. MKC-242, a novel 5-HT1A receptor agonist, facilitates cortical acetylcholine release by a mechanism different from that of 8-OH-DPAT in awake rats. Neuropharmacology 1997; 36(11–12):1733–1739.
60. Risch SC, Cohen PM, Janowsky DS, et al. Physostigmine induction of depressive symptomatology in normal human subjects. Psychiatry Res 1981; 4:89–94.
61. Mohs R, Hollander E, Haroutunian V, et al. Cholinomimetics in Alzheimer's Disease. Int J Neurosci 1987; 32:775–776.
62. Tamminga C, Smith RC, Change S, et al. Depression associated with oral choline. Lancet 1976; 2:905.
63. Casey DE. Mood alterations during deanol therapy. Psychopharmacology 1979; 187–191.
64. Bajada S. A trial of choline chloride and physostigmine in Alzheimer's dementia. In: Corkin S, Davis K, Growden J, eds. Alzheimer's Disease: A Report of Progress. New York: Raven Press, 1982:427–432.
65. Oppenheimer G, Ebstein R, Belmaker R. Effects of lithium on the physostigmine-induced behavioral syndrome and plasma cyclic GMP. J Psychiatry Res 1979; 14: 133–139.
66. Edelstein P, Schulz JF, Hirschowitz J, et al. Physostigmine and lithium response in schizophrenia. Am J Psychiatry 1981; 138:1078–1081.
67. Steinberg BJ, Trestman R, Mitropoulou V, et al. Depressive response to physostigmine challenge in borderline personality disorder patients. Neuropsychopharmacology 1997; 17(4):264–273.
68. Nurnberger JL Jr, Jimerson DC, Simmons-Alling S. Behavioral, physiological and neuroendocrine response to arecoline in normal twins and well state bipolar patients. Psychiatry Res 1983a; 9:191–200.
69. Nurnberger JL Jr, Sitaram N, Gershon ES, et al. A twin study of cholinergic REM induction. Biol Psychiatry 1983b; 18:1161–1173.
70. Bidzinski A, Puzynski S, Mrozek S. Choline transport in erythrocytes of healthy controls and patients with endogenous major depression. New Trends Exp Clin Psychiatry 1989; 5:179–185.
71. Stoll A, Cohen BM, Hanin I. Erythrocyte choline concentrations in psychiatric disorders. Biol Psychiatry 1991; 29:309–321.
72. Hankura H, Kato T, Murashita J, et al. Quantitative proton magnetic resonance spectroscopy of the basal ganglion in patients with affective disorder. Eur Arch Psychiatry Clin Neuroscience 1998; 248(1):53–58.
73. Hobson JA, McCarley RW, Wyzinski PW. Sleep cycle oscillation: reciprocal discharge by two brainstem neuronal groups. Science 1975; 89:55–58.
74. Gillin JC, Sutton L, Ruiz C, et al. The cholinergic rapid eye movement induction test with arecoline in depression. Arch Gen Psychiatry 1991; 48:264–270.
75. Berger M, Riemann D, Hochli D, et al. The cholinergic rapid eye movement sleep induction test with RS-86. Arch Gen Psychiatry 1989; 46:421–428.

76. Gann H, Riemann D, Hohagen F, et al. The sleep structure of patients with anxiety disorders in comparison to that of healthy controls and depressive patients under baseline conditions and after cholinergic stimulation. J Affect Dis 1992; 26:179–190.
77. Riemann D, Hohagen F, Krieger S, et al. Cholinergic REM induction test: muscarinic supersensitivity underlies polysomnographic findings in both depression and schizophrenia. J Psychiatr Res 1994; 28(3):195–210.
78. Dube S, Kuman N, Ettedgui A, et al. Cholinergic REM induction response: separation of anxiety and depression. Biol Psychiatry 1985; 20:408–418.
79. Dahl RE, Ryan ND, Perel J, et al. Cholinergic REM induction test with arecoline in depressed children. Psychiatry Res 1994; 51(3):269–282.
80. Poland RE, McCracken JT, Lutchmansingh P, et al. Differential response of rapid eye movement sleep to cholinergic blockade by scopolamine in currently depressed, remitted, and normal control subjects. Biol Psychiatry 1997; 41(9):929–938.
81. Rao U, Lin KM, Schramm P, et al. REM sleep and cortisol responses to scopolamine during depression and remission in women. Int J Neurpsychopharmacol 2004; 7(3): 265–274.
82. Gillin JC, Sutton L, Ruiz C, et al. The effects of scopolamine on sleep and mood in depressed patients with a history of alcoholism and a normal comparison group. Biol Psychiatry 1991; 30:157–169.
83. Dressing H, Riemann D, Gann H, et al. The effects of biperiden on nap sleep after sleep deprivation in depressed patients. Neuropsychopharmacology 1992; 7:1–5.
84. Schreiber W, Lauer CJ, Krumrey K, et al. Cholinergic REM sleep induction test in subjects at high risk for psychiatric disorders. Biol Psychiatry 1992; 32:79–90.
85. Berger M, Lund R, Bronisch T, et al. REM latency in neurotic and endogenous depression and the cholinergic REM induction test. Psychiatry Res 1983; 10:113–123.
86. Lauriello J, Kenny WM, Sutton L, et al. The cholinergic REM sleep induction test with pilocarpine in mildly depressed patients and normal controls. Biol Psychiatry 1993; 33: 33–39.
87. Sitaram N, Jones D, Dube S, et al. Supersensitive ACh REM-induction as a genetic vulnerability marker. Int J Neurosci 1985; 32:777–778.
88. Sitaram N, Jones D, Dube S, et al. The association of supersensitive cholinergic REM-induction and affective illness within pedigrees. J Psychiatr Res 1987; 21: 487–497.
89. DeMet EM, Sokolski KN. Sodium valproate increases pupillary responsiveness to a cholinergic agonist in responders with mania. Biol Psychiatry 1999; 46 (3):432–436.
90. Sokolski KN, DeMet EM. Cholinergic sensitivity predicts severity of mania. Psychiatry Res 2000; 95(3):195–200.
91. Janowsky DS, Risch SC, Huey LY, et al. Effects of physostigmine on pulse, blood pressure and serum epinephrine levels. Am J Psychiatry 1985; 142:738–740.
92. Janowsky DS, Risch SC, Judd LL, et al. Brain cholinergic systems and the pathogenesis of affective disorders. In: Singh MM, Warburton DM, Lal H, eds. Central Cholinergic Mechanisms and Adaptive Dysfunction. New York: Plenum, 1985:309–353.
93. Janowsky DS, Risch SC, Ziegler MG, et al. Physostigmine-induced epinephrine release in patients with affective disorder. Am J Psychiatry 1986; 143(7): 919–992.
94. Kubo T. Cholinergic mechanism and blood pressure regulation in the central nervous system. Brain Res Bull 1998; 46(6):475–481.
95. Dinan TG. Psychoneuroendocrinology of depression. Growth hormone. Psychiatr Clin North Am 1998; 21(2):325–339.
96. Janowsky DS, Risch SC, Kennedy B, et al. Central muscarinic effects of physostigmine on mood, cardiovascular function, pituitary, and adrenal neuroendocrine release. Psychopharmacology 1986; 89:150–154.
97. Janowsky DS, Risch SC. Cholinomimetic and anticholinergic drugs used to investigate an acetylcholine hypothesis of affective disorder and stress. Drug Dev Res 1984; 4: 125–142.
98. O'Keane V, O'Flynn K, Lucey J, et al. Pyridostigmine-induced growth hormone responses in healthy and depressed subjects: evidence for cholinergic supersensitivity in depression. Psychol Med 1992; 22(1):55–60.

99. Dinan TG, O'Keane V, Thakore J. Pyridostigmine induced growth hormone release in mania: focus on the cholinergic/somatostatin system. Clin Endocrinol (Oxf) 1994; 40(1):93–96.
100. Lucey JV, Butcher G, Clare AW, et al. Elevated growth hormone responses to pyridostigmine in obsessive-compulsive disorder: evidence of cholinergic supersensitivity. Am J Psychiatry 1993; 150:961–962.
101. O'Keane V, Abel K, Murray RM. Growth hormone responses to pyridostigmine in schizophrenia: evidence for cholinergic dysfunction. Biol Psychiatry 1994; 36(9):582–588.
102. Cooney JM, Lucey JV, O'Keane V, et al. Specificity of the pyridostigmine/growth hormone challenge in the diagnosis of depression. Biol Psychiatry 1997; 42(9): 827–833.
103. Rubin RT, Abbasi SA, Rhodes ME, et al. Growth hormone responses to low-dose physostigmine administration: functional sex differences (sexual diergism) between major depressives and matched controls. Psychol Med 2003; 33(4):655–665.
104. Coplan JD, Wolk SI, Goetz RR, et al. Nocturnal growth hormone secretion studies in adolescents without major depression re-examined: integration of adult clinical follow-up data. Biol Psychiatry 2000; 47(7):594–604.
105. Amsterdam JD, Maislin G, Skolnick B, et al. The assessment of abnormalities in hormonal responsiveness at multiple levels of the hypothalamic-pituitary-adrenocortical axis in depressive illness. Biol Psychiatry 1989; 26:265–278.
106. Doerr P, Berger M. Physostigmine-induced escape from dexamethasone suppression in normal adults. Biol Psychiatry 1983; 18:261–268.
107. Janowsky DS, Risch SC, Ziegler M, et al. Cholinomimetic model of motion sickness and space adaptation syndrome. Aviat Space Eng 1984; 55:692–696.
108. Risch SC, Janowsky DS, Kalin NH, et al. Cholinergic beta endorphin hypersensitivity associated with depression. In: Hanin I, Usdin E, eds. Biological Markers in Psychiatry and Neurology. Oxford: Pergamon Press, 1982:269–278.
109. Risch SC, Janowsky DS, Gillin JC. Muscarinic supersensitivity of anterior pituitary ACTH and beta endorphin release in major depressive illness. Peptides 1981; 9: 789–792.
110. Raskind MA, Peskind ER, Veith RC, et al. Neuroendocrine response to physostigmine in Alzheimer's Disease. Arch Gen Psychiatry 1989; 46:535–540.
111. Rubin RT, Rhodes ME, O'Toole S, et al. Sexual diergism of hypothalamo-pituitary-adrenal cortical responses to low-dose physostigmine in elderly vs. young women and men. Neuropsychopharmacology 2002; 26(5):672–681.
112. Rubin RT, O'Toole SM, Rhodes ME, et al. Hypothalamo-pituitary-adrenal cortical responses to low-dose physostigmine and arginine vasopressin administration: sex differences between major depressives and matched control subjects. Psychiatry Res 1999; 89(1):1–20.
113. Gilad GM. The stress-induced response of the septo-hippocampal cholinergic system. A vectorial outcome of psychoneuroendocrinological interactions. Psychoneuroendocrinology 1987; 12(3):167–184.
114. Finkelstein Y, Koffler B, Rabey JM, et al. Dynamics of cholinergic synaptic mechanisms in rat hippocampus after stress. Brain Res 1985; 43:314–319.
115. Mizuno T, Kimura F. Attenuated stress response of hippocampal acetylcholine release and adrenocortical secretion in aged rats. Neurosci Lett 1997; 222:49–52.
116. Mark GP, Rada PV, Shorts TJ. Inescapable stress enhances extracellular acetylcholine in the rat hippocampus and prefrontal cortex but not the nucleus accumbens or amygdala. Neuroscience 1996; 74:767–774.
117. Kaufer D, Friedman A, Seidman S, et al. Acute stress facilitates long-lasting changes in cholinergic gene expression. Nature 1998; 393(6683):373–377.
118. Day JC, Koehl M, Deroche V, et al. Prenatal stress enhances stress- and corticotrophin releasing factor-induced stimulation of hippocampal acetylcholine release in adult rats. J Neurosci 1998; 18:1886–1892.
119. Day JC, Koehl M, LeMoal M, et al. Cortiotropin-releasing factor administered centrally, but not peripherally, stimulates hippocampal acetylcholine release. J Neurochem 1998; 71:622–629.

120. Comings DE, Wu S, Rostamkhani M, et al. Association of the muscarinic cholinergic 2 receptor (CHRM2) gene with major depression in women. Am J Med Genet 2002; 114(5):527–529.
121. Wang JC, Hinrichs AL, Stock H, et al. Evidence of common and specific genetic effects: association of the muscarinic acetylcholine receptor M2 (CHRM2) gene with alcohol dependence and major depressive syndrome. Hum Mol Genet 2004; 13:1903–1911.
122. Luo X, Kranzler HR, Zuo L, et al. CHRM2 gene predisposes to alcohol dependence, drug dependence and affective disorders: results from an extended case-control structured association study. Hum Mol Genet 2005; 14(16):2421–2434.
123. Zavitsanou K, Katsifis A, Yu Y, et al. M2/M4 muscarinic receptor binding in the anterior cingulate cortex in schizophrenia and mood disorders. Brain Res Bull 2005; 65(5): 397–403.
124. Katerina Z, Andrew K, Filomena M, et al. Investigation of m1/m4 muscarinic receptors in the anterior cingulate cortex in schizophrenia, bipolar disorder, and major depression disorder. Neuropsychopharmacology 2004; 29(3):619–625.
125. Cohen BM, Lipinski JF, Altesman RI. Lecithin in the treatment of mania: double-blind, placebo controlled trials. Am J Psychiatry 1982; 139(9):1162–1164.
126. Leiva DB. The neurochemistry of mania: a hypothesis of etiology and rationale for treatment. Prog Neuropsychopharmacol Biol Psychiatry 1990; 14(3):423–429.
127. Avissar S, Schreiber G. Muscarinic receptor subclassification and G-proteins: significance for lithium action in affective disorders and for the treatment of extrapyramidal side effects of neuroleptics. Biol Psychiatry 1989; 26:113–130.
128. Avissar S, Schreiber, G. The involvement of guanine nucleotide binding proteins in the pathogenesis and treatment of affective disorders. Biol Psychiatry 1992; 31:435–459.
129. Hankura H, Kato T, Murashita J, et al. Quantitative proton magnetic resonance spectroscopy of the basal ganglion in patients with affective disorder. Eur Arch Psychiatry Clin Neuroscience 1998; 248(1):53–58.
130. Jope RS. Lithium selectively potentiates cholinergic activity in rat brain. Prog Brain Res 1993; 98:317–322.
131. Janowsky D, Judd L. the effects of lithium on cholinergic mechanisms. In: Perris C, Struwe G, Janson B, eds. Biological Psychiatry. Elseiver: Amsterdam/North-Holland Biomedical Press, 1981:653–656.
132. Overstreet DH. The Flinders sensitive line rats: a genetic animal model of depression. Neurosci Biobehav Rev 1993; 17:51–68.
133. Martin JR, Driscoll P, Gentsch C. Differential response to cholinergic stimulation in psychogenetically selected rat lines. Psychopharmacology 1984; 83:262–267.

5 Serotonergic Dysfunction in Mood Disorders

J. John Mann and Dianne Currier
Department of Psychiatry, Columbia University, New York, New York, U.S.A.

INTRODUCTION

First purified from blood and named in 1948 (1), serotonin (5-hydroxytryptamine) has a wide range of effects including cardiovascular regulation and intestinal motility outside the brain, and within the brain it modulates: respiration, thermoregulation, circadian rhythm entrainment, sleep-wake cycle, appetite, aggression, mood, sexual behavior, sensorimotor reactivity, pain sensitivity, and learning (2). Dysfunction of the serotonergic system is thought to play a role in a variety of psychiatric disorders including mood disorders, generalized anxiety disorder, panic disorder, obsessive-compulsive disorder, social phobia, schizophrenia, anorexia nervosa, and Alzheimer's dementia (2). This chapter outlines research findings that provide the basis for our current understandings of the role of serotonin in bipolar disorder.

HISTORICAL VIEW

Serotonergic dysfunction has long been implicated in the etiology of depressive disorders. In 1955 it was observed that reserpine, an antihypertensive drug that precipitated depression, depleted serotonin in the brain (3). In 1958, demonstration of the antidepressant properties of imipramine and iproniazid was subsequently linked to their actions as a serotonin reuptake inhibitor and monoamine oxidase inhibitor respectively, and further suggested a role of serotonin in the pathophysiology of depressive disorders. These findings contributed to the formulation, by Coppen (4) and Lapin and Oxenkrug (5) of the indoleamine hypothesis of depression, wherein the vulnerability to major depression was related to low serotonergic activity, attributable to either less serotonin release, fewer serotonin receptors, or impaired serotonin receptor-mediated signal transduction. In 1974 Prange et al. (6) extended this hypothesis from major depression by proposing a permissive hypothesis of serotonin function in bipolar disorder in which both the manic and depressive phases of bipolar disorder are characterized by a deficit in central serotonergic neurotransmission.

Over the ensuing 30 years a variety of research strategies have been pursued to elucidate the role of serotonin in mania and depression, including studies of: serotonin uptake and transporter binding in platelets; 5-hydroxyindole-acetic acid (5-HIAA) the main serotonin metabolite in cerebrospinal fluid (CSF); neuroendocrine challenge tests that provoke release of serotonin or activate serotonin receptors or block reuptake; postmortem studies of serotonin and its metabolites; serotonergic receptor density; and signal transduction in the brain. More recently molecular genetic studies and in vivo brain imaging studies of neuroreceptors have been utilized to study the serotonin system.

SEROTONERGIC FUNCTION IN MAJOR DEPRESSION

Most studies of serotonergic function in depression are conducted in major depressive disorder (MDD), also known as unipolar depression, or in mixed groups of MDD and bipolar depressed patients. This section considers studies of depressed groups of any composition and a later section reviews studies of bipolar depression specifically.

Cerebrospinal Fluid Studies

Levels of the major metabolite of serotonin, 5-hydroxyindoleacetic acid (5-HIAA), in cerebrospinal fluid (CSF) have been studied as an index of central serotonin turnover in depressive and other psychiatric disorders. Such an approach assumes that CSF 5-HIAA is related to brain serotonin activity. This premise is supported by the rostral-caudal concentration gradient of CSF 5-HIAA and the observation in postmortem studies that CSF 5-HIAA correlates with levels of 5-HIAA in the prefrontal cortex (7). Those findings are supplemented by the observation of a positive correlation with the prolactin response to fenfluramine, indicating that CSF 5-HIAA is a reasonable index of prefrontal serotonin turnover (8), although methodological factors and differences in study population may underlie the disparate findings of studies of CSF 5-HIAA and depression. Some report lower CSF 5-HIAA in depressed patients compared with healthy volunteers (9–14) but others do not (15–18). The level of CSF 5-HIAA generally does not correlate with severity of depression. In contrast to studies of CSF 5-HIAA in depression, the evidence of a relationship is more robust for suicidality. The finding of lower brainstem serotonin and/or 5-HIAA in depressed suicides and lower CSF 5-HIAA in depressed suicide attempters has been consistently replicated (19). The suicide-related finding may reflect a deficit in serotonin input to the ventromedial prefrontal cortex considered to underlie a predisposition to act on powerful feelings (20,21). Consistent with this model, correlations are also reported between severity of lifetime externally directed aggressive/impulsive behavior and low CSF 5-HIAA (22,23). This is a biochemical trait that can predict future suicide risk (24).

Platelet Studies

Platelets have been used as a peripheral model for serotonin neurons in studies of serotonin function in mood disorders. Platelets are more accessible for study than obtaining CSF, and share many properties with central serotonin neurons, including uptake, storage, and release of serotonin, some serotonin receptors, and serotonin transporter binding sites (25,26). Lower serotonin uptake, due mainly to fewer uptake sites (V_{max}), in unmedicated depressed patients has been consistently reported (27–31).

The ligands [^{3}H]imipramine and [^{3}H]paroxetine have been used to assay serotonin transporter binding in platelets and postmortem brain. In the brain, [^{3}H]imipramine binds to at least two classes of sites: high- and low-affinity. The high-affinity, sodium-dependent binding was found to be associated with the serotonin transporter complex (32). Depressed unmedicated subjects have fewer transporter sites as measured by lower [^{3}H]imipramine binding (31,33–38), though not all studies agree (39–42). Meta-analysis of over 70 studies of [^{3}H] imipramine comprising more than 1900 depressed patients and slightly fewer controls found fewer binding sites in depressed patients compared with controls (43). Studies using the ligand [^{3}H]paroxetine, which is thought to be more selective, have also reported lower binding (35,44,45) or no differences (38,46–48).

The 5-HT$_{2A}$ receptor has also been assayed in platelets with mixed results. Many studies (49–54) but not all (55,56) report more 5-HT$_{2A}$ binding sites in depressed patients compared with healthy controls. During treatment with antidepressants normalization of receptor density has been observed with symptomatic improvement (57,58) but not always (59,60).

Neuroendocrine Challenge and Depletion Studies

Another indicator of central serotonin system function is a hormonal response to the serotonin releasing agent/uptake inhibitor fenfluramine, and other related direct and indirect serotonin agonists. Release of serotonin from raphe nuclei projections to the hypothalamus causes the release of prolactin, cortisol, and adrenocorticotropic hormone (ACTH) from the pituitary. Thus, prolactin response is an index of central serotonin responsivity (61). Fenfluramine challenge can measure net serotonin transmission, including elements of presynaptic and postsynaptic serotonergic functioning, which is not possible in cerebrospinal fluid and platelet studies (62).

Fenfluramine causes the release of serotonin and inhibits serotonin reuptake, thus the prolactin response to fenfluramine is an index of serotonin responsivity. A blunted prolactin response to fenfluramine, indicating less serotonin release and/or serotonin 5-HT$_{1A}$ or 5-HT$_{2A}$ receptor signal transduction, in depressed patients, has been reported by many (63–69) but not all studies (70–72). Remitted patients still have blunted response, the same as severely depressed patients (73). Tryptophan is a precursor of serotonin and intravenous administration increases prolactin secretion. Blunted prolactin response to tryptophan challenge has been reported in major depression (74–76).

Other methods of studying serotonergic function are through the depletion of serotonin by inhibition of tryptophan hydroxylase (TPH), the rate-limiting biosynthetic enzyme for serotonin, with parachlorophenylalanine (PCPA) (77), or via acute tryptophan depletion combined with competitive inhibition of brain uptake by flooding with a bolus of large neutral amine acids (78,79). In long-term remitted, medication-free depressed patients, depression recurs in hours after acute serotonin or tryptophan depletion (80,81). In remitted patients taking serotonin-related antidepressant medication, tryptophan depletion induces a transient and rapid return of depression symptoms, while in untreated depressed patients or those on noradrenergic antidepressants, there is no clear worsening (82).

Postmortem Studies

Postmortem studies have examined a variety of serotonergic indices. Most postmortem studies are of depressed suicide victims and, given the evidence that serotonergic anomalies characterize suicide across diagnostic categories, it can be difficult to ascertain if abnormalities observed are related specifically to depression or to suicide.

SEROTONIN AND 5-HIAA

Postmortem studies found no differences between depressed suicide victims and control subjects in levels of serotonin in the hippocampus (83,84), occipital cortex

(83), frontal cortex (84,85), temporal cortex (84,85), caudate (84), striatum (85), or hypothalamus (85).

Postmortem studies of 5-HIAA have produced varied findings. In a study of depressed nonsuicide deaths, 5-HIAA in frontal cortex tended to be lower than in controls (86). Moreover, it was lower in patients who had not been on antidepressants in the month before death. In suicides Crow et al. (87) observed no significant differences in 5-HIAA levels in the frontal cortex, occipital cortex, and hippocampus of depressed compared with nondepressed suicide victims. Owen et al. (83) reported no difference in 5-HIAA in the hippocampus when comparing depressed suicide victims to nondepressed subjects but modestly higher 5-HIAA compared with normal control subjects. Likewise Cheetham (84) found depressed suicides did not differ from controls in levels of 5-HIAA in cortical regions, hippocampus, amygdala, and caudate; however, 5-HIAA levels were higher in the amygdala but not the hippocampus in drug-free depressed suicides.

Serotonin Transporter

Stanley et al. (1982) were the first to report lower [^{3}H]imipramine binding in the prefrontal cortex of suicides compared with nonsuicides and this was soon confirmed in suicides with and without depression (88) compared with nonsuicides, although others found no difference (83). It was not clear whether this finding was associated with suicide or major depression. Depressed nonsuicides were reported to have less serotonin transporter binding in the hippocampus and the occipital cortex compared with normal controls (89). Others found no differences (86,90). This question was resolved when a large study examining transporter binding using [^{3}H]cyanoimipramine found lower postmortem binding throughout the prefrontal cortex of a group with a major depression (20) whereas in suicide lower transporter binding appeared localized to the ventromedial prefrontal cortex (20,21).

5-HT$_{1A}$ Receptor

Several postmortem studies have examined the role of 5-HT$_{1A}$ receptors in MDD and suicide. In the main, controls are compared with suicide victims with major depression. The only study of depressed individuals who died by means other than suicide reported a trend for higher 5-HT$_{1A}$ binding compared with controls (86). Several studies reported no differences between depressed suicides and controls in the prefrontal cortex (91–93), cortex (92,94,95), occipital cortex (95), temporal cortex (94), hippocampus (92,93,95), and amygdala (95). Others report higher binding in prefrontal cortex (21), and more rostral segments of raphe nuclei (96) and lower binding in more caudal raphe nuclei (97), hippocampus (98), prefrontal cortex, and temporal cortex. Arango et al. also report less gene expression in the dorsal raphe (97).

5-HT$_{2A}$ Receptor

Postmortem studies of 5-HT$_{2A}$ receptors in the prefrontal cortex have likewise produced conflicting results. The first study reported higher 5-HT$_{2A}$ binding in prefrontal cortex of suicides compared with nonsuicides (99). Other studies replicated these findings (100–102). Pandey et al. subsequently showed that protein levels were elevated and so was gene expression (102). A number of studies, using

[^{3}H]ketanserin, did not find 5-HT$_{2A}$ receptors significantly different in the prefrontal cortex in depressed suicides compared with controls (83,87,91,93,103–106). Cheetham et al. (104) found that antidepressant-free depressed suicides had lower 5-HT$_{2A}$ binding in the hippocampus than normal controls. Still other studies find higher binding. One study of the prefrontal cortex found more 5-HT$_{2A}$ receptors (107) while another reported a trend in that direction (108). A study of the prefrontal cortex and the amygdala in depressed suicides also reported more 5-HT$_{2A}$ receptors (109). Ferrier et al. (86), in a nonsuicide postmortem sample, noted a trend for 5-HT$_{2A}$ receptors to be elevated in the frontal cortex of patients with major depression compared with dysthymia and nondepressed controls with the trend more pronounced in subjects who had been depressed immediately prior to death. Depressed non-suicides on antidepressants at the time of death did not show differences in [^{3}H]ketanserin binding in the frontal cortex compared with controls (107), suggesting that antidepressant treatment may reduce 5-HT$_{2A}$ receptors density in major depressives (107).

Imaging Studies

Positron emission tomography (PET) studies of 5-HT$_{1A}$ receptor binding potential in depressed subjects with primary, recurrent, familial mood disorders report lower mean 5-HT$_{1A}$ receptor binding potential in the midbrain raphe, limbic, and neocortical areas in the frontal, mesiotemporal, occipital, and parietal cortices compared with healthy controls (110,111). In the Drevets et al. (110) studies, the effect was mainly seen in bipolar subjects or unipolar with bipolar first degree relatives. Parsey et al. (2006) found depressed subjects who were antidepressant naïve had greater binding potential than controls. Moreover, higher 5HT$_{1A}$ binding was associated with the higher expressing G allele of the functional 5HT$_{1A}$ G(-1019) promoter polymorphism, and that allele was more common in the depressed group (112). Imaging studies of 5-HT$_{2A}$ receptor binding in depressed subjects have reported conflicting results. Some studies observe higher 5-HT$_{2A}$ receptor binding in the prefrontal cortex; others report lower binding in the cingulate, insula, and inferior frontal cortex; and still others observe no alterations (113). Variation in results may be related to the use of different ligands, patient heterogeneity, or downregulation of 5-HT$_{2A}$ receptor binding by antidepressants (114,115). Lower 5-HT$_{2A}$ receptor binding in the hippocampus has also been reported in medication-free depressed patients (116), supporting the findings of some postmortem studies (104,106). A SPECT study using [2-^{123}I]ketanserin to label 5-HT$_{2A}$ receptors found higher uptake of the tracer in the parietal cortex of the patients, and a right greater than left asymmetry in the infero-frontal region of the depressed subjects compared with control subjects (117).

Other imaging studies report decreased uptake of 1-^{11}C-hydroxytryptophan across the blood-brain barrier in depressed subjects compared with normal subjects (118). This finding suggests decreased serotonergic functioning secondary to reduced availability of serotonin precursors such as 5-hydroxytryptophan. In a FDG PET study of medication free major depression and healthy controls undergoing fenfluramine challenge, depressed patients had blunted regional brain glucose utilization in response to serotonin release by fenfluramine compared with controls (119). Finally, imaging studies of transporter binding report significantly lower serotonin transporter binding in the brainstem and other brain regions of medication-free unipolar depressed patients compared with healthy volunteers (112,120), although one study found higher binding in the hypothalamus/midbrain in depressed children and adolescents (121).

SEROTONERGIC FUNCTION IN MANIA

There is evidence for the contribution of serotonin dysfunction to mania, and involvement of the serotonin system in the mechanism of action of mood stabilizers (122); however, it is much less extensive than for depression. Moreover, altered functioning of other neurotransmitters in mania such as norepinephrine, dopamine, acetylcholine, and GABA and their interaction with serotonin are involved in the pathogenesis of bipolar disorder. Differences in these neurotransmitter systems possibly underlie differences in the pathogenesis of depressive and manic episodes.

Cerebrospinal Fluid Studies

Manic patients are variously reported to have lower CSF 5-HIAA levels (9,10,123,124), no difference (125–132), or higher CSF 5-HIAA compared with healthy controls (133). The majority of studies (10,17,123,127,129, 132,134,135), though not all (126), also found no difference in CSF 5-HIAA levels in mania compared with depression, consistent with the permissive hypothesis for bipolar disorders (6). This hypothesis states that low serotonin function has a role in episodes of both depression and mania.

Probenecid blocks the active transport of 5-HIAA out of the CSF and was used in older studies to boost levels of 5-HIAA when assays lacked the sensitivity of current methods. It is potentially a more dynamic measure of central serotonergic activity than basal level when CSF 5-HIAA is measured before and after probenecid administration because by blocking egress of 5-HIAA, the rise in 5-HIAA reflects the rate of production and transport into CSF of 5-HIAA. However, studies almost uniformly did not use probenecid in this way and just made one measurement after it was administered. Such studies of CSF 5-HIAA after probenecid administration in manics, depressives, and controls have yielded mixed findings. Two studies found low CSF 5-HIAA in both manic and depressed patients compared with controls (136,137), one reported lower CSF 5-HIAA accumulation in mania than depression and controls (127), and another, with no control group, found similar levels in mania and depression (129). Overall CSF studies suggest both mania and depression are associated with impaired central serotonin function.

Platelet Studies

Studies of platelet serotonin uptake in mania have reported mixed results with two studies finding no differences from healthy controls (28,138), one study reporting greater serotonin uptake (139), and another less serotonin uptake compared with controls (140). In the latter two studies not all manic patients were medication-free, potentially confounding the findings. Two studies of transporter binding in mania compared with depression found higher binding (141,142) and two others found no difference (42,143). In all four studies binding in mania was not different from healthy controls, which is inconsistent with the robust reports of lower binding in major depression. Velayudhan et al. (144) used [^{125}I]-ketanserin to examine platelet 5-HT_{2A} receptor binding sites in drug-free manic patients and found no differences compared with healthy controls.

Serotonin-induced platelet calcium mobilization is a measure of 5-HT_{2A} receptor signal transduction (145,146) and more robust responses are reported in untreated mania compared with healthy controls and euthymic lithium-treated bipolar patients (147) and in both manic and depressive phases of bipolar disorder compared with normal controls and euthymic bipolar patients (148). Together these two studies suggest an increase in sensitivity of 5-HT_{2A} receptors in both mania and bipolar depression.

Challenge Studies

Thakore et al. (149) found a blunted prolactin response to D-fenfluramine in medicated manic subjects compared with healthy controls. That may indicate a deficit in serotonin release because direct receptor agonists report enhanced 5-HT_{1A} receptor-mediated responses. Yatham (150,151) found no differences in prolactin response to the 5-HT_{1A} agonist buspirone or D,L-fenfluramine in mania compared with healthy controls; however, using the more selective 5-HT_{1A} agonist ipsapirone indicated enhanced ACTH and cortisol responses in mania compared with controls (152). ACTH and cortisol responses are mediated by postsynaptic 5-HT_{1A} receptors (153–155), thus suggesting an increase in postsynaptic 5-HT_{1A} receptor sensitivity in mania. Enhanced cortisol response has also been reported in mania in response to 5-hydroxytryptophan (156), perhaps because the precursor corrects a deficit in serotonin, and reflects the receptor supersensitivity. Consistent with this hypothesis, both medicated and medication- free mania show enhanced plasma GH and cortisol response to oral 5-hydroxytryptophan administration compared with MDD (156–158).

Challenge studies provide information about the hypothalamic serotonin activity only and not limbic or other cortical areas. Taken together, they suggest central presynaptic serotonin activity is decreased and postsynaptic serotonin receptor sensitivity is increased in mania.

Imaging Studies

A PET study of 5-HT_{2A} receptors using [18F]-setoperone in manic patients before and after treatment with valproate found that although all patients experienced remission of manic symptoms after treatment, there was no difference in 5-HT_{2A} binding after treatment (159). This suggests that changes in brain 5-HT_{2A} receptors are not involved in the antimanic effects of this mood stabilizer. Given the efficacy of atypical antipsychotics in mania, it may be that the role of 5-HT_{2A} receptors in mania is via down-stream signaling pathways.

Medication Studies

Medication studies offer insight into serotonergic functioning in bipolar disorder or at least its role in the therapeutic action of these treatments. Lithium is effective in bipolar disorder for mania and depression and as a mood stabilizer. Lithium increases CSF 5-HIAA in mania (122,160–162) and in euthymic bipolar patients (163). Lithium's short-term effect on platelet serotonin uptake in mania is less clear (138,164), but longer-term lithium treatment increases serotonin uptake in mania (164). Lithium treatment increases cortisol response to 5 hydroxytryptophan in mania (156). There is no effect of lithium administration on 5-HT_{1A} receptor-mediated prolactin response to buspirone in manic patients but perhaps ACTH or cortisol would be better indices for future studies (150). Tryptophan depletion in recently remitted mania induces relapse of mania (165) but not in bipolar patients that have been stable on lithium for a longer period of time (166–168).

FOCUS ON STUDIES OF BIPOLAR DISORDER PATIENTS

Most studies of serotonergic function do not separate results for bipolar depression and MDD. As studies of bipolar mania have already been considered, this section reviews findings in bipolar depression and euthymic bipolar patients.

Cerebrospinal Fluid Studies

Some studies of CSF 5-HIAA and bipolar depression find lower CSF 5-HIAA levels compared with normal controls, but no difference compared to MDD (13,169). Low CSF 5-HIAA is reported in bipolar depression, mania, and MDD compared with controls, and little changed with clinical recovery (10). Others find no difference between MDD, bipolar depressed, bipolar manic, and controls (17); euthymic bipolars and controls (163); and MDD depressed, bipolar depressed, and controls (125).

Platelet Studies

In a study comparing depressed bipolar I and depressed bipolar II patients, platelet serotonin content was higher in both groups compared with normal controls; however, there was no difference between the two groups (170). Comparing serotonin platelet uptake in different phases of bipolar disorder, two studies found that medication-free bipolar subjects had higher platelet serotonin content than controls in both depressed and hypomanic phases (171,172), while another study found greater serotonin uptake in platelets only in depressed bipolar patients, with manic and euthymic patients not different from healthy controls (173). Others report low serotonin uptake in platelets in depressed bipolar patients, similar to depressed MDD patients (28).

While ^{3}H-imipramine studies of platelet serotonin transporters in depression consistently find lower binding, and are mixed in mania, other studies in bipolar disorder suggest that altered transporter function may be a trait characteristic of the disorder. Lower binding is reported in depression, mania, and euthymia in bipolar disorder compared with normal controls (174) and in euthymic bipolar patients who had between 3–15 years of lithium treatment (175). However, not all studies agree: Lewis and McChesney report lower binding in bipolar depression compared with controls and mania (141), and Muscettola in 1986 found binding in bipolar depression and hypomania no different from controls (42). Recent evidence suggests genetic involvement, finding that both bipolar patients and their unaffected relatives have fewer transporter binding sites compared with normal controls (176). These studies did not report on suicidal behavior, which appears to be associated with lower platelet paroxetine binding in bipolar suicide attempters compared with bipolar nonattempters, who were all on mood stabilizers, and healthy controls (177).

5-HT_{2A} receptor binding in platelets, assessed with lysergic acid diethylamide (LSD), was the same in manic and depressed bipolar subjects, and higher binding compared with normal controls (178). Moreover, binding was higher still in suicidal compared with nonsuicidal bipolar patients, as reported by others in MDD (178).

Challenge Studies

Medication-free bipolar depression has a blunted prolactin response to fenfluramine, similar to MDD, when compared with healthy controls (66,179), although others find no difference in prolactin response (180). Tryptophan challenge produced blunted cortisol and ACTH responses in bipolar depressed (181), but prolactin response was not different compared with depressed MDD or healthy controls (182). Depressed bipolar disorder has an enhanced cortisol response to hydroxytryptophan compared with depressed MDD (156,158), but not to ipsapirone, a selective partial agonist of pre- and postsynaptic 5-HT_{1A} receptors, in bipolar depression compared with healthy controls (183). In a tryptophan deletion study, a mood-lowering effect was observed in euthymic bipolar subjects and their unaffected first degree relatives compared with controls (176). Precursor challenges are not in agreement in bipolar depression as to whether serotonin function is impaired. Studies are lacking to determine whether 5-HT_{1A} and 5-HT_{2A} responses are supersensitive in bipolar depression, as has been reported in mania.

Postmortem Studies

There have been few studies examining serotonergic indices in postmortem bipolar depressed subjects. Lower 5-HIAA levels in the frontal and parietal cortex and lower 5-HT/5-HIAA ratios in temporal cortex are reported compared with postmortem control brains (184). Bipolar subjects, on various medications including antipsychotics, had lower serotonin transporter affinity in the stratum lacunosum-moleculare region of the hippocampus and no change in 5-HT_{1A} or 5-HT_{2A} receptor binding compared with controls (185). Wiste et al. found less postmortem tryptophan hydroxylase (TPH) immunoreactivity, the rate-limiting biosynthetic enzyme for serotonin, in serotonin nerve terminals in the locus coeruleus of depressed bipolar suicides compared with depressed unipolar suicides and nonsuicide, nonpsychiatric controls (186). The same study found less tyrosine hydroxylase immunoreactivity in depression, and one manic case had the highest level of any case. These results are consistent with the serotonin-permissive hypothesis and suggest that noradrenergic activity may be dependent on mood state.

Genetic Studies

Genetic factors are estimated to account for 60% to 85% of the liability for bipolar disorder (187,188) and in recent years genetic studies have investigated association between bipolar disorder and a number of serotonin-related genes including the serotonin transporter, various serotonin receptors, and tryptophan hydroxylase 1, reporting variable results.

In bipolar disorder, three meta-analyses report an association between bipolar and the serotonin gene promotor 5-HTTLPR genotype (189–191), one finds a nonsignificant trend for an association (192), and one finds no association (193). There have been conflicting reports of association with bipolar disorder for a second variant in the same gene, namely, a variable number of tandem repeat (194,190).

Meta-analyses find no evidence of an association of bipolar disorder with the 5-HT_{2A} receptor T102C polymorphism (193,190). Recent large European multicentre studies also found no association between the 5-HT_{2A} 1438G/A and the His452Tyr polymorphisms and bipolar disorders (195). Associations have been reported between bipolar disorder and the 5-HT_{3A} variant C178T in 156 patients, but not C195T (196), and with 5-HT_{4A} in a small Japanese sample (197). A 5-HT_{5A} receptor allelic association was reported between the -19G/C polymorphism and bipolar disorder (198), although others found no association between 12A/T and bipolar disorder (199). Vogt and associates reported an association between a 5-HT_6 (267C) polymorphism and bipolar disorder in a small European sample (200), however no association between 5-HT_6 polymorphism (C267T) was observed in a Taiwanese group (201). Vincent et al. found no association between a 5-HT_7 variant and bipolar disorder (202).

The tryptophan hydroxylase 1 A218C polymorphism was associated with a small increase in susceptibly to bipolar disorder in a European sample (203) that is not replicated in other studies (202,204–207) and in a large multi-center European study of bipolar and MDD (208).

CONTRAST WITH FINDINGS IN MAJOR DEPRESSIVE DISORDER

Bipolar depression and major depressive disorder differ modestly in terms of the clinical picture of depression. With respect to the serotonergic system there is considerable evidence indicating serotonergic dysfunction in both disorders. The biochemical differences that may exist are likely to relate to the predisposition to mania in bipolar disorders.

PERSPECTIVES

The observations of serotonergic abnormalities in manic, depressed, and euthymic bipolar patients, and the unaffected relatives of bipolar patients suggest that serotonergic dysfunction is a trait-related characteristic of bipolar disorder rather than a mood state-dependent characteristic. Further studies in well-defined bipolar cohorts, in different phases of the disorder that include comparisons with MDD as well as healthy controls, are required to determine the neurobiologic correlates that are related to mood state. Such studies must include a consideration of the interaction of the serotonergic system with other neurotransmitters and the HPA axis. Moreover, the role of genetic factors and their interaction with environment in the development of this disorder requires investigation.

REFERENCES

1. Rapport MM, Green AA, Page IH. Crystalline serotonin. Science 1948; 108:329–330.
2. Lucki I. The spectrum of behaviors influenced by serotonin. Biol Psychiatry 1998; 44:151–162.
3. Pletscher A, Shore PA, Brodie BB. Serotonin release as a possible mechanism of reserpine action. Science 1955; 122:374–375.
4. Coppen AJ. Biochemical aspects of depression. Int Psychiatry Clin 1969; 6:53–81.
5. Lapin IP, Oxenkrug GF. Intensification of the central serotoninergic processes as a possible determinant of the thymoleptic effect. Lancet 1969; i:132–136.
6. Prange AJ Jr, Wilson IC, Lynn CW, Alltop LB, Stikeleather RA. L-Tryptophan in mania. Contributions to a permissive hypothesis of affective disorders. Arch Gen Psychiatry 1974; 30:56–62.

7. Stanley M, Träskman-Bendz L, Dorovini-Zis K. Correlations between aminergic metabolites simultaneously obtained from human CSF and brain. Life Sci 1985; 37:1279–1286.
8. Mann JJ. Role of the serotonergic system in the pathogenesis of major depression and suicidal behavior. Neuropsychopharmacology 1999; 21:99S–105S.
9. Dencker SJ, Malm U, Roos B-E, Werdinius B. Acid monoamine metabolites of cerebrospinal fluid in mental depression and mania. J Neurochem 1966; 13: 1545–1548.
10. Mendels J, Frazer A, Fitzgerald RG, Ramsey TA, Stokes JW. Biogenic amine metabolites in cerebrospinal fluid of depressed and manic patients. Science 1972; 175: 1380–1382.
11. Ågren H. Symptom patterns in unipolar and bipolar depression correlating with monoamine metabolites in the cerebrospinal fluid. II. Suicide. Psychiatry Res 1980; 3:225–236.
12. Åsberg M, Thorén P, Träskman L, Bertilsson L, Ringberger V. Serotonin depression—a biochemical subgroup within the affective disorders? Science 1976; 191:478–480.
13. Åsberg M, Bertilsson L, Martensson B, Scalia Tomba GP, Thorén P, Träskman-Bendz L. CSF monoamine metabolites in melancholia. Acta Psychiatr Scand 1984; 69:201–219.
14. Gibbons RD, Davis JM. Consistent evidence for a biological subtype of depression characterized by low CSF monoamine levels. Acta Psychiatr Scand 1986; 74:8–12.
15. Mann JJ, Malone KM. Cerebrospinal fluid amines and higher-lethality suicide attempts in depressed inpatients. Biol Psychiatry 1997; 41:162–171.
16. Reddy PL, Khanna S, Subhash MN, Channabasavanna SM, Rao BSSR. CSF amine metabolites in depression. Biol Psychiatry 1992; 31:112–118.
17. Koslow SH, Maas JW, Bowden CL, David JM, Hanin I, Javaid J. CSF and urinary biogenic amines and metabolism in depression and mania. A controlled univariate analysis. Arch Gen Psychiatry 1983; 40:999–1010.
18. Geracioti TD Jr, Loosen PT, Ekhator NN, et al. Uncoupling of serotonergic and noradrenergic systems in depression: preliminary evidence from continuous cerebrospinal fluid sampling. Depression and Anxiety 1997; 6:89–94.
19. Åsberg M. Neurotransmitters and suicidal behavior. The evidence from cerebrospinal fluid studies. Ann NY Acad Sci 1997; 836:158–181.
20. Mann JJ, Huang YY, Underwood MD, et al. A serotonin transporter gene promoter polymorphism (5-HTTLPR) and prefrontal cortical binding in major depression and suicide. Arch Gen Psychiatry 2000; 57:729–738.
21. Arango V, Underwood MD, Gubbi AV, Mann JJ. Localized alterations in pre- and postsynaptic serotonin binding sites in the ventrolateral prefrontal cortex of suicide victims. Brain Res 1995; 688:121–133.
22. Linnoila M, Virkkunen M, Scheinin M, Nuutila A, Rimond R, Goodwin FK. Low cerebrospinal fluid 5-hydroxyindoleacetic acid concentration differentiates impulsive from non-impulsive violent behavior. Life Sci 1983; 33:2609–2614.
23. Lidberg L, Belfrage H, Bertilsson L, Evenden MM, Åsberg M. Suicide attempts and impulse control disorder are related to low cerebrospinal fluid 5-HIAA in mentally disordered violent offenders. Acta Psychiatr Scand 2000; 101:395–402.
24. Mann JJ, Currier D, Stanley B, Oquendo MA, Amsel LV, Ellis SP. Can biological tests assist prediction of suicide in mood disorders? Int J Neuropsychopharmacol 2006; 9:465–474.
25. Stahl SM. The human platelet. Diagnostic and research tool for the study of biogenic amines in psychiatric and neurologic disorders. Arch Gen Psychiatry 1977; 34:509–516.
26. Stahl SM. Peripheral models for the study of neurotransmitter receptors in man. Psychopharmacol Bull 1985; 21:663–671.
27. Tuomisto J, Tukiainen E, Ahlfors UG. Decreased uptake of 5-hydroxytryptamine in blood platelets from patients with endogenous depression. Psychopharmacol (Berlin) 1979; 65:141–147.
28. Meltzer HY, Arora RC, Baber R, Tricou B-J. Serotonin uptake in blood platelets of psychiatric patients. Arch Gen Psychiatry 1981; 38:1322–1326.
29. Kaplan RD, Mann JJ. Altered platelet serotonin uptake kinetics in schizophrenia and depression. Life Sci 1982; 3:583–588.

30. Coppen A, Swade C, Wood K. Platelet 5-hydroxytryptamine accumulation in depressive illness. Clin Chim Acta 1978; 87:165–168.
31. Suranyi-Cadotte BE, Quirion R, Nair NPV, Lafaille F, Schwartz G. Imipramine treatment differentially affects platelet ^{3}H-imipramine binding and serotonin uptake in depressed patients. Life Sci 1985; 36:795–799.
32. Hrdina PD. Imipramine binding sites in brain and platelets: role in affective disorders. Int J Clin Pharmacol Res 1989; 9:119–122.
33. Briley MS, Langer SZ, Raisman R, Sechter D, Zarifian E. Tritiated imipramine binding sites are decreased in platelets of untreated depressed patients. Science 1980; 209: 303–305.
34. Wagner A, Aberg-Wistedt A, Asberg M, Ekqvist B, Martensson B, Montero D. Lower 3H-imipramine binding in platelets from untreated depressed patients compared to healthy controls. Psychiatry Res 1985; 16:131–139.
35. Nemeroff CB, Knight DL, Franks J, Craighead WE, Krishnan KRR. Further studies on platelet serotonin transporter binding in depression. Am J Psychiatry 1994; 151: 1623–1625.
36. Raisman R, Briley MS, Bouchami F, Sechter D, Zarifian E, Langer SZ. 3H-imipramine binding and serotonin uptake in platelets from untreated depressed patients and control volunteers. Psychopharmacology (Berl) 1982; 77:332–335.
37. Poirier MF, Benkelfat C, Loo H, et al. Reduced Bmax of [3H]-imipramine binding to platelets of depressed patients free of previous medication with 5HT uptake inhibitors. Psychopharmacology (Berl) 1986; 89:456–461.
38. Rosel P, Arranz B, Vallejo J, et al. Altered [3H]imipramine and 5-HT2 but not [3H]paroxetine binding sites in platelets from depressed patients. J Affect Disord 1999; 52:225–233.
39. Lawrence KM, Falkowski J, Jacobson RR, Horton RW. Platelet 5-HT uptake sites in depression: three concurrent measures using [^{3}H] imipramine and [^{3}H] paroxetine. Psychopharmacol (Berlin) 1993; 110:235–239.
40. Whitaker PM, Warsh JJ, Stancer HC, Persad E, Vint CK. Seasonal variation in platelet 3H-imipramine binding: comparable values in control and depressed populations. Psychiatry Res 1984; 11:127–131.
41. Tang SW, Morris JM. Variation in human platelet 3H-imipramine binding. Psychiatry Res 1985; 16:141–146.
42. Muscettola G, Di Lauro A, Giannini CP. Platelet 3H-imipramine binding in bipolar patients. Psychiatry Res 1986; 18:343–353.
43. Ellis PM, Salmond C. Is platelet imipramine binding reduced in depression? A meta-analysis. Biol Psychiatry 1994; 36:292–299.
44. Owens MJ, Nemeroff CB. Role of serotonin in the pathophysiology of depression: Focus on the serotonin transporter. Clin Chem 1994; 40:288–295.
45. Alvarez JC, Gluck N, Arnulf I, et al. Decreased platelet serotonin transporter sites and increased platelet inositol triphosphate levels in patients with unipolar depression: effects of clomipramine and fluoxetine. Clin Pharmacol Ther 1999; 66:617–624.
46. D'haenen H, De Waele M, Leysen JE. Platelet 3H-paroxetine binding in depressed patients. Psychiatry Res 1988; 26:11–17.
47. D'Hondt P, Maes M, Leysen JE, Gommeren W, Scharpé S, Cosyns P. Binding of [^{3}H]paroxetine to platelets of depressed patients: Seasonal differences and effects of diagnostic classification. J Affect Disord 1994; 32:27–35.
48. Nankai M, Yamada S, Yoshimoto S, et al. Platelet ^{3}H-paroxetine binding in control subjects and depressed patients: relationship to serotonin uptake and age. Psychiatry Res 1994; 51:147–155.
49. Pandey GN, Pandey SC, Janicak PG, Marks RC, Davis JM. Platelet serotonin-2 receptor binding sites in depression and suicide. Biol Psychiatry 1990; 28: 215–222.
50. Biegon A, Essar N, Israeli M, Elizur A, Bruch S, Bar-Nathan AA. Serotonin 5-HT2 receptor binding on blood platelets as a state dependent marker in major affective disorder. Psychopharmacol (Berlin) 1990; 102:73–75.
51. Arora RC, Meltzer HY. Increased serotonin$_2$ (5-HT$_2$) receptor binding as measured by ^{3}H-lysergic acid diethylamide (^{3}H-LSD) in the blood platelets of depressed patients. Life Sci 1989; 44:725–734.

52. Mikuni M, Kusumi I, Kagaya A, Kuroda Y, Mori H, Takahashi K. Increased 5-HT-2 receptor function as measured by serotonin-stimulated phosphoinositide hydrolysis in platelets of depressed patients. Prog Neuropsychopharmacol Biol Psychiatry 1991; 15:49–61.
53. Hrdina PD, Bakish D, Chudzik J, Ravindran A, Lapierre YD. Serotonergic markers in platelets of patients with major depression: upregulation of 5-HT2 receptors. J Psychiatry Neurosci 1995; 20:11–19.
54. Sheline YI, Bardgett ME, Jackson JL, Newcomer JW, Csernansky JG. Platelet serotonin markers and depressive symptomatology. Biol Psychiatry 1995; 37: 442–447.
55. McBride PA, Brown RP, DeMeo M, Keilp JG, Mieczkowski T, Mann JJ. The relationship of platelet 5-HT_2 receptor indices to major depressive disorder, personality traits, and suicidal behavior. Biol Psychiatry 1994; 35:295–308.
56. Cowen PJ, Charig EM, Fraser S, Elliott JM. Platelet 5-HT receptor binding during depressive illness and tricyclic antidepressant treatment. J Affect Disord 1987; 13: 45–50.
57. Biegon A, Weizman A, Karp L, Ram A, Tiano S, Wolff M. Serotonin 5-HT2 receptor binding on blood platelets–a peripheral marker for depression? Life Sci 1987; 41:2485–2492.
58. Butler J, Leonard BE. The platelet serotonergic system in depression and following sertraline treatment. Int Clin Psychopharmacol 1988; 3:343–347.
59. Cowen PJ, Charig EM, Fraser S, Elliott JM. Platelet 5-HT receptor binding during depressive illness and tricyclic antidepressant treatment. J Affect Disord 1987; 13: 45–50.
60. Hrdina PD, Bakish D, Ravindran A, Chudzik J, Cavazzoni P, Lapierre YD. Platelet serotonergic indices in major depression: Up-regulation of 5-HT_{2A} receptors unchanged by antidepressant treatment. Psychiatry Res 1997; 66:73–85.
61. Mann JJ, McBride PA, Brown RP, et al. Relationship between central and peripheral serotonin indexes in depressed and suicidal psychiatric inpatients. Arch Gen Psychiatry 1992; 49(6):442–446.
62. Yatham LN, Steiner M. Neuroendocrine probes of serotonergic function: a critical review. Life Sci 1993; 53:447–463.
63. Siever LJ, Murphy DL, Slater S, de la Vega E, Lipper S. Plasma prolactin changes following fenfluramine in depressed patients compared to controls: an evaluation of central serotonergic responsivity in depression. Life Sci 1984; 34:1029–1039.
64. Coccaro EF, Siever LJ, Klar HM, et al. Serotonergic studies in patients with affective and personality disorders. Correlates with suicidal and impulsive aggressive behavior. Arch Gen Psychiatry 1989; 46:587–599.
65. O'Keane V, Dinan TG. Prolactin and cortisol responses to *d*-fenfluramine in major depression: evidence for diminished responsivity of central serotonergic function. Am J Psychiatry 1991; 148:1009–1015.
66. Lichtenberg P, Shapira B, Gillon D, et al. Hormone responses to fenfluramine and placebo challenge in endogenous depression. Psychiatry Res 1992; 43:137–146.
67. Mann JJ, McBride PA, Malone KM, DeMeo MD, Keilp JG. Blunted serotonergic responsivity in depressed patients. Neuropsychopharmacology 1995; 13:53–64.
68. Cleare AJ, Murray RM, O'Keane V. Reduced prolactin and cortisol responses to d-fenfluraminein depressed compared to healthy matched control subjects. Neuropsychopharmacology 1996; 14:349–354.
69. Cleare AJ, Murray RM, O'Keane V. Assessment of serotonergic function in major depression using *d*-fenfluramine: Relation to clinical variables and antidepression response. Biol Psychiatry 1998; 44:555–561.
70. Weizman A, Mark M, Gil-Ad I, Tyano S, Laron Z. Plasma cortisol, prolactin, growth hormone, and immunoreactive beta-endorphin response to fenfluramine challenge in depressed patients. Clin Neuropharmacol 1988; 11:250–256.
71. Asnis GM, Eisenberg J, van Praag HM, Lemus CZ, Harkavy Friedman JM, Miller AH. The neuroendocrine response to fenfluramine in depressives and normal controls. Biol Psychiatry 1988; 24:117–120.

72. Kavoussi RJ, Kramer J, Hauger RL, Coccaro EF. Prolactin response to D-fenfluramine in outpatients with major depression. Psychiatry Res 1998; 79:199–205.
73. Flory JD, Mann JJ, Manuck SB, Muldoon MF. Recovery from major depression is not associated with normalization of serotonergic function. Biol Psychiatry 1998; 43: 320–326.
74. Upadhyaya AK, Pennell I, Cowen PJ, Deakin JFW. Blunted growth hormone and prolactin responses to L-tryptophan in depression; a state-dependent abnormality. J Affect Disord 1991; 21:213–218.
75. Cowen PJ, Anderson IM, Gartside SE. Endocrinological responses to 5-HT. In: Whitaker-Azmitia PM, Peroutka SJ, eds. The Neuropharmacology of Serotonin. New York, The New York Academy of Sciences, 1990.
76. Heninger GR, Charney DS, Sternberg DE. Serotonergic function in depression. Prolactin response to intravenous tryptophan in depressed patients and healthy subjects. Arch Gen Psychiatry 1984; 41:398–402.
77. Shopsin B, Friedman E, Gershon S. Parachlorophenylalanine reversal of tranylcypromine effects in depressed patients. Arch Gen Psychiatry 1976; 33:811–819.
78. Young SN. Behavioral effects of dietary neurotransmitter precursors: basic and clinical aspects. Neurosci Biobehav Rev 1996; 20:313–323.
79. Delgado PL, Price LH, Miller HL, et al. Rapid serotonin depletion as a provocative challenge test for patients with major depression: relevance to antidepressant action and the neurobiology of depression. Psychopharmacol Bull 1991; 27:321–330.
80. Moreno FA, Gelenberg AJ, Heninger GR, et al. Tryptophan depletion and depressive vulnerability. Biol Psychiatry 1999; 46:498–505.
81. Smith KA, Fairburn CG, Cowen PJ. Relapse of depression after rapid depletion of tryptophan. Lancet 1997; 349:915–919.
82. Neumeister A. Tryptophan depletion, serotonin, and depression: where do we stand? Psychopharmacol Bull 2003; 37:99–115.
83. Owen F, Chambers DR, Cooper SJ, et al. Serotonergic mechanisms in brains of suicide victims. Brain Res 1986; 362:185–188.
84. Cheetham SC, Crompton MR, Czudek C, Horton RW, Katona CLE, Reynolds GP. Serotonin concentrations and turnover in brains of depressed suicides. Brain Res 1989; 502:332–340.
85. Cochran E, Robins E, Grote S. Regional serotonin levels in brain: a comparison of depressive suicides and alcoholic suicides with controls. Biol Psychiatry 1976; 11: 283–294.
86. Ferrier IN, McKeith IG, Cross AJ, Perry EK, Candy JM, Perry RH. Postmortem neurochemical studies in depression. Ann NY Acad Sci 1986; 487:128–142.
87. Crow TJ, Cross AJ, Cooper SJ, et al. Neurotransmitter receptors and monoamine metabolites in the brains of patients with Alzheimer-type dementia and depression, and suicides. Neuropharmacology 1984; 23:1561–1569.
88. Stanley M, Virgilio J, Gershon S. Tritiated imipramine binding sites are decreased in the frontal cortex of suicides. Sci 1982; 216:1337–1339.
89. Perry EK, Marshall EF, Blessed G, Tomlinson BE, Perry RH. Decreased imipramine binding in the brains of patients with depressive illness. Br J Psychiatry 1983; 142: 188–192.
90. Lawrence KM, De Paermentier F, Cheetham SC, Crompton MR, Katona CLE, Horton RW. Symmetrical hemispheric distribution of ^{3}H-paroxetine binding sites in postmortem human brain from controls and suicides. Biol Psychiatry 1990; 28: 544–546.
91. Arranz B, Eriksson A, Mellerup E, Plenge P, Marcusson J. Brain 5-HT1A, 5-HT1D, and 5-HT2 receptors in suicide victims. Biol Psychiatry 1994; 35:457–463.
92. Dillon KA, Gross-Isseroff R, Israeli M, Biegon A. Autoradiographic analysis of serotonin 5-HT$_{1A}$ receptor binding in the human brain postmortem: Effects of age and alcohol. Brain Res 1991; 554:56–64.
93. Stockmeier CA, Dilley GE, Shapiro LA, Overholser JC, Thompson PA, Meltzer HY. Serotonin receptors in suicide victims with major depression. Neuropsychopharmacology 1997; 16:162–173.

94. Cheetham SC, Crompton MR, Katona CLE, Horton RW. Brain 5-HT_1 binding sites in depressed suicides. Psychopharmacol (Berlin) 1990; 102:544–548.
95. Lowther S, De Paermentier F, Cheetham SC, Crompton MR, Katona CL, Horton RW. 5-HT1A receptor binding sites in post-mortem brain samples from depressed suicides and controls. J Affect Disord 1997; 42:199–207.
96. Stockmeier CA, Shapiro LA, Dilley GE, Kolli TM, Friedman L, Rajkowska G. Increase in serotonin-1A autoreceptors in the midbrain of suicide victims with major depression—postmortem evidence for decrease serotonin activity. J Neurosci 1998; 18:7394–7401.
97. Arango V, Underwood MD, Boldrini M, et al. Serotonin 1A receptors, serotonin transporter binding and serotonin transporter mRNA expression in the brainstem of depressed suicide victims. Neuropsychopharmacology 2001; 25:892–903.
98. Lopez JF, Chalmers DT, Little KY, Watson SJ. A.E. Bennett Research Award. Regulation of serotonin1A, glucocorticoid, and mineralocorticoid receptor in rat and human hippocampus: implications for the neurobiology of depression. Biol Psychiatry 1998; 43:547–573.
99. Stanley M, Mann JJ. Increased serotonin-2 binding sites in frontal cortex of suicide victims. Lancet 1983; i:214–216.
100. Mann JJ, Stanley M, McBride PA, McEwen BS. Increased $serotonin_2$ and b-adrenergic receptor binding in the frontal cortices of suicide victims. Arch Gen Psychiatry 1986; 43:954–959.
101. Arango V, Ernsberger P, Marzuk PM, et al. Autoradiographic demonstration of increased serotonin 5-HT2 and beta-adrenergic receptor binding sites in the brain of suicide victims. Arch Gen Psychiatry 1990; 47:1038–1047.
102. Pandey GN, Dwivedi Y, Rizavi HS, et al. Higher expression of serotonin 5-HT(2A) receptors in the postmortem brains of teenage suicide victims. Am J Psychiatry 2002; 159:419–429.
103. Owen F, Cross AJ, Crow TJ, et al. Brain 5-HT_2 receptors and suicide. Lancet 1983; ii:1256.
104. Cheetham SC, Crompton MR, Katona CLE, Horton RW. Brain 5-HT_2 receptor binding sites in depressed suicide victims. Brain Res 1988; 443:272–280.
105. Lowther S, De Paermentier F, Crompton MR, Katona CLE, Horton RW. Brain 5-HT_2 receptors in suicide victims: violence of death, depression and effects of antidepressant treatment. Brain Res 1994; 642:281–289.
106. Rosel P, Arranz B, San L, et al. Altered 5-HT_{2A} binding sites and second messenger inositol trisphosphate (IP_3) levels in hippocampus but not in frontal cortex from depressed suicide victims. Psychiat Res Neuroimag 2000; 99:173–181.
107. Yates M, Leake A, Candy JM, Fairbairn AF, McKeith IG, Ferrier IN. 5HT2 receptor changes in major depression. Biol Psychiatry 1990; 27:489–496.
108. McKeith IG, Marshall EF, Ferrier IN, et al. 5-HT receptor binding in post-mortem brain from patients with affective disorder. J Affect Disord 1987; 13: 67–74.
109. Hrdina PD, Demeter E, Vu TB, Sótónyi P, Palkovits M. 5-HT uptake sites and 5-HT_2 receptors in brain of antidepressant-free suicide victims/depressives: increase in 5-HT_2 sites in cortex and amygdala. Brain Res 1993; 614:37–44.
110. Drevets WC, Frank E, Price JC, et al. PET imaging of serotonin 1A receptor binding in depression. Biol Psychiatry 1999; 46:1375–1387.
111. Sargent PA, Kjaer KH, Bench CJ, et al. Brain serotonin1A receptor binding measured by positron emission tomography with [11C]WAY-100635: effects of depression and antidepressant treatment. Arch Gen Psychiatry 2000; 57: 174–180.
112. Parsey RV, Oquendo MA, Ogden RT, et al. Altered serotonin 1A binding in major depression: a [carbonyl-C-11]WAY100635 positron emission tomography study. Biol Psychiatry 2006. In Press.
113. Stockmeier CA. Involvement of serotonin in depression: evidence from postmortem and imaging studies of serotonin receptors and the serotonin transporter. J Psychiatr Res 2003; 37:357–373.
114. Attar-Levy D, Martinot JL, Blin J, et al. The cortical serotonin2 receptors studied with positron-emission tomography and [18F]-setoperone during depressive illness and antidepressant treatment with clomipramine. Biol Psychiatry 1999; 45:180–186.

115. Yatham LN, Liddle PF, Dennie J, et al. Decrease in brain serotonin 2 receptor binding in patients with major depression following desipramine treatment: a positron emission tomography study with fluorine-18-labeled setoperone. Arch Gen Psychiatry 1999; 56:705–711.
116. Mintun MA, Sheline YI, Moerlein SM, Vlassenko AG, Huang Y, Snyder AZ. Decreased hippocampal 5-HT2A receptor binding in major depressive disorder: in vivo measurement with [18F]altanserin positron emission tomography. Biol Psychiatry 2004; 55:217–224.
117. D'haenen H, Bossuyt A, Mertens J, Bossuyt-Piron C, Gijsemans M, Kaufman L. SPECT imaging of serotonin$_2$ receptors in depression. Psychiatry Res Neuroimaging 1992; 45:227–237.
118. Ågren H, Reibring L, Hartvig P, et al. Low brain uptake of L-[^{11}C]5-hydroxytryptophan in major depression: a positron emission tomography study on patients and healthy volunteers. Acta Psychiatr Scand 1991; 83:449–455.
119. Mann JJ, Malone KM, Diehl DJ, Perel J, Cooper TB, Mintun MA. Demonstration *in vivo* of reduced serotonin responsivity in the brain of untreated depressed patients. Am J Psychiatry 1996; 153:174–182.
120. Malison RT, Price LH, Berman R, et al. Reduced brain serotonin transporter availability in major depression as measured by [^{123}I]-2b-carbomethoxy-3b-(4-iodophenyl)-tropane and single photon emission computer tomography. Biol Psychiatry 1998; 44: 1090–1098.
121. Dahlstrom M, Ahonen A, Ebeling H, Torniainen P, Heikkila J, Moilanen I. Elevated hypothalamic/midbrain serotonin (monoamine) transporter availability in depressive drug-naive children and adolescents. Mol Psychiatry 2000; 5:514–522.
122. Shiah IS, Yatham LN. Serotonin in mania and in the mechanism of action of mood stabilizers: a review of clinical studies. Bipolar Disord 2000; 2:77–92.
123. Coppen A, Prange AJ Jr, Hill C, Whybrow PC, Noguera R. Abnormalities of indoleamines in affective disorders. Arch Gen Psychiatry 1972; 26:474–478.
124. Banki CM. Correlation between cerebrospinal fluid amine metabolites and psychomotor activity in affective disorders. J Neurochem 1977; 28:255–257.
125. Gerner RH, Fairbanks L, Anderson GM, et al. CSF Neurochemistry in depressed, manic, and schizophrenic patients compared with that of normal controls. Am J Psychiatry 1984; 141(12):1533–1540.
126. Ashcroft GW, Crawford TB, Eccleston D, et al. 5-hydroxyindole compounds in the cerebrospinal fluid of patients with psychiatric or neurological diseases. Lancet 1966; 2:1049–1052.
127. Bowers MB Jr, Heninger GR, Gerbode F. Cerebrospinal fluid 5-hydroxyindoleactiic acid and homovanillic acid in psychiatric patients. Int J Neuropharmacol 1969; 8: 255–262.
128. Wilk S, Shopsin B, Gershon S, Suhl M. Cerebrospinal fluid levels of MHPG in affective disorders. Nature 1972; 235:440–441.
129. Goodwin FK, Post RM, Dunner DL, Gordon EK. Cerebrospinal fluid amine metabolites in affective illness: the probenecid technique. Am J Psychiatry 1973; 130:73–79.
130. Sjostrom R, Roos BE. 5-Hydroxyindolacetic acid and homovanillic acid in cerebrospinal fluid in manic-depressive psychosis. Eur J Clin Pharmacol 1972; 4: 170–176.
131. Swann AC, Secunda S, Davis JM, et al. CSF monoamine metabolites in mania. Am J Psychiatry 1983; 140:396–400.
132. Ashcroft GW, Glen AI. Mood and neuronal functions: a modified amine hypothesis for the etiology of affective illness. Adv Biochem Psychopharmacol 1974; 11: 335–339.
133. Vestergaard P, Sorensen T, Hoppe E, Rafaelsen OJ, Yates CM, Nicolaou N. Biogenic amine metabolites in cerebrospinal fluid of patients with affective disorders. Acta Psychiatr Scand 1978; 58:88–96.
134. Sjöström R, Roos B-E. 5-hydroxyindolacetic acid and homovanillic acid in cerebrospinal fluid in manic-depressive psychosis. Eur J Clin Pharmacol 1972; 4:170–176.
135. Swann AC, Stokes PE, Secunda SK, et al. Depressive mania versus agitated depression: biogenic amine and hypothalamic-pituitary-adrenocortical function. Biol Psychiatry 1994; 35:803–813.

136. Ross BE, Sjostrom R. 5-Hydroxyindoleacetic acid (and homovanillic acid) levels in the cerebrospinal fluid after probenecid application in patients with manic-depressive psychosis. Pharmacology Clinic 1969; 1:153–155.
137. Sjostrom R. 5-Hydroxyindole acetic acid and homovanillic acid in cerebrospinal fluid in manic-depressive psychosis and the effect of probenecid treatment. Eur J Clin Pharmacol 1973; 6:75–80.
138. Scott M, Reading HW, Loudon JB. Studies on human blood platelets in affective disorders. Psychopharmacol (Berlin) 1979; 60:131–135.
139. Meagher JB, O'Halloran A, Carney PA, Leonard BE. Changes in platelet 5-hydroxytryptamine uptake in mania. J Affect Disord 1990; 19:191–196.
140. Marazziti D, Lenzi A, Galli L, San Martino S, Cassano GB. Decreased platelet serotonin uptake in bipolar I patients. Int Clin Psychopharmacol 1991; 6:25–30.
141. Lewis DA, McChesney C. Tritiated imipramine binding distinguishes among subtypes of depression. Arch Gen Psychiatry 1985; 42:485–488.
142. Ellis PM, Mellsop GW, Beeston R, Cooke RR. Platelet tritiated imipramine binding in patients suffering from mania. J Affect Disord 1991; 22:105–110.
143. Marazziti D, Lenzi A, Cassano GB. Serotoninergic dysfunction in bipolar disorder. Pharmacopsychiat 1991; 24:164–167.
144. Velayudhan A, Sunitha TA, Balachander S, Reddy JY, Khanna S. A study of platelet serotonin receptor in mania. Biol Psychiatry 1999; 45:1059–1062.
145. Kusumi I, Koyama T, Yamashita I. Effect of various factors on serotonin-induced Ca2+ response in human platelets. Life Sci 1991; 48:2405–2412.
146. Kagaya A, Mikuni M, Kusumi I, Yamamoto H, Takahashi K. Serotonin-induced acute desensitization of $serotonin_2$ receptors in human platelets via a mechanism involving protein kinase C. J Pharmacol Exp Ther 1990; 255:305–311.
147. Okamoto Y, Kagaya A, Shinno H, Motohashi N, Yamawaki S. Serotonin-induced platelet calcium mobilization is enhanced in mania. Life Sci 1994; 56:327–332.
148. Berk M, Bodemer W, Van Oudenhove T, Butkow N. The platelet intracellular calcium response to serotonin is augmented in bipolar manic and depressed patients. Human Psychopharmacology 1995; 10:189–193.
149. Thakore JH, O'Keane V, Dinan TG. *d*-fenfluramine-induced prolactin responses in mania: evidence for serotonergic subsensitivity. Am J Psychiatry 1996; 153: 1460–1463.
150. Yatham LN. Buspirone induced prolactin release in mania. Biol Psychiatry 1994; 35:553–556.
151. Yatham LN. Prolactin and cortisol responses to fenfluramine challenge in mania. Biol Psychiatry 1996; 39:285–288.
152. Yatham LN, Shiah IS, Lam RW, Tam EM, Zis AP. Hypothermic, ACTH, and cortisol responses to ipsapirone in patients with mania and healthy controls. J Affect Disord 1999; 54:295–301.
153. Bagdy G, Makara GB. Hypothalamic paraventricular nucleus lesions differentially affect serotonin-1A ($5\text{-}HT_{1A}$) and $5\text{-}HT_2$ receptor agonist-induced oxytocin, prolactin, and corticosterone responses. Endocrinology 1994; 134:1127–1131.
154. Gilbert F, Brazell C, Tricklebank MD, Stahl SM. Activation of the 5-HT1A receptor subtype increases rat plasma ACTH concentration. Eur J Pharmacol 1988; 147: 431–439.
155. Koenig JI, Gudelsky GA, Meltzer HY. Stimulation of corticosterone and beta-endorphin secretion in the rat by selective 5-HT receptor subtype activation. Eur J Pharmacol 1987; 137:1–8.
156. Meltzer HY, Umberkoman-Wiita B, Robertson A, Tricou BJ, Lowy M, Perline R. Effect of 5-hydroxytryptophan on serum cortisol levels in major affective disorders. I. Enhanced response in depression and mania. Arch Gen Psychiatry 1984; 41:366–374.
157. Takahashi S, Kondo H, Yoshimura M, Ochi Y, Yoshimi T. Growth hormone responses to administration of L-5-hydroxytryptophan (L-5-HTP) in manic-depressive psychoses. Folia Psychiatr Neurol Jpn 1973; 27:197–206.
158. Meltzer HY, Uberkoman Wiita B, Robertson A, Tricou BJ, Lowy M. Enhanced serum cortisol response to 5-hydroxytryptophan in depression and mania. Life Sci 1983; 33:2541–2549.

159. Yatham LN, Liddle PF, Lam RW, et al. A positron emission tomography study of the effects of treatment with valproate on brain 5-HT2A receptors in acute mania. Bipolar Disord 2005; 7(suppl 5):53–57.
160. Price LH, Charney DS, Delgado PL, Heninger GR. Lithium and serotonin function: implications for the serotonin hypothesis of depression. Psychopharmacology (Berl) 1990; 100:3–12.
161. Fyro B, Petterson U, Sedvall G. The effect of lithium treatment on manic symptoms and levels of monoamine metabolites in cerebrospinal fluid of manic depressive patients. Psychopharmacologia 1975; 44:99–103.
162. Bowers MB Jr, Heninger GR. Lithium: clinical effects and cerebrospinal fluid acid monoamine metabolites. Commun Psychopharmacol 1977; 1:135–145.
163. Berrettini WH, Nurnberger JI Jr, Scheinin M, et al. Cerebrospinal fluid and plasma monoamines and their metabolites in euthymic bipolar patients. Biol Psychiatry 1985; 20:257–269.
164. Meltzer HY, Arora RC, Goodnick P. Effect of lithium carbonate on serotonin uptake in blood platelets of patients with affective disorders. J Affect Disord 1983; 5:215–221.
165. Cappiello A, Sernyak MJ, Malison RT, McDougle CJ, Heninger GR, Price LH. Effects of acute tryptophan depletion in lithium-remitted manic patients: a pilot study. Biol Psychiatry 1997; 42:1076–1078.
166. Benkelfat C, Seletti B, Palmour RM, Hillel J, Ellenbogen M, Young SN. Tryptophan depletion in stable lithium-treated patients with bipolar disorder in remission. Arch Gen Psychiatry 1995; 52:154–156.
167. Hughes JH, Dunne F, Young AH. Effects of acute tryptophan depletion on mood and suicidal ideation in bipolar patients symptomatically stable on lithium. Br J Psychiatry 2000; 177:447–451.
168. Johnson L, El Khoury A, Aberg-Wistedt A, Stain-Malmgren R, Mathe AA. Tryptophan depletion in lithium-stabilized patients with affective disorder. Int J Neuropsychopharmacol 2001; 4:329–336.
169. Ågren H. Symptom patterns in unipolar and bipolar depression correlating with monoamine metabolites in the cerebrospinal fluid: I. General patterns. Psychiatry Res 1980; 3:211–223.
170. Shiah IS, Ko HC, Lee JF, Lu RB. Platelet 5-HT and plasma MHPG levels in patients with bipolar I and bipolar II depressions and normal controls. J Affect Disord 1999; 52:101–110.
171. Wirz-Justice A, Puhringer W. Increased platelet serotonin in bipolar depression and hypomania. J Neural Transm 1978; 42:55–62.
172. Zemishlany Z, Munitz H, Rotman A, Wijsenbeek H. Increased uptake of serotonin by blood platelets from patients with bipolar primary affective disorder-bipolar type. Psychopharmacology (Berl) 1982; 77:175–178.
173. Modai I, Zemishlany Z, Jerushalmy Z. 5-Hydroxytryptamine uptake by blood platelets of unipolar and bipolar depressed patients. Neuropsychobiology 1984; 12:93–95.
174. Jeanningros R, Gronier B, Azorin JM, Tissot R. Platelet [3H]-imipramine binding according to DSM-III subtypes of depression. Neuropsychobiology 1989; 22:33–40.
175. Baron M, Barkai A, Gruen R, Peselow E, Fieve RR, Quitkin F. Platelet [3H] imipramine binding in affective disorders: trait versus state characteristics. Am J Psychiatry 1986; 143:711–717.
176. Quintin P, Benkelfat C, Launay JM, et al. Clinical and neurochemical effect of acute tryptophan depletion in unaffected relatives of patients with bipolar affective disorder. Biol Psychiatry 2001; 50:184–190.
177. Marazziti D, Dell'Osso L, Rossi A, et al. Decreased platelet [3H]paroxetine binding sites in suicide attempters. Psychiatry Res 2001; 103:125–131.
178. Pandey GN, Pandey SC, Ren X, Dwivedi Y, Janicak PG. Serotonin receptors in platelets of bipolar and schizoaffective patients: effect of lithium treatment. Psychopharmacology (Berl) 2003; 170:115–123.
179. Sher L, Oquendo MA, Li S, et al. Prolactin response to fenfluramine administration in patients with unipolar and bipolar depression and healthy controls. Psychoneuroendocrinology 2003; 28:559–573.

180. Mitchell P, Smythe G. Hormonal responses to fenfluramine in depressed and control subjects. J Affect Disord 1990; 19:43–51.
181. Nurnberger JI Jr, Berrettini W, Simmons-Alling S, Lawrence D, Brittain H. Blunted ACTH and cortisol response to afternoon tryptophan infusion in euthymic bipolar patients. Psychiatry Res 1990; 31:57–67.
182. Price LH, Charney DS, Delgado PL, Heninger GR. Serotonin function and depression: neuroendocrine and mood responses to intravenous L-tryptophan in depressed patients and healthy comparison subjects. Am J Psychiatry 1991; 148:1518–1525.
183. Shiah IS, Yatham LN, Lam RW, Tam EM, Zis AP. Cortisol, hypothermic, and behavioral responses to ipsapirone in patients with bipolar depression and normal controls. Neuropsychobiology 1998; 38:6–12.
184. Young LT, Warsh JJ, Kish SJ, Shannak K, Hornykeiwicz O. Reduced brain 5-HT and elevated NE turnover and metabolites in bipolar affective disorder. Biol Psychiatry 1994; 35:121–127.
185. Dean B, Scarr E, Pavey G, Copolov D. Studies on serotonergic markers in the human hippocampus: changes in subjects with bipolar disorder. J Affect Disord 2003; 75: 65–69.
186. Wiste AK, Underwood M, Ellis SP, Mann JJ. Norepinephrine and serotonin imbalance in the locus coeruleus in bipolar disorder. Biological Psychiatry. 2007. In Press.
187. McGuffin P, Rijsdijk F, Andrew M, Sham P, Katz R, Cardno A. The heritability of bipolar affective disorder and the genetic relationship to unipolar depression. Arch Gen Psychiatry 2003; 60:497–502.
188. Smoller JW, Finn CT. Family, twin, and adoption studies of bipolar disorder. Am J Med Genet 2003; 123C:48–58.
189. Furlong RA, Ho L, Walsh C, et al. Analysis and meta-analysis of two serotonin transporter gene polymorphisms in bipolar and unipolar affective disorders. Am J Med Genet 1998; 81:58–63.
190. Anguelova M, Benkelfat C, Turecki G. A systematic review of association studies investigating genes coding for serotonin receptors and the serotonin transporter: I. Affective disorders. Mol Psychiatry 2003; 8:574–591.
191. Lasky-Su JA, Faraone SV, Glatt SJ, Tsuang MT. Meta-analysis of the association between two polymorphisms in the serotonin transporter gene and affective disorders. Am J Med Genet B Neuropsychiatr Genet 2005; 133:110–115.
192. Lotrich FE, Pollock BG. Meta-analysis of serotonin transporter polymorphisms and affective disorders. Psychiatr Genet 2004; 14:121–129.
193. Craddock N, Dave S, Greening J. Association studies of bipolar disorder. Bipolar Disord 2001; 3:284–298.
194. Bellivier F, Roy I, Leboyer M. Serotonin transporter gene polymorphisms and affective disorder-related phenotypes. Current Opinion in Psychiatry 2002; 15:49–58.
195. Etain B, Rousseva A, Roy I, et al. Lack of association between 5HT2A receptor gene haplotype, bipolar disorder and its clinical subtypes in a West European sample. Am J Med Genet 2004; 129B:29–33.
196. Niesler B, Flohr T, Nothen MM, et al. Association between the 5′ UTR variant C178T of the serotonin receptor gene HTR3A and bipolar affective disorder. Pharmacogenetics 2001; 11:471–475.
197. Ohtsuki T, Ishiguro H, Detera-Wadleigh SD, et al. Association between serotonin 4 receptor gene polymorphisms and bipolar disorder in Japanese case-control samples and the NIMH Genetics Initiative Bipolar Pedigrees. Mol Psychiatry 2002; 7:954–961.
198. Birkett JT, Arranz MJ, Munro J, Osbourn S, Kerwin RW, Collier DA. Association analysis of the 5-HT5A gene in depression, psychosis and antipsychotic response. Neuroreport 2000; 11:2017–2020.
199. Arias B, Collier DA, Gasto C, et al. Genetic variation in the 5-HT5A receptor gene in patients with bipolar disorder and major depression. Neurosci Lett 2001; 303:111–114.
200. Vogt IR, Shimron-Abarbanell D, Neidt H, et al. Investigation of the human serotonin 6 [5-HT6] receptor gene in bipolar affective disorder and schizophrenia. Am J Med Genet 2000; 96:217–221.

201. Hong CJ, Tsai SJ, Cheng CY, Liao WY, Song HL, Lai HC. Association analysis of the 5-HT(6) receptor polymorphism (C267T) in mood disorders. Am J Med Genet 1999; 88:601–602.
202. Vincent JB, Masellis M, Lawrence J, et al. Genetic association analysis of serotonin system genes in bipolar affective disorder. Am J Psychiatry 1999; 156:136–138.
203. Bellivier F, Leboyer M, Courtet P, et al. Association between the tryptophan hydroxylase gene and manic-depressive illness. Arch Gen Psychiatry 1998; 55:33–37.
204. Furlong RA, Ho L, Rubinsztein JS, Walsh C, Paykel ES, Rubinsztein DC. No association of the tryptophan hydroxylase gene with bipolar affective disorder, unipolar affective disorder, or suicidal behaviour in major affective disorder. Am J Med Genet 1998; 81:245–247.
205. Kirov G, Owen MJ, Jones I, McCandless F, Craddock N. Tryptophan hydroxylase gene and manic-depressive illness. Arch Gen Psychiatry 1999; 56:98–99.
206. Kunugi H, Ishida S, Kato T, et al. No evidence for an association of polymorphisms of the tryptophan hydroxylase gene with affective disorders or attempted suicide among Japanese patients. Am J Psychiatry 1999; 156:774–776.
207. McQuillin A, Lawrence J, Kalsi G, Chen A, Gurling H, Curtis D. No allelic association between bipolar affective disorder and the tryptophan hydroxylase gene. Arch Gen Psychiatry 1999; 56:99–101.
208. Souery D, Van Gestel S, Massat I, et al. Tryptophan hydroxylase polymorphism and suicidality in unipolar and bipolar affective disorders: a multicenter association study. Biol Psychiatry 2001; 49:405–409.

6 Cell Membrane and Signal Transduction Pathways—Implications for the Pathophysiology of Bipolar Disorders

Guang Chen and Husseini K. Manji
Laboratory of Molecular Pathophysiology, Mood and Anxiety Disorders Research Program, National Institute of Mental Health, Bethesda, Maryland, U.S.A.

OVERVIEW AND HISTORICAL BACKGROUND

Although genetic factors play a major, unquestionable role in the etiology of bipolar disorder, the biochemical abnormalities underlying the predisposition to and the pathophysiology of this complex and intriguing neuropsychiatric disorder remain to be fully elucidated. The brain systems that have heretofore received the greatest attention in neurobiologic studies of these illnesses have been the monoaminergic neurotransmitter systems, which were implicated by the following observations:

1. Effective antidepressant drugs exert their *primary* biochemical effects by regulating intrasynaptic concentrations of serotonin and norepinephrine
2. Antihypertensives that deplete these monoamines sometimes precipitate depressive episodes in susceptible individuals.
3. Psychostimulants and dopamine agonists have been shown to be capable of triggering manic episodes in susceptible individuals.

Furthermore, the monoaminergic systems are extensively distributed throughout the network of limbic, striatal, and prefrontal cortical neuronal circuits thought to support the behavioral and visceral manifestations of mood disorders (1). Clinical studies over the past 40 years have attempted to uncover the biological factors mediating the pathophysiology of bipolar disorder utilizing a variety of biochemical and neuroendocrine strategies. Indeed, assessments of cerebrospinal fluid (CSF) chemistry, neuroendocrine responses to pharmacological challenge, and neuroreceptor and transporter binding have, in fact, demonstrated a number of abnormalities of the serotonergic, noradrenergic, and other neurotransmitter and neuropeptide systems in mood disorders.

While such investigations have been heuristic over the years, they have been of limited value in elucidating the unique biology of this affective disorder that must include an understanding of the underlying basis for the predilection to episodic and often profound mood disturbance that can become progressive over time. Thus, bipolar disorder likely arises from the complex interaction of multiple susceptibility (and protective) genes and environmental factors, and the phenotypic expression of the disease includes not only episodic and often profound mood disturbance, but also a constellation of cognitive, motoric, autonomic, endocrine, and sleep/wake abnormalities. Furthermore, while most antidepressants exert their initial effects by increasing the intrasynaptic levels of serotonin and/or norepinephrine, their clinical antidepressant effects are only observed after chronic (days to weeks) administration, suggesting that a cascade of downstream effects are ultimately responsible for their therapeutic effects. These observations have led to the appreciation that while dysfunction within the monoaminergic neurotransmitter systems is likely to play important roles in mediating some facets of the pathophysiology of bipolar disorder, they likely represent the downstream effects of other, *more primary abnormalities* (2).

The subsequent challenge for the basic and clinical neuroscientist will be the integration of these molecular/cellular changes to the systems and ultimately to the behavioral level wherein the clinical expression of bipolar disorder becomes fully elaborated.

Despite these formidable obstacles, there has been considerable progress in our understanding of the underlying molecular and cellular basis of this unique affective disorder in recent years. In particular, recent evidence demonstrating that impairments of signaling pathways may play a role in the pathophysiology of bipolar disorder, and that mood stabilizers exert major effects on signaling pathways which regulate neuroplasticity and cell survival, have generated considerable excitement among the clinical neuroscience community, and are reshaping views about the neurobiological underpinnings of these disorders (3–5). In this chapter, we critically review and appraise these data and discuss their implications not only for changing existing conceptualizations regarding the pathophysiology of mood disorders, but also for the strategic development of improved therapeutics.

Signaling Networks: The Cellular Machinery Underlying Information Processing and Long-Term Neuroplastic Events

It is hardly surprising that abnormalities in multiple neurotransmitter systems and physiological processes have been found in a disorder as complex as bipolar disorder. Signal transduction pathways are in a pivotal position in the central nervous system (CNS), able to affect the functional balance between multiple neurotransmitter systems and may therefore play a role in mediating the more "downstream" abnormalities that likely underlie the pathophysiology of affective disorders. Moreover, as we discuss below, recent research has clearly identified signaling pathways as therapeutically relevant targets for our most effective pharmacological treatments. Indeed, the molecular and cellular targets underlying lithium and valproate's abilities to stabilize a dysregulation of the limbic system and limbic associated function strongly suggest that abnormalities in signaling pathways may also play a critical role in the pathophysiology of bipolar disorder.

Signal transduction pathways serve the critical roles of first amplifying and "weighting" numerous extracellularly generated neuronal signals and then transmitting these integrated signals to effectors, thereby forming the basis for a complex information processing network (6). The high degree of complexity generated by these signaling networks may be one mechanism by which neurons acquire the flexibility for generating the wide range of responses observed in the nervous system. These pathways are thus undoubtedly involved in regulating such diverse vegetative functions such as mood, appetite, and wakefulness and are therefore likely to be involved in the pathophysiology of bipolar disorder.

G-Proteins

Mammalian guanine nucleotide binding proteins (G proteins) can be categorized into two major groups: heterotrimeric G proteins and small G proteins. Heterotrimeric G proteins convey signaling from G-protein-coupled-receptors (GPCRs) to their effectors, which include adenylyl and guanylyl cyclases (AC and GC), phosphodiesterases (PDE), phopholipases A2 (PLA2), phospholipase C (PLC), phosphoinositide 3-kinases (PI3Ks), ion channels, and Rho-GEF (Table 1). These G proteins are composed of three distinct subunits, α, β, and γ. These trimeric complexes are loosely associated with the GPCRs. Upon receptor stimulation, receptors and G proteins undergo conformational changes that lead to the exchange of guanosine diphosphate (GDP) for guanosine triphosphate (GTP) at $G\alpha$ subunit and dissociation of $G\alpha$ and $G\beta\gamma$ subunits. Consequently, $G\alpha$ and $G\beta\gamma$ stimulate effector molecules. $G\alpha$ subunits process intrinsic GTPase activities, thereby hydrolyzing GTP to GDT, resulting in $G\alpha$ inactivation and reassociation of $G\alpha$ with $G\beta\gamma$. There are four $G\alpha$ families expressed in the brains: $G\alpha s$ ($G\alpha s1$–4 and $G\alpha olf$), Gi ($G\alpha i1$–3, $G\alpha o1$–2, $G\alpha t1$–2, $G\alpha gust$, and $G\alpha z$), $G\alpha q$ ($G\alpha q$, and $G\alpha 11$, 14–16), and $G\alpha 12$ ($G\alpha 12$–13) (Table 1).

Small G proteins are monomeric G proteins with molecular weight of 20 to 40 kDa. Like heterotrimeric G proteins, small G proteins bind to guanine nucleotides, process intrinsic GTPase activity, and cycle through GDP- and GTP-bond forms. Exchange of GDP for GTP causes major conformational changes of these G proteins and their affinities for effector molecules. More than 150 small G proteins have been identified. Small G proteins are classified into five families: Ras, Rho, Rab, Ran, and Arf (Table 1). These G proteins have higher affinities for GDP and GTP and lower intrinsic GTPase activity, and GDP to GTP exchange activities. The functioning of these G proteins is modulated by regulator proteins: guanine nucleotide exchange factors (GEFs), which promote exchange between GDP and GTP, and GTPase-activating proteins, (GAPs) which stimulate hydrolysis of the bound GTP. Rho and Rab proteins are regulated by a third class of proteins,

TABLE 1 Major Signaling GTPase G Proteins

Groups	Types	Effectors and effects
$G\alpha$ proteins	$G\alpha s$	ACs, increase in cAMP
	$G\alpha i$	AC, inhibits cAMP production Ion channel Phosphodiesterase Phospholipase
	$G\alpha q$	Phospholipase C-β, increase DAG and IPs, in turn release Ca^{2+} and activate PKC
	$G\alpha 12$	RhoGEFs, activates Rho
Small G proteins	Ras	Regulate diverse signaling cascades involving gene expression and cell proliferation, differentiation and survival
	Rho	Regulate diverse signaling cascades involving actin organization, cell cycle progression and gene expression
	Rab	Regulate intracellular vesicular transport and trafficking of protein between different organelles of the endocytic and secretory pathways
	Ran	Function in nucleocytoplasmic transport of RNA and proteins
	Arf	Regulate vesicular transport

Abbreviations: AC, adenylyl cyclase; cAMP, cyclic AMP; DAG, diacylglycerol; IP, inositol phosphotase; PKC, protein kinase C.

guanine nucleotide dissociation inhibitors (GDIs). The GAPs for heterotrimeric G proteins are termed as the intracellular regulator of G protein signaling (RGS) proteins. RGS protein activities are regulated by a complex web of intracellular factors.

G Protein Abnormalities in Bipolar Disorder

Likely due to the fact that lithium affects G proteins (discussed below), several independent laboratories have investigated the potential role of G protein abnormalities in the pathophysiology of bipolar disorder. Potential G protein dysfunction in bipolar etiology has been investigated. Young and associates reported increases in protein levels of Gαs, but not Gαi, Gαo, and Gβ1/2, in the frontal and occipital cortex of bipolar patients (7). Follow-up work from the same laboratory suggested that these increases in Gαs protein levels are likely due to changes in post-translational enzymatic modification such as endogenous ADP-ribosylation (8), rather than being due to change in Gαs gene transcription (9). Consistent with these findings, Wang and Friedman reported elevated levels of Gαs, but not Gαi, Gαo, Gαq, Gαz, and Gβ1/2, in frontal cortical membrane preparations of bipolar patients (10). In a recent study, Gonzalez-Maeso et al. found that the frontal cortical membrane levels of Gαs, Gαi1/2, Gαi3, and Gαo are not different between control subjects and suicide victims with mood disorders. To investigate the functional coupling of receptors to G proteins, Friedman and Wang utilized agonist-induced GTPγS. They found increases in basal GTPγS binding, and increased GTPγS binding to Gαs induced by isoprenaline, Gαi, Gαo, and Gαq induced by carbachol, and Gαs, Gαi, Gαo, and Gαq induced by serotonin in postmortem frontal cortical samples of bipolar patients (11). Gonzalez-Maeso et al. found selective increases in GTPγS binding sensitivity in response to stimulation of α-adrenoceptor 2A, but not 5-HT1a, υ-opioid, GABA-B, and muscarinic receptors in frontal cortical membrane samples of suicide victims with mood disorders (12). Despite the increases in Gαs protein, Young and associates' follow-up studies did not reveal any significant increase in cAMP production induced by GTPγS, although there was an increase in cAMP production induced by forskolin which stimulates AC directly (13). Jope and associates found significant reductions in GTPγS, NaF, GTPγS plus carbachol, and GTPγS plus serotonin-induced phosphatidylinositol hydrolysis in occipital, but not temporal and frontal, cortical samples of bipolar subjects (14). By contrast, Mathews and associates reported an increase in protein level of Gαq11 in occipital, but not frontal or temporal, cortex of bipolar patients (15). Finally, several studies have also found elevated Gαs protein levels and mRNA levels in peripheral circulating cells in bipolar disorder, although the dependency on clinical state remains unclear (16–19).

Lithium and G Proteins

There are two proposed magnesium sites on G proteins: high affinity sites (nM) essential for GTPase activity, and low affinity (mM) sites required for exchange GDP for GTP (20). In 1988, Avissar and associates reported that lithium at 0.6 mM concentration completely blocked isoprenaline (β-adrenoceptor agonist), and carbachol (muscarinic receptor agonist) induced [H3]GTP bindings in rat cerebral cortical membrane preparations, in vitro and in cerebral cortical membrane preparations, from rats chronically treated with lithium (21). However, while there is some additional data in support of a *direct effect* of lithium on G proteins (22), this has not been consistently found.

Moreover, the inhibitory effects of *chronic* lithium treatment on rat brain adenylyl cyclase AC are not reversed by Mg^{2+}, and still persist after washing of the membranes, but are reversed by increasing concentrations of GTP (23–27). These results suggest that the effects of chronic lithium (those which are more likely to be therapeutically relevant) may be exerted at the level of signal-transducing G proteins at a GTP responsive step (discussed below). These two distinct actions of lithium on the AC system may explain the differing results which have been obtained by investigators using rat membrane preparations from those using slice preparations (2). Overall, for both Gs and Gi, lithium's major effects in both humans and rodents are most compatible with a stabilization of the heterotrimeric, undissociated, inactive ($\alpha\beta\gamma$) conformation of the G protein (16–18,28).

cAMP Signaling Cascade and PKA

Cyclic AMP (cAMP) is a second messenger generated by adenylyl cyclase (AC) upon extracellular stimulation (Fig. 1). There are at least nine isoforms of membrane-bound ACs, most of them expressed in the brain (Table 2). All membrane-bound ACs can be activated by forskolin and Gαs proteins. In addition, Group I ACs can be stimulated by calmodulin in Ca^{2+}-dependent manner. Group II ACs can be conditionally activated by G$\beta\gamma$. Group III ACs are highly sensitive to Gαi inhibition and also inhibited by Ca^{2+} and protein kinase A (PKA). Group IV AC is less responsive to FSK. Protein kinase C (PKC) markedly activates AC2, AC5, and AC7 and inactivates AC4 and AC9. There is also a soluble (cytosol) AC. This AC does not respond to FSK and Gαs, but becomes activated in presence of bicarbonate.

cAMP-dependent protein kinase A (PKA) in an inactivated state is a tetramer, containing two of each regulatory and catalytic subunits. There are four regulatory subunit isoforms (RIα, RIβ, RIIα, and RIIβ) and two catalytic subunit isoforms (Cα and Cβ). All isoforms are expressed in the brain (29). When intracellular cAMP level rises, cAMP binds to the regulatory subunit of protein kinase A (PKA), and this causes dissociation of the tetramer into its component monomers—two regulatory subunits with cAMP attached, and two active catalytic subunit. The freed catalytic subunits phosphorylate its cellular or nuclear substrates such as transcription factor cAMP responding element binding protein (CREB). cAMP is converted to

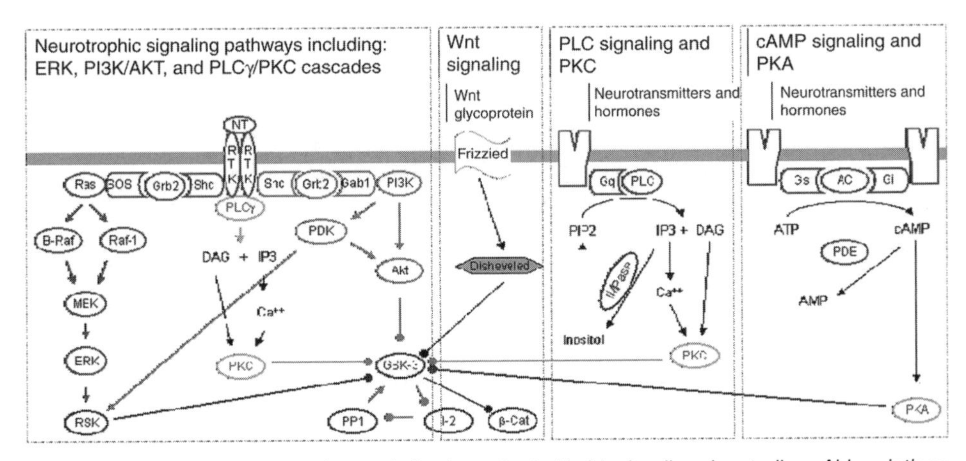

FIGURE 1 Some signaling pathways being investigated in bipolar disorder studies. *Abbreviations*: cAMP, cyclic AMP; DAG, diacyl glycerol; ERK, extracellular signal-regulated kinase; MEK, MAPK/ERK kinase; PLC, phospholipase C; PKC, protein kinase C; RSK, ribosomal S6 kinase.

TABLE 2 Membrane-Bound Adenylyl Cyclases

Group	Type	Distribution	FSK	$G\alpha s$	$G\alpha i/o$	$G\beta\gamma$	Ca^{2+}	PKA	PKC
I	AC1	Brain	↑	↑	↓ αi	↓	↑ CaM		↑ w
	AC3	Olfactory, pancreas	↑	↑	↓ w		↑ CaM		↑ w
	AC8	Brain, pancreas	↑	↑	↓ w		↑ CaM		
II	AC2	Lung, brain	↑	↑		↑ c			↑
	AC4	Widespread	↑	↑		↑ c			↓
	AC7	Widespread	↑	↑		↑ c			↑
III	AC5	Heart, striatum	↑	↑	↓		↓	↓	↑
	AC6	Widespread	↑	↑	↓		↓	↓	↓ and ↑ c
III	AC9	Widespread	↑ w	↑	↓ w		↓ Calcineurin		↓

Note: ↑, stimulation; ↓, inhibition; c, conditional action; w, weak action.

AMP by cyclic nucleotide phosphodiesterases (PDE); this event turns off the cAMP signaling at the second messenger level.

The cAMP/PKA Signaling Cascade in Bipolar Disorder

A limited number of postmortem brain studies suggest brain regional dysfunction of cAMP signaling at PKA, but not at AC, may play a role in bipolar disorder. Studies did not reveal any evidence for changes of basal and GTPγS-stimulated AC activities in a variety of brain regions including prefrontal, temporal, and occipital cortices, hippocampus, thalamus, and cerebellum in bipolar subjects (13,30,31). An initial report showed that forskolin-stimulated cAMP production was selectively increased in temporal and occipital cortexes of bipolar subjects (13), and later studies confirmed this finding, at least in temporal cortex (30,31). No differences were found between control and bipolar subjects in temporal cortex protein levels of AC1, AC4, and AC5/6. (30). The initial PKA postmortem brain study from the Warsh lab revealed widespread reductions in cytosolic, but not particulate, cAMP binding sites in bipolar subjects (32). In follow-up studies, the researchers found increases in PKA activities basal and cAMP-stimulated cytosolic PKA activities in temporal cortical samples. This higher PKA function seems due to reduction of regulatory subunit proteins (33), rather than due to changes in mRNA levels of PKA RIα, PKA RIIβ, and PKA Cα (34).

The data obtained from behavioral studies using transgenic and knockout mice indicate that selective components of cAMP-signaling cascade are required for regulation of behavior relevant to mood disorders. Mice lacking AC8 show reduced anxiety-like (or increased explorative) behaviors, which are more obvious after restraint stress (35). AC5 null mice exhibited enhanced sensitivity to D1 receptor agonist-induced locomotion and inverted locomotion response to D2 receptor antagonists (36). AC1 and AC8 double knockout mice showed signs of behavioral allodynia (37). Deletion of PKA RIIβ in mice resulted in lean body size (38), markedly increased home-cage activity (39,40), supersensitivity to amphetamine-induced locomotion (41), and more alcohol consumption and insensitivity to alcohol-induced sedation (42). Mice lacking RIb were indistinguishable from wild-type mice on indices of anxiety, exploration, and memory (43). Mice lacking PKA Cb also appear indistinguishable from wild-type mice in a variety of behavioral tests or tasks (43,44). PDE1B knockout mice possess baseline hyperlocomotion, enhanced locomotor response to psychostimulants, and spatial memory deficiencies (45). PDE4D, but not PDE4A and PDE4B, null mice showed antidepressant-like behavioral change in the forced swim and tail suspension tests (46).

Lithium and the cAMP/PKA Signaling Cascade

The effects of lithium on the cAMP system have been extensively studied, and reveal a complex, region-specific effect, likely due to lithium's effects at multiple levels of this signaling cascade. Studies in the early 70s showed that lithium inhibits membrane AC activities induced by GPCR and G protein stimulations (47–49), perhaps due to direct competition with magnesium required for AC catalytic activity. However, the clear inhibitory effects of lithium on AC are observed at concentrations above its therapeutic range. Lithium has been demonstrated to exert complex effects on the activity of AC, with the preponderance of the data demonstrating an elevation of basal AC activity, while attenuating a variety of receptor-mediated responses (16,50). Lithium in vitro inhibits the stimulation of AC by the poorly hydrolyzable analog of GTP, Gpp(NH)p, and also by Ca^{2+}/calmodulin, suggesting that lithium in vitro is directly able to inhibit the catalytic unit of AC (23–27). Since these inhibitory effects of lithium in vitro can be overcome by Mg^{2+}, they have been postulated to be mediated (at least in part) by a direct competition with magnesium (whose hydrated ionic radius is similar to that of lithium) for a binding site on the catalytic unit of AC (23–27). However, the inhibitory effects of *chronic* lithium treatment on rat brain AC are not reversed by Mg^{2+}, and still persist after washing of the membranes, but are reversed by increasing concentrations of GTP (23–27). This has led to an investigation of lithium's effects on the AC system in vivo, using microdialysis. These studies found that chronic lithium treatment produced a significant increase in basal and post-receptor stimulated (cholera toxin or forskolin) AC activity, while attenuating the β-adrenergic mediated effect (51,52). Interestingly, chronic lithium treatment resulted in an almost absent cAMP response to pertussis toxin, suggesting a lithium-induced attenuation of Gi function. It should be noted, however, that chronic lithium has also been found to increase not only cAMP levels (53), but also the levels of AC Type I and Type II mRNA and protein levels in frontal cortex (54,55), suggesting that lithium's complex effects on the system may represent the net effects of direct inhibition of AC, upregulation of AC subtypes, and effects on the stimulatory and inhibitory G proteins. Most recently, lithium's effects on the phosphorylation and activity of CREB have been examined in rodent brain and generally demonstrate a lithium-induced increase (56,57).

A series of studies have also examined lithium's effects on AC in humans. In a longitudinal study of healthy volunteers, two weeks of lithium administration was found to significantly increase platelet basal and postreceptor stimulated AC activity (58), effects which are strikingly similar to those observed in rodent brain. Consistent with a lithium-induced increase in basal cAMP and AC levels, a more recent study found that platelets obtained from lithium-treated euthymic bipolar enhanced basal and the cAMP-stimulated phosphorylation of Rap1 (a PKA substrate), as well as of a 38-kDa phosphoprotein (59). Somewhat surprisingly, these investigators did not find similar effects of lithium in healthy subjects.

Carbamazepine and the cAMP/PKA Signaling Cascade

Carbamazepine appears to dampen cAMP signaling. An earlier study showed that carbamazepine reduced cAMP levels in rabbit CSF (60), and attenuated basal- and stimulated-cAMP accumulation in cerebral tissue and cortical slices (61,62). Post and associates showed that carbamazepine reduced CSF cAMP levels in patients with mood disorders (63). In the follow-up studies, it was found that carbamazepine inhibits receptor- and forskolin-stimulated cAMP production in cultured cells and in membrane preparation made from cell and brain, and phosphorylation of CREB in cells, perhaps through direct interaction with adenylyl cyclase or proteins tightly associated with adenylyl cyclases (64).

Phosphoinositide Signaling Cascades and Protein Kinase C (PKC)
Phosphatidylinositol 4,5-bisphosphate [$PI(4,5)P_2$] is a minor membrane phospholipid. In response to extracellular stimulation, phosphoinositide-specific phospholipase C (PLC) catalyzes hydrolysis of $PI(4,5)P_2$ and results in the generation of two intracellular messengers, diacylglycerol (DAG) and inositol 1,4,5-trisphosphate ($I(1,4,5)P_3$) (Fig. 1) (65). PLC isozymes are divided into four types: PLCβ, PLCγ, PLCδ, and a new type of PLC, termed PLC-?. An isoform of PLC originally known as PLC-α is a proteolytic fragment of PLC-δ1. Gq family Gα proteins and G$\beta\gamma$ dimers activate PLCβ. Receptor protein tyrosine kinases and nonreceptor protein tyrosine kinases phosphorylate and activate PLCγ. GPCR stimulations can also activate PLCγ through nonreceptor protein tyrosine kinases. The molecular mechanisms underlying activations of PLCδ and PLC-? are not well established.

Once DAG is generated in response to extracellular signals, it activates protein kinase C (PKC). PKC is a large family of proteins divided into three subgroups: classic PKC (cPKC including α, βI, βII, and γ isoforms), new PKC (nPKC including ε, η, δ, θ isoform), and atypical PKC (aPKC including ζ, ι/λ isoform). cPKC isoforms, but not nPKC and aPKC isoforms, are also sensitive to Ca^{2+} ionophore. aPKC isoform does not respond to DAG, but are activated by other lipid mediators such as FFA, PS, and PIP3. PKC γ and ε are the isoforms more selectively expressed in the brain. Myristoylated alanine-rich C kinase substrate (MARCKS) is an acidic protein and the most prominent substrate for PKC in the brain.

IP3, another signal generated through PIP2 hydrolysis catalyzed by PLC, binds to the IP3 receptor, which functions as a calcium channel in the cell; binding IP3 to its receptor releases intracellular calcium reservoirs from the endoplasmic reticulum. IP3 binding sets forth downstream effects such as activation of calmodulins and calmodulin-dependent protein kinases and is then recycled back to PIP2 by the enzymes inositol monophosphatase phosphatase (IMPase, the rate limiting enzyme) and inositol polyphosphatase phosphatase (IPPase).

Lithium, at therapeutic concentration, directly inhibits IMPase and IPPase, therefore blocking conversion of IPs to inositols and consequently dampening PIP2 recycling and intracellular signaling requiring PIP2. The "inositol depletion hypothesis" posited that lithium, as an uncompetitive inhibitor IMPase, produced its therapeutic effects via a depletion of neuronal myo-inositol levels. Although this hypothesis has been of great heuristic value, numerous studies examined the effects of lithium on receptor-mediated PI responses, and although some report a reduction in agonist stimulated PIP2 hydrolysis in rat brain slices following acute or chronic lithium, these findings have often been small and inconsistent, and subject to numerous methodological differences (66,67). Most recently, a magnetic resonance spectroscopy study demonstrated that lithium-induced myo-inositol reductions are observed in the frontal cortex of BD patients after only five days of lithium administration, a time when the patients' clinical state is completely unchanged (68). Recent data suggest that valproate induces inositol depletion by blocking de novo inositol synthesis in yeast and cultured mammalian cells. These data provide additional support for the role of inositol depletion in the initiation of mood stabilizing effects. However, there is a lack of clear evidence for valproate-induced blockage of de novo inositol synthesis in the brain.

Given the predictable effects of inositol depletion on intracellular PIP2-dependent signaling, it has been argued that reducing myo-inositol levels per se is not associated with therapeutic response. This led to the working hypothesis that some of the initial actions of lithium may occur with a relative reduction of myo-inositol—that this reduction of myo-inositol initiates a cascade of secondary changes in the PKC signaling pathway and gene expression in the CNS, effects that are ultimately responsible for lithium's therapeutic efficacy.

Indeed, evidence accumulating from various laboratories clearly demonstrates that lithium, at therapeutically relevant concentrations, exerts major effects on the PKC-signaling cascade (2,67,69,70). The preponderance of the currently available data suggest that acute lithium may activate PKC, whereas chronic lithium exposure results in an attenuation of phorbol ester mediated responses, accompanied by a downregulation of PKC isozymes in the brain. Using quantitative autoradiographic techniques, it has also been demonstrated that chronic (five weeks) lithium administration results in a significant decrease in membrane-associated PKC in several hippocampal structures, most notably the subiculum and CA1 region, in the absence of any significant changes in the various other cortical and subcortical structures examined (71,72). Furthermore, immunoblotting using monoclonal anti-PKC antibodies revealed isozyme-specific decreases in PKC α and ε (which have been particularly implicated in facilitating neurotransmitter release), in the absence of significant alterations in PKCβ, PKCγ, PKCδ, or PKCζ. It is also noteworthy that exposure of immortalized hippocampal cells (67), neuroblastoma cells or PC12 cells (73) to lithium (1.0 mM) in vitro produces isozyme-selective decreases in PKCα, and (in the case of PC12 cells) PKC.

In view of lithium's significant effects on PKC as outlined above, the effects of valproate on various aspects of PKC functioning have also been investigated. It has been found that the structurally highly dissimilar agent, valproate, produces effects strikingly similar to lithium on the PKC signaling pathway (67,74). Interestingly, chronic lithium and valproate appear to regulate PKC isozymes by distinct mechanisms, with valproate's effects appearing to be largely independent of myo-inositol.

A major strategy used to investigate the downstream consequences of mood stabilizer-induced alteration in PKC isozymes is the examination of the effects of chronic lithium on endogenous PKC substrates in brain. The most prominent substrate for PKC in brain is an acidic protein, MARCKS (myristoylated alanine rich C kinase substrate), which is implicated in regulating long term neuroplastic events. Lenox and associates (1992) demonstrated that chronic lithium administration dramatically reduced MARCKS expression in hippocampus, effects that were not immediately reversed following lithium discontinuation (75). Subsequent studies carried out in immortalized hippocampal cells demonstrate that this action of chronic lithium on MARCKS regulation is dependent upon both the inositol concentration and the level of receptor-mediated activation of phosphoinositide hydrolysis (76). Recent studies provide evidence for regulation of transcription as a major site for the action of chronic lithium on MARCKS expression in brain (77). Since MARCKS may offer a specific target for pharmacologic agents targeting mood stabilizing drugs, the effects of the anticonvulsant valproate as well as lithium were studied in immortalized hippocampal cells, showing a time- and concentration-dependent reduction in MARCKS protein expression. The activity of valproate was observed within a concentration range and time course considered consistent with clinical studies of patients with bipolar disorder. Additionally, therapeutic concentrations of combined lithium and valproate induced an additive reduction in MARCKS also consistent with experimental findings that the two drugs work through different mechanisms on the PKC system, plus consistent with the clinical observation of the additivity of the two drugs in treatment responses (78).

Given that PKC signaling is a common target of lithium and valproate, and that this signaling pathway plays a pivotal role in the regulation of neuronal excitability, neurotransmitter release, and long term synaptic events, Manji and his associates postulated that the attenuation of PKC activity may mediate the antimanic effects of lithium and valproate. In a pilot study it was found that tamoxifen (a nonsteroidal antiestrogen known to be a PKC inhibitor at higher concentrations) may indeed possess antimanic efficacy (79). These data further support the potential role of the PKC pathway in mediating the pathophysiology of BD.

Phosphatidylinositol-3-Kinase (PI3K) Signaling Pathway

PI(4,5)P_2 is also an essential substrate for PI(3,4,5)P_3 production in the phosphatidylinositol-3-kinase (PI3K) signaling pathway (Fig. 1) (80). Activated receptor protein tyrosine kinases phosphorylate and activate phosphatidylinositol-3-kinase (PI3K). Class 1A PI3Ks, which are composed of heterodimers of an inhibitory adaptor/regulatory (p85α/p85β/p85γ) and a catalytic (p110α/p110β/p110δ) subunit, are initially purified from the brain tissue. PI3K catalyzes phosphorylation of PI(4,5)P_2. PI(3,4,5)P_3 facilitate the recruitment of AKTs (AKT1/AKT2/AKT3) [also known as protein kinase B (PKB)] to cell membrane. Akt is then phosphorylated by constitutively active phosphoinositide-dependent kinase 1 (PDK1) (at threonine 308 in AKT1) to stabilize the activation loop, and by PDK2 in the hydrophobic C-terminal domain (serine 473 in AKT1) for full activation. The downstream target AKT is glycogen synthase kinase-3 (GSK-3), which is phosphorylated by Akt at serine 21 of GSK-3α and serine 9 of GSK-3β, resulting in inactivation of its catalytic activity. PTEN dephosphorylates the 3 position of PtdIns(3,4,5)P_3 [and PtdIns(3,4)P_2] to reverse the action by PI3K (81).

Mood Stabilizers and PI3K Pathway

Although the inositol depletion theory predicts diminished PIP2-related intracellular signaling, recent data suggest an enhanced function of PI3K/AKT pathway by lithium and valproate treatments. Lithium is known to protect neurons against a variety of insults in cultured cells and in intact animals (82,83). The PI3K/AKT pathway is one of the signaling cascades that regulate cell survival. In search for the molecular mechanism by which lithium protects cerebellum granular cells against glutamate toxicity, Chuang and associates discovered that lithium activated the PI3K/AKT pathway. They revealed that lithium rapidly increases PI3K activity, phosphorylation of AKT1 at serine 473, and phosphorylation of GSK-3α at serine 21 (84). Soon after the initial lithium findings, Jope and associates reported that valproate also activates the PI3K/AKT pathway in cultured cells supported by coherent increases in phosphorylations of ATK1 at serine 473 and GSK-3β at serine 9 (85). Acute treatment with lithium increases phosphorylations of AKT-1 at threonin 308, but not serine 473, and of GSK-3 α and β at AKT sites (86). Studies also showed that chronic treatment with lithium and valproate increased GSK-3β serine 9 phosphorylations in frontal cortex and hippocampus (85,87), effects that are consistent with activation of the PI3K/AKT pathway by mood stabilizers. Other signaling cascades may also be involved (see below).

The PI3K/AKT pathway is also affected by antipsychotics, psychostimulants, and antidepressants; however, the behavioral and clinical relevance of these effects are unclear. It is reported that acute and chronic treatment with haloperidol increased phosphorylations of AKT1 at serine 473 and threonin 308 in whole brains of mice, and chronic treatment also increased phosphorylations of GSK-3β at serine 9 (88). Since chronic treatment with haloperidol is reported to decrease GSK-3β serine 9 phosphorylation (87), it appears these whole brain effects are not due to the changes in frontal cortex. In the same study, the investigators found that clozapine treatment increased GSK-3β serine 9 phosphorylation. Effects of psychostimulants on the AKT pathway are not consistent across studies. However, cocaine increased striatal levels of phosphorylated AKT1 at threonin 308 and seronine 485 (89), while amphetamine decreased striatal levels of phosphorylated AKT1 at threonine 308, GSK-3α at serine 21, and GSK-3β at seronine 9 (86). Imipramine treatment increased GSK-3α at serine 21 and GSK-3β at seronine 9 (90,91,92). Mice lacking functional AKT1 carry defects in both fetal and postnatal growth (93) and greater sensitivity to the sensorimotor gating-disruptive effects of amphetamine (88). However, mice with one copy of functional GSK-3β showed attenuated locomotor response to amphetamine (86). The genetic data on the AKT1 association with bipolar disorder is inconclusive (94).

The Extracellular Signal-Regulated Kinase Pathway

The extracellular signal-regulated kinase (ERK) pathway is one of the key signaling cascades mediating neurotrophic action and synaptic plasticity (Fig. 1). Neurotrophins such as nerve growth factor (NGF), brain-derived nerve growth factor (BDNF), and NT-3/4 bind to specific Trks. This binding triggers a cascade of events: Trk autophosphorylation and activation, recruitment of adaptor protein to activation site on the cell membrane, guanine nucleotide exchange and activation of small G protein Ras, Ras-facilitated dephosphorylation and phosphorylation reactions leading to Raf activation, and a chain of sequential phosphorylation-induced activations of MEK by Raf and ERK by MEK. ERK phosphorylates and activates RSK. ERK and RSK phosphorylate and thereby modulate activity of proteins with diverse functions including transcription factors, enzymes, ion channels, structure proteins, and receptors. In addition to Trk, NMDA receptors, PLC-coupled receptors, and adenylyl cyclase-coupled receptors also regulate the ERK pathway, exerting their actions on synaptic plasticity in the CNS.

The ERK pathway is activated by mood stabilizers in brain regions involved in mood regulation. Earlier works from several research groups showed that lithium and valproate induce function of transcription factors such as AP-1 and CREB and enhance expression of genes such as bcl-2 (95). These projects were led by Chen and associates who suspected that the ERK might be targeted by the mood stabilizers. They found that valproate activated the ERK pathway in cultured human neuroblastoma cells and promoted neuronal differentiation and survival of these cells (95). Later, they demonstrated that chronic treatments of rats with lithium and valproate activated the ERK pathway in prefrontal cortex and hippocampus (57,96). They showed that lithium and valproate promote neurogenesis in dentate gyrus of adult hippocampus, an effect mediated at least in part through their ERK pathway activating action (96,97). Lithium- and valproate-induced activation of the ERK pathway is also observed in other laboratories (98). ERK pathway-activating effects of carbamazepine (99), but not lamotrigine, have been reported from some studies using cells.

Studies on the effects of antidepressants on the ERK pathway have thus far yielded inconsistent results. Chronic treatment of rats with venlafaxine, a dual serotonin and norepinephrine reuptake inhibitor, did not induce significant changes on immunostaining of phosphorylated RSK in frontal cortex and hippocampus (100). Chronic treatment of rats with fluoxetine and reboxetine, a selective norepinephrine reuptake inhibitor, significantly reduced levels of phosphorylated ERK1, but not those of phosphorylated ERK2 in nuclear fraction of hippocampal tissue (101). The same treatments did not significantly alter levels of phosphorylated ERK1 and phosphorylated ERK2 in prefrontal cortex. Chronic treatment of rats with desipramine significantly elevated levels of phosphorylated ERK2 in nuclear fraction of hippocampal tissue, but decreased levels of phosphorylated ERK1 in nuclear fraction of prefrontal cortical tissue, without significant effects on levels of phosphorylated ERK1 in nuclear fraction of hippocampal tissue and levels of phospho-ERK2 in nuclear fraction of prefrontal cortical tissue (101). A recent study found that chronic treatment of rats with fluoxetine significantly reduced levels of phosphorylated ERK1 and phosphorylated ERK2 in nuclear fractions of hippocampal and frontal cortical tissues and reduced levels of phosphorylated ERK1 in hippocampal synaptosomes (102). However, chronic treatment with imipramine did not produce similar effects (102).

Studies conducted in cultured cells, and animals, show typical and atypical antipsychotics stimulate the ERK pathway. These effects, as suggested in some studies, are mediated through D2 and/or 5HT1A receptors and require pertussis toxin-sensitive Gi and Go proteins. Given that antipsychotics are highly effective in treatment of mania, it is reasonable to postulate that the ERK pathway activation may represent a convergent point of action of antimanic agents with varied chemical structures (98).

Inactivation of ERK pathway in CNS induces animal behavioral alterations reminiscent of manic symptoms; these complex behaviors likely depend on ERK's effects on discrete brain regions and the presence of other interacting molecules. U327 is a MEK inhibitor that can penetrate blood-brain barrier and attenuated ERK pathway activity. Chen and associates reported that systematic administration of U327 induced increases in locomotion and travel distance in the large open field test (57). The injection also reduced immobility and increased swimming time in the forced swim test. However, injection of its parent compound U0126, which cannot penetrate the blood-brain barrier, did not induce any significant effects on the outcomes of both tests, indicating the effects of U327 are due to inhibition of MEK in the CNS (57). Mice lacking one of two ERK subtypes are indistinguishable from wild-type mice in appearance and a variety of neurological and behavioral tests. However, these mice are hyperactive in novel environments; resistant to forced-swim induced immobility; supersensitive to rewarding property of morphine; over-engaged in wheel running, a naturally hedonic activity of rodents; and supersensitive to psychostimulant-induced locomotion (98). Direct infusion of a MEK inhibitor, or direct expression of functional null mutant ERK1 in left anterior cingulate cortex produce similar behavior types (98). Abnormal left anterior cingulate cortex in bipolar patients is a consistent finding from human brain imaging studies. The left anterior cingulate cortex is one of the brain regions repeatedly found altered in bipolar disorder in brain imaging and postmortem studies (98). These converging animal and human data support a critical role of the ERK pathway in therapeutic action of mood stabilizers. The roles of the ERK pathway in the susceptibility and pathophysiology of bipolar disorders are largely unknown.

Glycogen Synthase Kinase-3 (GSK-3)

Although it was originally identified as a key modulator of glycogen catabolism, GSK-3 plays critical roles in a variety of neuronal processes such as progenitor cell fate determination, neuronal survival and apoptosis, and synaptic plasticity (103–107). It has a wide substrate range and is a converging point of several signaling pathways described above (Fig. 1). Two GSK-3 subtypes have been identified, namely α and β, which share 97% sequence homology in their catalytic domains. Both subtypes are expressed in the brain and generally (but not always) have similar biological effects, while encoded by two separate genes. Both kinases are constitutively active. The constitutive activity arises from phosporylation of tyrosines 279 or 216 (α and β, respectively), although there is some evidence of active regulation of these sites in the brain (108). RSK, Akt, PKA, and PKC phosphorylate N-terminal serine residue of GSK-3 α and β (21 and 9, respectively) and the phosphorylation results in GSK-3 inhibition. Protein phosphatase 1 (PP1) and protein phosphatase 2A dephosphorylate GSK-3 (109–111). In addition, GSK-3 activity is regulated by binding proteins; for instance, in the Wnt signaling pathway, GSK-3-catalyzed phosphorylation of β-catenin is regulated by proteins adenomatous polyposis coli (APC), the scaffold protein Axin, and frequently rearranged in advanced T-cell lymphomas 1 (FRAT1) (106,107).

Seminal work by Klein and Melton (1996) led to the critical discovery that lithium inhibits GSK-3 activity through direct and indirect mechanisms. The initial in vitro data showed the IC50 of lithium to inhibit GSK-3 in test tube is about 1–2 or 2 mM (112,113), suggesting marginal direct inhibition of GSK-3 by lithium at therapeutically relevant serum levels of 0.6–1.2 mM. This direct inhibition is later identified to be through competition for magnesium (103,114).

The proposed indirect inhibition mechanisms include: (*i*) induction of GSK-3 phosphorylation-inactivation and (*ii*) inhibition of GSK-3 dephosphorylation-reactivation by mood stabilizers. The original work by Chuang and co-workers demonstrated that lithium activates the PI3K pathway and phosphorylations of GSK-3α at serine 21, the phosphorylations-inactivation site, in cultured cells (84). Jope et al. later demonstrated that lithium induced phosphorylation of GSK-3β at serine 9 in mouse brain after chronic treatment with lithium. In addition to the PI3K pathway, the effects of lithium on the ERK/RSK pathway and on PKC signaling could also contribute to phosphorylation-inactivation of GSK-3 (85).

Phosphatase inhibitor-2 (I-2) is an inhibitory component of PP1 complex. GSK-3 phosphorylates I-2 at threonine 72 (115), the phosphorylation alters interaction between I-2 and PP1 catalytic unit (116), resulting in increases in PP1 activity (116), and therefore PP1-induced dephosphorylation-reactivation of GSK-3 (111,117). According to this loop, an initial break on GSK-3 activity may cause further amplification of GSK-3 inhibitory signal. Although it is plausible, convincing direct evidence for amplification of GSK-3 inhibition through this feedback loop in the brain is lacking.

Chen and associates report that valproate, at therapeutic relevant concentrations, also inhibits GSK-3 activity in vitro (118). This finding has now been replicated by several, but not all, laboratories. Similar to the effects of lithium, studies also found valproate increased phosphorylation of GSK-3 at the inactivation sites (85), perhaps through its effects on PI3K (84,85) and ERK/RSK pathways (98,119). Taken together, these data suggest a therapeutic relevancy of GSK-3 inhibition in mood stabilization.

To further examine the therapeutic relevancy of mood stabilizer-induced GSK-3 inhibition, researchers evaluated the effects of GSK-3 blockage on rodent behavior. Reducing immobility in the forced swim test is one of the behavioral effects induced by acute administration of antidepressants. A recent study showed that lithium, similar to antidepressants, reduces the immobility in mice (120). GSK-3β deficient mice (120) and GSK-3 inhibitor-treated mice (121) and rats (122) exhibit similar reduction in immobility in the forced swim test, indicating GSK-3 inhibition mediates this behavioral effect of lithium. In support of such a contention, recent studies showed that antidepressant (90) and atypical antipsychotic (123), both agents known to reduce the immobility of animals in the forced swim test, increase GSK-3 phosphorylations at the inactivation site in several brain regions implicated in mood regulation. Lithium reduces holeboard exploration and amphetamine-induced locomotion in rodents (120). These phenomena are viewed as lithium-sensitive behavior related to antimanic action of this agent. GSK-3 heterozygous knockout mice showed reduced holeboard exploration (120) and rats that received GSK-3 inhibitor injection showed blunted response to amphetamine challenge (122). It appears that GSK-3 inhibition mediates antidepressant-like and antimanic agent-like behavioral effects (or mood stabilizing effect). However, it remains to be elucidated how the similar effects of mood stabilizers and antidepressants, two types of drugs with clearly different clinical profiles, especially for manic episodes of bipolar disorder, on brain GSK-3 phosphorylation at inactivation site result in different behavioral outcomes.

PROSPECTS AND FUTURE DIRECTIONS

Overall, it should be clear from the information reviewed in this chapter that bipolar disorder likely arises from abnormalities in cellular plasticity cascades, leading to aberrant information processing in synapses and circuits mediating affective, cognitive, motoric, and neurovegetative function. As we have discussed, there is a considerable body of evidence in support of abnormalities in the regulation of signaling as integral to the underlying neurobiology of bipolar disorder. Indeed, the role of cellular signaling cascades offers much explanatory power for understanding the complex neurobiology of manic-depressive illness:

- Signaling cascades regulate the multiple neurotransmitter and neuropeptide systems implicated in bipolar disorder.
- Abnormalities in cellular signaling cascades that regulate diverse physiologic functions likely explains the tremendous comorbidity with a variety of medical conditions (notably cardiovascular disease, diabetes mellitus, obesity, and migraine) and substance abuse.
- Signaling pathways are clearly major targets for hormones that have been implicated in the pathophysiology of manic-depressive illness, including gonadal steroids, thyroid hormones, and glucocorticoids.

- Cellular signal transduction cascades are clearly the targets for our most effective treatments for manic-depressive illness.
- Abnormalities in cellular plasticity cascades likely also represent the underpinnings of the impairments of structural plasticity seen in morphometric studies of bipolar disorder.

In conclusion, there have truly been tremendous advances in our understanding of the molecular and cellular underpinnings of bipolar disorder. Moreover, studies of cellular plasticity cascades in bipolar disorder are leading to a reconceptualization about the pathophysiology, course, and optimal long-term treatment of the illness. This data suggests that, while bipolar disorder is clearly not a classical neurodegenerative disease, it is in fact associated with impairments of cellular plasticity and resilience. As a consequence, there is a growing appreciation that optimal long-term treatment will likely be achieved by attempting to prevent the underlying disease progression and its attendant cellular dysfunction, rather than exclusively focusing on the treatment of signs and symptoms. There has, unfortunately, been little progress in developing truly novel drugs specifically for the treatment of manic-depressive illness, and most recent additions to the pharmacopeia are brain-penetrant drugs developed for the treatment of epilepsy or schizophrenia. This era may now be over as there are a number of pharmacologic "plasticity enhancing" strategies which may be of considerable utility in the treatment of bipolar disorder. This progress holds much promise for the development of novel therapeutics for the long term treatment of severe, refractory mood disorders, and for improving the lives of millions.

REFERENCES

1. Drevets WC. Neuroimaging studies of mood disorders. Biol Psychiatry 2000; 48(8):813–829.
2. Manji HK, Lenox RH. Signaling: cellular insights into the pathophysiology of bipolar disorder. Biol Psychiatry 2000; 48(6):518–530.
3. Manji HK, Drevets WC, Charney DS. The cellular neurobiology of depression. Nat Med 2001; 7(5):541–547.
4. Duman RS. Synaptic plasticity and mood disorders. Mol Psychiatry 2002; 7(suppl 1): S29–S34.
5. Nestler EJ, Barrot M, DiLeone RJ, Eisch AJ, Gold SJ, Monteggia LM. Neurobiology of depression. Neuron 2002; 34(1):13–25.
6. Bourne HR, Nicoll R. Molecular machines integrate coincident synaptic signals. Cell 1993; 72(suppl):65–75.
7. Young LT, Li PP, Kish SJ, Siu KP, Warsh JJ. Postmortem cerebral cortex Gs alpha-subunit levels are elevated in bipolar affective disorder. Brain Res 1991; 553(2):323–326.
8. Andreopoulos S, Li PP, Siu KP, Kish SJ, Warsh JJ. Altered CTX-catalyzed and endogenous [32P]ADP-ribosylation of stimulatory G protein alphas isoforms in postmortem bipolar affective disorder temporal cortex. J Neurosci Res 2003; 72(5):638–645.
9. Young LT, Asghari V, Li PP, Kish SJ, Fahnestock M, Warsh JJ. Stimulatory G-protein alpha-subunit mRNA levels are not increased in autopsied cerebral cortex from patients with bipolar disorder. Brain Res Mol Brain Res 1996; 42(1):45–50.
10. Wang HY, Friedman E. Enhanced protein kinase C activity and translocation in bipolar affective disorder brains. Biol Psychiatry 1996; 40(7):568–575.
11. Friedman E, Wang HY. Receptor-mediated activation of G proteins is increased in postmortem brains of bipolar affective disorder subjects. J Neurochem 1996; 67(3): 1145–1152.
12. Gonzalez-Maeso J, Rodriguez-Puertas R, Meana JJ, Garcia-Sevilla JA, Guimon J. Neurotransmitter receptor-mediated activation of G-proteins in brains of suicide

victims with mood disorders: selective supersensitivity of alpha(2A)-adrenoceptors. Mol Psychiatry 2002; 7(7):755–767.
13. Young LT, Li PP, Kish SJ, et al. Cerebral cortex Gs alpha protein levels and forskolin-stimulated cyclic AMP formation are increased in bipolar affective disorder. J Neurochem 1993; 61(3):890–898.
14. Jope RS, Song L, Li PP, et al. The phosphoinositide signal transduction system is impaired in bipolar affective disorder brain. J Neurochem 1996; 66(6):2402–2409.
15. Mathews R, Li PP, Young LT, Kish SJ, Warsh JJ. Increased G alpha q/11 immunoreactivity in postmortem occipital cortex from patients with bipolar affective disorder. Biol Psychiatry 1997; 41(6):649–656.
16. Manji HK, Potter WZ, Lenox RH. Signal transduction pathways. Molecular targets for lithium's actions. Arch Gen Psychiatry 1995; 52(7):531–543.
17. Emamghoreishi M, Schlichter L, Li PP, et al. High intracellular calcium concentrations in transformed lymphoblasts from subjects with bipolar I disorder. Am J Psychiatry 1997; 154(7):976–982.
18. Corson TW, Li PP, Kennedy JL, et al. Association analysis of G-protein beta 3 subunit gene with altered Ca(2+) homeostasis in bipolar disorder. Mol Psychiatry 2001; 6(2): 125–126.
19. Spleiss O, van Calker D, Scharer L, Adamovic K, Berger M, Gebicke-Haerter PJ. Abnormal G protein alpha(s) - and alpha(i2)-subunit mRNA expression in bipolar affective disorder. Mol Psychiatry 1998; 3(6):512–520.
20. Gilman AG. G proteins: transducers of receptor-generated signals. Annu Rev Biochem 1987; 56:615–649.
21. Avissar S, Schreiber G, Danon A, Belmaker RH. Lithium inhibits adrenergic and cholinergic increases in GTP binding in rat cortex. Nature 1988; 331(6155): 440–442.
22. Wang HY, Friedman E. Effects of lithium on receptor-mediated activation of G proteins in rat brain cortical membranes. Neuropharmacology 1999; 38(3):403–414.
23. Newman ME, Belmaker RH. Effects of lithium in vitro and ex vivo on components of the adenylate cyclase system in membranes from the cerebral cortex of the rat. Neuropharmacology 1987; 26(2–3):211–217.
24. Mork A, Geisler A. The effects of lithium in vitro and ex vivo on adenylate cyclase in brain are exerted by distinct mechanisms. Neuropharmacology 1989; 28(3): 307–311.
25. Mork A, Geisler A. Effects of GTP on hormone-stimulated adenylate cyclase activity in cerebral cortex, striatum, and hippocampus from rats treated chronically with lithium. Biol Psychiatry 1989; 26(3):279–288.
26. Mork A, Geisler A. Effects of lithium ex vivo on the GTP-mediated inhibition of calcium-stimulated adenylate cyclase activity in rat brain. Eur J Pharmacol 1989; 168(3):347–354.
27. Mork A, Geisler A, Hollund P. Effects of lithium on second messenger systems in the brain. Pharmacol Toxicol 1992; 71(suppl 1):4–17.
28. Stein MB, Chen G, Potter WZ, Manji HK. G-protein level quantification in platelets and leukocytes from patients with panic disorder. Neuropsychopharmacology 1996; 15(2): 180–186.
29. Cadd G, McKnight GS. Distinct patterns of cAMP-dependent protein kinase gene expression in mouse brain. Neuron 1989; 3(1):71–79.
30. Reiach JS, Li PP, Warsh JJ, Kish SJ, Young LT. Reduced adenylyl cyclase immunolabeling and activity in postmortem temporal cortex of depressed suicide victims. J Affect Disord 1999; 56(2-3):141–151.
31. Dowlatshahi D, MacQueen GM, Wang JF, Reiach JS, Young LT. G Protein-coupled cyclic AMP signaling in postmortem brain of subjects with mood disorders: effects of diagnosis, suicide, and treatment at the time of death. J Neurochem 1999; 73(3): 1121–1126.
32. Rahman S, Li PP, Young LT, Kofman O, Kish SJ, Warsh JJ. Reduced [3H]cyclic AMP binding in postmortem brain from subjects with bipolar affective disorder. J Neurochem 1997; 68(1):297–304.

33. Fields A, Li PP, Kish SJ, Warsh JJ. Increased cyclic AMP-dependent protein kinase activity in postmortem brain from patients with bipolar affective disorder. J Neurochem 1999; 73(4):1704–1710.
34. Chang A, Li PP, Warsh JJ. cAMP-Dependent protein kinase (PKA) subunit mRNA levels in postmortem brain from patients with bipolar affective disorder (BD). Brain Res Mol Brain Res 2003; 116(1–2):27–37.
35. Schaefer ML, Wong ST, Wozniak DF, et al. Altered stress-induced anxiety in adenylyl cyclase type VIII-deficient mice. J Neurosci 2000; 20(13):4809–4820.
36. Lee KW, Hong JH, Choi IY, et al. Impaired D2 dopamine receptor function in mice lacking type 5 adenylyl cyclase. J Neurosci 2002; 22(18):7931–7940.
37. Wei F, Qiu CS, Kim SJ, et al. Genetic elimination of behavioral sensitization in mice lacking calmodulin-stimulated adenylyl cyclases. Neuron 2002; 36(4): 713–726.
38. Cummings DE, Brandon EP, Planas JV, Motamed K, Idzerda RL, McKnight GS. Genetically lean mice result from targeted disruption of the RII beta subunit of protein kinase A. Nature 1996; 382(6592):622–626.
39. Nolan MA, Sikorski MA, McKnight GS. The role of uncoupling protein 1 in the metabolism and adiposity of RII beta-protein kinase A-deficient mice. Mol Endocrinol 2004; 18(9):2302–2311.
40. Newhall KJ, Cummings DE, Nolan MA, McKnight GS. Deletion of the RIIbeta-subunit of protein kinase A decreases body weight and increases energy expenditure in the obese, leptin-deficient ob/ob mouse. Mol Endocrinol 2005; 19(4):982–991.
41. Brandon EP, Logue SF, Adams MR, et al. Defective motor behavior and neural gene expression in RIIbeta-protein kinase A mutant mice. J Neurosci 1998; 18(10): 3639–3649.
42. Thiele TE, Willis B, Stadler J, Reynolds JG, Bernstein IL, McKnight GS. High ethanol consumption and low sensitivity to ethanol-induced sedation in protein kinase A-mutant mice. J Neurosci 2000; 20(10):RC75.
43. Huang YY, Kandel ER, Varshavsky L, et al. A genetic test of the effects of mutations in PKA on mossy fiber LTP and its relation to spatial and contextual learning. Cell 1995; 83(7):1211–1222.
44. Howe DG, Wiley JC, McKnight GS. Molecular and behavioral effects of a null mutation in all PKA C beta isoforms. Mol Cell Neurosci 2002; 20(3):515–524.
45. Reed TM, Repaske DR, Snyder GL, Greengard P, Vorhees CV. Phosphodiesterase 1B knock-out mice exhibit exaggerated locomotor hyperactivity and DARPP-32 phosphorylation in response to dopamine agonists and display impaired spatial learning. J Neurosci 2002; 22(12):5188–5197.
46. Zhang HT, Huang Y, Jin SL, et al. Antidepressant-like profile and reduced sensitivity to rolipram in mice deficient in the PDE4D phosphodiesterase enzyme. Neuropsychopharmacology 2002; 27(4):587–595.
47. Wolff J, Berens SC, Jones AB. Inhibition of thyrotropin-stimulated adenyl cyclase activity of beef thyroid membranes by low concentration of lithium ion. Biochem Biophys Res Commun 1970; 39(1):77–82.
48. Dousa T, Hechter O. Lithium and brain adenyl cyclase. Lancet 1970; 1(7651): 834–835.
49. Forn J, Valdecasas FG. Effects of lithium on brain adenyl cyclase activity. Biochem Pharmacol 1971; 20(10):2773–2779.
50. Wang JF, Asghari V, Rockel C, Young LT. Cyclic AMP responsive element binding protein phosphorylation and DNA binding is decreased by chronic lithium but not valproate treatment of SH-SY5Y neuroblastoma cells. Neuroscience 1999; 91(2): 771–776.
51. Masana MI, Bitran JA, Hsiao JK, Mefford IN, Potter WZ. Lithium effects on noradrenergic-linked adenylate cyclase activity in intact rat brain: an in vivo microdialysis study. Brain Res 1991; 538(2):333–336.
52. Masana MI, Bitran JA, Hsiao JK, Potter WZ. In vivo evidence that lithium inactivates Gi modulation of adenylate cyclase in brain. J Neurochem 1992; 59(1):200–205.
53. Wiborg O, Kruger T, Jakobsen SN. Region-selective effects of long-term lithium and carbamazepine administration on cyclic AMP levels in rat brain. Pharmacol Toxicol 1999; 84(2):88–93.

54. Colin SF, Chang HC, Mollner S, et al. Chronic lithium regulates the expression of adenylate cyclase and Gi-protein alpha subunit in rat cerebral cortex. Proc Natl Acad Sci U S A 1991; 88(23):10634–10637.
55. Jensen JB, Mork A. Altered protein phosphorylation in the rat brain following chronic lithium and carbamazepine treatments. Eur Neuropsychopharmacol 1997; 7(3): 173–179.
56. Ozaki N, Chuang DM. Lithium increases transcription factor binding to AP-1 and cyclic AMP-responsive element in cultured neurons and rat brain. J Neurochem 1997; 69(6):2336–2344.
57. Einat H, Yuan P, Gould TD, et al. The role of the extracellular signal-regulated kinase signaling pathway in mood modulation. J Neurosci 2003; 23(19): 7311–7316.
58. Risby ED, Hsiao JK, Manji HK, et al. The mechanisms of action of lithium. II. Effects on adenylate cyclase activity and beta-adrenergic receptor binding in normal subjects. Arch Gen Psychiatry 1991; 48(6):513–524.
59. Zanardi R, Racagni G, Smeraldi E, Perez J. Differential effects of lithium on platelet protein phosphorylation in bipolar patients and healthy subjects. Psychopharmacology (Berl) 1997; 129(1):44–47.
60. Myllyla VV. Effect of convulsions and anticonvulsive drugs on cerebrospinal fluid cyclic AMP in rabbits. Eur Neurol 1976; 14(2):97–107.
61. Lewin E, Bleck V. Cyclic AMP accumulation in cerebral cortical slices: effect of carbamazepine, phenobarbital, and phenytoin. Epilepsia 1977; 18(2):237–242.
62. Palmer GC, Jones DJ, Medina MA, Stavinoha WB. Anticonvulsant drug actions on in vitro and in vivo levels of cyclic AMP in the mouse brain. Epilepsia 1979; 20(2): 95–104.
63. Post RM, Ballenger JC, Uhde TW, Smith C, Rubinow DR, Bunney WE Jr. Effect of carbamazepine on cyclic nucleotides in CSF of patients with affective illness. Biol Psychiatry 1982; 17(9):1037–1045.
64. Chen G, Pan B, Hawver DB, Wright CB, Potter WZ, Manji HK. Attenuation of cyclic AMP production by carbamazepine. J Neurochem 1996; 67(5):2079–86.
65. Rhee SG. Regulation of phosphoinositide-specific phospholipase C. Annu Rev Biochem 2001; 70:281–312.
66. Jope RS, Williams MB. Lithium and brain signal transduction systems. Biochem Pharmacol 1994; 47(3):429–441.
67. Manji HK, Lenox RH. Ziskind-Somerfeld Research Award. Protein kinase C signaling in the brain: molecular transduction of mood stabilization in the treatment of manic-depressive illness. Biol Psychiatry 1999; 46(10):1328–1351.
68. Moore GJ, Bebchuk JM, Parrish JK, et al. Temporal dissociation between lithium-induced changes in frontal lobe myo-inositol and clinical response in manic-depressive illness. Am J Psychiatry 1999; 156(12):1902–1908.
69. Jope RS. Anti-bipolar therapy: mechanism of action of lithium. Mol Psychiatry 1999; 4(2):117–128.
70. Hahn CG, Friedman E. Abnormalities in protein kinase C signaling and the pathophysiology of bipolar disorder. Bipolar Disord 1999; 1(2):81–86.
71. Manji HK, Etcheberrigaray R, Chen G, Olds JL. Lithium decreases membrane-associated protein kinase C in hippocampus: selectivity for the alpha isozyme. J Neurochem 1993; 61(6):2303–2310.
72. Chen G, Masana MI, Manji HK. Lithium regulates PKC-mediated intracellular cross-talk and gene expression in the CNS in vivo. Bipolar Disord 2000; 2(3 Pt 2): 217–36.
73. Li X, Jope RS. Selective inhibition of the expression of signal transduction proteins by lithium in nerve growth factor-differentiated PC12 cells. J Neurochem 1995; 65(6): 2500–2508.
74. Chen G, Manji HK, Hawver DB, Wright CB, Potter WZ. Chronic sodium valproate selectively decreases protein kinase C alpha and epsilon in vitro. J Neurochem 1994; 63(6):2361–2364.
75. Lenox RH, Watson DG, Patel J, Ellis J. Chronic lithium administration alters a prominent PKC substrate in rat hippocampus. Brain Res 1992; 570(1–2):333–340.

76. Watson DG, Lenox RH. Chronic lithium-induced down-regulation of MARCKS in immortalized hippocampal cells: potentiation by muscarinic receptor activation. J Neurochem 1996; 67(2):767–777.
77. Wang L, Watson DG, Lenox RH. Myristoylation alters retinoic acid-induced down-regulation of MARCKS in immortalized hippocampal cells. Biochem Biophys Res Commun 2000; 276(1):183–188.
78. Watson DG, Watterson JM, Lenox RH. Sodium valproate down-regulates the myristoylated alanine-rich C kinase substrate (MARCKS) in immortalized hippocampal cells: a property of protein kinase C-mediated mood stabilizers. J Pharmacol Exp Ther 1998; 285(1):307–316.
79. Bebchuk JM, Arfken CL, Dolan-Manji S, Murphy J, Hasanat K, Manji HK. A preliminary investigation of a protein kinase C inhibitor in the treatment of acute mania. Arch Gen Psychiatry 2000; 57(1):95–97.
80. Hennessy BT, Smith DL, Ram PT, Lu Y, Mills GB. Exploiting the PI3K/AKT pathway for cancer drug discovery. Nat Rev Drug Discov 2005; 4(12):988–1004.
81. Cantley LC, Neel BG. New insights into tumor suppression: PTEN suppresses tumor formation by restraining the phosphoinositide 3-kinase/AKT pathway. Proc Natl Acad Sci USA 1999; 96(8):4240–245.
82. Chuang DM. The Antiapoptotic Actions of Mood Stabilizers: Molecular Mechanisms and Therapeutic Potentials. Ann N Y Acad Sci 2005; 1053:195–204.
83. Bachmann RF, Schloesser RJ, Gould TD, Manji HK. Mood stabilizers target cellular plasticity and resilience cascades: implications for the development of novel therapeutics. Mol Neurobiol 2005; 32(2):173–202.
84. Chalecka-Franaszek E, Chuang DM. Lithium activates the serine/threonine kinase Akt-1 and suppresses glutamate-induced inhibition of Akt-1 activity in neurons. Proc Natl Acad Sci USA 1999; 96(15):8745–8750.
85. De Sarno P, Li X, Jope RS. Regulation of Akt and glycogen synthase kinase-3 beta phosphorylation by sodium valproate and lithium. Neuropharmacology 2002; 43(7):1158–1164.
86. Beaulieu JM, Sotnikova TD, Yao WD, et al. Lithium antagonizes dopamine-dependent behaviors mediated by an AKT/glycogen synthase kinase 3 signaling cascade. Proc Natl Acad Sci USA 2004; 101(14):5099–5104.
87. Kozlovsky N, Amar S, Belmaker RH, Agam G. Psychotropic drugs affect Ser9-phosphorylated GSK-3beta protein levels in rodent frontal cortex. Int J Neuropsychopharmacol 2005:1–6.
88. Emamian ES, Hall D, Birnbaum MJ, Karayiorgou M, Gogos JA. Convergent evidence for impaired AKT1-GSK3beta signaling in schizophrenia. Nat Genet 2004; 36(2):131–137.
89. Brami-Cherrier K, Valjent E, Garcia M, Pages C, Hipskind RA, Caboche J. Dopamine induces a PI3-kinase-independent activation of Akt in striatal neurons: a new route to cAMP response element-binding protein phosphorylation. J Neurosci 2002; 22(20):8911–8921.
90. Roh MS, Eom TY, Zmijewska AA, De Sarno P, Roth KA, Jope RS. Hypoxia activates glycogen synthase kinase-3 in mouse brain in vivo: protection by mood stabilizers and imipramine. Biol Psychiatry 2005; 57(3):278–286.
91. Maragnoli ME, Fumagalli F, Gennarelli M, Racagni G, Riva MA. Fluoxetine and olanzapine have synergistic effects in the modulation of fibroblast growth factor 2 expression within the rat brain. Biol Psychiatry 2004; 55(11):1095–1102.
92. Chen MJ, Russo-Neustadt AA. Exercise activates the phosphatidylinositol 3-kinase pathway. Brain Res Mol Brain Res 2005; 135(1–2):181–193.
93. Cho H, Thorvaldsen JL, Chu Q, Feng F, Birnbaum MJ. Akt1/PKBalpha is required for normal growth but dispensable for maintenance of glucose homeostasis in mice. J Biol Chem 2001; 276(42):38349–38352.
94. Toyota T, Yamada K, Detera-Wadleigh SD, Yoshikawa T. Analysis of a cluster of polymorphisms in AKT1 gene in bipolar pedigrees: a family-based association study. Neurosci Lett 2003; 339(1):5–8.
95. Yuan PX, Huang LD, Jiang YM, Gutkind JS, Manji HK, Chen G. The mood stabilizer valproic acid activates mitogen-activated protein kinases and promotes neurite growth. J Biol Chem 2001; 276(34):31674–31683.

96. Hao Y, Creson T, Zhang L, et al. Mood stabilizer valproate promotes ERK pathway-dependent cortical neuronal growth and neurogenesis. J Neurosci 2004; 24(29): 6590–6599.
97. Chen G, Rajkowska G, Du F, Seraji-Bozorgzad N, Manji HK. Enhancement of hippocampal neurogenesis by lithium. J Neurochem 2000; 75(4):1729–1734.
98. Chen G, Manji HK. The extracellular signal-regulated kinase pathway: an emerging promising target for mood stabilizers. Curr Opin Psychiatry 2006; 19(3):313–323.
99. Mai L, Jope RS, Li X. BDNF-mediated signal transduction is modulated by GSK3beta and mood stabilizing agents. J Neurochem 2002; 82(1):75–83.
100. Khawaja XZ, Storm S, Liang JJ. Effects of venlafaxine on p90Rsk activity in rat C6-gliomas and brain. Neurosci Lett 2004; 372(1–2):99–103.
101. Tiraboschi E, Tardito D, Kasahara J, et al. Selective phosphorylation of nuclear CREB by fluoxetine is linked to activation of CaM kinase IV and MAP kinase cascades. Neuropsychopharmacology 2004; 29(10):1831–1840.
102. Fumagalli F, Molteni R, Calabrese F, Frasca A, Racagni G, Riva MA. Chronic fluoxetine administration inhibits extracellular signal-regulated kinase 1/2 phosphorylation in rat brain. J Neurochem 2005; 93(6):1551–1560.
103. Gurvich N, Klein PS. Lithium and valproic acid: parallels and contrasts in diverse signaling contexts. Pharmacol Ther 2002; 96(1):45–66.
104. Jope RS. Lithium and GSK-3: one inhibitor, two inhibitory actions, multiple outcomes. Trends Pharmacol Sci 2003; 24(9):441–443.
105. Jope RS, Johnson GV. The glamour and gloom of glycogen synthase kinase-3. Trends Biochem Sci 2004; 29(2):95–102.
106. Gould TD. Targeting glycogen synthase kinase-3 as an approach to develop novel mood-stabilising medications. Expert Opin Ther Targets 2006; 10(3):377–392.
107. Gould TD, Manji HK. Glycogen synthase kinase-3: a putative molecular target for lithium mimetic drugs. Neuropsychopharmacology 2005; 30(7):1223–1237.
108. Bhat RV, Shanley J, Correll MP, et al. Regulation and localization of tyrosine216 phosphorylation of glycogen synthase kinase-3beta in cellular and animal models of neuronal degeneration. Proc Natl Acad Sci USA 2000; 97(20):11074–11079.
109. Bennecib M, Gong CX, Grundke-Iqbal I, Iqbal K. Role of protein phosphatase-2A and -1 in the regulation of GSK-3, cdk5 and cdc2 and the phosphorylation of tau in rat forebrain. FEBS Lett 2000; 485(1):87–93.
110. Tanji C, Yamamoto H, Yorioka N, Kohno N, Kikuchi K, Kikuchi A. A-kinase anchoring protein AKAP220 binds to glycogen synthase kinase-3beta (GSK-3beta) and mediates protein kinase A-dependent inhibition of GSK-3beta. J Biol Chem 2002; 277(40): 36955–36961.
111. Zhang F, Phiel CJ, Spece L, Gurvich N, Klein PS. Inhibitory phosphorylation of glycogen synthase kinase-3 (GSK-3) in response to lithium. Evidence for autoregulation of GSK-3. J Biol Chem 2003; 278(35):33067–33077.
112. Klein PS, Melton DA. A molecular mechanism for the effect of lithium on development. Proc Natl Acad Sci USA 1996; 93(16):8455–8459.
113. Stambolic V, Ruel L, Woodgett JR. Lithium inhibits glycogen synthase kinase-3 activity and mimics wingless signalling in intact cells. Curr Biol 1996; 6(12): 1664–1668.
114. Ryves WJ, Harwood AJ. Lithium inhibits glycogen synthase kinase-3 by competition for magnesium. Biochem Biophys Res Commun 2001; 280(3):720–725.
115. Park IK, Roach P, Bondor J, Fox SP, DePaoli-Roach AA. Molecular mechanism of the synergistic phosphorylation of phosphatase inhibitor-2. Cloning, expression, and site-directed mutagenesis of inhibitor-2. J Biol Chem 1994; 269(2):944–954.
116. Park IK, DePaoli-Roach AA. Domains of phosphatase inhibitor-2 involved in the control of the ATP-Mg-dependent protein phosphatase. J Biol Chem 1994; 269(46): 28919–28928.
117. Szatmari E, Habas A, Yang P, Zheng JJ, Hagg T, Hetman M. A positive feedback loop between glycogen synthase kinase 3beta and protein phosphatase 1 after stimulation of NR2B NMDA receptors in forebrain neurons. J Biol Chem 2005; 280(45):37526–37535.
118. Chen G, Huang LD, Jiang YM, Manji HK. The mood-stabilizing agent valproate inhibits the activity of glycogen synthase kinase-3. J Neurochem 1999; 72(3):1327–1330.

119. Wang QM, Guan KL, Roach PJ, DePaoli-Roach AA. Phosphorylation and activation of the ATP-Mg-dependent protein phosphatase by the mitogen-activated protein kinase. J Biol Chem 1995; 270(31):18352–18358.
120. O'Brien WT, Harper AD, Jove F, et al. Glycogen synthase kinase-3beta haploinsufficiency mimics the behavioral and molecular effects of lithium. J Neurosci 2004; 24(30):6791–6798.
121. Kaidanovich-Beilin O, Milman A, Weizman A, Pick CG, Eldar-Finkelman H. Rapid antidepressive-like activity of specific glycogen synthase kinase-3 inhibitor and its effect on beta-catenin in mouse hippocampus. Biol Psychiatry 2004; 55(8):781–784.
122. Gould TD, Einat H, Bhat R, Manji HK. AR-A014418, a selective GSK-3 inhibitor, produces antidepressant-like effects in the forced swim test. Int J Neuropsychopharmacol 2004; 7(4):387–390.
123. Li X, Rosborough KM, Friedman AB, Zhu W, Roth KA. Regulation of mouse brain glycogen synthase kinase-3 by atypical antipsychotics. Int J Neuropsychopharmacol 2006:1–13.

7 Searching for a Cellular Endophenotype for Bipolar Disorder

Francine M. Benes
Department of Psychiatry, Harvard Medical School, Boston, Massachusetts, U.S.A.

INTRODUCTION

A critical question regarding the etiopathogenesis of bipolar disorder is whether there is a cellular endophenotype that can explain abnormalities at the level of dysfunctional neurons and circuits. A recent study has suggested that apoptosis could play a role in the pathophysiology of bipolar disorder. Is it possible that this abnormality is related to an endophenotype for this disorder? Fundamental differences in the genetic regulation of the apoptotic cascade, electron transport chain, and antioxidation enzymes are present and uniquely different from those seen in schizophrenia. These differences may reflect the cellular endophenotype for bipolar disorder and are reviewed below.

Postmortem evidence gathered over the past 20 years has suggested that bipolar disorder and, to a lesser extent, schizophrenia may involve apoptotic cell death (1,2). Several cell counting studies demonstrated a reduction of interneurons in the anterior cingulate cortex of schizophrenics, although this change has consistently shown a stronger covariation with affective disorder (3,4,5). Subjects with bipolar disorder, like those with schizophrenia, show decreases in the expression of mRNA for GAD65 (6,7) and GAD67 (7,8). Additionally, presynaptic axon terminals containing GAD65 have also been found to be significantly reduced in the anterior cingulate cortex of patients with bipolar disorder (9), suggesting that neuronal cell death may be a feature of this disorder. The idea that bipolar disorder involves cell death to a greater degree than schizophrenia is counterintuitive, because the latter disorder involves a characteristic deterioration in function that is not generally seen in affective disorder. This contrasts strikingly with the episodic nature of bipolar disorder and the characteristic return to normal baseline functioning following a manic or depressive episode.

Based on postmortem evidence, a reduction of GABAergic interneurons appears to be a more striking feature of bipolar disorder than schizophrenia (5). On this basis, it was postulated that apoptotic cell death may play a role in the pathophysiology of bipolar disorder to a greater extent than in schizophrenia (1,2). Interestingly, schizophrenics show a marked reduction of single-stranded DNA breaks, a marker for apoptosis, in the anterior cingulate cortex, whereas bipolars do not show this change (10). This pattern is potentially explained by the activation of a DNA "repair" mechanism that can offset the effects of the apoptotic cascade (11).

Research aimed at elucidating the underlying neurobiology and genetics of bipolar disorder, and factors associated with treatment response, have been limited by a heterogeneous clinical phenotype and lack of knowledge about its underlying diathesis (12). An important concept regarding the etiopathogenesis of bipolar disorder and schizophrenia that must be factored into this discussion involves the concept of the endophenotype, defined as the "measurable components unseen by the unaided eye along the pathway between disease and distal genotype" (13). Endophenotypes are quantifiable components in the genes-to-behaviors pathways, distinct from psychiatric symptoms, which make genetic and biological studies of etiologies for disease categories more manageable (14). It is no surprise then that the endophenotype concept has emerged as a strategic tool in neuropsychiatric research. The question that is explored in the discussion that follows is whether unique endophenotypes can be identified in bipolar disorder and schizophrenia using gene expression profiling data. The potential strength of this approach is the fact that databases obtained for these studies typically contain information for as many as 20,000–30,000 different genes that can be organized according to functionally relevant biopathways and/or clusters.

To explore this and other questions regarding the presence of a cellular endophenotype for bipolar disorder, gene expression profiling (GEP) has been used as a powerful screening tool, for the broad evaluation of molecular functions in subjects with bipolar disorder.

THE STUDY OF APOPTOSIS IN HUMAN HIPPOCAMPUS

Briefly, the cohort used in this study consisted of normal controls, schizophrenics, and bipolars that were matched for age and postmortem interval and have been previously described in detail (15). The normal control, schizophrenic, and bipolar subjects were reasonably well-matched for age, postmortem interval, gender, and hemisphere. All of the schizophrenic subjects were treated with neuroleptic agents and all of the bipolar subjects were treated with lithium carbonate and/or other mood stabilizing agents, including lithium carbonate, valproic acid, carbamazepine, and clonazepam, during the year prior to death.

In analyzing the data, initial tests of significance did not show appreciable changes in the apoptosis cascade in either group (16). Since the human hippocampus is composed of many different subregions, sublaminae, and cellular subtypes, it is relatively difficult to detect subtle changes in the expression of genes when whole extracts are analyzed, as they were in this study. To avoid Type II statistical errors, that is, suggesting that there were no significant changes, when in fact they might exist within discrete subregions, layers, or cellular subtypes within the hippocampus, we developed a post hoc method for analyzing gene expression profiling data that would minimize the risk of such errors. It is becoming broadly recognized that gene expression profiling and other genomics technology in the hunt for disease genes requires that we develop methods for evaluating associations between multiple and complicated factors (17) and assessing the reliability of microarray data (18,19). To overcome these limitations, we used a post hoc analysis (20) of the microarray data from bipolars and schizophrenics that evaluated functional clusters of genes using a low stringency approach (21).

An ad hoc metric was developed based on a combination of probability theory and two separate corrections for multiple comparisons. A requirement for such an analysis is that changes in the hybridization of any particular gene to its probe sets must occur independently of probe sets for other genes. If the probe sets for all of the genes in one particular biopathway were all clustered within a discrete sector of a microarray, then the hybridization of the respective mRNAs to their appropriate probe sets could not be assumed to be independent of one another and probability theory could not be used for the analysis.

To analyze the data, it was also important to examine changes in gene expression at the functional cellular level, so that a coherent understanding of the data could be obtained. In order to identify biologically relevant clusters of interrelated genes, GenMapp algorithms (www.genmapp.org) were used to relate the dChip findings to several different biochemical pathways and/or biologically related clusters of genes. The first stage of the analysis involved selecting genes according to high and low stringency criteria for inclusion in the GenMapp biopathways analysis. The inclusionary criterion was initially set at $p = 0.05$ and this was progressively increased to 0.1, 0.15, 0.2, and 0.25, respectively. When the inclusionary criterion was increased in a stepwise manner to $p = 0.25$, further increases in the number of genes entered into the GenMapp pathways were observed. Beyond this p-value, some of the GenMapp pathways that previously had not shown any activation began to show a random scatter of a small number of genes. This suggested that further increases of the inclusionary p-value would likely increase the background and compromise the signal-to-noise ratio of the post hoc analysis. When the post hoc composite probability (Pc) was calculated for the apoptotic cascade, it was 3×10^{-27} for bipolars and 4×10^{-9} for schizophrenics (Fig. 1).

As shown in Figure 1, there were striking differences in the changes in the expression of genes associated with apoptosis in schizophrenics versus bipolars. Some genes showed changes in the same direction in both groups, while most others occurred in opposite directions. For example, while caspase 2 was upregulated in both groups, the antiapoptotic factor, Bcl-2, showed decreased expression in bipolars, but increased expression in the schizophrenic group. The remainder of the apoptosis genes showed fundamental differences in regulation in the two disorders.

For the bipolars, 24 out of a total of 44 genes in the apoptosis pathways satisfied the low stringency criterion for inclusion in the analysis. As depicted in Figure 1, there were several upregulated proapoptotic genes, including FAS ligand, FAS receptor (22,23) perforin (24), TNFα (23,25), c-Jun (26), c-myc (27), BAK (28), APAF-1 and caspases 2 (29) and 8 (23,30). Other genes that are thought to inhibit apoptosis, such as TRAF1, IKK, IAP3, NF-kB (31), and Bcl-2 (32) also showed increased expression in the bipolar group, but these changes would tend to counteract the influence of the 10 upregulated proapoptotic genes, particularly when other key proapoptotic factors, such as JNKK and JNK (33) were found to be downregulated. The DNA repair enzyme, poly(adenosine diphosphate-ribosyl) polymerase, [PARP (11)], also showed a decrease in regulation and this would tend to increase the apoptotic potential of hippocampal cells in the bipolar group.

When the bipolars were broken down according to neuroleptic exposure (Table 1), mRNA expression for the proapoptotic factors, FAS ligand, RIP, BID, TRAF1, FADD, MDM-2, caspase 2, p53, and c-myc, as well as the antiapoptotic factors, NIK, IKK, IAP3, were all increased in the drug naïve bipolars, while the proapoptotic factors JNKK and JNK, as well as the antiapoptotic factors, IAP2, NF-kB-p105, and PARP, all showed decreased expression. The neuroleptic-free bipolar subjects also showed a decreased expression of PARP (34). For the neuroleptic-treated bipolars, proapoptotic factors, such as perforin, TNF-α, caspase 6, c-Jun, BAX, APAF-1, and caspase 2, all showed increased expression. Conversely, antiapoptotic factors, such as IKK, NF-kB-p105, NF-kB-p65, MCL-1, Bcl-2, and

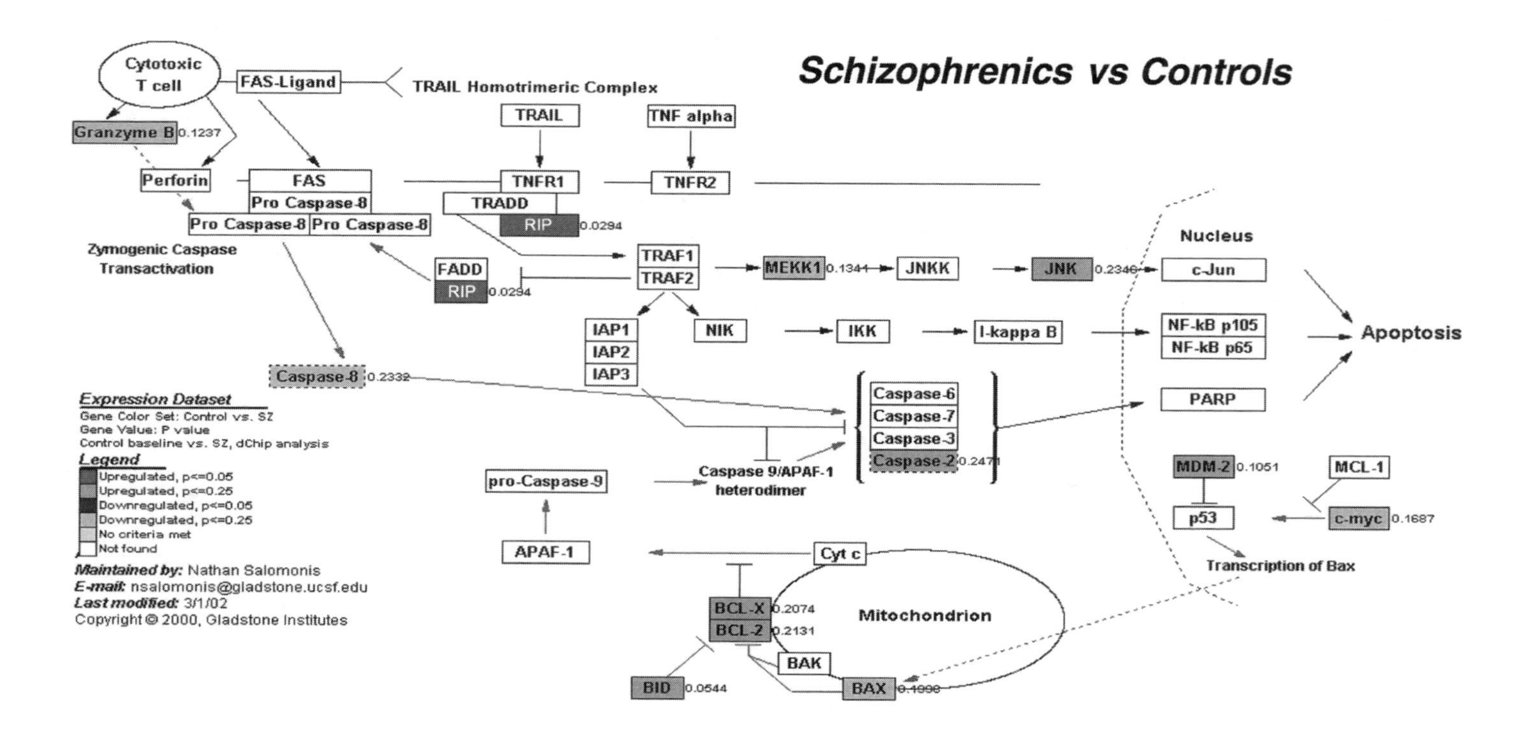
Schizophrenics vs Controls
Cytotoxic T cell
FAS-Ligand
TRAIL Homotrimeric Complex
Granzyme B 0.1237
TRAIL
TNF alpha
Perforin
FAS
Pro Caspase-8
TNFR1
TNFR2
TRADD
Pro Caspase-8
Pro Caspase-8
RIP 0.0294
Zymogenic Caspase Transactivation
TRAF1
TRAF2
FADD
RIP 0.0294
MEKK1 0.1341
JNKK
JNK 0.2346
Nucleus
c-Jun
IAP1
IAP2
IAP3
NIK
IKK
I-kappa B
NF-kB p105
NF-kB p65
Apoptosis
Caspase-8 0.2332
Caspase-6
Caspase-7
Caspase-3
Caspase-2 0.2471
PARP
Expression Dataset
Gene Color Set: Control vs. SZ
Gene Value: P value
Control baseline vs. SZ, dChip analysis
Legend
Upregulated, p<=0.05
Upregulated, p<=0.25
Downregulated, p<=0.05
Downregulated, p<=0.25
No criteria met
Not found
pro-Caspase-9
Caspase 9/APAF-1 heterodimer
MDM-2 0.1051
MCL-1
p53
c-myc 0.1687
APAF-1
Cyt c
Transcription of Bax
Maintained by: Nathan Salomonis
E-mail: nsalomonis@gladstone.ucsf.edu
Last modified: 3/1/02
Copyright © 2000, Gladstone Institutes
BCL-X 0.2074
BCL-2 0.2131
Mitochondrion
BAK
BID 0.0544
BAX

Apoptosis

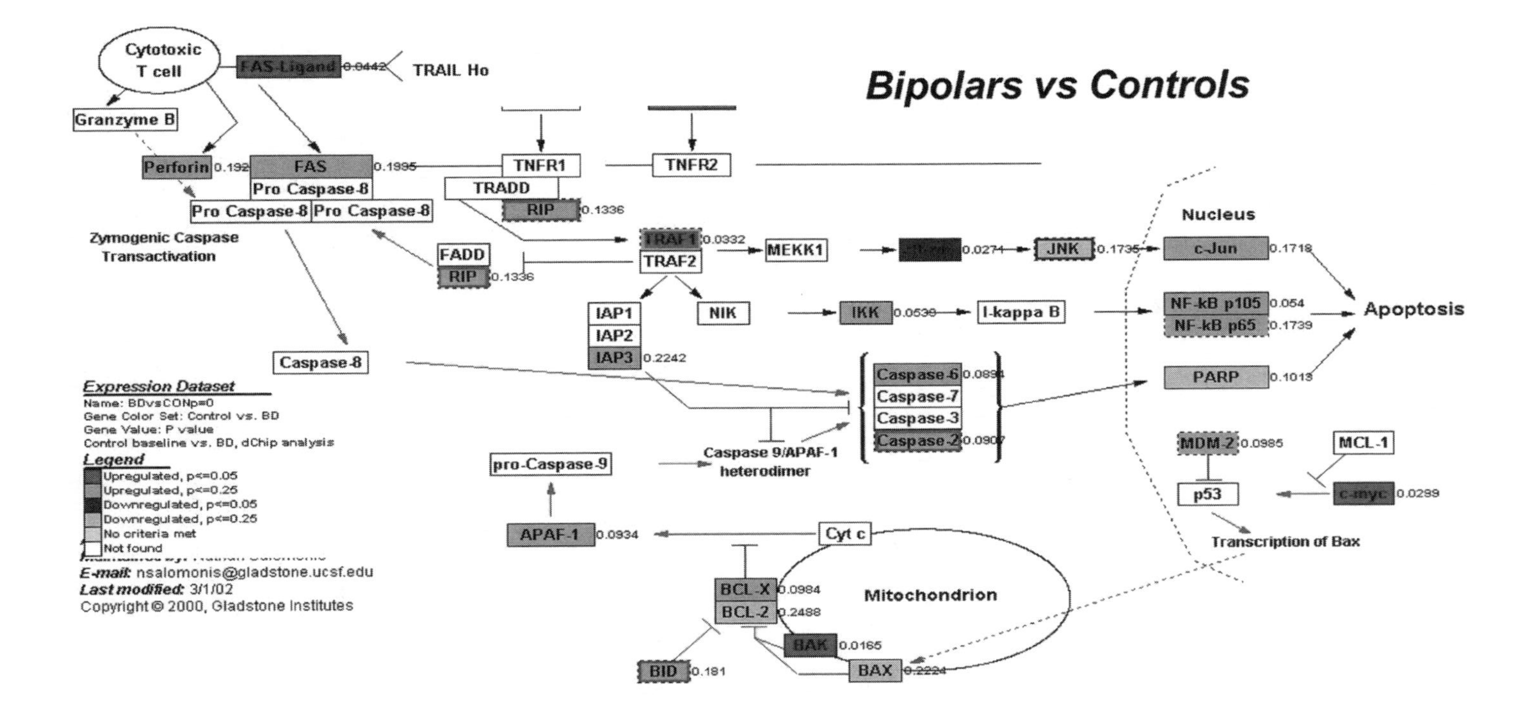

FIGURE 1 A set of diagrams showing the apoptosis pathways, schizophrenics versus normal controls (*upper*) and bipolar disorder versus normal controls (*lower*). The genes show either an upregulation (*dark gray*) or downregulation (*light gray*) and the intensity of the color varies according to whether the inclusion criterion for a particular gene was $p = 0.05$ (high stringency) or $p = 0.25$ (low stringency). The p-values are shown to the right for each gene. Using low stringency criteria, the apoptosis pathways for the bipolar subjects show 19 out of 44 genes with changes in expression. Of these 14 are upregulated and only five are downregulated. For the schizophrenics, eight genes were upregulated and five were downregulated. In both cases, pivotal genes at the beginning of the apoptosis cascade show changes in regulation. Overall, the bipolars show an increase of proapoptotic changes not seen in the schizophrenics.

TABLE 1 Proapoptotic and Prosurvival Changes in Gene Expression in Schizophrenics and Bipolars with and Without Neuroleptic Exposure

	Proapoptotic	Antiapoptotic	Total
Bipolars			
Neuroleptic-free	12 (66.7%)	6 (33.3%)	18
Neuroleptic-treated	14 (56.0%)	11 (44.0%)	25
Schizophrenics			
Low neuroleptic	9 (47.4%)	10 (52.6%)	19
High neuroleptic	10 (76.9%)	3 (23.1%)	13

Data represent the number of genes showing changes in expression in relation to neuroleptic drug exposure. *Proapoptotic* refers to genes associated with facilitation of the apoptosis cascade and cell death (e.g., BAX and APAF-1) that showed an increase of expression, or genes associated with an inhibition of apoptosis (e.g., Bcl-2 and MDM-2) that showed a decrease of expression. *Antiapoptotic* refers to an increased expression of genes that inhibit apoptosis and a decreased expression of genes that facilitate apoptosis.

Bcl-x, were all upregulated in the bipolars receiving neuroleptic. As shown in Table 1, there was an overall increase of antiapoptotic changes in gene expression in the bipolars, suggesting that neuroleptics may suppress apoptotic cell death in this disorder. The potential effect of mood-stabilizers on changes in apoptosis gene expression was also considered; however, all of the bipolar subjects were actively treated with these agents at the time of death.

In the schizophrenic group, significant changes in the expression of genes associated with apoptosis were also observed (post hoc $Pc = 4.3 \times 10^{-9}$), but there were differences with respect to the specific genes affected and the direction of the changes (Fig. 1). Four proapoptotic genes (RIP, BID, JNK, and caspase 2), and two prosurvival genes (Bcl-x and Bcl-2) showed an overall increase of expression. Other proapoptotic genes, such as granzyme B (24,35), caspase 8 (30,36), MEKK1 (27,37), and c-myc (27), showed decreased expression in the schizophrenic group. These latter changes would tend to suppress the apoptotic potential of hippocampal cells, as these four factors are believed to play a critical role in facilitating the progression of apoptosis. When the schizophrenic group was broken down according to "low" (CPZ-equivalent dose <500 mg/day; average $= 189 \pm 173$ mg/day) and "high" (CPZ-equivalent dose >500 mg/day; average $= 816 \pm 290$ mg/day) dose neuroleptic exposure, the proapoptotic factors Fas ligand, perforin, TRAIL, caspase 8, MEKK1, p53, c-myc, BAX, and BAK showed decreased expression in schizophrenics receiving low dose neuroleptic (Table 1). Some prosurvival genes, such as TRAF1 and MDM-2, showed decreased expression, whereas others, such as BCL-2, showed increased expression. In the "high" dose subgroup, the pattern observed was quite different. Several proapoptotic genes, including perforin, TRAIL, RIP, TNFα, caspase 2, and BAK, were upregulated, whereas anti-apoptotic genes, such as IAP3, MCL-1, and PARP, were downregulated. Unlike the subjects with bipolar disorder, the subjects with schizophrenia showed no difference in the number of genes showing proapoptotic changes in expression. On the other hand, the number of genes showing antiapoptotic changes in expression was markedly reduced, suggesting that neuroleptics might have some ability to increase apoptotic potential in hippocampal cells of schizophrenics. The DNA repair enzyme, PARP, showed a decrease of expression in the schizophrenics treated with high dose neuroleptic and this change could potentially increase the amount of DNA fragmentation present in these subjects (10). This observation further supports the view that the decrease of DNA damage in schizophrenic is probably not due to a neuroleptic effect (10).

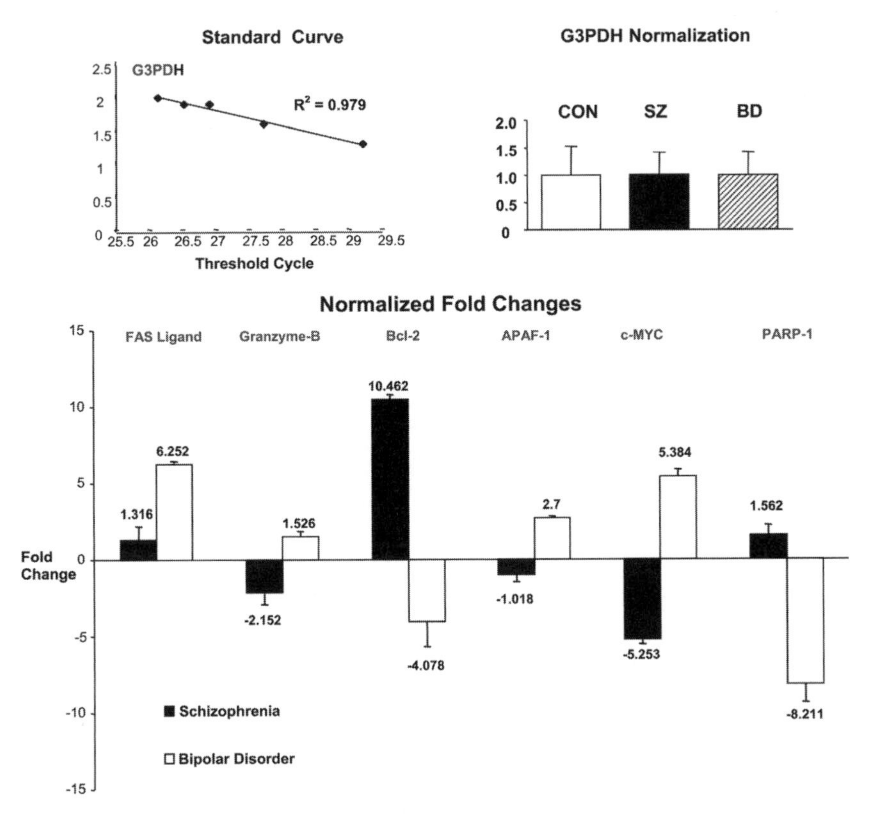

FIGURE 2 FRET-based quantitative RT-PCR validation of microarray analyses of apoptosis genes in normal controls, schizophrenics, and bipolars. Standard curves for log concentration versus threshold cycle, like that shown for G3PDH, were established for each of the genes evaluated with qRT-PCR. In addition, mRNA for glyceraldehyde-3-phosphate dehydrogenase (G3PDH), a "housekeeping" gene, was similar for the three groups (*upper right hand*) and this gene was used to normalize the data for the target genes (*lower tier*). As predicted from the microarray data, mRNA for c-MYC was markedly decreased in schizophrenics, but increased in bipolars. In contrast, PARP was strikingly reduced in the bipolars, while schizophrenics showed no differences.

As shown in Figure 2, the microarray data for several apoptotic genes, including FAS ligand, granzyme B, c-myc, PARP, BAK, Bcl-2, and APAF-1, were validated using qRT-PCR. In each case, the direction of change in expression for the control, schizophrenic, and bipolar groups were consistent with that seen using the microarray approach, except that the magnitude of these differences were generally much greater. For example, in the case of c-myc, the fold changes were −5.2 and +5.4, respectively, in the schizophrenics and bipolars. Although there was no change in PARP expression in the schizophrenics when neuroleptic effects were not considered, the bipolars showed a −8.2 fold decrease when compared to either the schizophrenic or control groups. Overall, the magnitude of the differences between the groups that were detected with qRT-PCR was much larger than those obtained with the microarrays and provided an important validation of the microarray results.

Numerous earlier studies from this laboratory had suggested that oxidative stress might occur in bipolar disorder and schizophrenia (38). Excessive amounts

TABLE 2 Comparison of Expression Profiling Results for Genes Associated with Antioxidant Reactions in Normal Controls vs. Bipolars and Schizophrenics

Bipolar disorder			
Lactoperoxidase	U39573	1.09	0.064
Heme oxygenease 1	Z82244	1.14	0.131
Superoxide dismutase 1	X02317	−1.21	0.081
Catalase	AL030579	−1.17	0.124
Glutathione peroxidase 2 (gastrointestinal)	X53463	1.11	0.041
Glutathione peroxidase 4 (phospholipid hydroperoxidase)	X71973	−1.23	0.014
Glutathione peroxidase 1	X13710	−1.14	0.134
Glyoxalase 1	NM006708	−1.32	0.014
Microsomal glutathione S-transferase 3	AF026977	−1.39	0.002
Hydroxyacyl glutathione hydrolase	X90999	−1.36	0.028
Glutathione S-transferase A4	AF025887	−1.24	0.046
Glutathione S-transferase M3 (brain)	AF043105	−1.19	0.198
Esterase D/formylglutathione hydrolase	AF112219	−1.31	0.044
Glutathione synthetase	U34683	−1.23	0.015
Glutathione S-transferase A2	M16594	1.16	0.045
Glutathione S-transferase M5	L02321	1.14	0.032
Glutathione-S-transferase like; glutathione transferase omega	U90313	−1.25	0.021
C-terminal PDZ domain ligand of neuronal nitric oxide synthase	AB007933	−1.12	0.090
Human neuronal nitric oxide synthase (*NOS1*) gene, exon 29, and complete cds	U17326	1.17	0.030
Human inducible nitric oxide synthase gene, promoter and exon 1	D29675	1.13	0.083
Nitric oxide synthase 3 (endothelial cell)	M93718	1.23	0.059
Schizophrenia			
Glutathione synthetase	U34683	−1.18	0.050

of glutamatergic activity acting upon GABAergic interneurons in both disorders (8) could potentially cause oxidative stress to these and other neuronal cells (9). In this setting, the antioxidation system, consisting of several different enzymes (Table 2) that are capable of clearing reactive oxygen species (ROS), could potentially be activated. If this occurs, it would tend to offset the effects of the apoptotic cascade that is driven by the accumulation of ROS intracellularly. As shown in Table 2, it is notable that several such genes were significantly downregulated in the bipolars, while the schizophrenics showed very little change in this system. Together with a highly significant reduction in the expression of the mitochondrial electron transport chain (39) that results in a decline of energy production, the reduced ability to clear ROS from cells in the hippocampus of bipolars could contribute to the activation of the apoptosis pathway in these subjects (40).

CONCLUSIONS

This study reports the results of a novel post hoc analysis of an extant microarray database, together with GenMapp biopathways and clusters, to obtain a more inclusive understanding of how complex aspects of transduction, signaling, and metabolism may be altered in schizophrenia and bipolar disorder. Indeed, with this new methodology, it has been possible to detect marked changes in the regulation of genes associated with the apoptosic cascade and would not have been detected, given the relative insensitivity of the standard approaches that employ an alpha level of p = 0.05 (39). The more sensitive analysis described above has revealed robust changes in the expression of apoptosis genes in hippocampal cells in both schizophrenic and bipolar subjects. Although there is some overlap in the genes showing differences in expression when compared to normal controls, the preponderance of such changes has been found to be remarkably different in schizophrenics when compared to bipolars. The hypothesis that apoptosis may play a role in the pathophysiology of schizophrenia and bipolar disorder can be traced to earlier microscopic studies in the anterior cingulate cortex (41,42) and hippocampus (43) that suggested a loss of interneurons occurs in both disorders. These changes were found to be far more striking in bipolars (30–35% reduction) when compared to schizophrenics (12–15% reduction) and it was postulated that there may be a more marked activation of apoptosis in affective disorder than in schizophrenia. A subsequent study demonstrated that there was a paradoxical reduction in the amount of DNA damage present in the anterior cingulate cortex of schizophrenic subjects (10). Indeed, the findings reported here are consistent with this latter observation, as there was a downregulation of several proapoptotic genes, such as granzyme B, caspase 8, c-myc, and BAX. Based on the results reported here, it seems more likely that GABA cell pathology in schizophrenia (4) may by related to an disturbance of intracellular signaling pathways, rather than to overt cell loss, as appears to be the case in bipolar disorder (34,44).

A noteworthy aspect of these findings is the observation that neuroleptic exposure was associated with a decrease of apoptotic potential in bipolars, but an increase in schizophrenics. Both typical (45) and atypical (46) antipsychotic drugs have been found to promote cell survival, although the atypical agents may be more effective in this regard (47). Indeed, several atypical antipsychotic drugs appear to protect against DNA damage (47). To date, only one neuroleptic, the typical agent perphenazine, has been associated with increased DNA fragmentation (48), although one study has reported that clozapine may act as a hapten and increase inflammatory potential (49). Contrary to the latter report, clozapine has also been found to activate Akt (50), a prosurvival factor that, in turn, inhibits glycogen synthase kinase-3β (GSK-3β), a protein that drives intracellular signaling toward cell death via the Wnt pathway. Lithium carbonate, a standard mood-stabilizing agent, is also believed to inhibit GSK-3β (51). Both lithium and valproate have also been found to be associated with increased expression of Bcl-2 (52,53) and ultimately influence both signal transduction (51,54,55) and intracellular signaling cascades that are fundamental to cell survival. Accordingly, the upregulation of antiapoptotic genes in neuroleptic-treated bipolar subjects may reflect the fact that all of these subjects were treated with mood-stabilizing agents during the year prior to death.

IS THERE A CELLULAR ENDOPHENOTYPE IN THE HIPPOCAMPUS OF BIPOLAR BRAIN?

Overall, the data reported here do support the hypothesis that there are fundamentally different patterns of gene expression in subjects with bipolar disorder when contrasted with those with schizophrenia (Fig. 3). These findings point to the possibility that the activity of the apoptotic cascade and L-type calcium channels are increased, while that for the electron transport chain and the

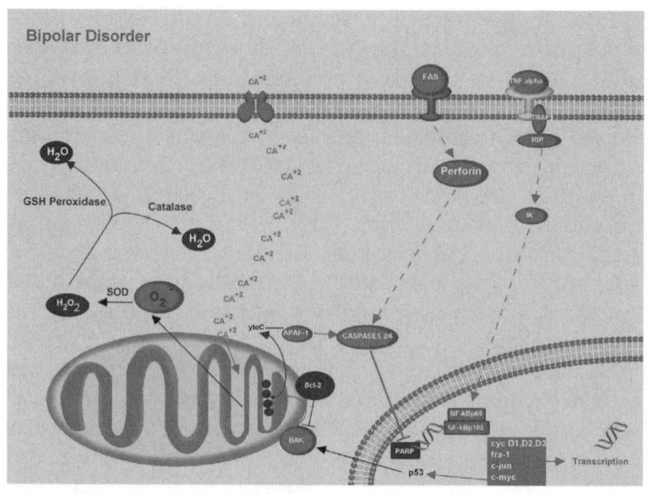

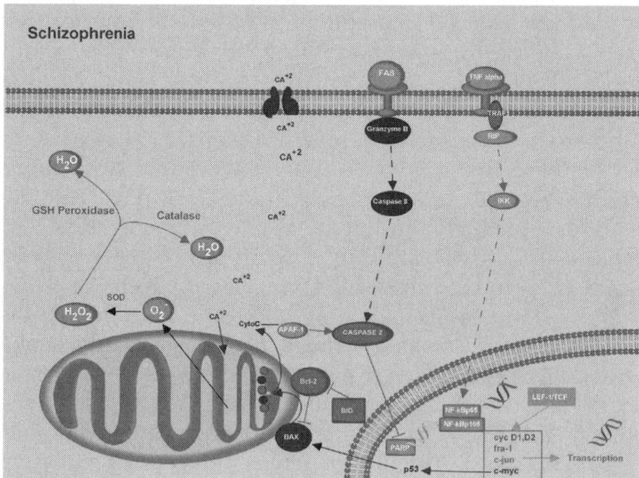

FIGURE 3 Schematic diagrams depicting genes associated with the apoptosis cascade that show changes in expression in bipolar disorder (*upper*) and schizophrenia (*lower*). In schizophrenia, as postulated in an earlier study (10), there is an overall downregulation of the apoptosis cascade, including granzyme B and caspase 8, in the cell periphery and the mitochondrial-associated antiapoptotic factor, Bcl-2. In bipolars, proapoptotic factors, including two death ligands, FAS and TNFα, as well as the FAS receptor, perforin, c-myc, and BAK are proapoptotic factors that are upregulated. NF-κBp65 and p105 are also upregulated, but they may play a role in stimulating the activity of the DNA repair enzyme, PARP-1, that is downregulated. Overall, the changes in bipolars would promote apoptotic injury or death, while those in schizophrenics would promote cell survival. Preliminary evidence has suggested that a voltage gated L-type calcium channel is downregulated in schizophrenics and upregulated in bipolars, a pattern consistent with the idea that oxidative stress may be contributing to the changes seen in the respective disorders.

anti-oxidation system are decreased in subjects with bipolar disorder. This distinctly proapoptotic pattern is in contradistinction to that seen in schizophrenics that is antiapoptotic in nature. It is important to emphasize that the interpretation of simultaneous changes in the expression of many genes that comprise a complex signaling pathway, such as apoptosis, is probably not straightforward, as it is quite possible that such changes may not be additive in nature, as the post-translational modification of their respective proteins may not result in corresponding changes.

Given this caveat, however, the detection of differences in the regulation of proapoptosis changes in gene expression in bipolar disorder and detection of the opposite changes in schizophrenia may provide some insight into the differential nature of cellular endophenotypes in the respective two disorders. What is currently lacking is a detailed understanding of how these changes in gene expression are determined at the cellular level. Are there genetic or epigenetic mechanisms involved in the control of apoptosis that are mutated? Is it possible that the neuronal activity mediated by these neurons contained within the hippocampus plays a role in determining the expression of apoptotic genes? Can environmental factors mediated via central mechanisms within the brain contribute to the abnormal regulation of apoptosis genes in bipolar subjects (56)? What is clear is that the answers to these questions, by their very nature, will bring us closer to understanding the nature of a cellular endophenotype in bipolar patients.

ACKNOWLEDGMENTS

This work was supported by grants from the National Institutes of Health (MH42261, MH62822, MH/NS31862) and the generosity of Menachem and Carmella Abraham, John and Virginia Taplin, and Anne Allen.

REFERENCES

1. Margolis RL, Chuang DM, Post RM. Programmed cell death: implications for neuropsychiatric disorders. Biol Psychiatry 1994; 35:946–956.
2. Jarskog LF, Gilmore JH, Selinger ES, Lieberman JA. Cortical bcl-2 protein expression and apoptotic regulation in schizophrenia. Biol Psychiatry 2000; 48:641–650.
3. Benes FM, McSparren J, Bird ED, Vincent SL, SanGiovanni JP. Deficits in small interneurons in prefrontal and anterior cingulate cortex of schizophrenic and schizoaffective patients. Arch Gen Psychiat 1991; 48:996–1001.
4. Benes FM, Berretta S. GABAergic interneurons: implications for understanding schizophrenia and bipolar disorder. Neuropsychopharmacology 2001; 25:1–27.
5. Todtenkopf MS, Vincent SL, Benes FM. A cross-study meta-analysis and three-dimensional comparison of cell counting in the anterior cingulate cortex of schizophrenic and bipolar brain. Schizophr Res 2005; 73:79–89.
6. Guidotti A, Auta J, Davis JM, et al. Decrease in Reelin and glutamate acid decarboxylase$_{67}$ (GAD$_{67}$) expression in schizophrenia and bipolar disorder. Arch Gen Psychiatry 2000; in press.
7. Heckers S, Stone D, Walsh J, Shick J, Koul P, Benes FM. Differential hippocampal expression of glutamic acid decarboxylase 65 and 67 messenger RNA in bipolar disorder and schizophrenia. Arch Gen Psychiatry 2002; 59:521–529.
8. Woo TU, Walsh JP, Benes FM. Density of glutamic acid decarboxylase 67 messenger RNA-containing neurons that express the *N*-methyl-D-aspartate receptor subunit NR2A in the anterior cingulate cortex in schizophrenia and bipolar disorder. Arch Gen Psychiatry 2004; 61:649–657.
9. Benes FM, Todtenkopf MS, Logiotatos P, Williams M. Glutamate decarboxylase(65)-immunoreactive terminals in cingulate and prefrontal cortices of schizophrenic and bipolar brain. J Chem Neuroanatomy 2004; 20:259–269.
10. Benes FM, Walsh J, Bhattacharyya S, Sheth A, Berretta S. DNA fragmentation decreased in schizophrenia but not bipolar disorder. Arch Gen Psychiatry 2003; 60:359–364.

11. Bouchard VJ, Rouleau M, Poirier GG. PARP-1, a determinant of cell survival in response to DNA damage. Exp Hematol 2003; 31:446–454.
12. Hasler G, Drevets WC, Gould TD, Gottesman, II, Manji HK. Toward constructing an endophenotype strategy for bipolar disorders. Biol Psychiatry 2006; 60:93–105.
13. Gottesman II, Gould TD. The endophenotype concept in psychiatry: etymology and strategic intentions. Am J Psychiatry 2003; 160:636–645.
14. Gould TD, Gottesman II. Psychiatric endophenotypes and the development of valid animal models. Genes Brain Behav 2006; 5:113–119.
15. Konradi C, Eaton M, MacDonald ML, Walsh J, Benes FM, Heckers S. Molecular evidence for mitochondrial dysfunction in bipolar disorder. Arch Gen Psychiatry 2004; 61:300–308.
16. Benes FM, Matzilevich D, Burke RE, Walsh J. The expression of proapoptosis genes is increased in bipolar disorder, but not in schizophrenia. Mol Psychiatry 2006; 11:241–251.
17. Tsunoda T, Yamada R, Tanaka T, Ohnishi Y, Kamatani N. Environmental factor dependent maximum likelihood method for association study targeted to personalized medicine. Genome Informatics 2000; 11:96–105.
18. Asyali MH, Shoukri MM, Demirkaya O, Khabar KSA. Assessment of reliability of microarray data and estimation of signal thresholds using mixture modeling. Nucleic Acids Res 2004; 32:2323–2335.
19. Raffelsberger W, Dembele D, Neubauer MG, Gottardis MM, Gronemeyer H. Quality indicators increase the reliability of microarray data. Genomics 2002; 80:385–394.
20. Benes FM, Burke RE, Walsh JP, Berretta S, Minns M, Konradi C. An upregulation of multiple monoamine and peptide G-coupled protein receptors in rat hippocampus in response to amygdalar activation. Molecular Psychiatry 2004; in press.
21. Konradi C, Eaton M, Walsh J, Benes FM, Heckers S. Molecular evidence for mitochondrial dysfunction in bipolar disorder. Archives of General Psychiatry 2004; in press.
22. Shin SW, Park JW, Suh MH, Suh SI, Choe BK. Persistent expression of FAS/FASL mRNA in the mouse hippocampus after a single NMDA injection. J Neurochem 1998; 71:1773–1776.
23. Hu WH, Johnson H, Shu HB. Activation of NF-κB by FADD, Casper, and caspase-8. J Biol Chem 2000; 275:10838–10844.
24. Ohara T, Morishita T, Suzuki H, Masaoka T, Ishii H. Perforin and granzyme B of cytotoxic T lymphocyte mediate apoptosis irrespective of Helicobacter pylori infection: possible act as a trigger of peptic ulcer formation. Hepatogastroenterology 2003; 50:1774–1779.
25. Thome M, Hofmann K, Burns K, et al. Identification of CARDIAK, a RIP-like kinase that associates with caspase-1. Curr Biol 1998; 8:885–888.
26. Mielke K, Brecht S, Dorst A, Herdegen T. Activity and expression of JNK1, p38 and ERK kinases, c-Jun N-terminal phosphorylation, and c-jun promoter binding in the adult rat brain following kainate-induced seizures. Neuroscience 1999; 91:471–483.
27. Alarcon-Vargas D, Tansey WP, Ronai Z. Regulation of c-myc stability by selective stress conditions and by MEKK1 requires aa 127–189 of c-myc. Oncogene 2002; 21:4384–4391.
28. Viktorsson K, Ekedahl J, Lindebro MC, et al. Defective stress kinase and BAK activation in response to ionizing radiation but not cisplatin in a non-small cell lung carcinoma cell line. Exp Cell Res 2003; 289:256–264.
29. Ferrer I, Lopez E, Blanco R, Rivera R, Krupinski J, Marti E. Differential c-Fos and caspase expression following kainic acid excitotoxicity. Acta Neuropathol (Berl) 2000; 99:245–256.
30. Northington FJ, Ferriero DM, Martin LJ. Neurodegeneration in the thalamus following neonatal hypoxia-ischemia is programmed cell death. Dev Neurosci 2001; 23:186–191.
31. Micheau O, Lens S, Gaide O, Alevizopoulos K, Tschopp J. NF-κB signals induce the expression of c-FLIP. Mol Cell Biol 2001; 21:5299–5305.
32. Adams JM, Cory S. The BCL-2 protein family: arbiters of cell survival. Science 1998; 281:1322–1326.
33. Yang DD, Kuan CY, Whitmarsh AJ, Rincon M, Zheng TS, Davis RJ, Rakic P, Flavell RA. Absence of excitotoxicity-induced apoptosis in the hippocampus of mice lacking the JNK3 gene. Nature 1997; 389:865–870.
34. Buttner EN, Bhattacharyya S, Walsh JP, Benes FM (2004) DNA fragmentation is increased in non-GABAergic neurons in bipolar disorder. Submitted.

35. Sun J, Bird CH, Thia KY, Matthews AY, Trapani JA, Bird PI. Granzyme B encoded by the commonly-occurring human RAH allele retains pro-apoptotic activity. J Biol Chem 2001; 279(17):16907–16911.
36. Wang CY, Guttridge DC, Mayo MW, Baldwin AS Jr. NF-κB induces expression of the BCL-2 homologue A1/Bfl-1 to preferentially suppress chemotherapy-induced apoptosis. Mol Cell Biol 1999; 19:5923–5929.
37. Boldt S, Weidle UH, Kolch W. The kinase domain of MEKK1 induces apoptosis by dysregulation of MAP kinase pathways. Exp Cell Res 2003; 283:80–90.
38. Benes FM. The role of stress and dopamine-GABA interactions in the vulnerability for schizophrenia. J Psychiatr Res 1997; 31:257–275.
39. Benes FM, Burke RE, Berretta S, Walsh JP, Minns M, Konradi C. Amygdalar activation induces an upregulation of multiple monoamine and peptide G-coupled protein receptors occurs in rat hippocampus. Molecular Psychiatry 2004; in press.
40. Pollack M, Leeuwenburgh C. Apoptosis and aging: role of the mitochondria. J Gerontol A Biol Sci Med Sci 2001; 56:B475–B482.
41. Benes FM, McSparren J, Bird ED, SanGiovanni JP, Vincent SL. Deficits in small interneurons in prefrontal and cingulate cortices of schizophrenic and schizoaffective patients. Arch Gen Psychiatry 1991; 48:996–1001.
42. Benes FM, Vincent SL, Todtenkopf M. The density of pyramidal and nonpyramidal neurons in anterior cingulate cortex of schizophrenic and bipolar subjects. Biol Psychiatry 2001; 50:395–406.
43. Benes FM, Kwok EW, Vincent SL, Todtenkopf MS. A reduction of nonpyramidal cells in sector CA2 of schizophrenics and manic depressives [see comments]. Biol Psychiatry 1998; 44:88–97.
44. Woo TW, Walsh JP, Benes FM. Density of glutamate acid decarboxylase 67 messenger RNA-containing neurons that express the *N*-methyl-D-aspartate subunit NR2a is decreased in the anterior cingulate cortex in schizophrenia and bipolar disorder. Arch Gen Psychiatry 2004.
45. Achour A, Lu W, Arlie M, Cao L, Andrieu JM. T cell survival/proliferation reconstitution by trifluoperazine in human immunodeficiency virus-1 infection. Virology 2003; 315:245–258.
46. Wei Z, Bai O, Richardson JS, Mousseau DD, Li XM. Olanzapine protects PC12 cells from oxidative stress induced by hydrogen peroxide. J Neurosci Res 2003; 73:364–368.
47. Qing H, Xu H, Wei Z, Gibson K, Li XM. The ability of atypical antipsychotic drugs vs. haloperidol to protect PC12 cells against MPP^+-induced apoptosis. Eur J Neurosci 2003; 17:1563–1570.
48. Gil-ad I, Shtaif B, Shiloh R, Weizman A. Evaluation of the neurotoxic activity of typical and atypical neuroleptics: relevance to iatrogenic extrapyramidal symptoms. Cell Mol Neurobiol 2001; 21:705–716.
49. Haack MJ, Bak ML, Beurskens R, Maes M, Stolk LM, Delespaul PA. Toxic rise of clozapine plasma concentrations in relation to inflammation. Eur Neuropsychopharmacol 2003; 13:381–385.
50. Kang UG, Seo MS, Roh MS, Kim Y, Yoon SC, Kim YS. The effects of clozapine on the GSK-3-mediated signaling pathway. FEBS Lett 2004; 560:115–119.
51. Li X, Bijur GN, Jope RS. Glycogen synthase kinase-3β, mood stabilizers, and neuroprotection. Bipolar Disord 2002; 4:137–144.
52. Manji HK, Moore GJ, Chen G. Clinical and preclinical evidence for the neurotrophic effects of mood stabilizers: implications for the pathophysiology and treatment of manic-depressive illness. Biol Psychiatry 2002; 48:740–754.
53. Manji HK, Moore GJ, Rajkowska G, Chen G. Neuroplasticity and cellular resilience in mood disorders. Mol Psychiatry 2002; 5:578–593.
54. Manji HK, Etcheberrigaray R, Chen G, Olds JL. Lithium decreases membrane-associated protein kinase C in hippocampus: selectivity for the alpha isozyme. J Neurochem 1993; 61:2303–2310.
55. Chen G, Masana MI, Manji HK. Lithium regulates PKC-mediated intracellular cross-talk and gene expression in the CNS in vivo. Bipolar Disord 2000; 2:217–236.
56. Benes FM. Emerging principles of altered neural circuitry in schizophrenia. Brain Res Brain Res Rev 2000; 31:251–269.

8 The Hypothalamic–Pituitary–Adrenal Axis in Bipolar Disorder

David J. Bond and Allan H. Young
Mood Disorders Centre of Excellence, University of British Columbia, Vancouver, British Columbia, Canada

INTRODUCTION

One of the most consistent findings in biological psychiatry is derangement of the hypothalamic-pituitary-adrenal (HPA) axis in patients with severe mood disorders. The HPA axis consists of the hypothalamus, pituitary, and adrenal glands, various hormones and releasing factors, and regulatory neural inputs (1). It regulates the body's acute response to stress, and its actions in this regard include mobilizing energy reserves through increased gluconeogenesis, lipolysis, and protein degradation (2). It also plays a role in long-term adaptive changes to physiological functions, for example, by modulating immune responses, facilitating learning, and activating the sympathetic nervous system (1). The purpose of this review is to examine the physiology of the HPA axis; outline the evidence for its implication in the pathogenesis of depression and bipolar disorder; summarize treatment options that may be efficacious in restoring normal HPA axis functioning, and thereby effective in treating mood disorders; and to suggest avenues for further research.

PHYSIOLOGY OF THE HPA AXIS

The hypothalamus receives input from afferent catecholaminergic nerve fibers, including noradrenergic fibers originating in the nucleus of the solitary tract and the locus coeruleus (3), as well as from limbic structures, including the amygdala, the hippocampus, and the stria terminalis. The medial parvicellular division of the paraventricular nucleus of the hypothalamus produces corticotropin-releasing hormone (CRH), a 41-amino acid peptide, and arginine vasopressin (AVP), a nonapeptide, which are released into the portal circulation. CRH interacts with the CRH1 receptor on the anterior surface of the pituitary gland, stimulating the release of adrenocortcotropic hormone (ACTH) from the corticotropes of the anterior pituitary. There is some evidence that AVP plays a role as well, such that CRH and AVP act synergistically in regulating ACTH release (3). ACTH enters the systemic circulation and binds to receptors on the adrenal cortex, inducing the synthesis and release of the steroid hormone cortisol (4). Once released into the serum, approximately 80% of cortisol molecules are bound to cortisol-binding globulin (5). Cortisol itself plays a vital role in regulating the HPA axis, through a negative feedback loop in which it interacts with receptors in the pituitary and hypothalamus to inhibit the synthesis and release of CRH and ACTH.

The actions of cortisol, including negative feedback, are mediated through two distinct cytoplasmic corticosteroid receptors, referred to as the type I or mineralocorticoid receptor (MR), and the type II or glucocorticoid receptor (GR) (1). The mineralocorticoid receptor has a high affinity for endogenous corticosteroids, and regulates cortisol levels during normal physiological conditions, such as normal circadian fluctuations in hormone levels. The glucocorticoid receptor has a high affinity for synthetic corticosteroids such as dexamethasone, but its affinity for cortisol is approximately one-tenth that of the MR (6). It is believed to mediate cortisol activity during periods when high concentrations are present, such as during the stress response (2).

The glucocorticoid receptor is present in almost all somatic cells, though the relative number of receptors varies between cell types. The receptor resides in the cytoplasm in an inactive form, as part of a complex with numerous other peptides, including heat shock proteins, phosphatase PP5, and immunophilins FKBP5 and Cyp40. When a substrate, such as cortisol or dexamethasone, binds to the ligand-binding domain at the carboxy-terminal end of the glucocorticoid receptor, the receptor complex is activated, and is translocated into the cell nucleus (2). Two projections at the centre of the receptor molecule allow the complex to bind to DNA. On binding, the receptor complex undergoes an allosteric change, permitting a second receptor complex to bind and dimerization of the complexes to occur. Once dimerization is complete, the cortisol-receptor complex induces transcriptional *trans*-activation or repression of selected genes, via interaction of two domains, an *N*-terminal glucocorticoid-independent domain and a C-terminal glucocorticoid-dependent domain, with glucocorticoid receptor elements in promoter regions (2).

The functional integrity of the HPA axis may be assessed by challenging it with synthetic or natural hormones or glucocorticoids. The dexamethasone suppression test (DST) measures serum cortisol levels following administration of the synthetic glucocorticoid dexamethasone. In subjects with an intact HPA axis, dexamethasone acts through the negative feedback loop outlined above to inhibit the release of CRH, ACTH, and cortisol. Escape from cortisol suppression is indicative of an abnormally functioning HPA axis. Recently, a refined version of this test, the combined dexamethasone/CRH test (7) has become widely utilized. The Dex/CRH test involves administering a dose of human CRH following pretreatment with dexamethasone. In patients with mood disorders, dexamethasone fails to prevent a substantial release of ACTH and cortisol following CRH administration. The Dex/CRH test has greater sensitivity in detecting abnormalities in the HPA axis in patients with mood disorders compared to the DST (8).

HPA FUNCTION IN PATIENTS WITH MAJOR DEPRESSION

Most studies investigating alterations in HPA functioning in patients with mood disorders have been carried out in patients with major depressive disorder. As they have been important in guiding subsequent studies in bipolar patients, and in understanding possible biological mechanisms of action, they are reviewed briefly here. Abnormalities in the HPA axis were reported in depressed patients more than 50 years ago (9). The most widely reported finding, that a significant percentage of patients with mood disorders hypersecrete cortisol, and do not suppress cortisol production in response to administration of dexamethasone, a state akin to a chronic sustained stress response, was first described 30 years ago (10).

Since this initial work, numerous studies have demonstrated other abnormalities in HPA axis functions. CRH levels have been reported as elevated in patients with major depression (11,12). The Dex/CRH test has revealed that depressed patients do not inhibit ACTH production to exogenous administration of CRH stimulus following pre-treatment with dexamethasone (11,13). There is less consistency in findings from studies of baseline levels of ACTH in depressed patients, with some reporting elevated levels of ACTH (14–18), and others describing normal or low ACTH levels (19–25). Enlarged pituitary and adrenal glands have been reported in studies of depressed patients (26–29), though these may be a state phenomenon, as adrenal gland volume has been reported to return to normal following remission of depression (30). Evidence for increased cortisol production includes elevated levels in serum, urine, and cerebrospinal fluid.

One theory to explain the discrepant findings regarding ACTH levels in depressed patients suggests that the physiology of hypercortisolism changes over the course of depressive illness (4,31). According to this hypothesis, elevated CRH levels lead to increased production of ACTH early in the course of depression. Over time, however, the sensitivity of the pituitary gland to CRH attenuates, and ACTH production decreases. As ACTH release drops, elevated levels of cortisol are maintained due to increased adrenal cortical sensitivity to ACTH. Supporting evidence for this theory includes the observations that depressed patients with hypercortisolemia display attenuated release of ACTH when administered exogenous CRF (30,32–34) and that the greatest attenuation of ACTH release occurs in the most severely depressed patients (23). As well, it has been demonstrated that administration of exogenous ACTH leads to a long-lasting hyper-responsiveness to ACTH in healthy individuals (35,36).

Successful treatment of depression is generally, though not invariably, associated with normalization of HPA axis functioning. Lack of resolution of HPA axis abnormalities in successfully treated patients has been shown to be predictive of a poor prognosis (37–39). Such patients in one study had a six-fold greater risk of relapse in the first six months after discharge from hospital (39). Recurrence of HPA abnormalities during remission from depression is similarly predictive of relapse (37), although whether this indicates a causal link, or simply that HPA axis abnormalities are one of the first signs of relapse, is not clear.

HPA FUNCTION IN PATIENTS WITH BIPOLAR DISORDER

Evidence of hypercortisolemia and other HPA axis abnormalities have been repeatedly demonstrated in all phases of bipolar illness. A review of 17 studies of the DST in manic patients reported an average nonsuppression rate of 39% across studies (40). Elevated baseline serum cortisol levels (41), nonsuppression of cortisol on the DST (42–45) and elevated cortisol release during the Dex/CRH test (46) have been frequently reported in bipolar depression. Rybakowski and Twardowska (46) also described a positive correlation between cortisol levels following CRH administration and severity of depression. While the relative rates of HPA axis hyperactivity in the manic and depressive phases of bipolar disorder have not been systematically investigated, one small study utilizing the DST in rapid cycling patients (47) detected higher levels of urinary free cortisol during depression than mania.

Elevated cortisol levels (48,49) and nonsuppression of cortisol in response to dexamethasone (41,50,51) have also been reported in patients experiencing mixed manic episodes. Studies that have compared patients with mixed mania to "pure" manic patients (40,48,50) have consistently reported greater HPA axis abnormalities in patients with mixed episodes, though possibly due to small sample sizes, the differences were generally not statistically significant. One study (48) did report significantly greater morning plasma cortisol, post-dexamethasone plasma cortisol, and CSF cortisol in patients with mixed mania compared to those with pure mania.

Two small studies (total $N = 11$) have reported on HPA axis functioning in bipolar patients with a rapid cycling course, with conflicting results. Watson et al. (52) found that cortisol response to the Dex/CRH was stable over two visits in their sample, and was independent of mood state. Tomitaka et al. (47) found greater cortisol levels during the DST in the depressed than the manic phase of illness.

The debate over whether HPA axis abnormalities in bipolar patients are dependent on illness state, or are trait features and so indicative of underlying pathophysiologic processes, has received some attention in the literature. A recent report on 53 bipolar subjects (8) found that remitted subjects and those with ongoing depressive symptoms had similarly enhanced cortisol response to the Dex/CRH test compared to normal controls. Another study (53) reported only partial normalization of cortisol response to the Dex/CRH test following remission from a manic episode. Cortisol levels during remission in this sample were still greater than those in normal controls. Interestingly, during serial testing of euthymic bipolar patients over four years (54), most patients displayed intermittently positive DSTs. This finding could be consistent with either the influence of state factors, or an inherently oscillating biological process. Other studies have reported elevated cortisol responses to the Dex/CRH test (55) and elevated salivary cortisol levels (56) in first degree relatives of bipolar patients. Finally, expression of glucocorticoid receptor alpha mRNA was noted to be reduced in affectively ill and remitted bipolar patients, as well as their first degree relatives, compared to a control group with no personal or family history of psychiatric illness (57). Thus, while some of the results are preliminary, the preponderance of evidence appears to suggest that HPA axis abnormalities in bipolar patients persist into euthymia, and are present to some extent in unaffected family members of bipolar probands.

Some evidence points toward distinct patterns of HPA axis abnormalities in major depressive disorder and bipolar disorder. For example, while a number of studies report a decreased density of brain GR receptors in both conditions, the affected brain regions appear to be unique to the illnesses (see subsequently). As well, reductions in GR receptors may be more pronounced in patients with major depression. A direct comparison of lymphocyte GR numbers between patients with major depressive disorder and bipolar manic patients described significantly lower mean GR numbers in the depressed patients (58). Depressed or euthymic bipolar patients were not studied, so the possibility that the difference was related to mood state in the bipolar patients cannot be ruled out. Interestingly, nonsuppression of cortisol during the DST appears to be more common in bipolar illness than in major depressive disorder (46,53). Finally, pituitary gland volume, which has been demonstrated to be increased in patients with major depression (27), was reported in one study to be reduced in bipolar patients compared to both depressed patients and healthy controls (59).

As DST nonsuppression has been described in many axis I psychiatric disorders, including major depressive disorder, schizophrenia, and eating disorders

(60), its usefulness as a diagnostic instrument for bipolar illness is limited. Nonetheless, as in patients with major depressive disorder, persistence of HPA axis abnormalities into remission in bipolar patients may have prognostic significance. In one study, Vieta et al. (61) administered 100 micrograms of CRH to 42 lithium-treated bipolar I patients in remission, and 21 healthy controls. The bipolar patients displayed higher baseline and peak ACTH concentrations than the control subjects. A higher area under the ACTH concentration curve in bipolar subjects predicted relapse at both 6 and 12 months. When bipolar subjects who subsequently relapsed were removed from the analysis, the difference between bipolar subjects and controls with respect to ACTH levels disappeared.

POSSIBLE MECHANISMS OF HPA AXIS ABNORMALITIES IN PATIENTS WITH MOOD DISORDERS

Research on the mechanism of HPA axis abnormalities in patients with major depressive disorder has focused on alterations in GR number or functioning (2). Research investigating whether the number of CNS or peripheral GR receptors is decreased in depressed patients has been contradictory (62), with some studies suggesting a decreased number of receptors (58,63–66), while others have not found differences between depressed patients and normal controls, or patients with other psychiatric illnesses (67–74). Studies in patients with bipolar disorder have also produced inconsistent results. One study of lymphocyte GR numbers (75) actually reported an increase in bipolar patients compared to healthy controls. Most other studies have used mRNA as a surrogate marker for receptors, and have reported decreased GR mRNA in lymphocytes (51) and postmortem brain samples (66,76). GR mRNA appears to be decreased in characteristic brain regions in specific mental illnesses. For example, Webster et al. (77) reported reduced GR mRNA in the entorhinal cortex, subiculum, and CA3 and CA4 of the hippocampus in bipolar patients, a pattern different from that seen in patients with major depression or schizophrenia. Perlman et al. (76) described decreased GR mRNA in the basolateral/lateral nuclei of the amygdala in patients with bipolar disorder and schizophrenia, but not patients with major depressive disorder.

The common clinical observation that depressed patients do not display the physical signs of hypercortisolism typical of Cushing's Syndrome suggests hypofunctionality of GRs (78). Many studies have demonstrated decreased GR functioning in patients with major depressive disorder, such as decreased translocation into the cell nucleus following ligand binding (69,79,80), or diminished inhibition of lymphocyte proliferation (71,80–83). Fewer studies have been carried out in patients with bipolar illness. One paper (84) reported that GR binding to DNA was decreased in depressed bipolar patients compared to healthy controls, and was at an intermediate level in euthymic bipolar patients. This was despite the fact that the number of nuclear GR receptors was increased in patients, suggesting normal or increased translocation into the nucleus.

Moutsatsou et al. (85) were unable to detect any mutations in GR DNA in 15 bipolar patients, using agarose gel electrophoresis, heteroduplex analysis, and DNA sequencing. This highlights the fact that physiologic derangements other than those directly involving GR structure could also contribute to the HPA axis abnormalities that have been observed in patients with mood disorders. These include changes in cortisol-binding globulin levels which alter bioavailability of cortisol (86); decreased entry or increased removal of glucocorticoids from cells

(87); alterations in intracellular metabolism of glucocorticoids (88); post-translational modifications such as phosphorylation (89); and expression or functioning of other transcription factors (75), or components of the receptor complex other than the GR (90,91). For example, Binder et al. (90) reported that overexpression of the receptor complex component FKBP5 was associated with decreased GR function in squirrel monkeys. Also, depressed patients exhibit reduced activity of cyclic-AMP-dependent protein kinase A (92), which normally plays a significant role in increasing GR activity.

There is evidence that AVP activity is increased in patients with mood disorders, and may play a role in maintaining ACTH release despite downregulation of CRH1 receptors and likely CRH activity. AVP has been demonstrated to have ACTH-releasing properties when administered to human subjects (93), through interaction with the V3 (also known as the V1b) receptor on the anterior surface of the pituitary gland (3). Studies employing stress paradigms in rats have demonstrated elevated AVP synthesis (94). Also, while CRH mRNA and CRH1-receptor mRNA levels are reduced by elevated glucocorticoid levels (95), V3 receptor mRNA levels are increased. Studies in patients with mood disorders are highly suggestive of increased AVP activity. A mixed sample of patients with major depressive disorder or bipolar disorder, depressed phase, exhibited a greater cortisol response when administered AVP than did control subjects (96). Watson et al. (52) reported that AVP levels after pre-treatment with dexamethasone were higher in bipolar patients and patients with major depressive disorder than in controls, suggesting a lack of negative feedback from glucocorticoids on AVP release.

The role of limbic structures, such as the amygdala and hippocampus, and brainstem regions such as the locus coeruleus, in mediating HPA axis abnormalities in patients with mood disorders has received surprisingly little attention. Several lines of evidence are suggestive in this regard. Electrical stimulation of the hippocampus has been demonstrated to decrease plasma corticosteroid levels in several species, including primates (97), and lesions of the hippocampus result in glucocorticoid hypersecretion during activation of the HPA through stress (98). Electrical stimulation of the amygdala in rats has been correlated with decreased release of CRH into the hypophyseal portal bloodstream (99). As well, depletion of biogenic amines in rats by reserpine is associated with decreased density of GRs in the hippocampus, frontal cortex, pituitary, and lymphoid tissues (100). A recent report in euthymic bipolar patients described increased amygdala volume with HPA axis abnormalities (101). A corticotropin-releasing hormone receptor, CHR2, is primarily expressed in the limbic brain, suggesting a bidirectional relationship between this region and the HPA axis (102). Studies in this area are clearly hindered by the complex interrelationships between limbic system structures, neurotransmitter systems, and the HPA axis (103), though electrical stimulation or neurotransmitter depletion studies such as those described above may prove to be fruitful areas of research.

CONSEQUENCES OF HYPERCORTISOLEMIA

Glucocorticoid and mineralocorticoid receptors are distributed differentially throughout the central nervous system. MRs are localized exclusively in the limbic system, particularly the hippocampus, parahippocampal gyrus, and the entorhinal and insular cortices. GRs have a wider distribution, and are found in limbic system structures, various nuclei of the hypothalamus, including the paraventricular nucleus, and particularly in the frontal cortex (6). The hippocampus plays an important role in the formation of episodic or declarative memories, while the frontal cortex mediates executive functioning and working memory.

Hypercortisolemia has been investigated in animal models, healthy human volunteers, patients with endocrine diseases that cause hypercortisolism, such as Cushing's Syndrome, and in psychiatric patients. Investigations in rodents indicate that chronic exposure to high levels of glucocorticoids is associated with cognitive impairment. Young rats so exposed develop memory loss and hippocampal atrophy similar to that seen in aged rats (104,105). Conversely, young rats that undergo adrenalectomy, and are kept alive with low levels of exogenous glucocorticoids, do not experience these changes as they age (105). Hippocampal changes appear to be a result of dendritic atrophy of CA3 pyramidal cells (106). Further studies have suggested that the ratio of MRs to GRs is important in mediating structural and cognitive changes associated with hypercortisolism (6), and that excitatory amino acids, serotonin, and NMDA also play roles in the development of cognitive impairment and structural brain changes related to hypercortisolism (107).

Impairments in both declarative and working memory have also been observed after glucocorticoid administration in healthy volunteers (108–110). For instance, in a placebo-controlled crossover trial involving 20 healthy male subjects, hydrocortisone 20 mg twice daily administered for 10 days was associated with impairments on tests of cognitive functioning sensitive to frontal lobe dysfunction. These deficits were reversible after discontinuation of hydrocortisone administration.

A relationship between chronically elevated levels of endogenous corticosteroids and cognitive impairment has also been unambiguously demonstrated. Starkman et al. (111) detected a significant relationship between memory impairment, high cortisol levels, and decreased hippocampal formation volume in a sample of patients with Cushing's syndrome. Lupien et al. (6) reported that long-term elevation of endogenous corticosteroids in elderly human subjects was associated with significantly impaired declarative memory, but not with changes in non-declarative memory. Furthermore, MRI scans revealed that subjects with chronically elevated cortisol levels had a 14% smaller hippocampal volume when compared to subjects with normal cortisol levels over time. Whether such cognitive impairment is reversible is not clear. Interestingly, however, loss of brain volume in 38 patients with Cushing's syndrome was reversed after hypercortisolemia was corrected (112).

Many studies document the existence of cognitive deficits in patients with bipolar disorder. The data from these reports suggest that this impairment is a trait phenomenon that persists even during euthymic intervals (113,114). One report has correlated neuropsychological impairments in bipolar patients with glucocorticoid receptor function, as measured by the DST (115). In this study, 17 euthymic bipolar patients and 16 controls participated in tests of verbal declarative and working memory. The patients made significantly more errors of omission and commission on tests of working memory, and also displayed impaired verbal recognition memory. Commission errors on tests of working memory were highly correlated with nonsuppression of cortisol on the DST ($r = 0.64$, $P < 0.0006$).

In addition to its effects on the central nervous system, cortisol is also known to antagonize the effects of insulin and to raise blood pressure, thereby increasing the risk of diabetes, hypertension, and coronary artery disease (6). This is particularly noteworthy, given the known increase in early mortality in patients suffering from bipolar disorder, particularly related to cardiovascular causes.

In summary, then, sustained hypercortisolism is associated with structural changes in the hippocampus, as well as deficits in working memory and declarative memory, functions known to be mediated by the hippocampus and the frontal lobes, areas of the brain which are rich in glucocorticoid receptors. The effects of hypercortisolism may also increase the risk of cardiovascular disease and diabetes.

EFFECTS OF SOMATIC TREATMENTS

Research suggests that antidepressant treatment leading to the resolution of depressive symptoms generally normalizes HPA axis functioning in patients with major depression (116–119). HPA axis abnormalities frequently persist, however, following unsuccessful antidepressant treatment (120). In contrast, there is evidence that lithium therapy actually accentuates abnormalities on the DST and Dex/CRH test, at least in patients with major depressive disorder. Bschor et al. studied depressed patients who were refractory to at least four weeks of adequate antidepressant therapy. Significant increases in serum cortisol and ACTH were noted with both the DST (121) and the Dex/CRH test (122) after lithium was added as an augmenting agent. Similar results were observed in both responders and nonresponders, and were in fact more pronounced in responders, suggesting that restitution of normal HPA axis functioning is not a prerequisite for treatment response. The mechanism by which lithium exerts this effect is not clear. Serotoninergic axons are known to project to both the hippocampus and the paraventricular nucleus of the hypothalamus (123), and it has been speculated that enhanced serotonin-mediated release of CRH as a result of lithium treatment may underlie the HPA effects of lithium (122).

Investigations on the effect of lithium on the HPA axis in bipolar patients are relatively lacking. One recent study reported that response to the Dex/CRH test in bipolar patients taking lithium was no different than that observed in patients receiving other medications, providing preliminary evidence that lithium may not potentiate HPA axis abnormalities in bipolar patients (8). Few other investigations on the effects of mood stabilizing medications have been reported in mood disorders patients. Studies utilizing the Dex/CRH test in euthymic patients with major depressive disorder (39) and bipolar disorder (8) reported that patients treated with carbamazepine experienced a greater rise in serum cortisol levels following CRH administration than did patients receiving other medications. However, this is likely related to a pharmacokinetic drug interaction rather than a specific effect on the HPA axis, as carbamazepine is known to induce CYP3A4, the primary enzyme responsible for dexamethasone metabolism. Given the resulting decrement in dexamethasone suppression of cortisol, CRH acts relatively unopposed in elevating ACTH and cortisol levels. Confirming this hypothesis, Watson et al. (8) found that patients taking carbamazepine had lower dexamethasone levels ($P < 0.0005$) than patients receiving other medications.

Research regarding the effects of lithium on the HPA axis has been carried out in animals and healthy human subjects. Many (124–126) but not all (127) studies indicate that rats receiving lithium over two to four weeks display increased GR numbers in regions of the brain such as the hippocampus and the paraventricular nucleus of the hypothalamus. A report on the effect of lithium on GR numbers in mouse lymphoma cells, however, was negative. Two studies (128,129) suggest that lithium diminishes GR functioning in rodents, possibly by way of interactions with the GR receptor complex, particularly BAG-1, a protein constituent of the complex (129). A study involving human subjects (130), however, reported that lithium, in contrast to antidepressant medications, did not induce translocation of the GR into the lymphocyte nucleus. If lithium does indeed inhibit GR functioning, this may explain its effect on accentuating HPA hyperactivity, a finding that has been replicated in animals receiving lithium (131–133).

The effects of sodium valproate and carbamazepine on the HPA axis have been studied in rodents and human volunteers. Valproate treatment for one week led to decreased CRH concentrations in several brain regions, including the amygdala and the paraventricular nucleus of the hypothalamus (134). However, neither acute (135) nor chronic (136) treatment with valproate decreased ACTH or cortisol levels induced by hypoglycemia in human subjects, nor did acute treatment lead to decreased release of ACTH stimulated by naloxone (137). Both valproate and carbamazepine, like lithium, were reported to impair GR functioning in rodents (128). Valproate, unlike the other two mood stabilizers, was also associated with decreased nuclear and cytoplasmic GR numbers. In a study in healthy human subjects, carbamazepine was associated with HPA axis hyperactivity, including robust ACTH response to CRH infusion, despite the presence of hypercortisolemia (138).

DIRECTIONS FOR FUTURE RESEARCH

Based on the currently available data, it is not possible to conclude with certainty that HPA axis abnormalities observed in patients with mood disorders have etiological significance, or whether they are a consequence of other pathophysiologic processes. Recently, researchers have begun to focus on modulating HPA axis functioning in patients with mood disorders as a possible treatment strategy. If these investigations bear fruit, not only will they broaden the armamentarium of available treatment options for major depression and bipolar illness, but they will also provide evidence of a primary pathophysiologic basis for HPA axis dysfunction.

Preliminary evidence exists for a number of different approaches, including the use of the adrenal steroid dehydroepiandrosterone (139), which is known to have antiglucocorticoid properties, and steroid synthesis inhibitors such as ketokonazone, aminoglutethimide, and metyrapone (140,141). Interestingly, both glucocorticoid receptor agonists such as dexamethasone (142,143) and antagonists such as mifepristone (144) may have some efficacy in the treatment of mood disorders. Activation of GRs leading to enhanced negative feedback at the level of the pituitary is hypothesized to underlie the efficacy of GR agonists, while GR antagonists are believed to have acute antiglucocorticoid activity, as well as leading to upregulation in GR number and consequently improved negative feedback. A recent small double-blind placebo-controlled crossover trial of mifepristone versus placebo in patients with bipolar disorder (144) provides preliminary evidence that this treatment yields benefits in both mood and cognition. Additional studies are needed to adequately assess the efficacy of these treatment approaches.

Further research is also needed to clarify other aspects of HPA axis functioning in patients with mood disorders. The roles of AVP, as well as neuroregulatory input from higher brain centers, such as limbic structures, in HPA axis dysfunction have yet to be fully elucidated. Little research has focused on the role of the mineralocorticoid receptor in mediating HPA axis abnormalities, although one recent study demonstrated deficits in mineralocorticoid receptor mRNA in the frontal cortex of patients with bipolar disorder and schizophrenia (143). Finally, most investigations of HPA axis function have focused on patients with Bipolar I Disorder, and there is little or no information specifically regarding Bipolar II patients.

REFERENCES

1. Daban C, Vieta E, Mackin P, Young AH. Hypothalamic-pituitary-adrenal axis and bipolar disorder. Psychiatri Clin North Am 2005; 28:469–480.
2. Pariante CM. Glucocorticoid receptor function in vitro in patients with major depression. Stress 2004; 7:209–219.
3. Dinan TG, Scott LV. Anatomy of melancholia: focus on hypothalamic-pituitary-adrenal axis overactivity and the role of vasopressin. J Anat 2005; 207:259–264.
4. Parker KJ, Schatzberg AF, Lyons DM. Neuroendcrine aspects of hypercortisolism in major depression. Horm Beh 2003; 43:60–66.
5. Brien TG. Human corticosteroid binding globulin. Clin endocrinol 1981; 14:193–212.
6. Lupien SJ, Fiocco A, Wan N, et al. Stress hormones and human memory function across the lifespan. Psychoneuroendocrinology 2005; 30:225–242.
7. Heuser I, Yassouridis A, Holsboer F. The combined dexamethasone/CRH test: a refined laboratory test for psychiatric disorders. J Psychiatr Res 1994; 28:341–356.
8. Watson S, Gallagher P, Ritchie JC, Ferrier N, Young AH. Hypothalamic-pituitary-adrenal axis function in patients with bipolar disorder. British J Psychiatry 2004; 184: 496–502.
9. Board F, Persky H, Hamburg DA. Psychological stress and endocrine functions. Psychosom Med 1956; 18:324–333.
10. Stokes PE, Pick GR, Stoll PM, et al. Pituitary-adrenal function in depressed patients: resistance to dexamethasone suppression. J Psychiatry Res 1975; 12:271–281.
11. Nemeroff CB, Widerlov E, Bissette G, et al. Elevated concentrations of CSF corticotropin-releasing factor-like immunoreactivity in depressed patients. Science 1984; 226: 1342–1344.
12. Arborelius L, Owens MJ, Plotsky PM, Nemeroff CB. The role of corticotropin releasing factor in depression and anxiety disorders. J Endocrinol 1999; 160:1–12.
13. Holsboer F. The corticosteroid receptor hypothesis of depression. Neuropsychopharmacology 2000; 23:477–501.
14. Kalin NH, Weiler SJ, Shelton SE. Plasma ACTH and cortisol concentrations before and after dexamethasone. Psychiatry Res 1982; 7:87–92.
15. Reus VI, Joseph MS, Dallman MF. ACTH levels after the dexamethasone suppression test in depression. N Engl J Med 1982; 306:238–239.
16. Pfohl B, Sherman B, Schlecte J, Winokur G. Differences in plasma ACTH and cortisol between depressed patients and normal controls. Biol Psychiatry 1985; 20:1055–1072.
17. Deuschle M, Schweiger U, Weber B, et al. Diurnal activity and pulsatility of the hypothalamic-pituitary-adrenal system in male depressed patients and healthy controls. J Clin Endocrinol Metab 1997; 82:234–238.
18. Young EA, Carlson MS, Brown MB. Twenty-four-hour ACTH and cortisol pulsatility in depressed women. Neuropsychopharmacology 2001; 25:267–276.
19. Fang VS, Tricou BJ, Robertsone A. Plasma ACTH and cortisol levels in depressed patients: relation to dexamethasone suppression test. Life Sci 1981; 29: 931–938.
20. Yerevanien BI, Woolf PD. Plasma ACTH levels in primary depression: relationship to the 24-hour dexamethasone suppression test. Psychiatr Res 1983; 9:45–51.
21. Sherman BM, Pfohl B, Winokur G. Correspondence of plasma ACTH and cortisol before and after dexamethasone in healthy and depressed patients. Psychiatr Med 1985; 3:41–52.
22. Linkowski P, Mendelwicz J, Leclercq R, et al. The 24-hour profile of adrenocorticotropin and cortisol in major depressive illness. J Clin Endocrinol Metab 1985; 61:429–438.
23. Gold PW, Loriaux DL, Roy A, et al. Responses to corticotrophin-releasing hormone in the hypercortisolism of depression and Cushing's disease. N Engl J Med 1986; 314: 1329–1335.

24. Murphy BEP. Steroids and depression. J Steroid Biochem Mol Biol 1991; 38: 537–559.
25. Posener JA, DeBattista C, Williams GH, Kraemer HC, Kalehzan BM, Schatzberg AF. 24-h monitoring of cortisol and corticotrophin secretion in psychotic and non-psychotic major depression. Arch Gen Psychiatry 2000; 57:755–760.
26. Amsterdam JD, Marinelli DL, Arger P, Winokur A. Assessment of adrenal gland volume by computed tomography in depressed patients and healthy volunteers: a pilot study. Psychiatr Res 1987; 21:189–197.
27. Krishnan KR, Doraiswamy PM, Lurie SN, et al. Pituitary size in depression. J Clin Endocrin Metab 1991; 72:256–259.
28. Nemeroff CB, Krishnan KRR, Reed D, Leder R, Beam C, Dunnick NR. Adrenal gland enlargement in major depression: a computed tomography study. Arch Gen Psychiatry 1992; 49:384–387.
29. Rubin RT, Phillips JJ, McCracken JT, Sadow TF. Adrenal gland volume in major depression: relationship to basal and stimulated pituitary-adrenal cortical axis function. Biol Psychiatry 1996; 40:89–97.
30. Rubin RT, Phillips JJ, Sadow TF, McCracken JT. Adrenal gland volume in major depression: increase during the depressive episode and decrease with successful treatment. Arch Gen Psychiatry 1995; 52:213–218.
31. Watson S, Gallagher P, Del-Estal D, Hearn A, Ferrier IN, Young AH. Hypothalamic-pituitary-adrenal function in patients with chronic depression. Psychol Med 2002; 32:1021–1028.
32. Amsterdam JD, Maislin G, Winokur A, Berwish N, Kling M, Gold P. The oCRH test before and after clinical recovery from depression. J Affect Disord 1988; 14:231–222.
33. Holsboer FA, Von Bardelbeden U, Gerken A, Staala GK, Muller OA. Blunted corticotrophin and normal cortisol response to human corticotropin-releasing factor in depression. N Engl J Med 1984; 311:1127–1128.
34. Holsboer FA, Gertken A, Stalla GK, Muller OA. Blunted aldosterone and corticotrophin release after human corticotrophin releasing hormone in depression. Am J Psychiatry 1987; 144:229–231.
35. Kolankowski J, Jeanjean M, Crabbe J. Response of human adrenal cortex to corticotrophin: potentialization in spite of a time lapse between two stimulations. Ann Endocrinol 1969; 30:857–864.
36. Kolankowski J, Pizzaro MA, Crabbe J. Potentiation of adrenocortical response upon intermittent stimulation with corticotrophin in normal subjects. J Clin Endocrinol Metab 1975; 41:453–465.
37. Holsboer F. The rationale for corticotrophin-releasing hormone receptor (CRH-R) antagonists to treat depression and anxiety. J Psychiatr Res 1999; 33:181–214.
38. Zobel AW, Yassouridis A, Frieboes RM, Holsboer F. Prediction of medium-term outcome by cortisol response to the combined dexamethasone/CRH test in patients with remitted depression. Am J Psychiatry 1999; 156:949–951.
39. Zobel AW, Nickel T, Sonntag A, et al. Cortisol response in the combined dexamethasone/CRH test as a predictor of relapse in patients with remitted depression: a prospective study. J Psychiatr Res 2001; 35:83–94.
40. Cassidy F, Ritchie JC, Carroll BJ. Plasma dexamethasone concentration and cortisol response during manic episodes. Biol Psychiatry 1998; 43:747–754.
41. Cervantes P, Gelber S, Kin FNKNY, Nair VNP, Schwartz G. Circadian secretion of cortisol in bipolar disorder. J Psychiatry Neuroscience 2001; 26:411–416.
42. Godwin CD, Greenberg LB, Shukla S. Consistent dexamethasone suppression test results with mania and depression in bipolar illness. Am J Psychiatry 1984; 141:1263–1265.
43. Bagdy G, Rihmer Z, Frecska E, Szadoczky E, Arato M. Platelet MAO activity and the dexamethasone suppression test in bipolar depression. Psychoneuroendocrinology 1986; 11:117–120.
44. Rush AJ, Giles DE, Schlesser MA, et al. Dexamethasone response thyrotropin releasing hormone stimulation rapid eye movement latency, and subtypes of depression. Biol Physchiatry 1997; 41:915–928.

45. Zhou D, Shen Y, Shu L, Lo H. Dexamthasone suppression test and unrinary MHPG-SO4 determination in depressive disorders. Biol Psychiatry 1987; 22: 883–891.
46. Rybakowsii JR, Twardowska K. The dexamethasone/corticotropin-releasing hormone test in depression in bipolar and unipolar affective illness. J Psychiatric Res 1999; 33:363–370.
47. Tomitaka S, Sakamoto K, Kojima I, Fujita H. Serial dexamethasone suppression tests by measuring urinary free cortisol among rapidly cycling patients. Biol Psychiatry 1995; 38:128–130.
48. Swann AC, Stokes PE, Casper R, et al. Hypothalamic-pituitary-adrenocortical function in mixed and pure mania. Acta Psychiatr Scand 1992; 85:270–274.
49. Swann AC, Stokes PE, Secunda SK, et al. Depressive mania versus agitated depression: biogenuc amine and hypothalamic-pituitary-adrenocortical function. Biol Psychiatry 1994; 35:803–813.
50. Evans DL, Nemeroff CB. The dexamethasone suppression test in bipolar disorder. Am J Psychiatry 1983; 140:615–617.
51. Krishnan RR, Maltbie AA, Davidson JRT. Abnormal cortisol suppression in bipolar patients with simultaneous manic and depressive symptoms. Am J Psychiatry 1983; 140:203–205.
52. Watson S, Thompson JM, Malik N, Ferrier IN, Young AH. Temporal stability of the Dex/CRH test in patients with rapid-cycling bipolar I disorder: a pilot study. Aus NZ J Psychiatry 2005; 39:244–248.
53. Schmider J, Lammers C-H, Gotthardt U, Dettling M, Holsboer F, Heuser I. Combined dexamethasone/corticotropin-releasing hormone test in acute and remitted manic patients, in acute depression, and in normal controls: I. Biol Psychiatry 1995; 38: 797–802.
54. Deschauer D, Grof E, Alda M, Grof P. Patterns of DST posotovity in remitted affective disorders. Biol Psychiatry 1999; 45:1023–1029.
55. Krieg J-C, Lauer CJ, Schrieber W, Modell S, Holsboer F. Neuroendocrine, polysomnographic, and psychometric observations in healthy subjects at high familial risk for affective disorders: the current state of the "Munich Vulnerability Study." J Affective Disord 2001; 62:33–27.
56. Ellenbogen MA, Hodgins S, Wlaker C-D. High levels of cortisol among adolescent offspring of parents with bipolar disorder: a pilot study. Psychoneuroendocrinology 2004; 29:99–106.
57. Matsubara T, Funato H, Kobayashi A, Nobumoto M, Watanabe Y. Reduced glucocorticoid receptor alpha expression in mood disoerder patients and first-degree relatives. Biol Psychiatry 2006; 59:689–695.
58. Yehuda R, Boisoneau D, Mason JW, Giller EL. Glucocorticoid receptro number and cortisol excretion in mood, anxiety, and psychotic disorders. Biol Psychiatry 1993; 34:18–25.
59. Sassi RB, Micoletti M, Brambilla P, Harenski K, Mallinger AG, Frank E, Kupfer DJ, Keshavan MS, Soares JC. Decreased pituitary volume in patients with bipolar disorder. Biol Psychiatry 2001; 50:271–280.
60. Krishnan KRR, Davidson JRT, Rayasam K, Tanas KS, Shope FS, Pelton S. Diagnostic utility of the DST. Biol Psychiatry 1987; 22:618–628.
61. Vieta E, Martinez-De-Osaba MJ, Colom F, Martinez-Aran A, Benabarre A, Gasto C. Enhanced corticotrophin response to corticotrophin-releasing hormone as a predictor of amnia in euthymic bipolar patients. Psychological Medicine 1999; 29: 971–978.
62. Pariante CM, Miller AH. Glucocrticoid receptors in major depression: relevance to pathophysiology and treatment. Biol Psychiatry 2001; 49:391–404.
63. Gormley GJ, Lowy MT, Reder AT, Hospelhorn VD, Antel JP, Meltzer HY. Glucocorticoid receptors in depression: relationship to the dexamethasone suppression test. Am J Psychiatry 1985; 142:1278–1284.
64. Whalley LJ, Borthwick N, Copolov D, Dick H, Christie JE, Fink G. Glucocorticoid receptors and depression. Br Med J 1986; 292:859–861.

65. Sallee FR, Nesbitt L, Dougherty D, Nandagopal VS, Sethuraman G. Lymphocyte glucocorticoid receptor: predictor of sertraline response in adolescent major depressive disorder. Pharmacol Bull 1995; 31:339–345.
66. Webster JC, Carlstedt-Duke J. Involvement of multi-drug resistance proteins (MDR) in the modulation of glucocorticoid response. J Steroid Biochem Mol Biol 2002; 82:277–288.
67. Schlecte JA, Sherman B. Lymphocyte glucocorticoid receptor binding in depressed patients with hypercortisolemia. Psychoneuroendocrinology 1985; 10:469–474.
68. Hunter R, Dick H, Christie JE, Goodwin GM, Fink G. Lymphocyte glucocorticoid receptor binding in depression: normal values following recovery. J Affect Disord 1988; 14:155–159.
69. Wassef A, Smith EM, Rose RM, Gardner R, Nguyen H, Meyer WJ. Mononuclear leukocyte glucocorticoid receptor binding characteristics and down-regulation in major depression. Psychoneuroendocrinology 1990; 15:59–68.
70. Rupprecht R, Korhhuber J, Wodarz N, et al. Disturbed glucocorticoid receptor autoregulation and corticotropin response to dexamethasone in depressives pretreated with metyrapone. Biol Psychiatry 1991a; 29:1099–1109.
71. Rupprecht R, Kornhuber J, Wodarz N, et al. Lymphocyte glucocorticoid receptor binding during depression and after clinical recovery. J Affect Disord 1991b; 22:31–35.
72. Wassef A, O'Boyle M, Gardner R, et al. Glucocorticoid receptor binding in three different cell types in major depressive disorder: lack of evidence of receptor binding defect. Prog Neuropsychopharmacol Biol Psychiatry 1992; 16:65–78.
73. Maguire TM, Thakore JU, Dinan TG, Hopwood S, Breen KC. Plasma sialyltransferase levels in psychiatric disorders as a possible indicator of HPA function. Biol Psychiatry 1997; 41:1131–1136.
74. Lopez JF, Chalmers DT, Little KY, Watson SJ. AE Bennett Research Award. Regulation of Serotonin 1A, glucocorticoid, and mineralocorticoid receptors in rat and human hippocampus: implications for the neurobiology of depression. Biol Psychiatry 1998; 43: 547–573.
75. Spiliotaki M, Salpeas V, Malitas P, Alevizos V, Moutsatsou P. Altered glucocorticoid receptor signalling cascade in lymphocytes of bipolar disorder patients. Psychoneuroendocrinology 2006; 31:748–760.
76. Perlman WR, Webster MJ, Kleinman JE, Weickert CS. Reduced glucocorticoid and estrogen receptor alpha messenger ribonucleic acid levels in the amygdale of patients with major mental illness. Biol Psychiatry 2004; 56:844–852.
77. Webster MJ, Knable MB, O'Grady J, Orthmann J, Weickert CS. Regional specificity of brain glucocorticoid receptor mRNA alterations in subjects with schizophrenia and mood disorders. Mol Psychiatry 2002; 7:985–994.
78. Lowy MT, Gormley GJ, Reder AT, et al. Immune function, glucocorticoid receptor regulation and depression. In: Miller AH, ed. Depressive Disorders and Immunity Washington DC: American Psychiatric Association, 1989:105–113.
79. Yehuda R, Halligan SL, Grossman R, Golier JA, Wong C. The cortisol and glucocorticoid receptor response to low dose dexamethasone administration in aging combat veterans and holocaust survivors with and without posttraumatic stress disorder. Biol Psychiatry 2002; 52:393–403.
80. Lowy MT, Reder AT, Gormley GJ, Meltzer HY. Comparison of in vivo and in vitro glucocorticoid sensitivity in depression: relationship to the dexamethasone suppression test. Biol Psychiatry 1988; 24:619–630.
81. Calfa G, Kademian S, Ceschin D, Vega G, Rabinovich GA, Volosin M. Characterization and functional significance of glucocorticoid receptors in patients with major depression: modulation by antidepressant treatment. Psychoneuroendocrinology 2003; 28:687–701.
82. Lowy MT, Reder AT, Antel JP, Meltzer HY. Glucocorticoid resistance in depression: the dexamethasone suppression test and lymphocyte sensitivity to dexamethasone. Am J Psychiatry 1984; 141:1365–1370
83. Wodarz N, Rupprecht R, Kornhuber J, et al. Normal lymphocyte responsiveness to lectins but impaired sensitivity to in vitro glucocorticoids in major depression. J Affect Disord 1991; 22:241–248.

84. Spiliotaki M, Salpeas V, Malitas P, Alevizos V, Moutsatsou P. Altered glucocorticoid receptor signaling cascade in lymphocytes of bipolar disorder patients. Psychoneuroendocrinology 2006; 31:748–760.
85. Moutsatsou P, Kazazoglou T, Fleischer-Lambropoulos H, Glucocorticoid receptor alpha and beta isoforms are not mutated in bipolar affective disorder. Mol Psychiatry 2000; 5(2):196–202.
86. Rosner W. Plasma steroid binding proteins. Endocrinol Metab Clin North Am 1991; 20:697–720.
87. deKloet ER, Vreugdenhil E, Oitzl MS, Joels M. Brain corticosteroid receptor balance in health and disease. Endocr Rev 1998; 19:269–301.
88. Seckl JR, Walker BR. Minireview: 11beta-hydroxysteroid dehydrogenase type I—a tissue-specific amplifier of glucocorticoid action. Endocrinology 2001; 142: 1371–1376.
89. Webster JC, Jewell CM, Bodwell JE, Munck A, Sar M, Cidlowski JA. Mouse glucocorticoid receptor phosphorylation status influences multiple functions of the receptor protein. J Biol Chem 1997; 272:9287–9293.
90. Binder EB, Salyakina D, Lichtner P, et al. Polymorphisms in FKBP5 are associated with increased recurrence of depressive episodes and rapid response to antidepressant treatment. Nat Genet 2004; 36:1319–1325.
91. vanRossum EF, Lamberts SW. Polymorphisms in the glucocorticoid receptor gene and their associations with metabolic parameters and body composition. Recent Prog Horm Res 2004; 59:333–357.
92. Manier DH, Shelton RC, Ellis TC, Peterson CS, Eiring A, Sulser F. Human fibroblasts as a relevant model to study signal transduction in affective disorders. J Affect Disord 2000; 61:51–58.
93. Salata RA, Jarrett DB, Verbalis JG, Robinson AG. Vasopressin stimulation of adrenocorticotropin hormone (ACTH) in humans. In vivo bioassay of corticotropin-releasing factor (CRF) which provides evidence for CRF mediation of the variation of the diurnal rhythm of ACTH. J Clin Invest 1988; 81:766–774.
94. DeGoeji D, Djikstra H, Tliders F. Chronic psychosocila strress enhances vasopressin but not corticotropin-releasing factor, in the external zone of the median eminence of male rats: relationship to subordinate status. Endocrinology 1992; 131:847–853.
95. Zhou JN, Hoffman MA, Swaab DF. Morphometric analysis of vasopressin and vasoactive intestinal polypeptide neurons in the human suprachiasmatic nucleus: influence of microwave treatment. Brain Res 1996; 742:334–338.
96. Meller WH, Kathol RC, Jaeckel RS, Lopez JF. Stimualtion of the pituitary-adrenal axis with arginie vasopressin in patients with depression. J Psychiat Res 1987; 21: 269–277.
97. Jacobson L, Sapolsky R. The role of the hippocampus in feedback regulation of the hypothalamic-pituitary-adrenocortical axis. Endocrine Rev 1991; 12:118–130.
98. Brown ES, Rush AJ, McEwen BS. Hippocampal remodelling and damage by corticosteroids: implications for mood disorders. Neuropsychopharmacology 1999; 21:474–484.
99. Tannahill LA, Sheward WJ, Robinson IC, Fink G. Corticotrophin-releasing factor-41, vasopressin and oxytocin release into hypophyseal portal blood in the rat: effects of electrical stimulation of the hypothalamus, amygdale and hippocampus. J Endocrinol 1991; 129:99–107.
100. Lowy MT. Reserpine induced decrease in type I and II corticosteroid receptors in neuronal and lymphoid tissues in adrenalectomized rats. Neuroendocrinology 1990; 51:190–196.
101. Lloyd AJ, Frangon S, Moore PB, et al. Increased amygdale size in euthymic bipolar patients on voxel based analysis. In press.
102. Holsboer F, Barden N. Antidepressants and hypothalamic-pituitary-adrenocortical regulation. Endocrine Rev 1996; 17:187–205.
103. Porter RJ, Gallagher P, Watson S, Young AH. Corticosteroid–serotonin interactions in depression: a review of the human evidence. Psychopharmacology 2004; 173: 1–17.
104. Landfield P, Waymire J, Lynch G. Hippocampal aging and adrenocorticoids: a quantitative correlation. Science 1978; 202:1098–1101.

105. Landfield P, Baskin RK, Pitler TA. Brain aging correlates: retardation by hormonal-pharmacological treatments. Science 1981; 214:581–583.
106. Vyas A, Mitra R, Shankaranarayana R, Chatharji S. Chronic stress induce contrasting patterns of dendritic remodelling in hippocampal and amygdalid nurons. J. Neurosci 2002; 22:10810–10818.
107. McEwen BS. The neurobiology of stress: from serendipity to clinical relevance. Brain Res 2000; 886:172–189.
108. Young AH, Sahakian BJ, Robbins TW, Cowen PJ. The effects of chronic administration of hydrocortisone on cognitive function in normal male volunteers. Psychopharmacology 1999; 145:260–266.
109. Wolkowitz OM, Reus VI, Weingartner H, et al. Cognitive deficits of corticosteroids. Am J Psychiatry 1990; 147:1297–1303.
110. Newcomber JW, Craft S, Hershey T, Askins K, Bardgett ME. Glucocorticoid induced impairment in declarative memory performance in adult humans. J Neurosci 1994; 14:2047–2053.
111. Starkman MN, Gebarski SS, Berent S, Schteingart De. Hippocampal formation volume, memory dysfunction, and cortisol levels in patients with Cushing's Syndrome. Biol Psychiatry 1992; 32:756–765.
112. Bourdeau I, Bard C, Noel B, et al. Loss of brain volume in endogenous Cushing's syndrome and its reversibility after correction of hypercortisolism. J Clin Endocrin Metab 2002; 87:1949–1954.
113. Robinson LJ, Thompson JM, Gallagher P, et al. A meta-analysis of cognitive deficits in euthymic patients with bipolar disorder. J Affect Disord 2006; 93(1–3):105–115.
114. Thompson JM, Gallagher P, Hughes JM, et al. Neurocognitive impairment in euthymic patients with bipolar affective disorder. Br J Psychiatry 2005; 186:32–40.
115. Watson S, Thompson JM, Ritchie JC, Ferrier IN, Young AH. Neuropsychological impairment in bipolar disorder: the relationship with glucocorticoid receptor function. Bipolar Disord 2006; 8:85–90.
116. Linkowski P, Mendlewicz J, Kerkhofs M, et al. 24-hour profiles of adrenocorticotropin, cortisol, and growth hormone in major depressive illness: effect of antidepressant treatment. J Clin Endocrinol Metab 1987; 65:141–152.
117. Heuser IJ, Schweiger U, Gotthardt U, et al. Pituitary-adrenal-system regulation and psychopathology during amitryptiline treatment in elderly depressed patients and normal comparison subjects. Am J Psychiatry 1996; 153:93–99.
118. Kunugi H, Ida I, Kimura M, et al. Assessment of the dexamthasone/CRH test as a state-dependent marker for hypothalamic-pituitary-adrenal (HPA) axis abnormalities in major depressive episode: a multicentre study. Neuropsychopharmacology 2006; 31:212–220.
119. Yuuki N, Ida A, Kumano H, et al. HPA axis normalization estimated by DEX/CRH test, but less alteration on cerebral glucose metabolism in depressed patients receiving ECT after medication treatment failures. Acta Psychiatrica Scandinavia 2005; 112: 257–265.
120. Kunzel HE, Binder EB, Nickel T, et al. Pharmacological and nonpharmacological factors influencing hypothalamic-pituitary-adrenocortical axis reactivity in acutely depressed psychiatric in-patients, measured by the DEX/CRH test. Neuropsychopharmacology 2003; 28:2169–2178.
121. Bschor T, Baethge C, Adli M, et al. Lithium augmentation increases post-dexamethasone cortisol in the dexamethasone suppression test in unipolar major depression. Depress Anx 2003; 17:43–48.
122. Bschor T, Adli M, Baethge C, et al. Lithium augmentation increase the ACTH and cortisol responses in the combined DEX/CRH test in unipolar major depression. Neuropsychopharmacology 2002; 27:470–478.
123. Jacobs BL, Azmitia EC. Structure and function of the brain serotonin system. Physiol Rev 1992; 72:165–229.
124. Budziszewska B, Lason W. Pharmacological modulation of glucocorticoid and mineralocorticoid receptors in the rat central nervous system. Pol J Pharmacol 1994; 46: 97–102.

125. Peiffer A, Veilleux S, Barden N. Antidepressant and other centally acting drugs regulate glucocorticoid receptor messenger RNA levels in rat brain. Psychoneuroendocrinology 1991; 16:505–515.
126. Semba J, Watanabe H, Suhara T, Akanuma N. Chronic lithium chloride injection increases glucocorticoid receptor but not mineralocorticoid receptor mRNA expression in rat brain. Neuroscience Res 2000; 38:313–319.
127. McQuade R, Leitch MM, Gartside SE, Young AH. Effect of chronic lithium treatment on glucocorticoid and 5HT-1A receptor messenger RNA in hippocampal and dorsal raphe nucleus regions of the rat brain. J Psychopharm 2004; 18:496–501.
128. Basta-Kaim A, Budziszewska B, Jaworska-Fail L, et al. Mood stabilizers inhibit glucocorticoid receptor function in LMCAT cells. European J Pharmacol 2004; 495:103–110.
129. Zhou R, Gray NA, Yuan P, et al. The antiapoptotic, glucocorticoid receptor cochaperone protein BAG-1 is a long-term target for the actions of mood stabilizers. J Neurosci 2005; 25:4493–4502.
130. Okuyama-Tamura M, Mikuni M, Kojima I. Modulation of human glucocorticoid receptor function by antidepressant compounds. Neurosci Letters 2003; 342: 206–210.
131. Vatal M, Aiyar AS. Some aspects of corticosterone metabolism in lithium treated rats. Chem Biol Interact 1983; 45:277–282.
132. Storlien LH, Higson FM, Gleeson RM, Smythe GA, Atrens DM. Effects of chronic lithium, amitryptiline, and mianserin on glucoregulation, corticosterone and energy balance in the rat. Pharmacol Biochem Behav 1985; 22:119–125.
133. Sugawara M, Hasimoto K, Hattori T, Takao T, Suemaru S, Ota Z. Effects of lithium on the hypothalamo-pituitary-adrenal axis. Endocrinol Jpn 1988; 35:655–663.
134. Stout SC, Owens MJ, Lindsey KP, Knight DL, Nemeroff CB. Effects of sodium valproate on corticotropin releasing factor systems in rat brain. Neuropsychopharmacology 2001; 24:624–631.
135. Petraglia F, Bakalakis S, Facchinetti F, Volpe A, Muller EE, Genazzani AR. Effects of sodium valproate and diazepam on beta-endorphin, beta-lipotropin and cortisol secretion induced by hypoglycemic stress in humans. Neuroendocrinology 1986; 44: 320–325.
136. Abraham RR, Dornhorst A, Wynn V, et al. Corticotropin, cortisol, prolactin and growth hormaone responses to insulin induced hypoglycemiain normal subjects given sodium valproate. Clin Endocrinol 1985; 22:639–644.
137. Torpy DJ, Grice JE, Hockings GI, Crosbie GV, Walters MM, Jackson RV. Effect of sodium valproate on naloxone-stimulated ACTH and cortisol release in humans. Clin Exp Pharmacol Physiol 1995; 22:441–443.
138. Perini GI, Devinsky O, Hauser P, et al. Effects of carbamazepine on pituitary-adrenal function in healthy volunteers. J Clin Endocrinol Metab 1992; 74:406–412.
139. Wolkowitz OM, Reus VI, Keebler A, et al. Double-blind treatment of major depression with dehydroepiandrosterone. Am J Psychiatry 1999a; 156:646–649.
140. Wolkowitz OM, Reus VI, Chan T, et al. Antiglucocorticoid treatment of depression: double-blind ketokonazole. Biol Psychiatry 1999b; 45:1070–1074.
141. Malison RT, Anand A, Pelton GH, et al. Limited efficacy of ketokonazole in treatment-refractory major depression. J Clin Psychopharm 1999; 19:466–470.
142. Arana GW, Santos AB, Laraia MT, et al. Dexamethasone for the treatment of depression: a randomized, placebo-controlled, double-blind trial. Am J Psychiatry 1995; 152: 265–267.
143. Bodani M, Sheehan B, Philpott M. The use of dexamethasone in elderly patients with antidepressant-resistant depressive illness. J Psychopharm 1999; 13:196–197.
144. Young AH, Gallagher P, Watson S, Del-Estal D, Owen BM, Ferrier IN. Improvement in neurocognitive functioning and mood following adjunctive treatment with mifepristone (RU486) in bipolar disorder. Neuropsychopharmacology 2004; 29:1538–1545.
145. Xing G-Q, Russell S, Webster MJ, Post RM. Decreased expression of mineralocorticoid receptor mRNA in the prefrontal cortex in schizophrenia and bipolar disorder. Int J Neuropsychopharmacol 2004; 7:143–153.

9 Brain Imaging Studies in Bipolar Disorder

E. Serap Monkul
Mood Disorders Clinical Neurosciences Program, Department of Psychiatry, University of Texas Health Science Center at San Antonio, San Antonio, Texas, U.S.A., and Department of Psychiatry, Dokuz Eylül University School of Medicine, Izmir, Turkey

Paolo Brambilla
Department of Pathology and Experimental and Clinical Medicine, Section of Psychiatry, University of Udine, Udine, and Scientific Institute IRCCS E. Medea, Bosisio Parini, Italy

Fabiano G. Nery
Mood Disorders Clinical Neurosciences Program, Department of Psychiatry, University of Texas Health Science Center at San Antonio, and Audie L. Murphy Division, South Texas Veterans Health Care System, San Antonio, Texas, U.S.A., and Institute of Psychiatry, University of São Paulo School of Medicine, São Paulo, Brazil

John P. Hatch
Mood Disorders Clinical Neurosciences Program, Department of Psychiatry, and Department of Orthodontics, University of Texas Health Science Center at San Antonio, San Antonio, Texas, U.S.A.

Jair C. Soares
Center of Excellence for Research and Treatment of Bipolar Disorders, Department of Psychiatry, University of North Carolina School of Medicine, Chapel Hill, North Carolina, U.S.A.

HISTORICAL BACKGROUND

The research on neurobiological markers of bipolar disorder has progressed rapidly over the past two decades. The rapid progress started with the early computed tomography studies in the 1980s, followed by studies with newer brain imaging techniques, in particular magnetic resonance imaging (MRI). These studies have focused mainly on brain structures that form a limbic–thalamic–cortical circuit and a limbic–striatal–pallidal–thalamic circuit, which are thought to play a role in the pathophysiology of mood disorders (1–4) (Fig. 1). The prefrontal cortex, amygdala-hippocampus complex, thalamus, basal ganglia, and connections among these areas comprise neuroanatomic brain circuits involved in mood regulation (1). Other areas, not spatially related to this fronto-limbic circuit, including the cerebellum, with its vermal connections to limbic structures, also participate in the regulation of mood (3). Functional, neurochemical and anatomical abnormalities in these brain regions have been examined with several imaging tools including structural MRI, functional MRI (fMRI), magnetic resonance spectroscopy (MRS), single photon emission computerized tomography (SPECT), and positron emission tomography (PET). These studies provide evidence of abnormalities in the prefrontal cortex, medial temporal lobe structures, striatum, and cerebellum of patients with bipolar disorder. Some of the suggested abnormalities include the following: (*i*) Decreased *N*-acetyl-aspartate levels, which is possibly an early marker of neuronal impairment and that could possibly be abnormal even before detectable anatomical MRI changes take place. (*ii*) Cerebellar vermis size decreases with repeated illness episodes, suggesting a neurodegenerative change. (*iii*) Amygdala volumes are directly correlated with age in adolescent bipolar patients, suggesting abnormal neurodevelopmental processes affecting the medial temporal lobe structures during adolescence. (*iv*) Callosal white matter density seems to be reduced in bipolar disorder, which may affect inter-hemispheric connections (5).

It is not known whether such abnormalities have a neurodevelopmental origin or if they result from neurodegenerative or compensatory mechanisms. In this chapter, we review existing neuroimaging literature on bipolar disorder, speculate about their possible origins vis-à-vis a neurodevelopmental versus neurodegenerative hypothesis, and assess emerging new findings that suggest future directions for this research.

STRUCTURAL NEUROIMAGING

Total Brain Volume

Studies evaluating the total brain volume of bipolar patients have produced interesting findings. A meta-analysis by McDonald and colleagues (6) including 404

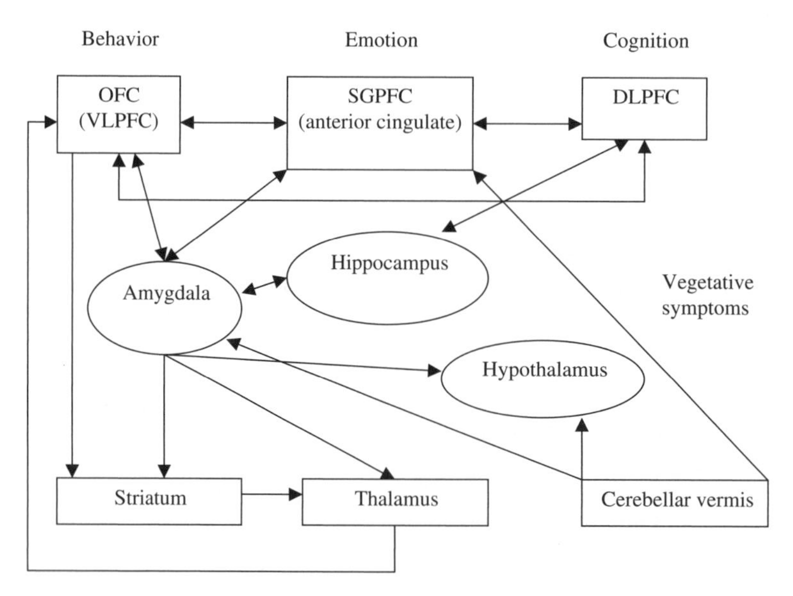

FIGURE 1 Neural circuitry related to the regulation of mood. *Abbreviations*: DLPFC, dorsolateral prefrontal cortex; OFC, orbitofrontal cortex; SGPFC, subgenual prefrontal cortex; VLPFC, ventrolateral prefrontal cortex. *Source*: modified from Ref. 99.

bipolar patients from 26 studies showed that global cerebral volume was preserved in bipolar disorder with the right lateral ventricular enlargement as the only significant abnormality, although the authors noted high levels of heterogeneity among studies. A recent MRI study, using voxel-based morphometry to assess abnormalities of total gray and white matter volumes throughout the entire brain, showed no significant differences in gray matter volume between patients with bipolar I disorder (n = 37, 31 on mood stabilizers, one on olanzapine, five untreated, eight on mood stabilizers plus antipsychotics) and healthy controls. In contrast, bipolar subjects showed frontal and parietal white matter deficits that overlapped anatomically with findings for schizophrenic patients (7). This finding was confirmed and extended in a longitudinal study by Farrow and colleagues (8), who followed first-episode psychosis patients (25 with schizophrenia and 8 with bipolar disorder) for 2.5 years after the initial scan. The authors noted that at the initial assessment, bipolar patients had gray matter deficits in regions differing from schizophrenic patients, localized to bilateral inferior temporal gyri. Also, at follow-up, bipolar patients showed progressive gray matter reduction in the anterior cingulate gyrus, which was not an area that initially differentiated them from healthy controls, pointing to the need for longitudinal anatomical MRI studies in bipolar disorder in order to clarify a possible neurodegenerative component. Three other studies also showed that bipolar patients have smaller cortical gray matter volumes and a more-pronounced age-related decline in total brain gray matter compared to healthy controls (9–11) suggesting some degree of neurodegeneration involved in the pathophysiology of bipolar disorder.

Interestingly, chronic lithium use may be protective against this gray matter loss. One longitudinal (12) study showed a 3% increase in total brain gray matter in eight out of ten bipolar patients treated with four weeks of lithium, while there were no changes in the white matter. Another study, although cross-sectional, showed that bipolar patients treated with lithium had greater total gray matter volumes than untreated bipolar patients and healthy controls, while there were no significant differences in total white matter volumes across three groups (13). These findings are thought to reflect neurotrophic effects of lithium, possibly with resulting neuropil increase. The possibility of mood stabilizers (particularly lithium) preventing or reversing gray matter changes should be kept in mind when interpreting anatomical MRI findings with medicated bipolar patients [i.e. 22 out of 37 bipolar patients in the McDonald et al. (7) study were on lithium]. Further support for the neural plasticity effects of lithium comes from Beyer and colleagues (14), showing enlarged left hippocampus volumes in older (n = 36, mean age = 58 years) bipolar patients compared to healthy controls. The authors thought this increase in hippocampal volume might be associated with lithium use. Future larger longitudinal studies may shed a light on the putative neurotrophic/neuroprotective effects of lithium and valproate in bipolar patients with first- and multiple-episodes, and the relationship of the brain changes with treatment response.

Prefrontal Cortex

The prefrontal cortex, including subgenual prefrontal cortex, anterior cingulate and dorsolateral prefrontal cortex (DLPFC), is a key region in the neuroanatomic model of mood regulation. The subgenual prefrontal cortex is the part of the anterior cingulate situated ventral to the genu of corpus callosum, corresponding to Brodmann area 24. Two studies (15,16) showed smaller left subgenual prefrontal cortex volumes in bipolar patients with a familial form of the disorder and one study (17) noted smaller right subgenual prefrontal cortex gray matter volumes in a combined sample of familial and non-familial bipolar patients. Drevets and colleagues also conducted a postmortem study and showed that the anatomical reduction of subgenual prefrontal cortex gray matter was associated with a selective and significant reduction in glial cell number and density, but not neuronal density or number, in a small group of familial unipolar (n = 6) and bipolar (n = 4) patients compared with healthy controls (n = 11) (18). In contrast, Brambilla and colleagues (19) did not show any volumetric abnormality in this brain region in familial or non-familial bipolar patients. Also, no abnormalities in the subgenual prefrontal cortex volume were found in a study of children and adolescents with bipolar disorder (n = 15) (20). These varying results could be due to different study samples, that is, including patients with bipolar disorder I and II, with first-episode versus multiple-episodes, and especially having a familial affective disorder history (Table 1). Moreover, the neurotropic effects of mood stabilizers should also be considered as a potential confounder in these results, as only the Drevets et al. (15) study had a mostly unmedicated sample.

Another region of the prefrontal cortex that has received much attention lately is the anterior cingulate (Brodmann's areas 24, 25, and 33). An early study (21) using relatively thick MRI slices (6 mm) did not demonstrate any cingulate abnormality in 17 unmedicated patients with bipolar disorder with first-episode mania. However, recent studies using thinner MRI slices show that left anterior cingulate gray matter volume is smaller in adult (22) and juvenile (23) bipolar patients compared to healthy controls. Complementary to the region-of-interest (ROI) volumetric studies, research employing a statistical parametric mapping approach revealed reduced gray matter density in the fronto-limbic cortex, particularly in the cingulate (24,25) (Table 2). In Lyoo et al.'s study (25), the greater number of prior manic episodes was correlated with a greater decrease in the gray matter density in left medial frontal and right inferior frontal gyri, suggesting a neurodegenerative effect over repeated episodes. As noted above, one of these studies showed a possible preventive or reversing effect of lithium on gray matter content, as the cingulate volumes of lithium-treated bipolar patients did not differ from those of healthy comparison subjects, while the untreated bipolar patients had smaller left anterior cingulate volumes compared to healthy controls (22).

DLPFC (Brodmann's areas 9 and 46) is a region with a main role in executive functions, working memory, and regulation of emotion. One study showed significantly smaller gray matter volumes in the left middle and superior, and the right middle and inferior prefrontal regions in 17 bipolar patients hospitalized for a manic episode and receiving various psychotropic medications (9). Patients in this study also presented a smaller left inferior prefrontal gray matter with increasing illness duration and smaller right inferior prefrontal gray matter related to antidepressant exposure. Although based on a cross-sectional design and in a small sample, the authors underline the need to further explore this possible antidepressant-related volume reduction as a possible way to understand the cycle acceleration induced by antidepressants in bipolar disorder. Consistent with the finding of reduced gray matter volume in prefrontal cortex in adult patients with bipolar disorder, a recent study using voxel-based morphometry demonstrated decreased gray matter volume in the left DLPFC in pediatric bipolar patients (n = 20, mean age 13.4 $\pm$ 2.5 years, bipolar I and II disorder, 19 of the patients on psychotropic medications) (26).

TABLE 1 Studies Measuring the Subgenual Prefrontal Cortex in Bipolar Disorder

Authors	Patients with bipolar disorder (age)	Healthy comparisons (age)	Field strength	Findings
Drevets et al., 1997	21 BP (35 ± 8.2) 6 medicated and 15 drug-free All with familial BD	21 (34 ± 8.2)	1.5 T	Left subgenual prefrontal cortex gray matter volume 39% smaller in patients
Hirayasu et al., 1999	24 patients with first-episode affective psychosis (23.7 ± 5.1) 21 manic, 3 depressive episode 14 with familial BD on various antipsychotics	20 (24.0 ± 4.3)	1.5 T	Left subgenual prefrontal cortex gray matter volume smaller in patients with a familial form of the disorder
Brambilla et al., 2002	27 BP (35 ± 11) 21 BP I, 6 BP II 12 with familial BD 11 unmedicated 16 on lithium monotherapy	38 (37 ± 10)	1.5 T	No significant differences in subgenual prefrontal cortex volumes
Sharma et al., 2003	12 BP I (38.3 ± 6.2), 6 with familial BD on various mood stabilizers	8 (38.3 ± 6.9)	4.0 T	Right subgenual prefrontal cortex volume reduction
Sanches et al., 2005	15 BP (15.5 ± 3.5) 11 BP I 3 BP II 1 BP NOS all with familial BD 13 on mood stabilizers	21 (16.9 ± 3.8)	1.5 T	No significant differences in subgenual prefrontal cortex volumes

Abbreviations: BD, Bipolar disorder; BP, Bipolar patients; NOS, Not otherwise specified.

TABLE 2 Studies Measuring the Cingulate in Bipolar Disorder

Authors	Patients with bipolar disorder (age)	Healthy comparisons (age)	Field strength	Findings
Strakowski et al., 1993	17 BP I (28.4 ± 6.8) first-episode mania Unmedicated	16 (30.9 ± 7.3)	1.5 T	No significant differences in the cingulate volume
Sassi et al., 2004	27 BP (35.1 ± 10) 21 BP I, 6 BP II 11 untreated 16 on lithium monotherapy 14 with familial BD	39 (37 ± 10)	1.5 T	Untreated BP had decreased left anterior cingulate volumes compared with HC and lithium-treated BP, lithium-treated BP not significantly different from HC
Doris et al., 2004	11 BP I (40.5 ± 11.6), on various medications	15 (39.1 ± 10.8)	2.0 T	Decreased gray matter density in fronto-limbic regions, mainly the cingulate (more common in the right hemisphere)
Lyoo et al., 2004	39 BP I (38.3 ± 11.6) 15 unmedicated, 24 on mood stabilizers 11 drug-naïve	43 (35.7 ± 10.1)	1.5 T	Decreased left anterior cingulate and left medial frontal gyrus gray matter density
Kaur et al., 2005	15 BP (15.5 ± 3.5) 11 BP I, 3 BP II, 1 BP NOS all with familial BD 13 on mood stabilizers	21 (16.9 ± 3.8)	1.5 T	Left anterior and posterior, and right posterior cingulate smaller in BP

Abbreviations: BD, Bipolar disorder; BP, Bipolar patients; NOS, Not otherwise specified.

Available studies report smaller cingulate and subgenual prefrontal cortex volumes in bipolar patients. DLPFC is an understudied region in MRI studies, possibly due to the lack of a sound manual ROI tracing method delineating Brodmann's areas 9 and 46. But with the availability of automated methods like voxel-based morphometry, future MRI studies can be directed to further clarify the anatomical changes in the DLPFC in bipolar patients.

White Matter and Corpus Callosum

One important aspect of the pathophysiology of bipolar disorder may be the loss of effective connectivity between prefrontal areas and limbic regions, as evidenced by white matter changes. One of the most consistently reported anatomical finding in bipolar disorder is the "white matter hyperintensities" (27,28). These lesions represent a change in the water content in the brain, and their etiology is unknown, although some associations with age, vascular pathology, and inflammatory response in the anterior cingulate have been described (29,30). White matter hyperintensities have been associated with some clinical characteristics of the disease, such as poor long-term outcome, female gender, multiple psychiatric admissions, and better response to lithium treatment (27,31,32) and are present even in bipolar adolescents (29).

The relationship between white matter hyperintensities, elements of one-carbon cycle metabolism (including serum folate, vitamin B_{12}, and homocysteine levels), and the outcome of antidepressant treatment was investigated in a relatively young sample of outpatients (mean age 40.6 ± 10.3 years) with major depressive disorder (33,34). It was shown that hypofolatemic and hypertensive patients had more severe white matter hyperintensities than normal controls. Also, the severity of subcortical white matter hyperintensities and the presence of hypofolatemia independently predicted lack of clinical response to antidepressant treatment. The authors concluded that hypofolatemia and hypertension might represent modifiable risk factors to prevent the occurrence of white matter hyperintensities.

Structural brain imaging studies also have revealed various other white matter abnormalities in bipolar patients. White matter density reductions in the anterior limb of internal capsule, connecting anterior thalamic nucleus to the frontal lobe, were found in patients with familial bipolar I disorder, suggesting a fronto-thalamic connection abnormality, common to both schizophrenia and bipolar disorder (35). Studies have also shown white matter volume deficits in bipolar I patients and their unaffected co-twins (in the left hemisphere) (36), familial male bipolar I patients (37), first-episode bipolar patients (21), and children and adolescents with bipolar disorder (in left superior temporal gyrus) (38). Presence of white matter and axonal disorganization in frontal and prefrontal regions was also shown in adult (39) and young first-episode (40) bipolar patients, by diffusion tensor imaging studies.

The corpus callosum is the major white matter commissure connecting the cerebral hemispheres. It plays a crucial role in inter-hemispheric communication (41) and higher cognitive processes, such as sensory-motor integration, attention, arousal, language, and memory (42,43). Corpus callosum is also an important structure for the development of structural brain asymmetry, resulting in cerebral functional lateralization in humans. Anatomical asymmetry is believed to be the substrate of the functional lateralization, and corpus callosum size reflects inter-hemispheric connectivity, suggesting that cerebral asymmetry occurs with decreased callosal connectivity (44). Inter-hemispheric information exchange decreases with increasing cerebral asymmetry, leading to hemispheric independence/dominance. Therefore, corpus callosum abnormalities may also play a major role in affecting the development of structural and functional lateralization in bipolar disorder, ultimately resulting in abnormal inter-hemispheric communication.

Corpus callosum area is reduced in adult bipolar patients (45,46). Abnormally reduced corpus callosum signal intensity was found in adult bipolar patients (47), and lower circularity of the splenium was observed in young (mean age 16) patients with bipolar disorder (48), suggesting abnormalities in corpus callosum white matter in bipolar patients, possibly due to altered myelination, which may lead to impaired inter-hemispheric communication.

In conclusion, it is likely that white matter hyperintensities are involved in the pathophysiology of bipolar disorder by interrupting essential neural pathways in the brain that are involved in mood regulation, as put forth by Soares and Mann (1). Researchers also suggested that left hemisphere white matter changes may reflect genetic factors predisposing to bipolar disorder (36), and that glial loss might underlie both the white matter hyperintensities and volume reductions observed in bipolar patients (49). Evidence from brain imaging studies point to the importance of examining white matter changes in bipolar disorders.

Medial Temporal Structures

Three different research groups (50–52) reported amygdala enlargement in adult bipolar patients, suggesting that hypertrophy of this region might reflect dysfunction underlying the mood lability of bipolar disorder. In contrast, two studies (53,54) found smaller amygdala sizes in adult bipolar patients. In children and adolescents with bipolar disorder, there is more consistent evidence of smaller amygdala volumes (54–59). In Chen and colleagues' (55) cross-sectional study, pediatric bipolar outpatients ($n = 16$, 14 on mood stabilizers, 11 with comorbid Axis I disorders) showed a direct correlation between left amygdala volume and age, in contrast to healthy comparisons, who showed the expected inverse relationship. The tendency for larger amygdala volumes to occur in the older bipolar children may be the result of some compensatory mechanisms of developmental abnormalities. One recent longitudinal study (59) rescanned the same 10 bipolar I outpatients (7 with various Axis I comorbidities, all exposed to various psychotropics between first and second scans) after an average of 2.5 ± 0.4 years. The authors reported the persistence of decreased amygdala volumes in adolescents and young adults with bipolar disorder, with healthy controls having an increase in the left amygdala volume and bipolar patients having a decrease in the right amygdala volume, although both small and nonsignificant (59). Preliminary evidence suggests that lithium and valproate treatment is associated with increased amygdala volume in pediatric bipolar patients (58). In Chang et al.'s study (58), bipolar patients with past lithium or valproate exposure tended to have greater amygdala gray matter volumes than bipolar patients without such exposure, and the authors concluded that prolonged medication exposure to lithium or valproate may account for findings of increased amygdala volumes relative to healthy controls in adults with bipolar disorder. Further support for this possible neurotropic effect of mood stabilizers comes from a neuropathological study (60) reporting higher glial numbers in the amygdala of 10 bipolar patients who were chronically treated with either lithium and valproate compared with 2 bipolar patients who did not receive lithium and/or valproate and 10 healthy subjects, although this needs to be replicated.

Hippocampus is another temporal lobe structure with important mood-regulating functions. One study, using 0.5 T field strength and 10 mm image slices, found smaller right hippocampus in 48 bipolar patients (more pronounced in males) (61), but this finding was not replicated in several other studies with moderate sample sizes (~25) (50,51,62–65). Only in one study (65) did the authors compare drug-free (n = 9) and lithium-treated bipolar patients (n = 15), and they did not find any significant differences of hippocampal measurements between the two groups. One recent study (66) reported a trend toward smaller left hippocampal volumes in psychotic bipolar patients (n = 38, 23 with psychosis) compared to healthy subjects, suggesting that future anatomical studies measuring hippocampus in bipolar disorder should also take into account psychosis. Moderate to severe forms of hippocampal shape anomaly (characterized by a rounded hippocampus) were observed in patients with familial schizophrenia, whereas patients with familial bipolar disorder did not have such anomaly (67). The findings in children and adolescents are different from the adults, as two studies found smaller hippocampal volumes in children and adolescents with bipolar disorder relative to healthy subjects (54,68). Two possible explanations for this discrepancy between adult and pediatric findings are that age at onset may differentially affect brain structure, or that childhood- or adolescent-onset bipolar disorder may be a distinct disorder from adult-onset illness with a different set of neuroanatomic correlates, as put forth by Frazier and colleagues (68).

In conclusion, there is very credible evidence for smaller amygdala volumes in pediatric bipolar disorder (69), and there is also some evidence from adult studies to suggest enlargement of this structure. Brambilla and colleagues (5) suggested that abnormal pruning mechanisms in childhood and adolescence or, alternatively, compensatory mechanisms might lead to enlargement of the amygdala in adulthood. An alternative explanation for a larger amygdala in adulthood, as put forth by Chang and colleagues (58), is prolonged exposure to lithium or valproate. Longitudinal studies, in children and adolescents with bipolar disorder, in high-risk populations, and also with controlled medication treatment, are needed to define the nature of amygdala abnormalities in bipolar disorder. The hippocampal volume changes in bipolar disorder are not well established. Also, antidepressant and antipsychotic treatments, as well as mood stabilizers, may affect hippocampal volume, and so far no study has examined the hippocampus in a sizable sample of untreated bipolar individuals.

Thalamus

The thalamus is another key structure in brain anatomic circuits involved in the pathophysiology of mood disorders. ROI-based MRI studies yielded conflicting results concerning the thalamus in bipolar patients. Two studies with sizable samples (i.e., ≥24) found larger thalamic volumes in bipolar patients (51,70). Nonetheless, six other studies, with two conducted in young bipolar patients (68,71), failed to replicate these findings (21,72–74). Also, no significant difference was found in the thalamic volumes of first- versus multiple-episode adult bipolar patients (73) and healthy controls. Young and colleagues (75) reported that neuron numbers and volumes in limbic thalamic nuclei were normal in patients with schizophrenia and bipolar disorder. However, a recent study utilizing voxel-based morphometry in patients with schizophrenia and bipolar disorder and their unaffected relatives demonstrated gray matter reductions in the anterior thalamic nucleus (receiving major input from the hippocampus and mamillary bodies) in all groups compared to healthy controls, suggesting that this abnormality might be a marker of liability for both disorders (76) and especially for psychosis, but this finding needs to be replicated.

In conclusion, there are no consistent findings about thalamic anatomical changes in bipolar disorder. Again, the distinction between psychotic and non-psychotic bipolar disorder may be playing a role here, and needs to be further investigated, preferably with more advanced techniques that can measure specific thalamic nuclei.

Basal Ganglia

The basal ganlia are part of the neural networks underlying mood regulation, with extensive connections to prefrontal areas. The earlier anatomical studies about the basal ganglia in bipolar disorder did not find any significant abnormalities (21,61,70,77). However, a more recent study by Brambilla and colleagues (78) reported that age and length of illness may have significant effects on basal ganglia structures in bipolar patients (more pronounced among bipolar I patients). In this study, the length of illness predicted smaller left putamen volumes, and older patients (>36 years) had a significantly larger left globus pallidus than younger ones (≤36 years). Also, two studies in adult (51,79) and one in adolescent (57) bipolar patients reported enlargement of basal ganglia structures. Recently, caudate size was also shown to be decreased in older (mean age 58) bipolar patients (14). The age-related decline in some basal ganglia structures is also documented in children and adolescents with bipolar disorder by Sanches and colleagues (80). They found a significant inverse correlation between age and the volumes of left caudate, right caudate, and left putamen in bipolar patients, not present in healthy controls. These findings suggest that bipolar patients may have more pronounced age-related changes in basal ganglia structures than healthy individuals (81).

Cerebellum

The cerebellar hemispheres do not seem to be affected in bipolar disorder, as shown by several MRI reports (82–85). However, there may be anatomical changes in the cerebellar vermis that originate from neurodegenerative processes during the course of the illness. Vermis area V3 (the inferior posterior and flocculonodular lobes, lobules VIII-X) seems to be smaller in patients with multiple episodes than in first-episode bipolar patients and healthy volunteers (82,83,85). Another recently conducted volumetric study showed that patients with multiple episodes had smaller V2 volumes compared to first-episode patients and healthy controls (84). In bipolar children and adolescents with familial mood disorder, smaller vermis V2 area (superior posterior vermis, neocerebellar vermal lobules VI-VII) was reported (85). Taken together, these findings suggest that cerebellar vermal abnormalities occur in bipolar disorder, and that these are likely neurodegenerative and progressive.

FUNCTIONAL NEUROIMAGING

Blood Flow and Metabolism Studies

Earlier positron emission tomography (PET) studies (86–90) showed that bipolar patients had decreased cerebral blood flow (i.e., hypofrontality) in prefrontal cortices in the resting state. Martinot and colleagues also reported that successful medication treatment with tricyclic antidepressants could reduce the left-right prefrontal asymmetry, but not the hypofrontality and whole-cortex hypometabolism that was found in the patients in the depressed state (90), suggesting that this was more a trait-like abnormality. Patients with bipolar affective illness (n = 16) also had significantly lower metabolic rates in their basal ganglia in comparison to normal controls, by PET using [18F]2-deoxyglucose as a tracer (88).

These earlier SPECT and PET studies were followed by newer studies using more sophisticated techniques to examine the effect of mood state and drug treatment. Gyulai and colleagues (91) examined the same patients (n = 12) in depressed/dysphoric, manic/hypomanic, and euthymic states and concluded that there may be a state-dependent dysfunction in the temporal lobe, as evidenced by an asymmetric distribution of I-123 iofetamine in depression but not in euthymia, in bipolar patients with rapid cycling. One SPECT study with *n*-isopropiliodoamphetamine showed increased blood flow in the temporal lobe and basal ganglia in manic state (n = 11) (92). Recently, mood state-dependent alterations in blood flow and metabolism in other brain regions were also reported. Blumberg and colleagues reported greater manic versus euthymic state-related activity in the anterior cingulate and caudate (93), and decreased activity in the orbitofrontal cortex (94), although the sample sizes were small (5 manic and 6 euthymic patients in both studies, on various psychotropics). The findings of Blumberg and colleagues' studies also converge with Drevets and colleagues' (15) report that bipolar patients in the manic phase (n = 4) have a higher metabolism in the subgenual prefrontal cortex, compared to healthy subjects and bipolar depressed patients. Goodwin and colleagues (95) reported decreased anterior cingulate and right caudate blood flow after lithium withdrawal, whereas development of mania on lithium withdrawal was associated with increased anterior cingulate blood flow. One explanation put forth by Blumberg and colleagues (93) for the heightened anterior cingulate and caudate activity was a compensatory mechanism for the increased demand to limit "manic-type distractibility and maladaptive excessive behaviors" under scan conditions. Cingulate hypermetabolism as an adaptive response to depression, and its failure underlying a poor outcome to antidepressant treatment, were also reported by Mayberg and colleagues (96). Also, an fMRI study by Caligiuri and colleagues (97) showed that affective state in bipolar disorder might be related to a disturbance of inhibitory regulation within the basal ganglia and that antipsychotics and/or mood stabilizers might normalize cortical and subcortical (basal ganglia and thalamus) hyperactivity.

Recently, Drevets and colleagues (98), using higher-resolution glucose metabolism images and MRI-based ROI analysis, showed that the amygdala metabolism in remitted bipolar patients (n = 8) was intermediate between that of healthy controls (n = 12) and depressed bipolar subjects (n = 7), although not significantly different from either. The left and right amygdala metabolism was increased in the remitted bipolar patients who were not taking mood stabilizers relative to the patients who were taking one of these agents and healthy comparisons. The authors concluded that chronic mood stabilizer treatment might normalize the increased amygdala metabolism, and thus reduce and prevent pathological mood episodes (98).

In conclusion, there may be an elevated limbic activity with deficient frontal modulation, underlying the illness recurrence in mood disorders (98,99).

Activation Studies

Functional imaging studies coupled with cognitive testing provide a helpful tool to understand the neurophysiology of bipolar disorder. Bipolar patients have attention, memory, and executive deficits that are apparent during affective episodes (72), but also persist in euthymia (100). Results from studies (100–103) indicate that euthymic bipolar patients have poorer performance than healthy comparisons on working memory tasks, and they exhibit increasing impairment as the task becomes more difficult, and suggest that memory impairments might be more a trait marker than state in bipolar disorder (104). One recent study examined verbal learning in eight euthymic, remitted bipolar I patients (7 of them unmedicated) with ^{15}O PET scanning while doing a verbal learning paradigm. Bipolar patients had more difficulties learning the lists of words compared with the control subjects, and they had a blunted regional cerebral blood flow increase in the left DLPFC during encoding the words (104).

The cognitive deficits that persist during euthymia were hypothesized to arise from an over-reactive anterior limbic network (including prefrontal regions, thalamus, striatum, amygdala, and the cerebellar vermis), which affects the functioning of the cognitive network through reciprocal connections. Strakowski and colleagues reported evidence of anterior limbic overactivation during a simple attentional task in euthymic, unmedicated bipolar patients (101). In this study, bipolar patients' performance on the task was similar to that of healthy subjects, but they exhibited activation in different cortical regions, and the authors interpreted this as a compensation for interference from emotional brain networks in order to maintain task performance. As a next step, they incorporated a more challenging attention task in which compensatory mechanisms would be overwhelmed in patients, thereby further differentiating brain activation patterns between the groups (105). Activation of the anterior cingulate (Brodmann's area 24/32) and DLPFC (Brodmann's area 9) while performing the Stroop task were not significantly different between groups, but patients demonstrated a pattern of lower activation in temporal regions, midline cerebellum, ventrolateral prefrontal cortex, and putamen. Failure to activate secondary brain regions involved in this task may have contributed to impaired task performance in bipolar patients, as put forth by Strakowski and colleagues (105).

Results of a recent SPECT study exploring the possible correlation of neuropsychological functioning and cerebral blood flow in patients with bipolar disorder ($n = 30$, all unmedicated, 7 manic, 8 hypomanic, 12 depressed, and 3 euthymic) showed that worse performance on memory, executive, and attention tasks was related to a greater perfusion in the striatum (106). In this study, patients with poor performance on tests of executive functions also showed low perfusion in the frontal region and cerebellum and high perfusion in the cingulate. In another study, bipolar patients showed activation in the thalamus and amygdala in addition to cortical activation in regions implicated in the cognitive processing of affective stimuli (107). The authors concluded that the bipolar brain manifested a different perfusion pattern than the healthy brain, and perhaps an abnormal cognitive emotional processing through ventral prefrontal cortex dysfunction (103,107). A recent ^{18}F-fluorodeoxyglucose-PET study, conducted in a small group ($n = 8$) of unmedicated depressed bipolar patients using the continuous performance test to assess sustained attention and vigilance, provided additional evidence that decreased prefrontal activity may represent failure to activate some areas of inhibitory control (108). This study showed that decreased subgenual prefrontal cortex metabolism was related to decreased attention, and decreased DLPFC metabolism was related to decreased inhibitory control.

Krüger and colleagues (109) assessed bipolar risk and resilience in a very small group of lithium-responsive bipolar patients ($n = 9$) and their healthy siblings ($n = 9$) using an emotional challenge test. They reported that both patients and their siblings had rCBF increases in the anterior cingulate and decreases in the orbitofrontal and inferior temporal cortices. In addition, the siblings had an increased metabolism in the medial frontal cortex, a finding that was not observed in the patients, which the authors commented as being a compensatory response, and possibly conferring resilience to these high-risk individuals. However, the lack of a healthy control group in this study limits the interpretation of these findings. The authors also compared lithium-responsive patients to valproate-responsive bipolar patients from an earlier study by their group, and concluded that changes specific to the DLPFC and rostral anterior cingulate distinguished these patient groups from each other.

In conclusion, both resting blood flow and glucose metabolism studies using SPECT and PET and studies using cognitive tasks during functional imaging complement results from other imaging modalities showing that limbic dysregulation by prefrontal regions is a key element in the pathophysiology of bipolar disorder.

Neurochemical Brain Imaging

PET and SPECT Neuroreceptor Studies

Dopaminergic System

Dopaminergic system abnormalities are thought to underlie mania and depression, and especially the psychosis associated with these conditions. Two PET studies measuring dopamine D_2 receptor density, as measured by D_2 binding potential with [^{11}C]raclopride (110), and [^{11}C]*N*-methylspiperone as a ligand (111) showed that D_2 receptors were unaltered in acute non-psychotic mania. Also, presynaptic dopamine function as reflected by [^{18}F]DOPA uptake does not seem to be altered in medication-naïve first-episode bipolar patients with non-psychotic mania (112). In contrast, patients with psychotic mania have a higher D_2 receptor density compared with healthy comparisons, and the authors concluded that the higher density of D_2 receptors might be related to psychosis and not to manic symptoms (111). Yatham and colleagues (110) also showed that although divalproex sodium treatment did not seem to affect D_2 receptor availability, presynaptic dopamine function in manic patients was lower after treatment with divalproex sodium as measured by the [^{18}F]6-fluoro-L-dopa ([^{18}F]DOPA) uptake in the striatum (112). The authors concluded that divalproex sodium might exert its antimanic effects via reducing the rate of dopamine synthesis (110,112).

In vivo brain intrasynaptic release of dopamine after amphetamine release can be quantified by currently available neuroimaging methods. A SPECT study investigating the differences in amphetamine-induced dopamine release in euthymic bipolar disorder patients and healthy comparison subjects found no differences in baseline D_2 receptor binding between the two groups (113). Although there was a greater behavioral response to an amphetamine challenge in euthymic bipolar disorder patients, this was not accompanied by a significant difference in decrease in [^{123}I]IBZM binding between the two groups. As a decrease in [^{123}I]IBZM binding is related to the amount of striatal dopamine release, the authors concluded that mania-like symptoms induced by amphetamine may not be related to increased dopamine release.

Suhara and colleagues (114) studied D_1 binding potential in 10 bipolar patients in various mood states, and found that the binding potentials for the frontal cortex were significantly lower in patients than healthy controls, whereas those for striatum were not significantly different, suggesting that D_1 dopamine receptors in the frontal cortex may be in a different state in patients with bipolar disorder, but this finding needs to be replicated.

In conclusion, very few in vivo brain-imaging studies examined dopamine receptors in bipolar patients, and they suggested that D_2 receptor abnormalities might be related to psychosis in bipolar patients, and dopamine receptor blockade might contribute to the anti-manic property of medications. The levels of dopamine receptors in brain regions other than striatum should be assessed in future studies. Also, future research should be directed to examining dopamine transporters in medication-free bipolar individuals.

Serotonergic System

Involvement of the serotonergic system in the pathophysiology of mood disorders is suggested by PET studies conducted mostly in patients with unipolar depression. One study showed decreased mid-brain serotonin transporter (5-HTT) density (115) in drug-free unipolar depressed patients. Another study conducted with 13 drug-free male patients with mood disorders (seven with major depressive disorder and six with bipolar disorder) showed no significant change in 5-HTT binding in the midbrain, but there was elevated thalamic binding in unipolar patients compared to bipolar patients and healthy controls (116).

$5HT_{1A}$ receptor binding potential was also studied in vivo, and found to be reduced in the raphe and mesiotemporal cortex, most prominently in bipolar depressive patients ($n = 4$) and unipolar depressive patients with bipolar relatives ($n = 4$) (117). Sargent and colleagues (118) reported widespread reduction in cortical postsynaptic $5\text{-}HT_{1A}$ receptor binding potential measured by positron emission tomography (PET) with [^{11}C]WAY-1006357 in unmedicated depressed patients, with a similar reduction in binding of [^{11}C]WAY-100635 to $5\text{-}HT_{1A}$ autoreceptors in the raphe nuclei in the midbrain. These reductions persisted in patients who responded clinically to treatment with selective serotonin reuptake inhibitors, and the authors hypothesized that lowered post-synaptic $5\text{-}HT_{1A}$ receptor binding availability might be a trait abnormality and a vulnerability factor in recurrent unipolar depression. They tested this hypothesis recently with 14 male euthymic unipolar patients, who had been free of psychotropic medications for at least six months, and reported a persistent dysfunction in cortical $5\text{-}HT_{1A}$ binding potential (119), supporting their hypothesis. Further studies are clearly needed to determine the origin of this reduction in cortical $5\text{-}HT_{1A}$ receptor availability in bipolar and unipolar patients.

Brain $5\text{-}HT_2$ receptors have also been examined in vivo in unipolar disorder patients. A PET study (120) reported a decrease in uptake of [^{18}F]altanserin in the right anterior part of insular cortex and right posterolateral orbitofrontal cortex in untreated unipolar depressed patients, consistent with 5-HT-2A receptor levels. Three other PET studies, that used [^{18}F]setoperone as a tracer, reported a significant decrease in [^{18}F]setoperone binding in the frontal cortex (121) and frontal, temporal, parietal, and occipital cortical regions (122) of depressed patients compared with control subjects, or no difference between the two groups (123). Another recent PET study using (^{18}F)-setoperone and conducted in seven bipolar patients in a manic episode (drug naïve or psychotropic medication free for at least one week) and treated with valproate (n = 7) and lithium (n = 1) for 3 to 5 weeks, showed that treatment with valproate alone or in combination with lithium had no significant effect on brain 5-HT2A receptor binding (124). 5-HT_{2A} receptors seem to be altered in major depression, and some researchers suggested that the reduction in brain 5-HT_{2A} receptor density might be the result of a homeostatic mechanism by the brain to compensate for depression (122). Studying 5-HT_2 receptors in bipolar patients is an intriguing area of research, as atypical antipsychotics that are effective in the treatment of acute mania and as mood-stabilizers have greater affinity for 5-HT_{2A} receptors than dopamine receptors (125).

Biochemical Findings

Magnetic resonance spectroscopy (MRS) is unique among in vivo brain imaging modalities as being a noninvasive way to measure levels of different chemicals in the brain. ^{1}H-MRS can be used to quantitate neurochemicals such as *N*-acetyl-aspartate (NAA), choline, myo-inositol, creatine, glutamate, glutamine, and GABA. ^{31}P-MRS provides information about phosphorus-based membrane processes and metabolism, by estimation of brain levels of phosphomonoesters, phosphodiesters, and pH.

NAA is an amino acid found in high concentrations in mature neurons and considered to be a marker of neuronal integrity, the formation and maintenance of myelin and mitochondrial energy production (126,127). Lower dorsolateral prefrontal cortex NAA levels were reported in adult (n = 20, euthymic, unmedicated) (128) and pediatric (129) bipolar patients, and also in both treatment-naïve (130) and medicated (131) adolescent patients with bipolar disorder. Decreased NAA levels were reported in other brain regions including orbital frontal gray matter (132) and hippocampus (133,134) in adult patients with bipolar disorder. This decrease in NAA may be the result of neurodegenerative changes of long-standing illness duration, as a recent study by Gallelli and colleagues (135) in 60 children of parents with bipolar I or II disorder (32 with bipolar disorder and 28 with subsyndromal symptoms of bipolar disorder) did not reveal any significant changes in dorsolateral prefrontal NAA levels. Two other studies did not find significant differences of dorsolateral prefrontal cortical NAA between mostly medicated adult bipolar patients and healthy volunteers (133,136). There are other negative studies where no abnormalities of NAA were found in other brain regions such as the basal ganglia (133,137–139), cingulate (133), and thalamus (133). However, these studies included mostly medicated subjects, and it is possible that any regional abnormalities in NAA levels that were present and related to the illness were altered by neurotrophic/neuroprotective effects of mood stabilizers. Chronic lithium treatment was shown to increase NAA concentrations in total brain (140), temporal lobe (141), and basal ganglia (139), providing indirect evidence of its proposed neurotrophic/neuroprotective effects (142). Also, Brambilla and colleagues (136) reported significantly higher NAA/PCr + Cr ratios in lithium-treated bipolar patients compared to unmedicated ones and healthy controls. This NAA increase has not been demonstrated with sodium valproate (141).

Choline is another neurochemical that can be measured via ^{1}H-MRS; it is a membrane component and considered a potential biomarker for the status of membrane phospholipid metabolism. Choline levels (measured as a ratio to creatine) in the basal ganglia were elevated in euthymic (138,139) and depressive (137) bipolar patients. Lithium and valproate treatment do not alter basal ganglia choline resonance in bipolar patients (143–145) or healthy volunteers (146). Thus, the increased choline in basal ganglia may not be related to medication exposure, and may represent an important aspect of the pathophysiology of bipolar disorder, although future studies with drug-free patients and technologies that would allow the measurement of free choline are warranted.

Glutamate, glutamine, and GABA (Glx peak) are also of interest, as antiglutamatergic and GABAergic anticonvulsants are useful in treating bipolar disorder and glutamatergic abnormalities may be involved in neurotoxicity that is potentially responsible for specific brain insults present in bipolar disorder. In a group of depressed bipolar adolescents, Castillo and colleagues (147) reported a bilateral increase in Glx in frontal cortex and basal ganglia. Another study that examined medication-free patients with bipolar disorder found elevated gray matter Glx and lactate concentrations, which were postulated to reflect bioenergetic alterations (148). In an extension of the previous study, the authors showed that the gray matter Glx elevation was reduced with lithium treatment, but not with valproate treatment (145).

Myo-inositol (mI) is another important metabolite in proton spectroscopy that is a substrate for phosphoinositide cycle and a marker for glia, as it is actively transported into astrocytes (149). Lithium inhibits inositol monophosphatase and polyphosphate-1-phosphatase that are involved in recycling inositol mono- and poly-phosphates to myo-inositol. Sodium valproate also decreases myo-inositol concentration and increases the concentration of inositol monophosphates in rat brain (150). In unmedicated euthymic bipolar patients, myo-inositol levels appear unchanged in the frontal lobe (128). Moore and colleagues (151) found decreased myo-inositol in the right frontal lobe of depressed bipolar subjects following acute (5–7 days) lithium administration, which persisted through one month of treatment. However, the patients' clinical state was clearly unchanged at this time, supporting the hypothesis that the initial actions of lithium may occur with a reduction of myo-inositol, and that this reduction initiates a cascade of secondary changes that are ultimately responsible for the therapeutic efficacy of lithium. Another study did not find any differences in myo-inositol levels in the cingulate cortex of depressed bipolar patients and healthy controls (152). However, Davanzo and colleagues (153) observed a significant decrease in anterior cingulate mI/Cr ratios following seven days of lithium therapy in children and adolescents with early-onset bipolar disorder, and children in the manic phase of bipolar disorder had elevated mI/Cr levels within the anterior cingulate cortex.

Another metabolite that is closely related to myo-inositol and the phosphatidylinositol cycle is phosphomonoester (PME) level. PME can be quantified using ^{31}P-MRS and it represents precursors of membrane phospholipid metabolism. In a comprehensive series of studies, a group of investigators (154–156) demonstrated that frontal lobe PME levels vary with mood state in bipolar patients. Deicken and colleagues (157) also reported significantly reduced PME in both the right and left temporal lobes in unmedicated, euthymic bipolar patients compared with healthy controls. An increase in PME concentration with 7 and 14 days of lithium administration in the human brain in vivo was observed (158), and as patients in the above-mentioned studies were mostly on lithium or off lithium for short periods of time, increased levels of PME could reflect medication effects. This is significant as lithium inhibits inositol monophosphatase, producing increased levels of PME, which would be consistent with increased membrane anabolism.

^{1}H-MRS studies have been using the creatine + phosphocreatine peak as an unchanging internal standard, an application that has recently been challenged by several studies showing that both creatine and phosphocreatine peaks change in bipolar disorder. Hamakawa and colleagues (159) showed that creatine concentrations were reduced in the frontal lobe of patients with bipolar depression, whereas in another study euthymic male bipolar I patients had higher creatine concentrations in the thalamus (160). Medication treatment also may modify creatine levels, as O'Donnell and colleagues (161) showed that both lithium and valproate changed creatine concentrations. Thus, with higher field strength magnets becoming more available, researchers are now more inclined to quantify absolute metabolite concentrations or brain water concentration as the reference peak.

In conclusion, MRS studies have identified changes in cerebral concentrations of NAA, glutamate/glutamine, choline, myo-inositol, lactate, phosphocreatine, phosphomonoesters, and intracellular pH in bipolar patients. A recent review of MRS studies in bipolar disorder, focusing on the dysfunction of cellular energy metabolism, proposed a hypothesis of mitochondrial dysfunction in bipolar disorder that involves impaired oxidative phosphorylation, a resultant shift toward glycolytic energy production, a decrease in total energy production and/or substrate availability, and altered phospholipid metabolism (127).

The MRS studies carried out in bipolar patients generally have used lower field strength magnets, which did not allow optimal resolution of many metabolites of interest, a problem that has recently been targeted by improved MRS methods. Also, longitudinal MRS investigations with larger samples of untreated patients are needed to further explore the putative neuroprotective effects of lithium and their relationship to treatment response in bipolar disorder.

Contrast with Findings in Unipolar Disorder

Strakowski et al. (162) reviewed magnetic resonance imaging studies of mood disorders and concluded that bipolar and unipolar disorders may share the pathology of reduced volumes of prefrontal structures, causing loss of cortical modulation over limbic structures. In regards to subcortical brain regions, the most striking difference between unipolar and bipolar disorders seems to be in medial temporal lobe structures, when the hippocampus and amygdala are measured as discrete structures. Magnetic resonance imaging studies demonstrate that hippocampal volume is decreased in patients with recurrent depression, whereas hippocampal volume is preserved in bipolar disorder (6,163). The cellular basis for this observed reduction in hippocampal volume in unipolar depression may be a significant reduction in neuropil (the lattice of glial cells and their processes, dendrites, and proximal axons surrounding neuron cell bodies), as detected in a recent postmortem study by Stockmeier and colleagues (164).

The volumes of the basal ganglia are also different in unipolar and bipolar disorders. In contrast to most of the studies of bipolar disorder, decreased volumes of basal ganglia have been reported in unipolar depression, although it is not clear whether medication effects are ruled out (81). Future studies comparing unmedicated bipolar and unipolar patients for this specific brain region are needed.

Bipolar disorder and unipolar depression also differ regarding the corpus callosum abnormalities. In contrast to bipolar disorder, in unipolar depression the anterior and posterior quarters of the corpus callosum were larger in depressed patients than in controls in an earlier study (165). Consistent with this study, Lacerda and colleagues (166) also found significantly increased anterior (middle genu) and posterior (anterior and middle splenium) callosal sub-divisions in patients with familial major depression. Although the functional significance of the observed callosal changes and their roles in the pathophysiology of mood disorders remain unclear, future neuroimaging and neuropsychological studies with larger patient samples will possibly clarify the difference between bipolar and unipolar disorders regarding this brain structure.

As for the results of postmortem studies, there may be regional "disease-specific" differences in glial pathology between unipolar and bipolar mood disorders. These differences are more apparent ventrally in the orbitofrontal cortex and subgenual anterior cingulate, and less pronounced in dorsal regions such as the dorsolateral prefrontal cortex and the supracallosal part of the anterior cingulate (167). Rajkowska and colleagues (167) suggested that these disease-specific changes might be a reflection of different circuits involved in the neuropathology of these two disorders and/or a consequence of different medication treatments. When abnormal prefrontal cortical markers of neural plasticity, neurotransmission (dopaminergic D2 receptor RNA, glutamate receptor dysfunction), signal transduction, glial cells, and GABA containing interneuron function were analyzed in schizophrenia, bipolar disorder and non-psychotic depression, bipolar disorder was found to be more similar to schizophrenia than depression (168).

There also are differences between unipolar and bipolar disorders regarding MRS findings. Decreased GABA levels were reported in occipital cortex in nonmedicated depressed unipolar patients (169), and occipital cortex GABA concentrations after SSRI treatment were significantly higher than the pretreatment concentrations (170). Bipolar patients do not have such reductions in GABA levels (171).

SUMMARY OF MAIN FINDINGS

Available findings from anatomical MRI studies implicate key regions involved in mood regulation, such as anterior cingulate, subgenual prefrontal cortex, amygdala, basal ganglia, and corpus callosum in adult and pediatric bipolar patients. Both adult and pediatric bipolar patients seem to lose more brain gray matter by aging than healthy controls, and this "cortical thinning" may be related to impairment of emotional, cognitive, and sensory processing in bipolar disorder (11). There seem to be regionally-specific brain abnormalities before the onset of disease, and illness-specific, possibly neurodegenerative, changes taking place with repeated episodes during the illness course. It is, however, not an easy task to differentiate illness-related changes from medication treatment-related changes. Also, whether these volume changes are related to changes in neuropil, neuronal size, or dendritic or axonal consolidation and/or pruning, is not yet very clear, although findings point to glial changes taking place in the early phases of the disease. The active episodes seem to have a neurotoxic effect, and those patients who show progressive changes with recurring episodes may also be suffering from a more severe, recurrent form of bipolar disorder, whereas mood stabilizer treatment seems to prevent and/or reverse some of the illness-associated changes (172). Psychotropic medications affect volumes of certain brain structures that are thought to play a role in mood disorders (58,173,174), and some of the volumetric and neurochemical changes seem to be reversed by mood-stabilizer treatment, such as lithium and valproate. This points to the importance of keeping in mind the previous and present exposure to psychotropic medications when interpreting the volumetric and neurochemical findings.

Volume reduction in various brain regions is also corroborated by decreased levels of *N*-acetyl-aspartate in DLPFC, which is the most consistently reported finding in MRS studies, in adults as well as children and adolescents with bipolar disorder, possibly developing only after the onset of bipolar disorder. In addition, phosphomonoesther concentrations seem to be altered in different mood states, reduced/unchanged in euthymic bipolar patients, and increased in manic and hypomanic patients, suggesting that changes in phosphoinositol activity occur in frontal and temporal regions.

PET and SPECT studies characterizing receptor changes in the brain show serotonin receptors may be involved in disease vulnerability and prevention or relief from mood disorder symptoms. These studies also provide insight that increased dopamine levels may be important in psychosis associated with mood disorders, and reduction of dopamine synthesis may play a role in anti-manic efficacy.

Future longitudinal studies that prospectively examine the effects of medication exposure and other factors associated with illness chronicity are needed. Additionally, examining bipolar patients at the onset of illness would help control for treatment effects and other variables associated with illness chronicity, and also confounding variables, such as medication history, electroconvulsive therapy, and substance abuse. For future research, some of the above-mentioned structural changes may qualify as putative brain structure endophenotypes, as suggested by Hasler and colleagues (175), and to understand the potential relevance of these brain changes to genetic susceptibility to the disorder.

ACKNOWLEDGMENTS

This work was partly supported by grants MH 068662, MH 068766, MH069774, RR020571, UTHSCSA GCRC (M01-RR-01346), NARSAD, the Krus Endowed Chair in Psychiatry (UTHSCSA), and the Veterans Administration (VA Merit Review).

REFERENCES

1. Soares JC, Mann JJ. The anatomy of mood disorders–review of structural neuroimaging studies. Biol Psychiatry 1997; 41:86–106.
2. Sheline YI. Neuroimaging studies of mood disorder effects on the brain. Biol Psychiatry 2003; 54:338–352.
3. Strakowski SM, DelBello MP, Adler C, et al. Neuroimaging in bipolar disorder. Bipolar Disord 2000; 2:148–164.
4. Sheline YI. 3D MRI studies of neuroanatomic changes in unipolar major depression: the role of stress and medical comorbidity. Biol Psychiatry 48(8):791–800, 2000.
5. Brambilla P, Glahn DC, Balestrieri M, et al. Magnetic resonance findings in bipolar disorder. Psychiatr Clin N Am 2005; 28:443–467.
6. McDonald C, Zanelli J, Rabe-Hesketh S, et al. Meta-analysis of magnetic resonance imaging brain morphometry studies in bipolar disorder. Biol Psychiatry 2004; 56:411–417.

7. McDonald C, Bullmore E, Sham P, et al. Regional volume deviations of brain structure in schizophrenia and psychotic bipolar disorder: computational morphometry study. Br J Psychiatry 2005; 186:369–377.
8. Farrow TF, Whitford TJ, Williams LM, et al. Diagnosis-related regional gray matter loss over two years in first episode schizophrenia and bipolar disorder. Biol Psychiatry 2005; 58(9):713–723.
9. Lopez-Larson MP, DelBello MP, Zimmerman ME, et al. Regional prefrontal gray and white matter abnormalities in bipolar disorder. Biol Psychiatry 2002; 52(2):93–100.
10. Brambilla P, Harenski K, Nicoletti M, et al. Differential effects on brain gray matter in bipolar patients and healthy individuals. Neuropsychobiology 2001; 43:242–247.
11. Lyoo IK, Sung YH, Dager SR, et al. Regional cerebral cortical thinning in bipolar disorder. Bipolar Disorder 2006; 8(1):65–74.
12. Moore GJ, Bebchuk JM, Wilds IB, et al. Lithium-induced increase in human brain grey matter. Lancet 2000; 356:1241–1242.
13. Sassi RB, Nicoletti M, Brambilla P, et al. Increased gray matter volume in lithium-treated bipolar disorder patients. Neurosci Lett 2002; 329:243–245.
14. Beyer JL, Kuchibhatla M, Payne M, et al. Caudate volume measurement in older adults with bipolar disorder. Int J Geriatr Psychiatry 2004; 19:109–114.
15. Drevets WC, Price JL, Simpson JR Jr, et al. Subgenual prefrontal cortex abnormalities in mood disorders. Nature 1997; 386(6627):824–827.
16. Hirayasu Y, Shenton ME, Salisbury DF, et al. Subgenual cingulate cortex volume in first-episode psychosis. Am J Psychiatry 1999; 156(7):1091–1093.
17. Sharma V, Menon R, Carr TJ, et al. An MRI study of subgenual prefrontal cortex in patients with familial and non-familial bipolar I disorder. J Affect Disord 2003; 77(2):167–171.
18. Ongur D, Drevets WC, Price JL. Glial reduction in the subgenual prefrontal cortex in mood disorders. Proc Natl Acad Sci U S A 1998; 95(22):13290–13295.
19. Brambilla P, Nicoletti MA, Harenski K, et al. Anatomical MRI study of subgenual prefrontal cortex in bipolar and unipolar subjects. Neuropsychopharmacology 2002; 27:792–799.
20. Sanches M, Sassi RB, Axelson D, et al. Subgenual prefrontal cortex of child and adolescent bipolar patients: a morphometric magnetic resonance imaging study. Psychiatry Res 2005; 138(1):43–49.
21. Strakowski SM, Wilson DR, Tohen M, et al. Structural brain abnormalities in first-episode mania. Biol Psychiatry 1993; 33:602–609.
22. Sassi RB, Brambilla P, Hatch JP, et al. Reduced left anteriot cingulate volumes in untreated bipolar patients. Biol Psychiatry 2004; 56(7):467–475.
23. Kaur S, Sassi RB, Axelson D, et al. Cingulate cortex anatomical abnormalities in children and adolescents with bipolar disorder. Am J Psychiatry 2005; 162(9): 1637–1643.
24. Doris A, Belton E, Ebmeier KP, et al. Reduction of cingulate gray matter density in poor outcome bipolar illness. Psychiatry Res 2004; 130:153–159.
25. Lyoo IK, Kim MJ, Stoll AL, et al. Frontal lobe gray matter density decreases in bipolar I disorder. Biol Psychiatry 2004; 55:648–651.
26. Dickstein DP, Milham MP, Nugent AC, et al. Frontotemporal alterations in pediatric bipolar disorder: results of a voxel-based morphometry study. Arch Gen Psychiatry 2005; 62:734–741.
27. Altshuler LL, Curran JG, Hauser P, et al. T2 hyperintensities in bipolar disorder: magnetic resonance imaging comparison and literature meta-analysis. Am J Psychiatry 1995; 152:1139–1144.
28. Moore PB, Shepherd DJ, Eccleston D, et al. Cerebral white matter lesions in bipolar affective disorder: relationship to outcome. Br J Psychiatry 2001; 178:172–176.
29. Pillai JJ, Friedman L, Stuve TA, et al. Increased presence of white matter hyperintensities in adolescent patients with bipolar disorder. Psychiatry Res 2002; 114:51–56.
30. Thomas AJ, Davis S, Ferrier IN, et al. Elevation of cell adhesion molecule immunoreactivity in the anterior cingulate cortex in bipolar disorder. Biol Psychiatry 2004; 55: 652–655.

31. Silverstone T, McPherson H, Li Q, et al. Deep white matter hyperintensities in patients with bipolar depression, unipolar depression and age-matched control subjects. Bipolar Disord 2003; 5:53–57.
32. Kato T, Fujii K, Kamiya A, et al. White matter hyperintensity detected by magnetic resonance imaging and lithium response in bipolar disorder: A preliminary observation. Psychiatry Clin Neurosci 2000; 54:117–120.
33. Iosifescu DV, Papakostas GI, Lyoo IK, et al. Brain MRI white matter hyperintensities and one-carbon cycle metabolism in non-geriatric outpatients with major depressive disorder (Part I). Psychiatry Res 2005; 140(3):291–299.
34. Papakostas GI, Iosifescu DV, Renshaw PF, et al. Brain MRI white matter hyperintensities and one-carbon cycle metabolism in non-geriatric outpatients with major depressive disorder (Part II). Psychiatry Res 2005; 140(3):301–307.
35. McIntosh AM, Job DE, Moorhead WJ, et al. White matter density in patients with schizophrenia, bipolar disorder and their unaffected relatives. Biol Psychiatry 2005; 58:254–257.
36. Kieseppa T, van Erp TG, Haukka J, et al. Reduced left hemispheric white matter volume in twins with bipolar I disorder. Biol Psychiatry 2003; 54(9):896–905.
37. Davis KA, Kwon A, Cardenas VA, et al. Decreased cortical gray and cerebral white matter in male patients with familial bipolar I disorder. J Affect Disord 2004; 82:475–485.
38. Chen HH, Nicoletti MA, Hatch JP, et al. Abnormal left superior temporal gyrus volumes in children and adolescents with bipolar disorder: a magnetic resonance imaging study. Neurosci Lett 2004; 363:65–68.
39. Adler CM, Holland SK, Schmithorst V, et al. Abnormal frontal white matter tracts in bipolar disorder: a diffusion tensor imaging study. Bipolar Disord 2004; 6:197–203.
40. Adler CM, Adams J, DelBello MP, et al. Evidence of white matter pathology in bipolar disorder adolescents experiencing their first episode of mania: a diffusion tensor imaging study. Am J Psychiatry 2006; 163(2):322–324.
41. Gazzaniga MS. Cerebral specialization and interhemispheric communication: does the corpus callosum enable the human condition? Brain 2000; 123:1293–1326.
42. Rueckert L, Levy J. Further evidence that the callosum is involved in sustaining attention. Neuropsychologia 1996; 34(9):927–935.
43. Levy J, Trevarthen C. Perceptual, semantic and phonetic aspects of elementary language processes in split-brain patients. Brain 1977; 100(Pt 1):105–118.
44. Rosen GD, Sherman GF, Galaburda AM. Interhemispheric connections differ between symmetrical and asymmetrical brain regions. Neuroscience 1989; 33:525–533.
45. Coffman JA, Bornstein RA, Olson SC, et al. Cognitive impairment and cerebral structure by MRI in bipolar disorder. Biol Psychiatry 1990; 27:1188–1196.
46. Brambilla P, Nicoletti MA, Sassi RB, et al. Magnetic resonance imaging study of corpus callosum abnormalities in patients with bipolar disorder. Biol Psychiatry 2003; 54:1294–1297.
47. Brambilla P, Nicoletti M, Sassi RB, et al. Corpus callosum signal intensity in patients with bipolar and unipolar disorder. J Neurol Neurosurg Psychiatry 2004; 75:221–225.
48. Yasar AS, Monkul ES, Sassi RB, et al. MRI study of corpus callosum in children and adolescents with bipolar disorder. Psychiatry Res 2006; 146(1):83–85.
49. Hajek T, Carrey N, Alda M. Neuroanatomical abnormalities as risk factors for bipolar disorder. Bipolar Disord 2005; 7:393–403.
50. Altshuler LL, Bartzokis G, Grieder T, et al. An MRI study of temporal lobe structures in men with bipolar disorder or schizophrenia. Biol Psychiatry 2000; 48:147–162.
51. Strakowski SM, DelBello MP, Sax KW, et al. Brain magnetic resonance imaging of structural abnormalities in bipolar disorder. Arch Gen Psychiatry 1999; 56:254–260.
52. Brambilla P, Harenski K, Nicoletti M, et al. MRI investigation of temporal lobe structures in bipolar patients. J Psychiatr Res 2003; 37:287–295.
53. Pearlson GD, Barta PE, Powers RE. Medial and superior temporal gyral volumes and cerebral asymmetry in schizophrenia versus bipolar disorder. Biol Psychiatry 1997; 41:1–4.
54. Blumberg HP, Kaufman J, Martin A, et al. Amygdala and hippocampal volumes in adolescents and adults with bipolar disorder. Arch Gen Psychiatry 2003; 60(12):1201–1208.

55. Chen BK, Sassi R, Axelson D, et al. Cross-sectional study of abnormal amygdala development in adolescents and young adults with bipolar disorder. Biol Psychiatry 2004; 56(6):399–405.
56. Caetano SC, Olvera R, Hunter K, et al. Abnormal amygdala volumes in pediatric bipolar disorder. Biol Psychiatry 2004; 55:111S.
57. DelBello MP, Zimmerman ME, Mills NP, et al. Magnetic resonance imaging analysis of amygdala and other subcortical brain regions in adolescents with bipolar disorder. Bipolar Disord 2004; 6:43–52.
58. Chang K, Karchemskiy A, Barena-Goraly N, et al. Reduced amygdalar gray matter volume in familial pediatric bipolar disorder. J Am Acad Child Adolesc Psychiatry 2005; 44(6):565–573.
59. Blumberg HP, Fredericks C, Wang F, et al. Preliminary evidence for persistent abnormalities in amygdala volumes in adolescents and young adults with bipolar disorder. Bipolar Disord 2005; 7(6):570–576.
60. Bowley MP, Drevets WC, Ongur D, et al. Low glial cell numbers in the amygdala in major depressive disorder. Biol Psychiatry 2002; 52:404–412.
61. Swayze VWII, Andreasen NC, Alliger RJ, et al. Subcortical and temporal structures in affective disorder and schizophrenia: a magnetic resonance imaging study. Biol Psychiatry 1992; 31:221–240.
62. Hauser P, Altshuler LL, Berrettini W, et al. Temporal lobe measurement in primary affective disorder by magnetic resonance imaging. J Neuropsychiatry Clin Neurosci 1989; 1:128–134.
63. Hauser P, Matochik J, Altshuler LL, et al. MRI-based measurements of temporal lobe and ventricular structures in patients with bipolar I and bipolar II disorders. J Affect Disord 2000; 60:25–32.
64. Pearlson GD, Barta PE, Powers RE. Medial and superior temporal gyral volumes and cerebral asymmetry in schizophrenia versus bipolar disorder. Biol Psychiatry 1997; 41:1–4.
65. Brambilla P, Harenski K, Nicoletti M, et al. MRI investigation of temporal lobe structures in bipolar patients. J Psychiatr Res 2003; 37:287–295.
66. Strasser HC, Lilyestrom J, Ashby ER, et al. Hippocampal and ventricular volumes in psychotic and nonpsychotic bipolar patients compared with schizophrenia patients and community control subjects: a pilot study. Biol Psychiatry 2005; 57(6):633–639.
67. Connor SEJ, Ng V, McDonald C, et al. A study of hippocampal shape anomaly in schizophrenia and in families multiply affected by schizophrenia or bipolar disorder. Neuroradiology 2004; 46:523–534.
68. Frazier JA, Chiu S, Breeze JL, et al. Structural brain magnetic resonance imaging of limbic and thalamic volumes in pediatric bipolar disorder. Am J Psychiatry 2005; 162(7):1256–1265.
69. Caetano SC, Olvera RL, Glahn D, et al. Fronto-limbic brain abnormalities in juvenile onset bipolar disorder. Biol Psychiatry 2005; 58(7):525–531.
70. Dupont RM, Jernigan TL, Heindel W, et al. Magnetic resonance imaging and mood disorders. Localization of white matter and other subcortical abnormalities. Arch Gen Psychiatry 1995; 52:747–755.
71. Monkul ES, Spence D, Sassi RB, et al. Magnetic resonance imaging study of thalamus volumes in adolescent bipolar patients. Depression and Anxiety 2006; 23(6):347–352.
72. Sax KW, Strakowski SM, Zimmerman ME, et al. Frontosubcortical neuroanatomy and the continuous performance test in mania. Am J Psychiatry 1999; 156:139–141.
73. Strakowski SM, DelBello MP, Zimmerman ME, et al. Ventricular and periventricular structural volumes in first- versus multiple-episode bipolar patients. Am J Psychiatry 2002; 159:1841–1847.
74. Caetano SC, Sassi R, Brambilla P, et al. MRI study of thalamic volumes in bipolar and unipolar patients and healthy individuals. Psychiatry Res 2001; 108:161–168.
75. Young KA, Holcomb LA, Yazdani U, et al. Elevated neuron number in the limbic thalamus in major depression. Am J Psychiatry 2004; 161(7):1270–1277.

76. McIntosh AM, Job DE, Moorhead WJ, et al. Voxel-based morphometry of patients with schizophrenia or bipolar disorder and their unaffected relatives. Biol Psychiatry 2004; 56:544–552.
77. Harvey I, Persaud R, Ron MA, et al. Volumetric MRI measurements in bipolars compared with schizophrenics and healthy controls. Psychol Med 1994; 24:689–699.
78. Brambilla P, Harenski K, Nicoletti MA, et al. Anatomical MRI study of basal ganglia in bipolar disorder patients. Psychiatry Res 2001; 106:65–80.
79. Aylward EH, Roberts-Twillie JV, Barta PE, et al. Basal ganglia volumes and white matter hyperintensities in patients with bipolar disorder. Am J Psychiatry 1994; 151:687–693.
80. Sanches M, Roberts RL, Sassi RB, et al. Developmental abnormalities in striatum in young bipolar patients: a preliminary study. Bipolar Disord 2005; 7(2):153–138.
81. Bonelli RM, Kapfhammer HP, Pillay SS, et al. Basal ganglia volumetric studies in affective disorder: what did we learn in the last 15 years? J Neural Transm 2006; 113(2):255–268.
82. DelBello MP, Strakowski SM, Zimmerman ME, et al. MRI analysis of the cerebellum in bipolar disorder: a pilot study. Neuropsychopharmacology 1999; 21:63–68.
83. Brambilla P, Harenski K, Nicoletti M, et al. MRI study of posterior fossa structures and brain ventricles in bipolar patients. J Psychiatr Res 2001; 35:313–322.
84. Mills NP, DelBello MP, Adler CM, et al. MRI analysis of cerebellar vermal abnormalities in bipolar disorder. Am J Psychiatry 2005; 162(8):1530–1532.
85. Monkul ES, Sassi RB, Axelson D, et al. MRI study of the cerebellum and vermis in children and adolescents with bipolar disorder. Biol Psychiatry 55:136S, 2004.
86. Baxter LR Jr, Schwartz JM, Phelps ME, et al. Reduction of prefrontal cortex glucose metabolism common to three types of depression. Arch Gen Psychiatry 1989; 46(3):243–250.
87. al-Mousawi AH, Evans N, Ebmeier KP, et al. Limbic dysfunction in schizophrenia and mania. A study using 18F-labelled fluorodeoxyglucose and positron emission tomography. Br J Psychiatry 1996; 169(4):509–516.
88. Buchsbaum MS, Wu J, DeLisi LE, et al. Frontal cortex and basal ganglia metabolic rates assessed by positron emission tomography with [18F]2-deoxyglucose in affective illness. J Affect Disord 1986; 10(2):137–152.
89. Cohen RM, Semple WE, Gross M, et al. Evidence for common alterations in cerebral glucose metabolism in major affective disorders and schizophrenia. Neuropsychopharmacology 1989; 2(4):241–254.
90. Martinot JL, Hardy P, Feline A, et al. Left prefrontal glucose hypometabolism in the depressed state: a confirmation. Am J Psychiatry 1990; 147(10):1313–1317.
91. Gyulai L, Alavi A, Broich K, et al. I-123 iofetamine single-photon computed emission tomography in rapid cycling bipolar disorder: a clinical study. Biol Psychiatry 1997; 41(2):152–161.
92. O'Connell RA, Van Heertum RL, Luck D, et al. Single-photon emission computed tomography of the brain in acute mania and schizophrenia. J Neuroimaging 1995; 5(2):101–104.
93. Blumberg HP, Stern E, Martinez D, et al. Increased anterior cingulate and caudate activity in bipolar mania. Biol Psychiatry 2000; 48(11):1045–1052.
94. Blumberg H, Stern E, Ricketts S, et al. Rostral and orbital prefrontal cortex dysfunction in the manic state of bipolar disorder. Am J Psychiatry 1999; 156:1986–1988.
95. Goodwin GM, Cavanagh JTO, Glabus MF, et al. Uptake of 99mTc-exametazime shown by single photon emission computed tomography before and after lithium withdrawal in bipolar patients: associations with mania. Br J Psychiatry 1997; 170: 426–430.
96. Mayberg HS, Brannan SK, Mahurin RK, et al. Cingulate function in depression: a potential predictor of treatment response. Neuroreport 1997; 8(4):1057–1061.
97. Caligiuri MP, Brown GG, Meloy MJ, et al. An fMRI study of affective state and medication on cortical and subcortical brain regions during motor performance in bipolar disorder. Psychiatry Res 2003; 123(3):171–182.

98. Drevets WC, Price JL, Bardgett ME, et al. Glucose metabolism in the amygdala in depression: relationship to diagnostic subtype and plasma cortisol levels. Pharmacol Biochem Behav 2002; 71(3):431–447.
99. Strakowski SM, DelBello MP, Adler CM. The functional neuroanatomy of bipolar disorder: a review of neuroimaging findings. Molec Psychiatry 2005; 10:105–116.
100. Bearden CE, Hoffman KM, Cannon TD. The neuropsychology and neuroanatomy of bipolar affective disorder: a critical review. Bipolar Disord 2001; 3:106–150.
101. Strakowski SM, Adler CM, Holland SK, et al. A preliminary fMRI study of sustained attention in unmedicated, euthymic bipolar disorder. Neuropsychopharmacology 2004; 29:1734–1740.
102. Adler CM, Holland SK, Schmithorst V, et al. Changes in neuronal activation in patients with bipolar disorder during performance of a working memory task. Bipolar Disord 2004; 6:540–549..
103. Blumberg HP, Leung HC, Skudlarski P, et al. A functional magnetic resonance imaging study of bipolar disorder: state- and trait-related dysfunction in ventral prefrontal cortices. Arch Gen Psychiatry 2003; 60:601–609.
104. Deckersbach T, Dougherty DD, Savage C, et al. Impaired recruitment of the dorsolateral prefrontal cortex and hippocampus during encoding in bipolar disorder. Biol Psychiatry 2006; 59(2):138–146.
105. Strakowski SM, Adler CM, Holland SK, et al. Abnormal FMRI brain activation in euthymic bipolar disorder patients during a counting Stroop interference task. Am J Psychiatry 2005; 162(9):1697–1705.
106. Benabarre A, Vieta E, Martinez-Aran A, et al. Neuropsychological disturbances and cerebral blood flow in bipolar disorder. Aust N Z J Psychiatry 2005; 39(4):227–234.
107. Malhi GS, Lagopoulos J, Ward PB, et al. Cognitive generation of affect in bipolar depression. An fMRI study. Eur J Neurosci 2004; 19:741–754.
108. Brooks JO 3rd, Wang PW, Strong C, et al. Preliminary evidence of differential relations between prefrontal cortex metabolism and sustained attention in depressed adults with bipolar disorder and healthy controls. Bipolar Disord 2006; 8(3):248–254.
109. Krüger S, Alda M, Young LT, et al. Risk and resilience markers in bipolar disorder: brain responses to emotional challenge in bipolar patients and their healthy siblings. Am J Psychiatry 2006; 163(2):257–264.
110. Yatham LN, Liddle PF, Lam RW, et al. PET study of the effects of valproate on dopamine D(2) receptors in neuroleptic- and mood-stabilizer-naive patients with nonpsychotic mania. Am J Psychiatry 2002; 159(10):1718–1723.
111. Pearlson GD, Wong DF, Tune LE, et al. In vivo D_2 dopamine receptor density in psychotic and nonpsychotic patients with bipolar disorder. Arch Gen Psychiatry 1995; 52: 471–477.
112. Yatham LN, Liddle PF, Shiah IS, et al. PET study of [(18)F]6-fluoro-L-dopa uptake in neuroleptic- and mood-stabilizer-naive first-episode nonpsychotic mania: effects of treatment with divalproex sodium. Am J Psychiatry 2002; 159(5):768–774.
113. Anand A, Verhoeff P, Seneca N, et al. Brain SPECT imaging of amphetamine-induced dopamine release in euthymic bipolar disorder patients. Am J Psychiatry 2000; 157: 1108–1114.
114. Suhara T, Nakayama K, Inoue O, et al. D1 dopamine receptor binding in mood disorders measured by positron emission tomography. Psychopharmacology (Berl) 1992; 106(1):14–18.
115. Malison RT, Price LH, Berman R, et al. Reduced brain serotonin transporter availability in major depression as measured by [123I]-2 beta-carbomethoxy-3 beta-(4-iodophenyl)-tropane and single photon emission computed tomography. Biol Psychiatry 1998; 44(11):1090–1098.
116. Ichimiya T, Suhara T, Sudo Y, et al. Serotonin transporter binding in patients with mood disorders: a PET study with [11C](+)McN5652. Biol Psychiatry 2002; 51(9): 715–722.
117. Drevets WC, Frank E, Price JL, et al. PET imaging of serotonin 1A receptor binding in depression. Biol Psychiatry 1999; 46(10):1375–1387.

118. Sargent PS, Kjaer KH, Bench CJ, et al. Brain serotonin1A receptor binding measured by positron emission tomography with [11C]WAY-100635: effects of depression and antidepressant treatment. Arch Gen Psychiatry 2000; 57(2):174–180.
119. Bhagwagar Z, Rabiner EA, Sargent PA, et al. Persistent reduction in brain serotonin1A receptor binding in recovered depressed men measured by positron emission tomography with [11C]WAY-100635. Mol Psychiatry 2004; 9(4):386–392.
120. Biver F, Wikler D, Lotstra F, et al. Serotonin 5-HT_2 receptor imaging in major depression: focal changes in orbito-insular cortex. Br J Psychiatry 1997; 171:444–448.
121. Attar-Levy D, Martinot JL, Blin J, et al. The cortical serotonin2 receptors studied with positron-emission tomography and [^{18}F]-setoperone during depressive illness and antidepressant treatment with clomipramine. Biol Psychiatry 1999; 45:180–186.
122. Yatham LN, Liddle PF, Shiah IS, et al. Brain serotonin2 receptors in major depression: a positron emission tomography study. Arch Gen Psychiatry 2000; 57(9):850–858.
123. Meyer JH, Kapur S, Houle S, et al. Prefrontal cortex 5-HT_2 receptors in depression: an [^{18}F]setoperone PET imaging study. Am J Psychiatry 1999; 156:1029–1034.
124. Yatham LN, Liddle PF, Lam RW, et al. A positron emission tomography study of the effects of treatment with valproate on brain 5-HT2A receptors in acute mania. Bipolar Disord 2005; 7(suppl 5):53–57.
125. Yatham LN. Translating knowledge of genetics and pharmacology into improving everyday practice. Bipolar Disord 2005; 7(suppl 4):13–20.
126. Baslow MH. Functions of *N*-acetyl-*L*-aspartate and *N*-acetyl-*L*-aspartylglutamate in the vertebrate brain: role in glial cell-specific signaling. J Neurochem 2000; 75(2):453–459.
127. Stork C, Renshaw PF. Mitochondrial dysfunction in bipolar disorder: evidence from magnetic resonance spectroscopy research. Mol Psychiatry 2005; 10(10):900–919.
128. Winsberg ME, Sachs N, Tate DL, et al. Decreased dorsolateral prefrontal *N*-acetyl aspartate in bipolar disorder. Biol Psychiatry 2000; 47:475–481.
129. Chang KD, Adleman N, Dienes K, et al. Decreased N-acetyl aspartate in children with familial bipolar disorder. Biol Psychiatry 2003; 53:1059–1065.
130. Kusumakar V, MacMaster FP, Sparkes S. Dorsolateral prefrontal cortex *N*-acetyl-aspartate in treatment naïve adolescent mood disorders. Biol Psychiatry 2002; 51:12S.
131. Sassi RB, Stanley JA, Axelson D, et al. Reduced NAA levels in the dorsolateral prefrontal cortex of young bipolar patients. Am J Psychiatry 2005; 162(11):2109–2115.
132. Cecil KM, DelBello MP, Morey R, et al. Frontal lobe differences in bipolar disorder as determined by proton MR spectroscopy. Bipolar Disord 2002; 4:357–365.
133. Bertolino A, Frye M, Callicott JH, et al. Neuronal pathology in the hippocampal area of patients with bipolar disorder: a study with proton magnetic resonance spectroscopic imaging. Biol Psychiatry 2003; 53(10): 906–913.
134. Deicken RF, Pegues MP, Anzalone S, et al. Proton MRSI evidence for hippocampal neuronal pathology in familial bipolar I disorder. Biol Psychiatry 2002; 51:84S.
135. Gallelli KA, Wagner CM, Karchemskiy A, et al. N-acetylaspartate levels in bipolar offspring with and at high-risk for bipolar disorder. Bipolar Disord 2005; 7(6):589–597.
136. Brambilla P, Stanley JA, Nicoletti MA, et al. 1H magnetic resonance spectroscopy investigation of the dorsolateral prefrontal cortex in bipolar disorder patients. J Affect Disord 2005; 86(1):61–67.
137. Hamakawa H, Kato T, Murashita J, et al. Quantitative proton magnetic resonance spectroscopy of the basal ganglia in patients with affective disorders. Eur Arch Psychiatry Clin Neurosci 1998; 248:53–58.
138. Kato T, Hamakawa H, Shioiri T, et al. Choline-containing compounds detected by proton magnetic resonance spectroscopy in the basal ganglia in bipolar disorder. J Psychiatry Neurosci 1996; 21:248–254.
139. Sharma R, Venkatasubramanian PN, Barany M, et al. Proton magnetic resonance spectroscopy of the brain in schizophrenic and affective patients. Schizophrenia Res 1992; 8:43–49.
140. Moore GJ, Bebchuk JM, Hasanat K, et al. Lithium increases N-acetyl-aspartate in the human brain: in vivo evidence in support of bcl-2's neurotrophic effects? Biol Psychiatry 2000; 48:1–8.

141. Silverstone PH, Wu RH, O'Donnell T, et al. Chronic treatment with lithium, but not sodium valproate, increases cortical *N*-acetyl-aspartate concentrations in euthymic bipolar patients. Int Clin Psychopharmacol 2003; 18(2):73–79.
142. Manji HK, Moore GJ, Rajkowska G, et al. Neuroplasticity and cellular resilience in mood disorders. Mol Psychiatry 2000; 5(6):578–593.
143. Stoll AL, Renshaw PF, Sachs GS, et al. The human brain resonance of choline-containing compounds is similar in patients receiving lithium treatment and controls: an in vivo proton magnetic resonance spectroscopy study. Biol Psychiatry 1992; 32: 944–949.
144. Wu RH, O'Donnell T, Ulrich M, et al. Brain choline concentrations may not be altered in euthymic bipolar disorder patients chronically treated with either lithium or sodium valproate. Ann Gen Hosp Psychiatry 2004; 3(1):13.
145. Friedman SD, Dager SR, Parow A, et al. Lithium and valproic acid treatment effects on brain chemistry in bipolar disorder. Biol Psychiatry 2004; 56(5):340–348.
146. Silverstone PH, Hanstock CC, Rotzinger S. Lithium does not alter the choline/creatine ratio in the temporal lobe of human volunteers as measured by proton magnetic resonance spectroscopy. J Psychiatry Neurosci 1999; 24:222–226.
147. Castillo M, Kwock L, Courvoisie H, et al. Proton MR spectroscopy in children with bipolar affective disorder: preliminary observations. Am J Neuroradiol 2000; 21: 832–838.
148. Dager SR, Friedman SD, Parow A, et al. Brain metabolic alterations in medication-free patients with bipolar disorder. Arch Gen Psychiatry 2004; 61(5):450–458.
149. Wolfson M, Bersudsky Y, Hertz E, et al. A model of inositol compartmentation in astrocytes based upon efflux kinetics and slow inositol depletion after uptake inhibition, Neurochem Res 2000; 25:977–982.
150. O'Donnell T, Rotzinger S, Nakashima TT, et al. Chronic lithium and sodium valproate both decrease the concentration of *myo*-inositol and increase the concentration of inositol monophosphates in rat brain. Brain Res 2000; 880:84–91.
151. Moore GJ, Bebchuk JM, Parrish JK, et al. Temporal dissociation between lithium-induced changes in frontal lobe myo-inositol and clinical response in manic-depressive illness. Am J Psychiatry 1999; 156:1902–1908.
152. Moore CM, Breeze JL, Gruber SA, et al. Choline, myo-inositol and mood in bipolar disorder: a proton magnetic resonance spectroscopic imaging study of the anterior cingulate cortex. Bipolar Disord 2000; 2:207–216.
153. Davanzo P, Thomas MA, Yue K, et al. Decreased anterior cingulate myo-inositol/creatine spectroscopy resonance with lithium treatment in children with bipolar disorder. Neuropsychopharmacology 2001; 24:359–369.
154. Kato T, Takahashi S, Shioiri T, et al. Alterations in brain phosphorous metabolism in bipolar disorder detected by in vivo 31P and 7Li magnetic resonance spectroscopy. J Affect Disord 1993; 27:53–59.
155. Kato T, Shioiri T, Murashita J, et al. Phosphorus-31 magnetic resonance spectroscopy and ventricular enlargement in bipolar disorder. Psychiatry Res 1994; 55:41–50.
156. Deicken RF, Fein G, Weiner MW. Abnormal frontal lobe phosphorous metabolism in bipolar disorder. Am J Psychiatry 1995; 152:915–918.
157. Deicken RF, Weiner MW, Fein G. Decreased temporal lobe phosphomonoesters in bipolar disorder. J Affect Disord 1995; 33:195–199.
158. Yıldız A, Demopulos CM, Moore CM, et al. Effect of lithium on phosphoinositide metabolism in human brain: a proton decoupled ^{31}P magnetic resonance spectroscopy study. Biol Psychiatry 2001; 50:3–7.
159. Hamakawa H, Kato T, Shioiri T, et al. Quantitative proton magnetic resonance spectroscopy of the bilateral frontal lobes in patients with bipolar disorder. Psychol Med 1999; 29:639–644.
160. Deicken RF, Eliaz Y, Feiwell R, et al. Increased thalamic N-acetylaspartate in male patients with familial bipolar I disorder. Psychiatry Res 2001; 106:35–45.
161. O'Donnell T, Rotzinger S, Nakashima TT, et al. Chronic lithium and sodium valproate both decrease the concentration of *myo*-inositol and increase the concentration of inositol monophosphates in rat brain. Brain Res 2000; 880:84–91.

162. Strakowski SM, Adler CM, DelBello MP. Volumetric MRI studies of mood disorders: do they distinguish unipolar and bipolar disorder? Bipolar Disord 2002; 4(2):80–88.
163. Campbell S, Marriott M, Nahmias C, et al. Lower hippocampal volume in patients suffering from depression A meta-analysis. Am J Psychiatry 2004; 161:598–607.
164. Stockmeier CA, Mahajan GJ, Konick LC, et al. Cellular changes in the postmortem hippocampus in major depression. Biol Psychiatry 2004; 56(9):640–650.
165. Wu JC, Buchsbaum MS, Johnson JC, et al. Magnetic resonance and positron emission tomography imaging of the corpus callosum: size, shape and metabolic rate in unipolar depression. J Affect Disord 1993; 28(1):15–25.
166. Lacerda AL, Brambilla P, Sassi RB, et al. Anatomical MRI study of corpus callosum in unipolar depression. J Psychiatr Res 2005; 39(4):347–354.
167. Rajkowska G. Cell pathology in bipolar disorder. Bipolar Disord 2002; 4(2):105–116.
168. Knable MB, Torrey EF, Webster MJ, et al. Multivariate analysis of prefrontal cortical data from the Stanley Foundation Neuropathology Consortium. Brain Res Bull 2001; 55(5):651–659.
169. Sanacora G, Mason GF, Rothman DL, et al. Reduced cortical gamma-aminobutyric acid levels in depressed patients determined by proton magnetic resonance spectroscopy. Arch Gen Psychiatry 1999; 56:1043–1047.
170. Sanacora G, Mason GF, Rothman DL, et al. Increased occipital cortex GABA concentrations in depressed patients after therapy with selective serotonin reuptake inhibitors. Am J Psychiatry 2002; 159:663–665.
171. Mason GF, Sanacora G, Anand A, et al. Cortical GABA reduced in unipolar but not bipolar depression. Biol Psychiatry 2000; 47:92S.
172. Monkul ES, Malhi GS, Soares JC. Anatomical MRI abnormalities in bipolar disorder: do they exist and do they progress? Aust N Z J Psychiatry 2005; 39(4):222–226.
173. Schmitt A, Weber S, Jatzko A, et al. Hippocampal volume and cell proliferation after acute and chronic clozapine or haloperidol treatment. J Neural Transm 2004; 111: 91–100.
174. Tebartz van Elst L, Baumer D, Ebert D, et al. Chronic antidopaminergic medication might affect amygdala structure in patients with schizophrenia. Pharmacopsychiatry 2004; 37(5):217–220.
175. Hasler G, Drevets WC, Gould TD, et al. Toward constructing an endophenotype strategy for bipolar disorders. Biol Psychiatry 2006; 60(2):93–105.

10 Sleep and Biological Rhythms Abnormalities in the Pathophysiology of Bipolar Disorders

Stephany Jones and Ruth M. Benca
Department of Psychiatry, University of Wisconsin–Madison, Madison, Wisconsin, U.S.A.

INTRODUCTION

Bipolar illness is, by definition, a cycling disorder. The longitudinal course is characterized by recurrent episodes of depression and mania or hypomania, with intervening periods of euthymia (1). The fact that cyclicity is a salient feature of the clinical phenotype has led many investigators to speculate that abnormalities in biological rhythms might prove etiologically or pathophysiologically significant in the disease process, and several lines of evidence support a relationship between biological rhythm disturbances in the onset and maintenance of bipolar episodes. Notably, physiological and behavioral timekeeping processes are often altered in bipolar patients, and there is evidence to suggest that a susceptibility to biological rhythm instability may be a factor in the onset and maintenance of affective episodes. For some patients, interventions that serve to stabilize or resynchronize sleep and biological rhythms prove therapeutically effective. Furthermore, sleep propensity is markedly altered during different phases of the disorder, and sleep disturbance is perhaps the most potent predictor of mood deterioration. Lastly, although quite preliminary, several recent studies have attempted to identify mutations in genes involved in circadian clock regulation that might influence the disease process in a subgroup of patients. The prevalence of biological rhythm dysfunction in bipolar disorder, the cyclical presentation of symptoms, and the strong link between sleep disturbance and episodes of mania and depression suggest that these features are pathophysiologically important in the disease process and might inform treatment development for this illness. This chapter will review these areas of research in bipolar illness.

BIOLOGICAL RHYTHM DYSFUNCTION

Irregularities in the Timing of Activity–Rest

A number of studies have used actigraphy to record the onset and amplitude of activity rhythms in bipolar patients across periods of clinical illness and euthymia. Given that increased energy and activity, combined with decrements in perceived need for sleep, are hallmarks of a manic episode, it is perhaps not surprising that several studies have reported an overall increase in motor activity during the manic phase of bipolar illness relative to either the euthymic or depressed phases (2,3). In contrast, both diminished activity and a delay in the timing of normal activity onset is reported in the depressed phase of bipolar disorder relative to the activity level of patients with a purely unipolar course (4–6). In addition to providing evidence of activity irregularities during periods of clinical illness, behavioral rhythm alterations also serve as prognosticators of an approaching clinical episode, particularly an episode of mania. Patients who experienced relapse into mania during lithium discontinuation were reported to have higher baseline levels of daytime motor activity preceding the onset of mania. In contrast, those patients who managed to remain euthymic showed no such increases in levels of daily activity (7). Changes in the duration of sleep also provide an index of changing clinical state. Using actigraphy to longitudinally record sleep and activity over 18 months in 11 bipolar patients, Leibenluft reported that reduced sleep duration and earlier awakening time were reliably prognostic for the onset of mania or hypomania the following day (8). Similarly, in a recent meta-analysis of bipolar prodromes, Lam and Wong reported that early predictors of an emerging episode of mania were recognized by 97% of patients and of these, 77% reported sleep disturbance, particularly sleep reduction and fragmentation, as the most prominent early symptom of mania (9).

Although it is not entirely clear if rhythm and sleep disruption are causally related to mood pathology, both clinical and experimental evidence support the idea that disruptions of daily routine and activity patterns can lead to an episode of clinical illness (10). Stressful life events, in particular, are often associated with mood deterioration, and this effect is particularly marked when those stressful life events result in abrupt changes in daily patterns of sleep timing and duration (11–14). Malkoff-Schwartz et al. analyzed the association between stressful life events and the onset of an affective episode in 39 bipolar patients and found that life events characterized as subjectively stressful, which consequently led to a disruption in normal social routines, were significantly predictive of the onset of mania (15). Moreover, although clinicians have long recognized that reductions in sleep duration are correlated with the onset of mania, the importance of sleep loss as a trigger for mania is empirically supported by studies employing sleep deprivation as a therapeutic intervention for depression. In a meta-analysis based on 10 studies of sleep deprivation effects, Wu and Bunney found a 30% switch rate into the manic or hypomanic state following one night of enforced sleep deprivation as a treatment for depression (16). Although estimates of state switching following sleep deprivation vary, the clear relationship between mania and sleep loss has led some to speculate that sleep loss is "a final common pathway in the genesis of mania" (17). This data suggests that at least a subset of patients suffering from bipolar disorder may have a particular sensitivity to biological rhythm disruption. A closer analysis of the subset of patients who are susceptible to the effects of rhythm disruption may offer insight into whether they show genetic abnormalities in the circadian pacemaker.

Evidence of Circadian Abnormalities: Endocrine Variables

Based on the alterations in behavioral rhythms demonstrated by many patients, Goodwin and Wirz-Justice hypothesized that patients with mood disorders, including bipolar disorder, may have an endogenous circadian period that is significantly shorter or longer than 24 hours. The alteration in the period of the circadian clock is thought to lead to an abnormal phase relationship between the individual and the environment, and this lack of synchrony is thought to contribute to the onset and maintenance of affective episodes (18,19). The notion that desynchrony between the internal and external environment might somehow contribute to affective deterioration is intuitively appealing given that a synchronous relationship between the endogenous rhythm of the clock and the external environment is critical to well-being; the adverse affects on mood, cognition, and physiology that result from jet-lag, shift work, or other conditions that produce desynchrony are pronounced (20). A number of empirical studies have sought to substantiate this hypothesis by attempting to characterize endogenous circadian pacemaker abnormalities in bipolar patients. These attempts, however, have produced largely contradictory and inconclusive findings due, in part, to the methodological difficulties inherent in accurately measuring human clock function.

Mammalian circadian rhythms are governed by a master oscillator in the suprachiasmatic nucleus (SCN) of the hypothalamus. The SCN regulates a number of physiological variables in humans, including temperature, certain elements of the sleep–wake cycle such as rapid-eye movement sleep propensity, as well as the secretion of a variety of hormones. Measurements of these variables can, theoretically, be used to assess the period and amplitude of the endogenous circadian rhythm. The oscillation and amplitude of most output rhythms, however, is masked by both behavior and environment (21). Protocols specialized for disentangling the true period of the endogenous pacemaker from the environment can be used to measure the period and phase of the endogenous clock accurately, but unfortunately, few studies have used such rigorous experimental control in studies of bipolar patients. As a result, most studies published to date have been confounded by the masking effects of activity and light exposure on the rhythms studied. In addition to these methodological difficulties, other factors such as small sample size, psychotropic medication use by patients, and diagnostic inconsistency conspire to limit the ability of a given study to detect statistically significant differences between groups.

Dysregulation of the hypothalamic pituitary axis (HPA) is a highly replicated finding in at least a subset of patients with mood disorders. Frequently reported findings include elevated levels of cortisol and corticotropin-releasing hormone (CRH), nonsuppression of cortisol on the dexamethasone suppression test, and/or a blunted adrenocorticotropic hormone (ACTH) response to CRH (22). Given that the profile of cortisol secretion is regulated in large measure by the circadian system, substantial research has focused on identifying abnormalities in the circadian regulation of the HPA axis in depression, but support for abnormalities in either phase or amplitude of circadian rhythms have been mixed. Linkowski, for example, reported a significant phase advance of the serum cortisol and ACTH rhythms in unipolar depressed patients, but although several of the patients with a bipolar course also showed this phase advance, the effect did not reach significance at the group level. A later study by this same group characterized sleep and 24-hour profiles of cortisol in 14 patients during the manic state and reported an elevation of nocturnal cortisol levels and an earlier occurrence of the nadir of the circadian secretory profile relative to controls (23,24). In contrast, Cervantes and colleagues analyzed the amplitude and phase of the cortisol profile in 18 bipolar patients in either the depressed, hypomanic, or euthymic phase relative to a control group, and although they did report hypersecretion of serum cortisol in the depressed and hypomanic phase, they found no evidence for a circadian phase shift in either phase of the illness relative to controls (25). Sleep loss and changes in sleep timing, both of which invariably accompany depression and mania, are known to alter both the timing and amplitude of the cortisol secretory profile and could thus easily account for the conflicting findings. In general, mania-associated phase or amplitude abnormalities in the cortisol rhythm have not been reported consistently.

Attempts to characterize the phase and period of the endogenous clock in patients with affective disorders have also been based on peripheral measurements of the hormone melatonin. The pineal gland, which is responsible for the production and secretion of melatonin, is directly controlled by the SCN. Since melatonin production is inhibited by light exposure, serial measurement under dim light conditions has historically been thought to provide a valid estimate of circadian phase. As a result, melatonin output has been widely used to assess the period of the internal clock. Potential confounds to this measurement have been suggested by more recent data, however; for instance, melatonin production is likely affected by even dim light as well as postural position. In addition, the timing of the melatonin secretory profile is further altered by sleep loss and hypercortisolemia, both of which invariably accompany clinical illness (26). These confounds ultimately complicate the interpretation of the many early studies that used melatonin output to assess clock function.

A recent longitudinal analysis comparing the melatonin secretory profile of bipolar patients in all phases (depressed, manic, and euthymic) of the disorder relative to controls reported no evidence for circadian phase differences during any illness period, although the amplitude of melatonin secretion was diminished in bipolar patients during all phases relative to controls. The authors speculated that this diminished amplitude in melatonin production might result from an abnormal response to light in bipolar patients, and suggested that this might provide the mechanism for an abnormal coupling of the endogenous pacemaker and the environment (27).

As a consequence, a number of recent analyses of melatonin abnormalities have focused on the effects of light on melatonin suppression in bipolar patients. Some, but not all studies support an increased sensitivity to light-induced suppression of melatonin in a subgroup of bipolar patients, and it has been hypothesized that this sensitivity might result in circadian phase instability (28–30). Lewy et al. compared the effects of light exposure on melatonin secretion between the hours of 2 and 4 am in euthymic bipolar patients who were age and gender-matched with controls. Bipolar patients were reported to have a 61.5% suppression of melatonin after light exposure relative to a 28% suppression of melatonin in controls (31). Similarly, in a recent comparison of euthymic bipolar, unipolar, and control subjects, Nurnberger and colleagues reported enhanced melatonin suppression by light in bipolar I patients relative to controls (32). This same group also reported evidence of a hypersensitivity to light-induced melatonin suppression in the offspring of bipolar disorder patients (33). The data do raise the possibility that any circadian phase instability seen in bipolar patients may not be a dysfunction in the clock itself, but may instead be a consequence of a more peripheral abnormality, which consequently results in an improper alignment between the clock and the environment.

Seasonality in Bipolar Disorder

Most epidemiological studies support a seasonal component for bipolar disorder, although there is some disagreement regarding the peak incidence for various types of episodes, likely due in part to clinical heterogeneity. Nevertheless, there is evidence for a significant peak of episodes of mania in the spring, with some studies reporting a minor or major peak in the late summer or fall as well (34–37). If all episode types are included, there is a more clear spring-fall pattern, probably related to a fall increase in depressive or mixed episodes (38,39).

Seasonality of mood disorders is also evident in seasonal affective disorder (SAD), a condition originally described by Rosenthal to characterize a bipolar patient who developed seasonally recurrent depressions (40). Although SAD has since been reconceptualized as primarily a disorder in which patients are afflicted by winter-onset depression, there are those patients who present with seasonally induced mania in addition to winter depressions. Although the shorter photoperiod during winter is thought to be etiologically responsible for SAD, studies attempting to show an increased prevalence at more northern latitudes have not been conclusive; some have been equivocal (41,42), whereas others suggest increased prevalence in regions with reduced winter light (43). Although the results of epidemiological studies have been inconsistent, SAD patients have been reported to demonstrate unusual responses to seasonal lighting changes.

The duration of the melatonin signal is directly related to the length of the dark period, and in this way melatonin secretion serves as the signal of photoperiod length for many mammals. Wehr et al. have recently helped to characterize the nature of circadian and circannual abnormalities in SAD. They reported that SAD patients generate a seasonal change in the duration of the dim-light melatonin secretion profile, specifically that they produced a longer duration of nocturnal melatonin secretion in winter relative to summer, whereas control subjects fail to show such seasonal alteration in melatonin secretion (44). These data suggest that SAD patients, in contrast to normal subjects, exhibit a change-of-season signal similar to that used by mammals that show significant seasonal changes in behavior and physiology. Although this finding does not explain how abnormal responses to the changing photoperiod lead to affective deterioration, it provides further support for the idea that patients with bipolar disorder may have particular sensitivities to biological rhythm disruption.

Cyclicity in bipolar illness not related to seasonality has primarily been described in terms of rapid cycling over the course of hours or days (2); it is likely that many bipolar patients may have underlying cycling with periods ranging from days to weeks to months. In summary, mood disorders, in particular bipolar disorder and SAD, are characterized by cycling that is often seasonally related, suggesting a role for circadian and/or seasonal pacemakers in the onset and maintenance of these disorders.

SLEEP ABNORMALITIES IN BIPOLAR DISORDER

Psychiatrists have been interested in the relationship between sleep and mental illness since the time of Sigmund Freud when dream analysis was a major component of the treatment for mental illness. After the discovery of rapid eye movement (REM) sleep in 1953, psychiatrists, spurred on by the seeming similarities between the hallucinations associated with psychosis and the hallucinogenic quality of dreams, began to speculate that schizophrenia might represent a disorder of REM sleep. Although this speculation did not prove accurate, an interest in the relationship between specific sleep abnormalities and psychiatric disorders has persisted. Sleep in depression has been studied more thoroughly that any other disorder, and as early as the 1950s abnormalities in both slow wave sleep (SWS) and REM sleep were reported in depressed patients. Coincident with technological advances in sleep science, a deeper understanding of the neural control of sleep was developing. Given the widespread and robust changes in sleep architecture demonstrated by patients with affective disorders, it was hoped that an understanding of the neural control of sleep might offer insight into the neurochemical abnormalities of patients with mood disorders.

Overview of Sleep Regulation

Sleep is modulated by the endogenous circadian clock as well as by the accumulation of sleep need over the course of wakefulness. The daily temporal modulation of sleep and wakefulness is described by the two-process mathematical model of sleep regulation, first articulated by Borbely, where the circadian pacemaker (process C) interacts with a sleep dependent homeostatic process (process S) to regulate sleep onset and offset. The homeostatically regulated process S reflects an increasing sleep drive, estimated from slow-wave analysis of the electroencephalogram (EEG), and increases as duration of wakefulness increases. Process C is thought to be largely governed by the circadian clock, and is subjectively measured by self-report assessments of sleepiness during sleep-deprivation studies. REM sleep onset and offset is governed largely, but not completely, by the circadian clock (45,46).

The transition from wakefulness to sleep is regulated by a complex interaction between circadian and homeostatic processes and by myriad neurochemical changes. The idea that the rostral brainstem reticular formation plays a central role in arousal has not changed in over 50 years, but our understanding of the complex neural circuits and multiple neuromodulatory systems involved in sleep–wakefulness has grown considerably. Arousal is mediated by two cholinergic nuclei of the laterodorsal tegmental (LDT) and pedunculopontine tegmental (PPT) nuclei of the brainstem reticular formation. The dorsal pathway to the thalamus, which sends a broad glutamatergic influence to the cortex, is the classical substrate of cortical arousal. A ventral pathway projects to both the hypothalamus, which, in turn, sends a diffuse histaminergic projection to the cortex as well as to the basal forebrain. Cells of the basal forebrain synthesize both acetylcholine and GABA and project broadly to the cortex. Sleep is produced by structures distributed throughout the brain that exert inhibitory influences upon this arousal system. Neurons from the lower brainstem exert an inhibitory influence on the rostral reticular formation, and GABAergic neurons of the ventrolateral preoptic area of the hypothalamus and adjacent regions of the basal forebrain send inhibitory signals to both the histaminergic nucleus of the hypothalamus, as well as to the cholinergic nuclei of the reticular formation, dampening cortical and behavioral arousal (47,48).

The neural control of REM sleep, initially proposed by McCarley and Hobson, outlined a reciprocal relationship between so-called REM-off and REM-on neurons in the regulation of REM sleep (49). Experimental evidence identified brainstem cholinergic neurons of the LDT and PPT nuclei as the cells that control the onset of REM sleep. Noradrenergic cells of the locus coeruleus and serotonergic cells of the dorsal raphe nucleus project to the PPT and LDT and normally have an inhibitory influence on these cholinergic cells. Just prior to the onset of REM sleep, these aminergic cells (referred to as REM-on neurons) decrease firing, thus lifting their inhibitory influence on PPT and LDT cholinergic neurons. The firing of these neurons supports an increase in acetylcholine release and, ultimately, the cerebral arousal evidenced by the activated EEG of REM sleep (49).

Sleep Abnormalities in Bipolar Depression

Some of the best-documented clinical findings in depressed patients involve sleep disruption. Disturbed sleep, including prolonged sleep latency, early morning awakenings, and reduced sleep efficiency, is characteristic of depressive episodes.

Consistent changes in polysomnographic sleep characteristics of depressed patients have also been widely and consistently reported. Depressed patients show a relative loss of slow-wave sleep (SWS), including a reduction in the absolute number of EEG delta waves during the first nonrapid eye movement (NREM) period. Changes in REM sleep are perhaps the most specific to mood disorders. In healthy subjects, REM sleep propensity normally reaches its maximum in the latter portion of the sleep period. In contrast, depressed patients often exhibit abnormally early REM sleep onset (reduced REM latency), a larger proportion of REM sleep during the first third of the night, and increased eye movements during the REM sleep period (50). Although most of the studies on depression to date have likely included a mixture of unipolar and bipolar patients, a few studies have focused exclusively on the question of whether the sleep of bipolar depressed patients differs from that of unipolar depressed patients.

Riemann et al. analyzed the EEG of 27 bipolar and unipolar depressed patients age- and gender-matched with controls. Although, as expected, both the unipolar and bipolar depressed patients showed increased REM density, increased percentage of REM sleep throughout the night, and a loss of SWS, differences between the two patient groups were subtle. The authors reported a trend toward increased REM density in the first REM period in bipolar patients, but this effect did not reach significance (51). Similarly, Fossion et al., comparing EEG data of unipolar and bipolar depressed patients during an episode of major depression, reported a trend toward a longer REM sleep latency in the bipolar patients relative to the depressed patients, but this trend was only evident in the more severely afflicted bipolar patients. A total of five studies, in addition to the two reviewed, have directly compared the sleep EEG of bipolar and unipolar depressed patients (52). Although there have been some subtle differences between the sleep of unipolar and bipolar depressed patients, overall these studies have failed to demonstrate compelling differences between the two (53–58).

Sleep Abnormalities in Bipolar Mania

A subjectively reported decrease in the need for sleep is one of the most remarkable correlates of a manic episode. As mentioned early in the chapter, it is regularly reported in the clinical literature that sleep disruption not only often predicts but may also precipitate the onset of a manic episode. These striking clinical observations have naturally led to an interest in analyzing the sleep EEG in acutely manic patients, but studying sleep in mania beyond behavioral observation is challenging. As a result, only a few controlled EEG studies have been performed during bipolar mania.

Hudson et al., studying nine bipolar manic patients and nine age-matched healthy controls, reported an overall decrease in time spent asleep, reduced REM latency, and increased REM density in the manic patients relative to controls. In a later study, the same group performed a comparison between the sleep EEG in 19 bipolar manic patients, 19 age- and gender-matched depressed patients, and 19 healthy controls. As expected, both patient groups exhibited behavioral and EEG abnormalities relative to healthy controls, but although manic patients were reported to have a decreased total sleep time relative to the depressed patients, EEG recordings failed to distinguish manic from depressed patients on measures of REM latency and density (59). Similarly, Linkowski et al. found that EEG recordings did not distinguish bipolar manic patients from depressed patients or healthy controls, with the exception of longer sleep latencies and diminished total sleep time in the manic group (60).

Longitudinal Studies of Sleep in Bipolar Patients

Longitudinal studies have the potential to provide information that cannot be achieved through cross-sectional analyses. These studies, however, are difficult to conduct and the extant studies have produced conflicting data. Bunney et al. analyzing the EEG of one patient during three switches into mania and one switch out of mania, reported that REM sleep was uniformly decreased as patients entered into the manic state (61). The authors suggested that since noradrenergic cells of the locus coeruleus and serotonergic cells of the dorsal raphe nucleus are thought to inhibit REM sleep by suppressing cholinergic firing, the decrease in REM sleep preceding mania onset suggests the possibility of a decreasing cholinergic tone in mania. The authors claimed that the data was supported by an early hypothesis of mood disorders articulated by Janowsky, which argued for the pathophysiological importance of increased cholinergic activity in depression and increased aminergic activity in mania (62). In contrast to the Bunney data, however, Kupfer et al. reported a higher REM density directly preceding the onset of a manic episode in a rapid-cycling patient with a 48-hour cycle (63). Although longitudinal designs have the potential to provide informative clinical data across behavioral states, to date, these studies have proved inconsistent.

Sleep in Bipolar Remission

There is controversy over whether or not sleep abnormalities in bipolar patients normalize with remission, or if these alterations reflect enduring trait or vulnerability markers. State-dependent abnormalities, which resolve when symptoms of the affective disorder resolve, reflect alterations in neurobiological processes underlying an acute episode. Trait markers, in contrast, do not resolve with symptom remission and suggest an increased risk for occurrence of the disorder, presumably linked to genetic transmission. Several authors contend that some sleep alterations, in particular REM abnormalities, normalize in unipolar and bipolar patients during periods of euthymia, and may therefore be characterized as state-dependent. Buysse et al. reported a decrease in REM density and a decrease in REM latency during remission in depressed patients, and similar results have been reported in several longitudinal studies, supporting the hypothesis that some REM abnormalities may be episode-related biological features (64–67). A variety of other findings, however, do not support this view. Giles et al. reported a persistence of shortened REM latencies and enhanced REM density during remission in bipolar and unipolar patients, and longitudinal studies have further shown REM latency to be stable during depressive episodes and periods of euthymia (Rush, 68–70). Familial analyses provide further support for REM abnormalities, including increased REM density and decreased REM latency as stable traits in patients with mood disorders. (69,71). Lauer, investigating 54 high-risk probands by polysomnography, found that the EEG patterns of subjects without a personal history of depression, but with a strong familial history, showed reduced SWS and increased REM density in the first sleep cycle compared to control subjects with no personal or family history of affective disorders (72). A recent study, attempting to identify risk factors for the development of affective disorders similarly reported increased REM density in the first REM period in 82 subjects with a familial history of affective disorders (73). This group failed to find evidence of distinct differences between relatives of unipolar and bipolar patients. EEG abnormalities, particularly increased REM density, were similarly expressed in both first degree relatives of bipolar and unipolar patients, causing the authors to conclude that REM abnormalities might represent vulnerability markers for mood disorders in general (74).

In summary, sleep disturbances are an integral feature of bipolar disorder. Like the disorder itself, the associated sleep disturbances are heterogenous, ranging from hypersomnia to difficulty initiating or maintaining sleep (75). Although there is controversy over whether sleep abnormalities represent stable, trait-like features of depressive and bipolar disorders, most studies appear to support the notion that reduced REM latency and increased REM density, at least, are stable features. The persistence of these abnormalities during periods of euthymia does call into question whether or not these abnormalities are pathophysiologically significant or simply reflect a trait generally correlated with, but not integral to, the pathophysiology of bipolar disorder. However, a major goal of metal health research is to identify specific quantifiable, heritable markers of psychiatric research, and particular sleep abnormalities may represent such markers. Identifying discrete physiological features of bipolar disorder promises to not only improve diagnostic reliability and consistency, but this strategy offers hope for elucidating the genetic underpinnings of the disorder.

TREATMENTS FOR BIPOLAR DISORDER

Perhaps the most compelling demonstration for a significant role for sleep and rhythm disturbances in bipolar disorder comes from treatment studies. Manipulations of the sleep–wake cycle, including total sleep deprivation as well as phase advances in the timing of sleep onset, have been demonstrated to produce mood improvement in bipolar depressed patients and also to induce mania in susceptible patients. In addition, fostering sleep and stabilizing its timing has been shown to be helpful in decreasing the duration of a manic episode. The most widely prescribed mood stabilizer, lithium, has demonstrated effects on both sleep and circadian rhythmicity, as do most of the antidepressant medications prescribed for bipolar depression, although it is not known if their therapeutic efficacy is related to effects on circadian rhythms. How sleep and rhythm manipulations achieve therapeutic efficacy, or how, in some cases, these manipulations lead to mood symptoms, is unclear. An understanding of the mechanism(s) through which these interventions lead to mood deterioration or improvement may ultimately lead to the development of pharmaceutical agents that produce more rapid symptom alleviation than currently available pharmaceutical agents are able to do.

Manipulations of the Sleep–Wake Cycle: Sleep Deprivation

Total sleep deprivation for one night has been convincingly demonstrated to produce rapid improvement in mood, as well as in cognitive and motor functions, in approximately 60% of patients with unipolar depression. Late night sleep deprivation, where a patient is kept awake for the latter half of the nocturnal sleep period, has also been shown to produce significant symptom alleviation. Selective REM deprivation, a process in which the patient is awakened whenever he or she shows polysomnographic evidence of REM sleep, has also been shown to have some antidepressant effect, although there is a considerable latency between initiation of the REM deprivation protocol and mood improvement (76). Although the data supporting the therapeutic effect of sleep deprivation specifically in bipolar disorder patients is not as extensive as that in unipolar depressed patients, there is evidence to both support and contraindicate this therapeutic sleep intervention in bipolar patients. In terms of efficacy, the response of bipolar depressed patients to sleep deprivation appears to be comparable to that seen in unipolar depressed patients, with some studies suggesting a slightly better response in bipolar patients (77). Early studies indicated that up to 30% of bipolar patients were at risk for entering a manic or hypomanic episode following a night of sleep deprivation. More recent evidence, however, suggests that a switch into mania following a night of sleep loss may be more characteristic of bipolar patients with a rapid cycling course (78,79). A recent analysis of a large population of bipolar depressed patients revealed a rate of switch into mania of approximately 5% following one night of sleep deprivation, a rate comparable to that seen following initiation of antidepressant treatment (79,80). The precise estimate of the rate of switch from depression to mania is still controversial, and caution is warranted when sleep deprivation is employed as a treatment for bipolar depression.

Sleep Deprivation: Potential Mechanisms

Although the effect of sleep deprivation therapy is transient, with relapse typically occurring immediately following recovery sleep, it provides rapid symptom improvement. This rapid effect is in contrast to antidepressant medications, which typically require anywhere from several days to a week before symptom relief occurs. Elucidating the biological mechanisms through which this rapid therapeutic effect is achieved could thus aid in the development of antidepressants that are superior to existing agents, and, to that end, many investigators have attempted to characterize the mechanism of the therapeutic efficacy of sleep deprivation. Although the neurobiological mechanisms of depression and mania are complex, theoretical conceptualizations of depression have implicated monoaminergic signaling as a significant factor in mood pathology.

The role of serotonin in modulating sleep and arousal, and the clinical efficacy of antidepressant medications that enhance monoaminergic transmission in the central nervous system, provides an overt link between the systems regulating both mood and sleep. Selective serotonin uptake inhibitors (SSRIs) represent the most widely prescribed class of antidepressant medication and the most consistent effect of the SSRIs on sleep is a reduction of REM sleep. It has long been suggested that the antidepressant effect of SSRIs might be tied to this reduction of REM sleep; however, recently developed antidepressant drugs with therapeutic efficacy similar to that of the SSRIs are not REM-suppressive, casting doubt on the importance of REM reduction for mood elevation (81). Nevertheless, the serotonin system is still thought be involved in the response to sleep deprivation. The SSRI drug fluoxetine, the mixed serotonergic-noradrenergic drug amitriptyline, and the 5HT1-A-beta adrenoreceptor blocker drug pindolol have been demonstrated to enhance and sustain the response to sleep deprivation in depressed patients (82–84). In further support of a role for the serotonin system, a recent study reported that a polymorphism in the transcriptional control region of the serotonin transporter (5-HTT), a polymorphism which ultimately increases central serotonin production, is associated with a better response to sleep deprivation in bipolar patients relative to those patients homozygous for the short variant of this allele. Another group, however, failed to replicate this effect (85,86). Altered serotonin signaling clearly does not explain all of the therapeutic effects of sleep deprivation, however, since experimental depletion of tryptophan, which rapidly lowers brain serotonin levels and produces relapse in medicated depressed patients, does not reverse the antidepressant effects of sleep deprivation in unmedicated patients (87).

Dopamine has also received attention as a possible mediator of the effects of sleep derivation (88). Patients who show symptom improvement in response to sleep deprivation have been shown to have lower levels of cerebral spinal fluid homovanillic acid—a dopamine metabolite—before sleep deprivation and an increase in this metabolite following sleep deprivation, when compared to those patients who do not respond (88–90). Ebert et al., using SPECT to study receptor occupancy of the dopamine D2 receptor, reported that sleep deprivation responders had decreased occupancy of the D2 receptor following the sleep deprivation while nonresponders showed an increase in D2 binding, and the authors interpreted this result as an indication that sleep deprivation enhanced dopamine release in responders to sleep deprivation. Later studies, however, failed to replicate this finding and one group reported enhanced D2 binding in sleep deprivation responders (91,92). An additional piece of indirect evidence supporting the importance of dopaminergic signaling in the effects of sleep deprivation comes from studies of patients afflicted with Parkinson's disease. Interestingly, following one night of sleep deprivation, Parkinson's patients show marked improvement in both tremor and rigidity, suggesting a transient normalization of dopamine (93,94). Although dopamine signaling may be increased following sleep deprivation, its role in mood improvement is unclear.

Interestingly, recent data in animal models indicates that both antidepressant medication and sleep deprivation lead to increased expression of genes involved in synaptic potentiation and plasticity, suggesting a common mechanism for their effects on depression. The time course of these changes in gene expression parallels the time course of the antidepressant effect. Sleep deprivation results in immediate increases in plasticity-related genes in rats (95) and an immediate antidepressant effect in humans, while chronic (but not acute) administration of antidepressant medication results in increased expression of these same genes in rodents (96). It has been suggested that the antidepressant effect of sleep deprivation may be directly related to synaptic potentiation and the induction of such genes (97) and that "sleep deprivation may bring about its rapid antidepressant effects by activation of the robust expression of plasticity genes, such as *CREB* (cyclic AMP response element binding protein), *BDNF* (brain-derived neurotrophic factor) and *TkrB* (tyrosine kinase receptor beta), and consequently a rapid antidepressant response" (98).

Despite these intriguing data, the mechanism of the antidepressant action of sleep deprivation has not yet been identified. Nevertheless, a more complete understanding of the effects of sleep deprivation has the potential to lead to both a better understanding of the neurobiology of mood disorders as well as the development of improved pharmaceutical interventions.

Manipulations of the Sleep–Wake Cycle: Circadian Phase Advances

As described above, depressed patients often demonstrate a marked phase advance in the first REM sleep period of the night. Given that REM sleep timing is regulated largely by the circadian system, the appearance of REM sleep earlier in the night in depressed patients has been hypothesized to reflect abnormally advanced circadian rhythms relative to the sleep–wake cycle. Several investigators have therefore argued that depressed patients may thus sleep at the wrong circadian time, and have advocated shifting the sleep–wake period to earlier in the evening in an effort to synchronize out-of-phase rhythms with hopes of inducing mood improvement. Several studies employing sleep phase advance, in which a patient's sleep onset and offset time is moved to an earlier clock time, have been reported to alleviate depressive symptoms, although the time to improvement is considerably longer than that seen following one night of total sleep deprivation (99,100). Additionally, sleep phase advances have also been shown to successfully preserve the antidepressant effects of total sleep deprivation (101,102).

Although these data suggest that an abnormal phase relationship between sleep and the endogenous circadian rhythm or abnormal circadian control of sleep might be of pathophysiological significance, there is currently no evidence that these sleep–wake manipulations alter REM sleep latency or distribution (103). Reimann et al. recently examined REM sleep distribution of depressed patients following total sleep deprivation with a subsequent sleep phase advance. Even in those patients who responded positively to the therapeutic regimen, shortened REM latencies persisted (104). The failure to correct the purported circadian abnormality calls into question whether these REM sleep abnormalities are pathophysiologically significant or whether they are merely epiphenomena of the depressed state. Nevertheless, the efficacy of sleep phase advance has been shown to be superior to sleep phase delays in the treatment of depression, providing support for the idea that the timing of sleep and its interaction with other biological variables may be an essential element in symptom relief. Moreover, a recent study provided evidence for an interactive effect between sleep and the circadian system on mood regulation in normal subjects, further highlighting the importance of proper alignment of endogenous circadian phase and sleep–wakefulness timing for mood regulation (105).

Treating and Preventing Mania: Rhythm Stabilization

Although much of the emphasis on treating and preventing mania in bipolar patients is on pharmacotherapy, there is some evidence that rhythm stabilization might be effective in both shortening the duration of an acute episode of mania as well as in preventing episode recurrence. Several case studies have demonstrated that enforced sleep and extended darkness successfully reduces the duration of acute episodes of mania in rapid-cycling bipolar patients (106,107). Wehr and colleagues employed a protocol of 14 hours of enforced bed rest in complete darkness for a bipolar disorder patient with a long history of rapid cycling. This enforced sleep and darkness protocol effectively reduced the duration of an acute manic episode, and additionally served to dramatically stabilize the patient's rapid-cycling course (108). Similarly, behavioral interventions aimed at normalizing and structuring a regular pattern of sleep and daytime activity also appear to have some efficacy in both reducing the duration of a manic episode, as well as extending the time between episodes of clinical illness (109). As a result, Frank and Kupfer developed social rhythm therapy and they advocate its use as an adjuvant to medication treatment for bipolar patients. This form of psychotherapy, which augments the more conventional interpersonal therapy, employs behavioral strategies to stabilize daily routines and rhythms (110,111). Currently, results are anticipated from an ongoing large-scale clinical trial of social rhythm therapy at the University of Pittsburgh.

Medication for Mania: Effects on Sleep and Circadian Rhythms

Lithium carbonate is the first line of treatment for bipolar patients. It has established effects on both sleep and circadian rhythmicity, and these effects have been proposed to at least partially explain its clinical effect (112,113). Several early studies demonstrated that lithium remediated sleep abnormalities associated with depression in bipolar patients. A 150-night analysis of sleep in five bipolar patients reported that chronic lithium administration produced a significant decrease in both REM density as well as an increase in the latency to REM sleep (114). In addition to confirming this finding, a follow-up study by Mendels et al. reported that lithium administration also significantly increased stage 1, 3, and 4 sleep in bipolar patients. Similarly, Hudson et al., studying nine bipolar manic patients, reported evidence of both REM suppression and an increase in REM latency following initiation of lithium carbonate therapy (59).

Lithium also has demonstrated effects on measures of circadian physiology beyond REM sleep regulation in both healthy subjects and patients; it has been demonstrated to delay the sleep–wake rhythm by approximately 15 minutes in healthy human subjects (115). Campbell et al. examined the effects of lithium on the circadian rhythms of body temperature and REM sleep in a single patient with bipolar depression, and reported a phase delay of the body temperature rhythm of 74 minutes after one week of lithium treatment. The authors also reported a markedly increased REM sleep latency following lithium administration as well as a significant decrease in REM density (115). The data are consistent with the hypothesized phase-delaying properties of lithium, although it is not clear if these properties are related to therapeutic effect. Interestingly, however, recent data have demonstrated that both lithium and valproic acid, another widely prescribed mood stabilizer, are capable of reducing melatonin suppression by light in healthy controls. As discussed above, sensitivity of the pineal hormone melatonin to bright light suppression has been posited as a putative marker of affective disorders. Hallam et al. recently demonstrated that both lithium and valproic acid significantly reduce the sensitivity of nocturnal melatonin secretion to light in healthy volunteers (116). Both sets of data are in accord with a trend for unmediated bipolar patients to demonstrate greater sensitivity to the melatonin-suppressing effects of light relative to lithium-treated patients (32). The clinical relevance of this finding is not entirely clear, however.

GENES AND BIPOLAR DISORDER

Evidence from family, twin and adoption studies indicate that bipolar disorder has a genetic basis with heritability estimates as high as 80% (117). To date, however, neither genes of large or small effect have been identified. The search for genes conveying risk for bipolar illness is complicated by the fact that the disorder is both phenotypically and genetically heterogeneous (118). A number of genes interacting with each other, as well as with environmental factors, likely conspire to modulate susceptibility to the disease phenotype. Despite these complexities, a substantial research effort is underway to identify genetic factors that contribute to the bipolar disorder phenotype. In particular, several recent studies have attempted to link particular genes involved in the regulation of the circadian clock with a general susceptibility to bipolar disorder or with some specific feature of the disease process.

The last decade has witnessed a phenomenal growth in our understanding of the molecular basis of the circadian clock. In the 1980s, induced mutations in fruit flies led to the identification of the first circadian clock mutants period (Per) and frequency (Frq), and in the 1990s a similar mutagenesis approach in mice isolated the first mammalian circadian mutation CLOCK (119,120). Although most of our understanding of the molecular regulation of the circadian clock comes from animal studies, genes dedicated to the generation and regulation of circadian rhythms have also been identified in humans. Patients afflicted by Familial Advanced Sleep Phase Syndrome (ASPS) have a short circadian period with a four-hour advance of the daily sleep–wake cycle, and this trait has been linked to a missense mutation that replaces a serine for a glycine in the human Per2 gene (121), proving that, as is the case in lower animals, mutations in circadian clock regulatory genes have demonstrable effects on the circadian phenotype. Researchers have therefore begun to consider the role of circadian genes involved in other putative rhythmic disorders, including bipolar disorder. Although the results of these studies have thus far failed to identify definitively any gene or genes involved in susceptibility to bipolar disorder, our increasing understanding of the molecular regulation of the circadian system has the potential to identify circadian genes that might contribute to some element of the bipolar phenotype.

As noted above, lithium treatment is known to lengthen the period of circadian rhythms in mammals (122). Although the mechanism through which this effect is achieved is not entirely clear, lithium has been shown to directly inhibit the activity of glycogen synthase kinase-3 beta (GSK3-B), a serine/threonine kinase essential in a number of signaling pathways, which has recently been identified as a fundamental regulator of the mammalian circadian clock (123,124). The inhibition of GSK3-B is speculated to be the mechanism through which lithium lengthens the circadian period, an effect that has been linked to its therapeutic efficacy, making GSK3-B a plausible candidate gene for bipolar disorder (125). Three studies have attempted to link a particular allelic variant in the promoter of the gene encoding the GSK3-B protein with particular features of bipolar disorder. Benedetti et al. analyzed 185 bipolar disorder patients and reported that homozygotes for the wild-type variant (T/T) of the promoter had an earlier age of onset than the carriers of the mutant (T/C) allele. These authors further evaluated the association of GSK3-B with the response to both total sleep deprivation and to lithium treatment in bipolar patients. In both cases, a moderate association was made between the mutant allele (T/C) and an enhanced response to both therapeutic interventions relative to the (C/C) or (T/T) variants (126–128). Although these associations are intriguing, sample sizes were small and differences in allele frequency in different populations may introduce a bias. Importantly, there is currently no experimental evidence that the particular allelic variant in the GSK3-B promoter has any functional effect on GSK3-B gene expression or on the expression of any constituent of the circadian clock in humans. Both larger sample sizes as well as basic research on the functional role of GSK3-B variant are necessary before any firm conclusions can be drawn.

CONCLUSION

Although not all patients with bipolar illness show evidence of sensitivities to biological rhythm or sleep disruption, nor do all patients respond to treatments that normalize these rhythms, there is evidence that these abnormalities are important in the disease process for at least a subset of bipolar patients. Given the increasing recognition of etiological heterogeneity in bipolar illness, a critical next step is to identify particular groups of patients who present with these abnormalities and determine whether the abnormalities are merely epiphenomena of the disorder, or, rather, are causally related. Advances in basic science and technology, including the sequencing of the human genome and refinements in neuroimaging technology, will help to expand and refine our understanding of the central nervous system mechanisms disrupted in bipolar disorder.

REFERENCES

1. Goodwin FK. The biology of recurrence: new directions for the pharmacologic bridge. J Clin Psychiatry 1989; (50 suppl):40–44; discussion 45–47.
2. Koukopoulos A, Sani G, et al. Duration and stability of the rapid-cycling course: a long-term personal follow-up of 109 patients. J Affect Disord 2003; 73(1–2): 75–85.
3. MacKinnon DF, Zandi PP, et al. Association of rapid mood switching with panic disorder and familial panic risk in familial bipolar disorder. Am J Psychiatry 2003; 160(9):1696–1698.
4. Wolff EA III, Putnam FW, et al. Motor activity and affective illness. The relationship of amplitude and temporal distribution to changes in affective state. Arch Gen Psychiatry 1985; 42(3):288–294.
5. Teicher MH. Actigraphy and motion analysis: new tools for psychiatry. Harv Rev Psychiatry 1995; 3(1):18–35.
6. Ashman SB, Monk TH, et al. Relationship between social rhythms and mood in patients with rapid cycling bipolar disorder. Psychiatry Res 1999; 86(1):1–8.
7. Klein E, Lavie P, et al. Increased motor activity and recurrent manic episodes: predictors of rapid relapse in remitted bipolar disorder patients after lithium discontinuation. Biol Psychiatry 1992; 31(3):279–284.
8. Leibenluft E, Albert PS, et al. Relationship between sleep and mood in patients with rapid-cycling bipolar disorder. Psychiatry Res 1996; 63(2–3):161–168.
9. Lam D, Wong G. Prodromes, coping strategies and psychological interventions in bipolar disorders. Clin Psychol Rev 2005; 25(8):1028–1042.
10. Goodwin FK, Redfield Jamison K. Manic-Depressive Illness. New York: Oxford University Press, 1990.
11. Ehlers CL, Frank E, et al. Social zeitgebers and biological rhythms. A unified approach to understanding the etiology of depression. Arch Gen Psychiatry 1998; 45(10):948–952.
12. Mathew MR, Chandrasekaran R, et al. A study of life events in mania. J Affect Disord 1994; 32(3):157–161.
13. Johnson SL, Roberts JE. Life events and bipolar disorder: implications from biological theories. Psychol Bull 1995; 117(3):434–449.
14. Hammen C, Gitlin M. Stress reactivity in bipolar patients and its relation to prior history of disorder. Am J Psychiatry 1997; 154(6):856–857.
15. Malkoff-Schwartz S, Frank E, et al. Stressful life events and social rhythm disruption in the onset of manic and depressive bipolar episodes: a preliminary investigation. Arch Gen Psychiatry 1998; 55(8):702–707.
16. Wu JC, Bunney WE. The biological basis of an antidepressant response to sleep deprivation and relapse: review and hypothesis. Am J Psychiatry 1990; 147(1):14–21.
17. Wehr TA, Sack DA, et al. Sleep reduction as a final common pathway in the genesis of mania. Am J Psychiatry 1987; 144(2):201–204.
18. Goodwin FK, Wirz-Justice A, et al. Evidence that the pathophysiology of depression and the mechanism of action of antidepressant drugs both involve alterations in circadian rhythms. Adv Biochem Psychopharmacol 1982; 32:1–11.
19. Kripke DF. Critical interval hypotheses for depression. Chronobiol Int 1984; 1(1):73–80.
20. Wagner DR. Disorders of the circadian sleep-wake cycle. Neurol Clin 1996; 14(3):651–670.
21. Scheer FA, Czeisler CA. Melatonin, sleep, and circadian rhythms. Sleep Med Rev 2005; 9(1):5–9.
22. Gold PW, Chrousos GP. Organization of the stress system and its dysregulation in melancholic and atypical depression: high vs low CRH/NE states. Mol Psychiatry 2002; 7(3):254–275.
23. Linkowski P, Mendlewicz J, et al. The 24-hour profile of adrenocorticotropin and cortisol in major depressive illness. J Clin Endocrinol Metab 1985; 61(3):429–438.

24. Linkowski P, Mendlewicz J, et al. 24-hour profiles of adrenocorticotropin, cortisol, and growth hormone in major depressive illness: effect of antidepressant treatment. J Clin Endocrinol Metab 1987; 65(1):141–152.
25. Cervantes P, Gelber S, et al. Circadian secretion of cortisol in bipolar disorder. J Psychiatry Neurosci 2001; 26(5):411–416.
26. Czeisler CA, Klerman EB. Circadian and sleep-dependent regulation of hormone release in humans. Recent Prog Horm Res 1999; 54:97–130; discussion 130–132.
27. Leibenluft E, Feldman-Naim S, et al. Salivary and plasma measures of dim light melatonin onset (DLMO) in patients with rapid cycling bipolar disorder. Biol Psychiatry 1996; 40(8):731–735.
28. Whalley LJ, Perini T, et al. Melatonin response to bright light in recovered, drug-free, bipolar patients. Psychiatry Res 1991; 38(1):13–19.
29. Kennedy SH, Kutcher SH, et al. Nocturnal melatonin and 24-hour 6-sulphatoxymelatonin levels in various phases of bipolar affective disorder. Psychiatry Res 1996; 63(2–3):219–222.
30. Nathan PJ, Burrows GD, et al. Melatonin sensitivity to dim white light in affective disorders. Neuropsychopharmacology 1999; 21(3):408–413.
31. Lewy AJ, Nurnberger JI Jr, et al. Supersensitivity to light: possible trait marker for manic-depressive illness. Am J Psychiatry 1985; 142(6):725–727.
32. Nurnberger JI Jr, Adkins S, et al. Melatonin suppression by light in euthymic bipolar and unipolar patients. Arch Gen Psychiatry 2000; 57(6):572–579.
33. Nurnberger JI Jr, Berrettini W, et al. Supersensitivity to melatonin suppression by light in young people at high risk for affective disorder. A preliminary report. Neuropsychopharmacology 1988; 1(3):217–223.
34. Frangos E, Athanassenas G, et al. Seasonality of the episodes of recurrent affective psychoses. Possible prophylactic interventions. J Affect Disord 1980; 2(4): 239–247.
35. Carney PA, Fitzgerald CT, et al. Influence of climate on the prevalence of mania. Br J Psychiatry 1988; 152:820–823.
36. Kamo K, Tomitaka S, et al. Season and mania. Jpn J Psychiatry Neurol 1993; 47(2): 473–474.
37. Cassidy F, Carroll BJ. Seasonal variation of mixed and pure episodes of bipolar disorder. J Affect Disord 2002; 68(1):25–31.
38. D'Mello DA, McNeil JA, et al. Seasons and bipolar disorder. Ann Clin Psychiatry 1995; 7(1):11–18.
39. Suhail K, Cochrane R. Seasonal variations in hospital admissions for affective disorders by gender and ethnicity. Soc Psychiatry Psychiatr Epidemiol 1998; 33(5):211–217.
40. Rosenthal NE, Sack DA, et al. Seasonal affective disorder. A description of the syndrome and preliminary findings with light therapy. Arch Gen Psychiatry 1984; 41(1):72–80.
41. Magnusson A. An overview of epidemiological studies on seasonal affective disorder. Acta Psychiatr Scand 2000; 101(3):176–184.
42. Levitt AJ, Boyle MH. The impact of latitude on the prevalence of seasonal depression. Can J Psychiatry 2002; 47(4):361–367.
43. Hardin TA, Wehr TA, et al. Evaluation of seasonality in six clinical populations and two normal populations. J Psychiatr Res 1991; 25(3):75–87.
44. Wehr TA, Duncan WC, Jr, et al. A circadian signal of change of season in patients with seasonal affective disorder. Arch Gen Psychiatry 2001; 58(12):1108–1114.
45. Borbely AA. A two process model of sleep regulation. Hum Neurobiol 1982; 1(3): 195–204.
46. Achermann P, Dijk DJ, et al. A model of human sleep homeostasis based on EEG slow-wave activity: quantitative comparison of data and simulations. Brain Res Bull 1993; 31(1–2):97–113.
47. Jones BE. Arousal systems. Front Biosci 2003; 8:s438–s451.
48. Jones BE. From waking to sleeping: neuronal and chemical substrates. Trends Pharmacol Sci 2005; 26(11):578–586.
49. Hobson JA, McCarley RW, et al. Sleep cycle oscillation: reciprocal discharge by two brainstem neuronal groups. Science 1975; 189(4196):55–58.

50. Benca RM, Obermeyer WH, et al. Sleep and psychiatric disorders. A meta-analysis. Arch Gen Psychiatry 1992; 49(8):651–668; discussion 669–670.
51. Weske G, Berger M, et al. Neurobiological findings in a patient with 48-hour rapid cycling bipolar affective disorder. A case report. Nervenarzt 2001; 72(7):549–554.
52. Fossion P, Staner L, et al. Does sleep EEG data distinguish between UP, BPI or BPII major depressions? An age and gender controlled study. J Affect Disord 1998; 49(3):181–187.
53. Duncan WC Jr, Pettigrew KD, et al. REM architecture changes in bipolar and unipolar depression. Am J Psychiatry 1979; 136(11):1424–1427.
54. Giles DE, Rush AJ, et al. Sleep parameters in bipolar I, bipolar II, and unipolar depressions. Biol Psychiatry 1986; 21(13):1340–1343.
55. Kupfer DJ, Reynolds CF, et al. Comparison of automated REM and slow-wave sleep analysis in young and middle-aged depressed subjects. Biol Psychiatry 1986; 21(2):189–200.
56. Linkowski P, Van Cauter E, et al. Neuroendocrine rhythms in uni- and bipolar depressions. Neurophysiol Clin 1998; 18(2):141–151.
57. Lauer CJ, Wiegand M, et al. All-night electroencephalographic sleep and cranial computed tomography in depression. A study of unipolar and bipolar patients. Eur Arch Psychiatry Clin Neurosci 1992; 242(2–3):59–68.
58. Hubain P, Van Veeren C, et al. Neuroendocrine and sleep variables in major depressed inpatients: role of severity. Psychiatry Res 1996; 63(1):83–92.
59. Hudson JI, Lipinski JE, et al. Electroencephalographic sleep in mania. Arch Gen Psychiatry 1988; 45(3):267–273.
60. Linkowski P, Kerkhofs M, et al. Sleep during mania in manic-depressive males. Eur Arch Psychiatry Neurol Sci 1986; 235(6):339–341.
61. Bunney WE Jr, Murphy DL, et al. The "switch process" in manic-depressive illness. I. A systematic study of sequential behavioral changes. Arch Gen Psychiatry 1972; 27(3):295–302.
62. Janowsky DS, el-Yousef MK, et al. A cholinergic-adrenergic hypothesis of mania and depression. Lancet 1972; 2(7778):632–635.
63. Kupfer DJ, Heninger GR. REM activity as a correlate of mood changes throughout the night. Electroencephalographic sleep patterns in a patient with a 48-hour cyclic mood disorder. Arch Gen Psychiatry 1972; 27(3):368–373.
64. Lee JH, Reynolds CF, et al. Electroencephalographic sleep in recently remitted, elderly depressed patients in double-blind placebo-maintenance therapy. Neuropsychopharmacology 1993; 8(2):143–150.
65. Buysse DJ, Frank E, et al. Electroencephalographic sleep correlates of episode and vulnerability to recurrence in depression. Biol Psychiatry 1997; 41(4): 406–418.
66. Buysse DJ, Tu XM, et al. Pretreatment REM sleep and subjective sleep quality distinguish depressed psychotherapy remitters and nonremitters. Biol Psychiatry 1999; 45(2):205–213.
67. Buysse DJ, Hall M, et al. Sleep and treatment response in depression: new findings using power spectral analysis. Psychiatry Res 2001; 103(1):51–67.
68. Rush AJ, Erman MK, et al. Polysomnographic findings in recently drug-free and clinically remitted depressed patients. Arch Gen Psychiatry 1986; 43(9): 878–884.
69. Giles DE, Kupfer DJ, et al. Polysomnographic parameters in first-degree relatives of unipolar probands. Psychiatry Res 1989; 27(2):127–136.
70. Giles DE, Roffwarg HP, et al. A cross-sectional study of the effects of depression on REM latency. Biol Psychiatry 1990; 28(8):697–704.
71. Giles DE, Biggs MM, et al. Risk factors in families of unipolar depression. I. Psychiatric illness and reduced REM latency. J Affect Disord 1988; 14(1):51–59.
72. Lauer CJ, Schreiber W, et al. In quest of identifying vulnerability markers for psychiatric disorders by all-night polysomnography. Arch Gen Psychiatry 1995; 52(2):145–153.
73. Modell S, Ising M, et al. The Munich Vulnerability Study on Affective Disorders: stability of polysomnographic findings over time. Biol Psychiatry 2002; 52(5): 430–437.
74. Modell S, Huber J, et al. The Munich Vulnerability Study on Affective Disorders: risk factors for unipolarity versus bipolarity. J Affect Disord 2003; 74(2):173–184.

75. Harvey AG, Schmidt DA, et al. Sleep-related functioning in euthymic patients with bipolar disorder, patients with insomnia, and subjects without sleep problems. Am J Psychiatry 2005; 162(1):50–57.
76. Vogel GW. A review of REM sleep deprivation. Arch Gen Psychiatry 1975; 32:749–761.
77. Wirz-Justice A, Van den Hoofdakker RH. Sleep deprivation in depression: what do we know, where do we go? Biol Psychiatry 1999; 46(4):445–453.
78. Barbini B, Colombo C, et al. The unipolar-bipolar dichotomy and the response to sleep deprivation. Psychiatry Res 1998; 79(1):43–50.
79. Colombo C, Benedetti F, et al. Rate of switch from depression into mania after therapeutic sleep deprivation in bipolar depression. Psychiatry Res 1999; 86(3):267–270.
80. Grunze H. Reevaluating therapies for bipolar depression. J Clin Psychiatry 2005; (66 suppl 5):17–25.
81. Rush AJ, Armitage R, et al. Comparative effects of nefazodone and fluoxetine on sleep in outpatients with major depressive disorder. Biol Psychiatry 1998; 44(1):3–14.
82. Kuhs H, Farber D, et al. Amitriptyline in combination with repeated late sleep deprivation versus amitriptyline alone in major depression. A randomised study. J Affect Disord 1996; 37(1):31–41.
83. Benedetti F, Barbini B, et al. Sleep deprivation hastens the antidepressant action of fluoxetine. Eur Arch Psychiatry Clin Neurosci 1997; 247(2):100–103.
84. Smeraldi E, Benedetti F, et al. Sustained antidepressant effect of sleep deprivation combined with pindolol in bipolar depression. A placebo-controlled trial. Neuropsychopharmacology 1999; 20(4):380–385.
85. Benedetti F, Serretti A, et al. Influence of a functional polymorphism within the promoter of the serotonin transporter gene on the effects of total sleep deprivation in bipolar depression. Am J Psychiatry 1999; 156(9):1450–1452.
86. Baghai TC, Schule C, et al. No Influence of a functional polymorphism within the serotonin transporter gene on partial sleep deprivation in major depression. World J Biol Psychiatry 2003; 4(3):111–114.
87. Neumeister A, Praschak-Rieder N, et al. Effects of tryptophan depletion in drug-free depressed patients who responded to total sleep deprivation. Arch Gen Psychiatry 1998; 55(2):167–172.
88. Ebert D, Berger M. Neurobiological similarities in antidepressant sleep deprivation and psychostimulant use: a psychostimulant theory of antidepressant sleep deprivation. Psychopharmacology (Berl) 1998; 140(1):1–10.
89. Lerner P, Goodwin FK, et al. Dopamine-beta-hydroxylase in the cerebrospinal fluid of psychiatric patients. Biol Psychiatry 1978; 13(6):685–694.
90. Gerner RH, Post RM, et al. Biological and behavioral effects of one night's sleep deprivation in depressed patients and normals. J Psychiatr Res 1979; 15(1):21–40.
91. Ebert D, Feistel H, et al. Single photon emission computerized tomography assessment of cerebral dopamine D2 receptor blockade in depression before and after sleep deprivation—preliminary results. Biol Psychiatry 1994; 35(11):880–885.
92. Klimke AR, Larisch R, et al. Dopamine D2 receptor binding before and after treatment of major depression measured by [123I]IBZM SPECT. Psychiatry Res 1999; 90(2): 91–101.
93. Reist C, Sokolski KN, et al. The effect of sleep deprivation on motor impairment and retinal adaptation in Parkinson's disease. Prog Neuropsychopharmacol Biol Psychiatry 1995; 19(3):445–454.
94. Demet EM, Chicz-Demet A, et al. Sleep deprivation therapy in depressive illness and Parkinson's disease. Prog Neuropsychopharmacol Biol Psychiatry 1999; 23(5): 753–784.
95. Cirelli C, Gutierrez CM, et al. Extensive and divergent effects of sleep and wakefulness on brain gene expression. Neuron 2004; 41(1):35–43.
96. Nibuya M, Nestler EJ, et al. Chronic antidepressant administration increases the expression of cAMP response element binding protein (CREB) in rat hippocampus. J Neurosci 1996; 16(7):2365–2372.
97. Tononi G, Cirelli C. Sleep function and synaptic homeostasis. Sleep Med Rev 2006; 10(1):49–62.

98. Payne JL, Quiroz JA, et al. Timing is everything: does the robust upregulation of noradrenergically regulated plasticity genes underlie the rapid antidepressant effects of sleep deprivation? Biol Psychiatry 2002; 52(10):921–926.
99. Schilgen B, Bischofs W, et al. Total and partial sleep deprivation in the treatment of depression: preliminary communication. Arzneimittelforschung 1976; 26(6):1171–1173.
100. Schilgen B, Tolle R. Partial sleep deprivation as therapy for depression. Arch Gen Psychiatry 1980; 37(3):267–271.
101. Smeraldi E, Benedetti F, et al. Sustaining the effect of sleep deprivation. Am J Psychiatry 1998; 155(8):1134–1135.
102. Benedetti F, Barbini B, et al. Sleep phase advance and lithium to sustain the antidepressant effect of total sleep deprivation in bipolar depression: new findings supporting the internal coincidence model? J Psychiatr Res 2001; 35(6):323–329.
103. Riemann D, Voderholzer U, et al. Sleep and sleep-wake manipulations in bipolar depression. Neuropsychobiology 2002; 45(suppl 1):7–12.
104. Riemann D, Konig A, et al. How to preserve the antidepressive effect of sleep deprivation: a comparison of sleep phase advance and sleep phase delay. Eur Arch Psychiatry Clin Neurosci 1999; 249(5):231–237.
105. Boivin DB, Czeisler CA, et al. Complex interaction of the sleep-wake cycle and circadian phase modulates mood in healthy subjects. Arch Gen Psychiatry 1997; 54(2): 145–152.
106. Leibenluft E,Suppes T. Treating bipolar illness: focus on treatment algorithms and management of the sleep-wake cycle. Am J Psychiatry 1999; 156(12):1976–1981.
107. Wirz-Justice A, Quinto C, et al. A rapid-cycling bipolar patient treated with long nights, bedrest, and light. Biol Psychiatry 1999; 45(8):1075–1077.
108. Wehr TA, Turner EH, et al. Treatment of rapidly cycling bipolar patient by using extended bed rest and darkness to stabilize the timing and duration of sleep. Biol Psychiatry 1998; 43(11):822–828.
109. Malkoff-Schwartz S, Frank E, et al. Social rhythm disruption and stressful life events in the onset of bipolar and unipolar episodes. Psychol Med 2000; 30(5):1005–1016.
110. Frank E, Hlastala S, et al. Inducing lifestyle regularity in recovering bipolar disorder patients: results from the maintenance therapies in bipolar disorder protocol. Biol Psychiatry 1997; 41(12):1165–1173.
111. Frank E, Swartz HA, et al. Interpersonal and social rhythm therapy: managing the chaos of bipolar disorder. Biol Psychiatry 2000; 48(6):593–604.
112. Klemfuss H. Rhythms and the pharmacology of lithium. Pharmacol Ther 1992; 56(1): 53–78.
113. Ikonomov OC, Manji HK. Molecular mechanisms underlying mood stabilization in manic-depressive illness: the phenotype challenge. Am J Psychiatry 1999; 156(10): 1506–1514.
114. Mendels J, Chernik DA. The effect of lithium carbonate on the sleep of depressed patients. Int Pharmacopsychiatry 1973; 8(3):184–192.
115. Campbell SS, Gillin JC, et al. Lithium delays circadian phase of temperature and REM sleep in a bipolar depressive: a case report. Psychiatry Res 1989; 27(1):23–29.
116. Hallam KT, Olver JS, et al. Low doses of lithium carbonate reduce melatonin light sensitivity in healthy volunteers. Int J Neuropsychopharmacol 2005; l 8(2):255–259.
117. Cardno AG, Marshall EJ, et al. Heritability estimates for psychotic disorders: the Maudsley twin psychosis series. Arch Gen Psychiatry 1999; 56(2):162–168.
118. MacKinnon DF, Jamison KR, et al. Genetics of manic depressive illness. Annu Rev Neurosci 1997; 20:355–373.
119. Wager-Smith K, Kay SA. Circadian rhythm genetics: from flies to mice to humans. Nat Genet 2000; 26(1):23–27.
120. Allada R, Emery P, et al. Stopping time: the genetics of fly and mouse circadian clocks. Annu Rev Neurosci 2001; 24:1091–1119.
121. Toth LA. Identifying genetic influences on sleep: an approach to discovering the mechanisms of sleep regulation. Behav Genet 2001; 31(1):39–46.
122. Johnsson A, Engelmann W, et al. Influence of lithium ions on human circadian rhythms. Z Naturforsch [C] 1980; 35(5–6):503–507.

123. Grimes CA, Jope RS. The multifaceted roles of glycogen synthase kinase 3beta in cellular signaling. Prog Neurobiol 2001; 65(4):391–426.
124. Martinek S, Inonog S, et al. A role for the segment polarity gene shaggy/GSK-3 in the Drosophila circadian clock. Cell 2001; 105(6):769–779.
125. Iitaka C, Miyazaki K, et al. A role for glycogen synthase kinase-3beta in the mammalian circadian clock. J Biol Chem 2005; 280(33):29397–23402.
126. Benedetti F, Bernasconi A, et al. A single nucleotide polymorphism in glycogen synthase kinase 3-beta promoter gene influences onset of illness in patients affected by bipolar disorder. Neurosci Lett 2004; 355(1–2):37–40.
127. Benedetti F, Serretti A, et al. A glycogen synthase kinase 3-beta promoter gene single nucleotide polymorphism is associated with age at onset and response to total sleep deprivation in bipolar depression. Neurosci Lett 2004; 368(2):123–126.
128. Benedetti F, Serretti A, et al. Long-term response to lithium salts in bipolar illness is influenced by the glycogen synthase kinase 3-beta -50 T/C SNP. Neurosci Lett 2005; 376(1):51–55.

11 Hypothesis of an Infectious Etiology in Bipolar Disorder

Robert H. Yolken
Stanley Division of Developmental Neurovirology, Johns Hopkins University School of Medicine, Baltimore, Maryland, U.S.A.

E. Fuller Torrey
Stanley Medical Research Institute, Chevy Chase, Maryland, U.S.A.

INTRODUCTION

Until recently, the possibility of viruses or other infectious agents being involved in the etiology of bipolar disorder had not been seriously considered. For example, in their 782 page *Manic-Depressive Illness*, published in 1990, Goodwin and Jamison (1) covered viral factors in two pages, almost all of which was devoted to viral factors in unipolar depression, not bipolar disorder. A major reason for the neglect of a possible viral etiology of bipolar disorder has been the widespread assumption that it is predominantly or exclusively genetic in origin.

Findings suggest that viruses and other infectious agents should be considered as part of a multifactorial etiological pathway for at least some cases of bipolar disorder: viral infections are strongly influenced by genetic factors, and thus viral and genetic etiologies are not incompatible; viral central nervous system (CNS) infections may mimic bipolar disorder; and season-of-birth, urban birth, and perinatal studies all point toward an environmental risk factor in bipolar disorder that may be a virus. This chapter summarizes these three factors and then discusses recent findings of specific infectious agents and bipolar disorder. Immunological research, which is also consistent with an infectious etiology, is discussed elsewhere in this volume.

GENETIC FACTORS IN MICROBIAL INFECTION

At first glance, the postulate that infections and other environmental factors may play a role in the etiopathogenesis of bipolar disorder seems to contradict the large body of evidence presented elsewhere in this volume indicating that genetic factors are important predictors of disease susceptibility. However, it is becoming increasingly clear that host factors under genetic control are major determinants of host susceptibility to infectious agents and to the response to infection after it has occurred (Table 1).

The human genes with the clearest association with susceptibility to infection are those that encode components of the immune system. There are many single-gene disorders associated with a complete ablation of a major component of the immune system, such as major defects in T cells, B cells, macrophages, or complement cascade (2). These rare defects are generally associated with serious, often life-threatening, infection and are associated with a substantial rate of

TABLE 1 Genetic Determinants of Infection

Immunological activity
T cells—cell-mediated immunity
B cells—antibody generation
Macrophages
Complement
Cytokines
Interferons
Microbial receptors
Monokines
Lectins
Fusin
Mannose-binding proteins
Miscellaneous
Hemoglobin variants
Blood groups
Determinants of mucosal integrity

morbidity and mortality (3). However, there are also more subtle genetic alterations that are often specific for susceptibility or resistance to infection with defined pathogenic agents. These polymorphisms are generally more common in human populations than single-gene defects leading to complete immunodeficiency (4).

The human infectious agents that have been studied in most detail in terms of genetic patterns of susceptibility and resistance are the plasmodia that cause malaria. Interest in genetic interactions between malaria parasites and host genes was initiated by Haldane (5), who noted in 1948 that sickle cell disease, thalassemia, and other homoglobinopathies were most prevalent in areas of the world where malaria is endemic. Since this original observation, more than 120 mutations in the hemoglobin molecule have been identified that are associated with protection against malaria. The mechanism by which altered hemoglobin provides against malaria is not known with certainty but may be related to the increased clearance of parasites in altered erythrocytes or by decreased parasitic growth under conditions of lowered oxygen concentrations. Susceptibility to malaria has also been associated with genetic polymorphisms in other molecules, including HLA class I, HLA class II, and tumor necrosis factor alpha (TNF-α). Epidemiological studies have indicated additional, as yet unrecognized, factors that contribute to susceptibility to malaria. The large number of polymorphisms associated with susceptibility to malaria is a testimony to both the pervasiveness of malaria as a human pathogen and the diversity by which genetic mechanisms may determine susceptibility and resistance to an infectious agent (6).

The association between genes and susceptibility to malaria has led to the search for genes that confer susceptibility to other parasitic agents. For example, investigators have identified a region on chromosome 5 (5q31-33) that is associated with susceptibility to *Schistosoma mansoni*. Known genes in this region include a gene cluster that encodes a number of immunologically active molecules, including granulocyte-macrophage colony stimulating factor, immune regulatory factor 1, and interleukins 3, 4, 5, and 13 (7). An analysis of parasitic infections in Brazil identified additional regions that may determine susceptibility to leishmaniasis, including HLA II and HLA III, and a region on chromosome 17q that encodes a number of inducible cytokine molecules. Susceptibility to leishmaniasis may also be determined by polymorphisms in the gene encoding the promoter region of

TNF-α (8). Finally, a study of hookworm in Gambia concluded that approximately 37% of the susceptibility to this infection was related to genetic factors; however, the specific genetic factors have not yet been determined (9).

Genetic factors have also been recognized as important factors in susceptibility to bacterial infections. Although such factors have been recognized in terms of the response to pyogenic bacteria such as *Haemophilus influenzae* (10), most information in this area had been directed at slow-growing intracellular bacteria, particularly those of the family mycobacteraciae. For example, susceptibility to *Mycobacteria tuberculosis* has been linked to the Nramp locus on human chromosome 2. This locus appears to control the functioning of bactericidal activity within cells of the macrophage lineage and is thus central in determining the intracellular survival of mycobacteria after phagocytosis (11). This gene locus may also be a major determinant of disease response to other mycobacteria, including those that cause leprosy. The response to lower pathogenicity mycobacteria, such as bacillus C-G, is determined by another gene that encodes the receptor for TNF-α. Individuals with defects in this gene have deficient upregulation of TNF-α after infection with mycobacteria and subsequent deficiencies in bacterial clearance (12). The response to mycobacteria may also be determined by other genes such as the 5q31-33 and 17q loci described above. Genetic factors may also play a role in the response to other slowly growing bacteria. For example, susceptibility to *Helicobacter pylori*, the organism that causes intestinal ulcers, is determined, to a great extent, by DQA genes that are components of the human major histocompatibility system (13).

The response to viral infection is also determined, in part, by genetic factors. For example, infection with Epstein-Barr virus results in asymptomatic or mild infection in most individuals. However, some infected individuals will develop overwhelming infections or malignant tumors due to mutations involved in the immune response to infection (14). The response to other viral agents, such as those causing hepatitis, is also under the control of a number of genes, including those that encode the TNF-α promoter, mannose-binding proteins, and components of the histocompatibility locus (15). A striking example of genetically encoded protection against viral infection is provided by analyses of infection with human immunodeficiency virus type 1 (HIV-1). In this case, individuals with homozygous mutations in the viral coreceptor CCR5 are protected against infection despite exposure to high levels of infecting virus (16,17). Heterozygosity, although not associated with protection, may result in a slower course of disease progression in infected individuals (18). Interestingly, these mutations, which occur in 1% to 3% of individuals of European extraction, do not appear to have a deleterious effect on the host (19). The discovery of these protective genes has led to research directed at providing similar protective mechanisms to individuals who are not genetically endowed with this inherent mechanism of disease protection.

It can be seen from the above discussion that many genes determine the susceptibility to human infectious diseases. Furthermore, the clinical manifestations of these genes depend, to a great extent, on individual environmental exposures; most genes do not have any effect on individuals who are not exposed to the infectious agents. Genes of a similar nature could be operant in psychiatric diseases such as bipolar disorder in that they may confer susceptibility to infectious agents in the absence of other clinical manifestations. This concept is consistent with the fact that most studies of the genetics of bipolar disorder have identified multiple genomic regions of weak effect, none of which appear to be completely determinant for disease acquisition (20). It is of note in this regard that many regions associated with the acquisition of human psychiatric disease, such as 6p, contain several genes involved in the immune response to infectious agents (21). Further studies of the genetics of bipolar disorder are likely to identify additional regions involved in the immune response to infectious agents. However, the role of these genes in the pathogenesis of bipolar disorder is unlikely to be accurately evaluated without corresponding data related to exposure to infectious agents and other environmental stimuli of the immune response.

VIRAL INFECTIONS OF THE CNS AND BIPOLAR DISORDER

The epidemic of HIV infection has served as a reminder that many viruses can infect the central nervous system (CNS) and cause symptoms that are clinically identical to the symptoms of bipolar disorder. For example, Harris et al. (22) described 31 cases of new-onset psychosis in HIV-infected individuals. Of these, 25 had "mood and/or affective disturbance," including nine with depressions, three with euphoria or irritability, and two with both depression and euphoria. Many of these individuals responded to antipsychotic medication. The Epstein-Barr virus, which causes mononucleosis, can also cause bipolar-like symptoms. For example, Weinstein et al. (23) described a 22-year-old woman who developed "auditory hallucinations, pressure of speech and flight of ideas" one week after having been diagnosed with mononucleosis. Similarly, Goldney and Temme (24) described a 23-year-old woman with the onset of flight of ideas and mania six weeks after mononucleosis; her psychiatric symptoms were controlled with lithium but recurred when the lithium was stopped.

Koehler and Guth (25) described a 41-year-old man who had severe depression lasting three weeks and then abruptly switched to having classic mania with pressure of speech, flight of ideas, psychomotor agitation, and grandiosity. Symptoms of depression or mania lasted intermittently for six months, with a good response to antipsychotic medication. Because he had a headache and stiff neck at the onset of his illness, a lumbar puncture was done and revealed a diagnosis of underlying herpes simplex encephalitis.

A similar case, suspected of being caused by Coxsackie virus, was reported by Myers and Dunner (26). A 28-year-old woman developed increased energy, decreased need for sleep, elation, pressure of speech, hypersexuality, and disorganized behavior and was diagnosed as having bipolar disorder, manic type. She was treated with antipsychotic medication and lithium, with improvement of her symptoms. She then developed a fever, headache, and stiff neck; a lumbar puncture and electroencephalograph were therefore done and suggested a diagnosis of viral encephalitis. Her symptoms abated on medication, and she was maintained on lithium for six months, at which time it was discontinued. Six months later, she relapsed and had a second episode of mania.

There is, of course, no way to conclusively prove a cause-and-effect relationship in such cases and rule out a coincidental but unrelated onset of bipolar disorders and encephalitis. However, the co-occurrence of bipolar disorder symptoms with encephalitis has been repeatedly noted since the early 1900s and such descriptions were especially prominent after the influenza pandemic (27) and during the epidemic of encephalitis in the 1920s (28,29).

ENVIRONMENTAL RISK FACTORS FOR BIPOLAR DISORDER

Despite the fact that genes are known to play a prominent role in the etiology of bipolar disorder, it is also known that nongenetic environmental factors are also operant. The pairwise concordance rate for bipolar disorder among monozygotic twins was 56% (44/79 pairs) in five European studies (30), suggesting that environmental factors also play a significant role. Recent studies of these environmental factors are consistent with an etiological role for viruses.

One environmental factor is the season of birth of individuals who later develop bipolar disorder. Torrey et al. (31), using time series analysis in a study of 18,021 individuals in four states who were diagnosed with Diagnostic and Statistical Method, 3rd edition serial (DSM-III-R) bipolar disorder, reported a 5.8% excess of births in December through March compared with 27.3 million general births in the same states for the same years. Especially noteworthy were the results in North Carolina, which had an 18% excess of bipolar disorder births for February and a 22% excess for March. In this study, the 5.8% bipolar disorder winter birth excess was slightly greater than the 5.0% excess for undifferentiated schizophrenia for the same months. Individuals with severe depression showed a different birth pattern, with a 5.4% excess for March, April, and May. The results of this study of bipolar disorder births are consistent with four other studies of bipolar disorder with much smaller numbers and are also consistent with previous seasonal birth studies of manic-depressive psychosis and mania (32).

Urban birth is now a well-described risk factor for the development of later schizophrenia, but far fewer studies have been done on bipolar disorder. A study in The Netherlands reported some support for regarding urban birth as a risk factor for "affective psychosis" (33), but studies from Denmark, using the Danish Case Register, did not find an association for either affective psychosis or bipolar disorder (34).

Another environmental factor that predisposes individuals to the later development of bipolar disorder is having complications of pregnancy or delivery. Lewis and Murray (35), using retrospective maternal histories, rated definite obstetrical complications in individuals with one of eight different psychiatric diagnosis, including 110 with bipolar disorder. Individuals with schizophrenia (17%) and anorexia nervosa (16%) had the highest percentage of definite obstetrical complications, followed by bipolar disorder (11%), unipolar depression (10%), other psychosis (7%), personality disorder (6%), neurosis (5%), and alcoholism and drug dependence (3%).

Kinney et al. (36) compared obstetrical complications in 16 individuals with DSM-III-R bipolar disorder and 20 of their unaffected siblings by blindly rating their obstetrical records. The weighted-sum score for the affected individuals was 3.56 compared with 1.95 for their siblings (Wilcoxon test, $p = 0.01$). According to the authors, "the higher obstetrical complications (OC) scores in bipolar probands were due to moderate elevations in the rates of a variety of different OCs."

Viral Etiology in Bipolar Disorder

Using prospectively collected data from the 1958 British Perinatal Morality Survey, Sacker et al. (37) compared 44 individuals with affective psychosis as diagnosed by the Catego system with 16,812 control subjects. Among the 44 individuals, 15 had mania, nine had depressive psychosis, and 20 had retarded depression. Compared with control subjects, the group of individuals with affective psychosis had more mothers over age 34 at the time of delivery ($p < 0.05$) and more cesarean sections and forceps deliveries ($p < 0.05$), had been lighter at birth (<2500 g; $p < 0.01$), had been more premature (gestation <37 weeks; $p < 0.01$), and had been given vitamin K at birth more often because of a risk of bleeding ($p < 0.01$).

Recent research, however, has cast doubt on the importance of perinatal complications in bipolar disorder. Klaning et al. (38), using the Danish case register, compared the incidence of bipolar disorder among twins, which are known to have more perinatal complications, and singleton births; the incidence was found to be the same. Scott et al. (39) examined all the relevant studies in a meta-analysis and concluded that "there is no robust evidence that exposure to [obstetrical complications] increases the risk of later development of BP [bipolar disorder]."

Findings for Specific Infectious Agents

In the modern era, the first infectious agent to be tentatively linked to the etiology of bipolar disorder was the borna disease virus (BDV). Two studies in the 1980s reported increased BDV antibodies in patients with unipolar and bipolar depression (40,41). Interest in BDV increased in the 1990s when Salvatore et al. (42) reported finding BDV in two out of five postmortem brain samples from patients with bipolar disorder and Bode et al. (43) described a BDV-positive patient with chronic bipolar disorder whose symptoms improved dramatically when treated with amantadine, an antiviral medication.

Since that time, interest in BDV as a possible etiologic factor in bipolar disorder has waned. Studies of BDV antibody in the sera of bipolar patients by Fu et al. (44), Sauder et al. (45), Kim et al. (46), and Terayama et al. (47) have yielded negative findings; additional unpublished studies in our laboratory have been uniformly negative. Studies of BDV in patients with unipolar depression have been modestly more promising, and additional research may be warranted for that condition.

An infectious agent that has been intensively studied in relationship to its possible role in causing schizophrenia is the influenza virus. For bipolar disorder, there are fewer studies, and most of those combine bipolar disorder with unipolar depression, with or without psychotic features, thus making any definitive view of bipolar disorder problematic. In the study of Sacker et al. cited above (37), there was a weak trend ($p < 0.1$) for the mothers of the individuals with affective psychosis to have had more influenza during pregnancy. Machón et al. (48) in Finland reported a significant ($p < 0.002$) increase in the births of individuals with "unipolar forms of major affective disorder" and a trend ($p < 0.05$) for an increase of individuals with "bipolar forms of major affective disorder." Cannon et al. (49) in Ireland also reported a significantly ($p = 0.003$) increased risk of depressive disorder in the offspring of mothers exposed to influenza but no increase in bipolar disorder. Three other studies in England (50), Queensland (51), Western Australia (52) that examined the risk of affective disorders in offspring after maternal exposure to influenza during pregnancy reported negative results.

The most interesting infectious agent to emerge from recent research on bipolar disorder has been the herpes simplex virus, type 1 (HSV-1). This virus has long been of interest to psychiatric researchers because of its known propensity for neurotropism in general and its affinity for limbic tissue in particular, as well as for its ability to remain latent in the brain and periodically recur. Cases of known, HSV-2 encephalitis with clinical features of bipolar disorder have been well described.

In a series of studies Dickerson et al. (53,54) have shown that individuals with schizophrenia and those with bipolar disorder who also have past exposure to HSV-1, as demonstrated by antibodies, are significantly more likely to have greater cognitive dysfunction, especially recent memory deficits, as measured by the Repeatable Battery for the Assessment of Neuropsychological Status (RBANS) test. In the bipolar study 49 patients who had HSV-1 antibodies had an RBANS score of 77.5, compared with 68 patients who did not have antibodies who had an RBANS score of 90.3 ($p < 0.00002$). Among the patients who had the lowest RBANS scores (lowest quartile), 65 percent had HSV-1 antibodies; among patients who had the highest scores (highest quartile), only 14 percent had HSV-1 antibodies. Treatment with lithium, other mood stabilizers, or antidepressants and other possible confounding variables did not account for the association. There was also no significant association between RBANS-measured cognitive dysfunction and antibodies to other herpes family viruses, specifically HSV-2, Cytomegalo virus (CMV), Epstein-Barr virus (EBV), Varicella-Zoster virus (VZV), or human herpes virus-6 (HHV-6). The association between cognitive dysfunction and HSV-1 antibodies was found only in the patient population and did not exist among the normal controls, suggesting an interaction between the infectious agent and the disease.

Since it is known that genes play a significant role in the etiology of bipolar disorder, Dickerson and her colleagues, using the same patients as in the above study, examined a polymorphism on the catechol-O-methyltransferase gene at amino acid 158 (COMT Val 158 Met polymorphism) to ascertain whether it might explain the association of cognitive dysfunction and HSV-1 antibodies. They found that the polymorphism and HSV-1 antibodies were both independent risk factors for the cognitive dysfunction but that individuals with bipolar disorder who had both "were more than 85 times more likely to be in the lowest quintile of cognitive functioning as compared with the highest quintile when controlling for potential confounding variables such as symptom severity and education" (55). This strongly suggests an interaction between a genetic predisposition and an environmental (specifically, infectious) factor.

Finally, there is a preliminary study linking *Toxoplasma gondii* to bipolar disorder. *T. gondii* is a coccidian protozoa whose definitive host is felines. It is known to cause a congenital syndrome when transmitted from the mother to the fetus early in pregnancy, but later perinatal infection or postnatal infection have been thought to be relatively benign. In recent years, however, several research groups have reported significantly increased antibodies to *T. gondii* in individuals with schizophrenia (56). A study of 148 outpatients with bipolar disorder and 170 controls has also reported an increase in *T. gondii* antibodies (Odds Ratio 5.4; $p < 0.001$) (57).

CONCLUSIONS

Bipolar disorder is a complex human disease state that may represent the interaction of genetic susceptibilities and environmental risk factors. These environmental factors may include infections and the inflammatory response to infectious agents. It is of note in this regard that there are several pathways by which infectious agents can induce disease processes. In classic infectious diseases, there is a clear cause-and-effect relationship between acquisition of the infectious agent and the initiation of the disease process. This is the case for agents that conform to Koch's postulates of cause and effect. As depicted in Figure 1, this is the pattern followed by highly contagious infections such as plague, cholera, and

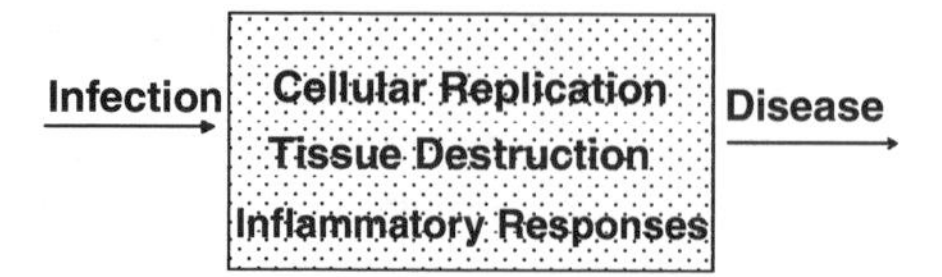

FIGURE 1 Model for infectious diseases for which there is a correspondence between a single infectious agent and a defined disease process. The mediators of the disease process are designated within the box.

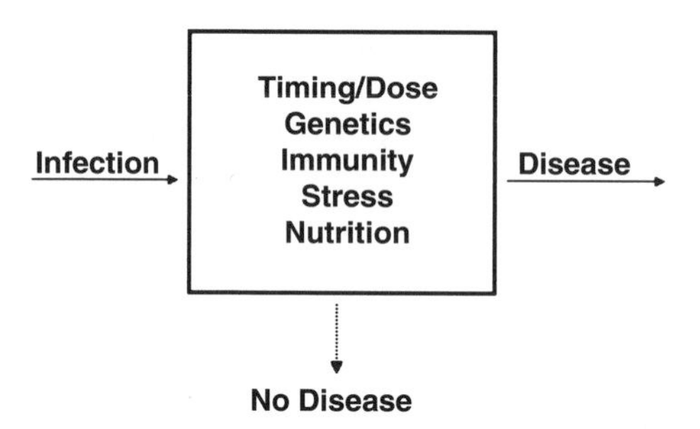

FIGURE 2 Model for infectious diseases where infection may or may not lead to a disease state, depending on the microbial and host factors designated within the box.

measles that cause disease in virtually all susceptible individuals who are exposed to a suitably large infectious dose. However, microbial agents can also induce disease by less direct pathogenic mechanisms. For example, there are agents that only cause disease in a subset of infected individuals. The factors that determine which individuals become infected are determined by both genetic susceptibilities and environmental factors such as nutritional status (Fig. 2). Finally, there are infectious agents that initiate disease processes that can also be initiated by noninfectious agents such as toxins. As depicted in Figure 3, these agents generally cause chronic diseases of organ systems. In light of the above discussion concerning genetic factors, it is likely that if infectious agents are involved in the pathogenesis of bipolar disorder and other human psychiatric diseases, they are operating following the model outlined in Figure 2 or 3. However, it is of note that for both

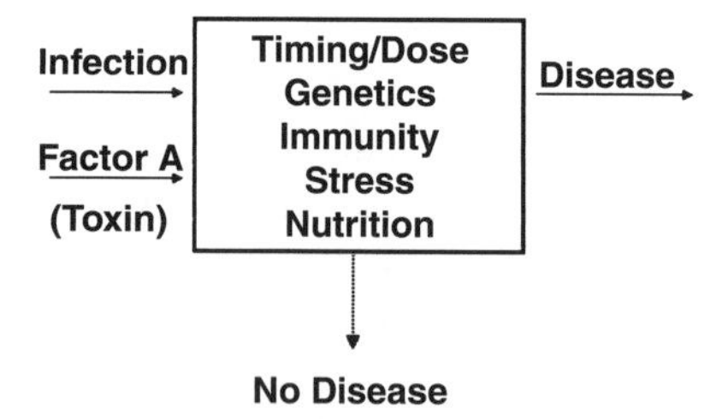

FIGURE 3 Model for infectious diseases in which infectious agents may combine with other environmental factors to initiate a disease process. The expression of the disease process is also modulated by the microbial and host factors designated in the box.

models the successful prevention or treatment of the infectious agent would result in a substantial decrease in both disease incidence and morbidity.

The identification of specific environmental factors related to bipolar disorder would result in the development of improved strategies for the diagnosis and management of this devastating disease.

REFERENCES

1. Goodwin FK, Jamison KR. Manic-Depressive Illness. New York: Oxford University Press, 1990.
2. Hill AVS. Genetics of infectious disease resistance. Curr Opin Genet Dev 1996; 6:348–353.
3. McLeoad R, Buschman E, Arbuckle LD, et al. Immunogenetics in the analysis of resistance to intracellular pathogens. Curr Opin Immunol 1995; 7:539–552.
4. Abel L, Dessein A. The impact of host genetics on susceptibility to human infectious diseases. Curr Opin Immunol 1997; 9:509–516.
5. Haldane JBS. The rate of mutation of human genes. Proceedings of the Eighth International Congress of Genetics and Heredity. Hereditas 1998; 35(suppl):267–273.
6. Weatherall DJ. Host genetics and infectious disease. Parasitology 1996; 112(suppl): S23–S29.
7. Marquet S, Abel L, Hillaire D, et al. Genetic Localization of a locus controlling the intensity of infection by Schistosoma mansoni on chromosome 5q31-q33. Nat Genet 1996; 14:181–184.
8. Blackwell JM, Black GF, Peacock CS, et al. Immunogenetics of leishmanial and mycobacterial infections: the Belem Family study. Phil Trans R Soc Lond 1997; 352:1331–1345.
9. Williams-Blangero S, Blangero J, Bradley M. Quantitative genetic analysis of susceptibility to hookworm infection in a population from rural Zimbabwe. Hum Biol 1997; 69:201–208.
10. Mäkelä PH, Takala AK, Peltola H, et al. Epidemiology of Haemophilus influenzae disease: epidemiology of invasive Haemophilus influenzae type b disease. J Infect Dis 1992; 165(suppl 1):S2–S6.
11. Vidal S, Gros P, Skamene E. Natural resistance to infection with intracellular parasites: molecular genetics identifies Nramp1 as the Bcg/Ity/Lsh locus. J Leuk Biol 1995; 58:382–390.
12. Newport MJ, Huxley CM, Huston, et al. A mutation in the interferon-gamma-receptor gene and susceptibility to mycobacterial infection. N Engl J Med 1996; 335:1941–1949.
13. Azuma T, Konishi J, Ito Y, et al. Genetic differences between duodenal ulcer patients who were positive or negative for Helicobacter pylori. J Clin Gastroenterol 1995; 21(suppl 1):5151–5154.
14. Andersson J. Clinical and immunological considerations in Epstein-Barr virus-associated diseases. Scand J Infect Dis 1996; 100:72–82.
15. Thursz MR, Thomas HC. Host factors in chronic viral hepatitis. Semin Liver Dis 1997; 17:345–350.
16. Dean M, Carrington M, Winkler C, et al. Genetic restriction of HIV-1 infection and progression to AIDS by a deletion allele of the CCR5 structural gene. Science 1996; 273:1856–1862.
17. Quillent C, Oberlin E, Braun J, et al. HIV-1-resistance phenotype conferred by combination of two separate inherited mutations of CCR5 gene. Lancet 1998; 351:14–18.
18. Samson M, Libert F, Doranz BJ, et al. Resistance to HIV-1 infection in Caucasian individuals bearing mutant alleles of the CCR-5 chemokine receptor gene. Nature 1996; 382:722–725.
19. Martison JJ, Chapman NH, Rees DC, et al. Global distribution of the CCR5 gene 32-basepair deletion. Nat Genet 1997; 16:100–103.
20. Moldin SO. The maddening hunt for madness genes. Nat Genet 1997; 17:127–129.
21. Maziade M, Bissonnette L, Rouillard E, et al. 6p24-22 region and major psychoses in the Eastern Quebec population, Le Groupe IREP. Am J Med Genet 1997; 74:311–318.

22. Harris MJ, Jeste DV, Gleghorn A, et al. New-onset psychosis in HIV-infected patients. J Clin Psychiatry 1991; 52:369–376.
23. Weinstein EA, Linn L, Kahn RL. Encephalitis with a clinical picture of schizophrenia. J Mt Sinai Hosp 1955; 21:341–354.
24. Goldney RD, Temme PB. Case report: manic depressive psychosis following infectious mononucleosis. J Clin Psychiatry 1980; 41:322–323.
25. Koehler K, Guth W. The mimicking of mania in "benign" herpes simplex encephalitis. Biol Psychiatry 1979; 14:405–411.
26. Myers K, Dunner DL. Acute viral encephalitis complicating a first manic episode. J Fam Pract 1984; 18:403–407.
27. Menninger KA. Influenza and schizophrenia: an analysis of post-influenzal "dementia precox," as of 1918, and five years later. Am J Psychiatry 1926; 5:469–527.
28. Kasanin J, Peterson JN. Psychosis as an early sign of epidemic encephalitis. J New Ment Dis 1926; 64:352–358.
29. Sands IJ. The acute psychiatric type of epidemic encephalitis. Am J Psychiatry 1928; 7:975–987.
30. Torrey EF. Are we overestimating the genetic contribution to schizophrenia? Schizophr Bull 1992; 18:159–170.
31. Torrey EF, Rawlings RR, Ennis JM, et al. Birth seasonality in bipolar disorder, schizophrenia, schizoaffective disorders and stillbirths. Schizophr Res 1996; 21:141–149.
32. Torrey EF, Miller J, Rawlings R, et al. Seasonality of births in schizophrenia and bipolar disorder: a review of the literature. Schizophr Res 1997; 28:1–38.
33. Marcelis M, Navarro-Mateu F, Murray R, et al. Urbanization and psychosis: a study of 1942–1978 birth cohorts in The Netherlands. Psychol Med 1998; 28:871–879.
34. Eaton WW, Mortensen PB, Frydenberg M. Obstetric factors, urbanization and psychosis. Schizophr Res 2000; 43:117–123.
35. Lewis S, Murray RM. Obstetric complications, neurodevelopmental deviance, and risk of schizophrenia. J Psychiatr Res 1987; 21:413–421.
36. Kinney DK, Yurgelun-Todd DA, Levy DL, et al. Obstetrical complications in patients with bipolar disorder and their siblings. Psychiatry Res 1993; 48:47–56.
37. Sacker A, Done DJ, Crow TJ, et al. Antecedents of schizophrenia and affective illness: obstetric complications. Br J Psychiatry 1995; 166:734–741.
38. Klaning U, Laursen TM, Licht RW, et al. Is the risk of bipolar disorder in twins equal to the risk in singletons? A nationwide register-based study. J Affect Disord 2004; 81:141–145.
39. Scott J, McNem Y, Cowanagh J, et al. Obstetric complication and bipolar disorders: a systematic review and meta-analysis of bipolar disorders compared with healthy controls, unipolar disorders and schizophrenia [abstract]. Eur Neuropsycopharm 2005; 15(suppl):S133–S134.
40. Rott R, Herzog S, Fleischer B, et al. Detection of serum antibodies to Borna disease virus in patients with psychiatric disorders. Science 1985; 228:755–756.
41. Amsterdam JD, Winokur A, Dyson W, et al. Borna disease virus. A possible etiologic factor in human affective disorders? Arch Gen Psychiatry 1985; 42:1093–1096.
42. Salvatore M, Morzunov S, Schwemmle M, et al. Borna disease virus in brains of North American and European people with schizophrenia and bipolar disorder. Bornavirus Study Group. Lancet 1997; 349:1813–1814.
43. Bode L, Dietrich DE, Stoyloff R, et al. Amantadine and human Borna disease virus in vitro and in vivo in an infected patient with bipolar depression. Lancet 1997; 349:958.
44. Fu ZF, Amsterdam JD, Kao M, et al. Detection of Borna disease virus-reactive antibodies from patients with affective disorders by western immunoblot technique. J Affect Disord 1993; 27:61–68.
45. Sauder C, Müller A, Cubitt B, et al. Detection of borna disease virus (BDV) antibodies and BDV RNA in psychiatric patients: evidence for high sequence conservation of human blood-derived BDV RNA. J Virol 1996; 70:7713–7724.
46. Kim Y-K, Kim S-H, Han C-S, et al. Borna disease virus and deficit schizophrenia. Acta Neuropsych 2003; 15:262–265.

47. Terayama H, Nishino Y, Kishi M, et al. Detection of anti-Borna Disease Virus (BDV) antibodies from patients with schizophrenia and mood disorders in Japan. Psychiatry Res 2003; 120:201–206.
48. Macón RA, Mednick SA, Huttunen MO. Adult major affective disorder after prenatal exposure to an influenza epidemic. Arch Gen Psychiatry 1997; 54:322–328.
49. Cannon M, Cotter D, Coffey VP, et al. Prenatal exposure to the 1957 influenza epidemic and adult schizophrenia: a follow-up study. Br J Psychiatry 1996; 168:368–371.
50. Takei N, O'Callaghan E, Sham PC, et al. Does prenatal influenza divert susceptible females from later affective psychosis to schizophrenia? Acta Psychiatr Scand 1993; 88:328–336.
51. Welham J, McGrath J, Pemberton M. Affective psychose and the influenza epidemics of 1954, 1957, and 1959 [abstract]. Schizophr Res 1998; 29:19.
52. Morgan V, Castle D, Page A, et al. Influenza epidemics and incidence of schizophrenia, affective disorders and mental retardation in Western Austria: no evidence of a major effect. Schizophr Res 1997; 26:25–39.
53. Dickerson FB, Boronow JJ, Stallings C, et al. Association of serum antibodies to herpes simplex virus 1 with cognitive deficits in individuals with schizophrenia. Arch Gen Psychiatry 2003; 60:466–472.
54. Dickerson FB, Boronow JJ, Stallings, et al. Infection with herpes simplex virus type 1 is associated with cognitive deficits in bipolar disorder. Biol Psychiatry 2004; 55:588–593.
55. Dickerson FB, Boronow JJ, Stallings C, et al. The cathechol O-methyltransferase Val158Met polymorphism and herpes simplex virus type 1 infection are risk factors for cognitive impairment in bipolar disorder: additive gene-environmental effects in a complex human psychiatric disorder. Bipolar Dis 2006; 8:124–132.
56. Torrey EF, Yolken RH. Toxoplasma gondii and schizophrenia. Emerg Infect Dis 2003; 9:1375–1380.
57. Dickerson F, Yolken R. Toxoplasma gondii: prevalence and correlates in individuals with schizophrenia and bipolar disorder [abstract]. Toxoplasmosis Conference, Annapolis, MD, 2005.

12 EEGs and ERPs in Bipolar Disorders

R. Hamish McAllister-Williams
School of Neurology, Neurobiology, and Psychiatry, University of Newcastle upon Tyne, Newcastle upon Tyne, U.K.

INTRODUCTION

The first report of an electroencephalographic (EEG) recording from a mammal was made more than a century ago (1) and the first human recording over 65 years ago (2). A search of PubMed for papers referring to EEG or event-related potentials (ERPs) and bipolar disorder (BD) flags nearly three hundred items (between 1966 and December 2005). However, of these papers, over two-thirds are more than 10 years old. EEG and ERP studies are not new but certainly are not increasing at the rate seen with other experimental methodologies in psychiatric illnesses, such as functional magnetic resonance imaging (fMRI) and molecular studies. Put bluntly, EEG is not a "sexy" topic. This stems largely from the technique not delivering important findings after years of study, with a few notable exceptions such as the diagnosis of epilepsy and sleep disorders. However, EEG and ERP techniques still have much to offer in the further elucidation of the pathophysiology of mood disorders if used in appropriate ways. In the past, much reliance has been placed on the visual inspection of paper and ink EEG recordings and rather simple ERP paradigms with no a priori hypothetical rationale behind their design. While this situation continues in clinical electrophysiology to a large extent, the increasing use of novel computational analysis of EEG and ERP data, often coupled with specific pharmacological or cognitive challenges, offers exciting prospects for the future. This chapter is a selective review of EEG and ERP data to illustrate this argument. It is not a comprehensive review of all of the EEG studies in BD. In particular, sleep EEG research is not covered at all. The chapter is written with the non-EEG specialist bipolar researcher in mind.

THE NORMAL EEG

The EEG [together with magneto-encephalography (MEG)] has an advantage over any other form of functional imaging, such as fMRI or positron emission tomography (PET), in that it records signals that are the direct result of electrical activity of neurons, rather than some "downstream" consequence such as changes in oxygen utilization or blood flow. Partly as a result of this, EEG and MEG have temporal resolutions that are generally an order of magnitude better than fMRI or PET. EEG also has an advantage of being relatively inexpensive, easy to conduct, and well-tolerated compared with both fMRI and PET and does not require the use of radio-ligands as in PET. This facilitates multiple recordings in subjects, such as in cross-over or longitudinal studies. However, the EEG does have some drawbacks, most notably problems of spatial resolution. This relates to how and what EEG electrodes record.

The EEG is generated by inhibitory and excitatory postsynaptic potentials at cortical neuronal synapses. These postsynaptic potentials summate in the cortex and can be recorded at the scalp as EEG. Action potentials probably do not contribute significantly to the EEG because there is little penetration of these into extracellular space and hence transmission to the scalp. Rather, scalp EEG electrodes record the summation of postsynaptic potentials of large areas of pyramidal neurons in the underlying cortex. For a signal to be detectable these neurons must be firing simultaneously with the cells oriented in parallel at 90° to the plane of the scalp. Knowing this, one might wonder that it is possible to record any electrical activity at the scalp. However, many cortical areas have intrinsic oscillatory activity that occurs synchronously across large groups of cells generating rhythmic EEG activity. Cortical neuronal firing is heavily influenced by the thalamus which also has intrinsic oscillatory activity. In addition, many systems project from the brainstem and basal forebrain areas and are able to alter cortical rhythmic activity such as by desynchronizing cell firing. However, it should be remembered that the EEG only directly records activity from superficial cortical areas. Another problem in interpreting EEG data is that the polarity of the signal is not determined simply by whether the postsynaptic potentials are excitatory or inhibitory, but also by the location and orientation of the neurons. As a result it is not possible to deduce an electrical response as reflecting activation or inhibition in a neuronal network from the polarity of the EEG. The electrical activity from neurons must pass through the tissue between them and the EEG electrodes (brain, CSF, skull, and scalp) and is attenuated and modified by this. The degree of attenuation is determined by the degree of synchronization of the postsynaptic potentials, the orientation of the neurons, and the size of the participating area of cortex. All of these factors mitigate against localizing the source of the electrical activity recorded at the scalp. In fact, there is no unique solution that can be calculated for any recording. Instead, source localization methods must make certain assumptions to be able to find a solution (e.g., that the activity originates in grey matter, that there are a limited number of electrical sources, and/or that activity at any particular location closely resembles neighboring locations). Different techniques make different assumptions and hence localize scalp EEG activity as having originated in different intracranial locations. This is a major weakness of EEG studies. For a full and detailed understanding of the spatial location of cortical networks involved in any particular task, as well as the temporal relationship between the various components of the network, both EEG (or MEG) need to be conducted alongside other neuroimaging techniques such as fMRI.

In the past, EEG was recorded from a few scalp electrodes, often placed on the midline of the head, with an analog signal passing to a pen and paper chart recorder. The resultant tracers were analyzed simply by inspection by experienced EEG analysts. Such individuals are able to identify aberrant activity, particularly that associated with epilepsy. They are also able to identify gross changes in the quantities of the common frequencies seen in the EEG (delta <4 Hz, theta 4–8 Hz, alpha 9–13 Hz, and beta >13 Hz). Nowadays, EEG recordings can be made from dozens, if not hundreds, of electrodes placed across the scalp. Signals are digitized, allowing computer-based analysis, such as Fast Fourier Transformations (FFTs) that calculate the "amount" of different frequencies in the whole EEG, and attempts at source localization. An additional method of obtaining extra information from the EEG is to record this while a subject is performing a particular cognitive task. The EEG record is subsequently divided into epochs time-locked to the presentation of a stimulus or the subject's response. Several epochs associated with particular stimuli and/or responses can then be averaged together. This is what is meant by "event-related potentials" (ERPs). By doing this, "noise" (unrelated EEG activity) is averaged out, leaving only the electrical activity that relates to the task the subject was performing. Over the years there have been a number of ERP paradigms developed using visual or verbal stimuli such as the "P300" and "missmatch negativity" (MMN). Unfortunately, many of these tasks depend to a large extent on a whole variety of psychological processes and so their interpretation is at times difficult.

CLINICAL EEG ABNORMALITIES IN BIPOLAR DISORDER

The obvious starting point for considering EEG abnormalities in BD patients are observations that patients with epilepsy have high rates of mood disorders (3) and bipolar patients are not infrequently treated with anticonvulsants (4). However, while epidemiological studies demonstrate that up to half of patients with epilepsy have clinically significant depression and/or anxiety (3), BD does not occur with a higher than expected frequency (5). Classic bipolar I disorder is rarely seen, and manic episodes occur almost exclusively in the setting of postictal psychosis or after epilepsy surgery (6,7). This raises the question whether affective disorders in epilepsy are different from those seen in nonepileptic patients. It is possible that epilepsy is associated with a specific interictal dysphoric syndrome related to limbic dysfunction (7) that differs from the pathophysiology underlying nonepilepsy related depression. Nevertheless, it is of interest to note that depression is more common in patients with left temporal lobe epilepsy (8), which does suggest that it is not simply psychosocial factors that account for the high rates of comorbidity. Intriguingly, the nature and even the direction of the causality between depression and epilepsy is not clear given findings that depressive illness is associated with a sixfold increase in the risk of the development of seizures in an elderly population (9). Perhaps the association between the two conditions may be merely a reflection of some common underlying pathology.

Despite the observation of no increase in the rate of BD in epilepsy, Post has previously suggested that a possible pathophysiological process in BD is similar to one seen in epilepsy, namely "kindling" (see Chapter 18) (10). This hypothesis is based on physiological findings showing that intermittent subthreshold electrical stimuli produce increasingly strong neuronal depolarization in the brain through a process of sensitization. It is suggested that processes such as kindling predict that in BD, psychosocial stressors would be more frequent earlier in the course of illness, with the severity of stressor needed to precipitate an episode of illness decreasing over time leading to the frequency of episodes increasing as originally observed by Kraeplin and subsequently by others (11). If this is the case, then it may be that some subictal electrical dysrhythmia may be associated with the development of BD and be evident in EEG recordings from patients.

Reviews of the EEGs of bipolar patients do reveal an abnormal slowing of delta and theta EEG activity (12) and the presence of various small spikes and focal slow waves, usually in the frequency range of 4 Hz to 7 Hz (13,14). Abnormalities are reported in 15% to 50% of patients (15–17). Most studies report that abnormalities are seen bilaterally, though reports of unilateral bias more commonly relate to the right (16,18,19) rather than the left hemispheres (17). As already stated, the EEG abnormalities may, however, simply reflect an underlying brain abnormality that predisposes both to BD and abnormal EEGs. There may also be a difference in the underlying pathophysiology of bipolar patients with and without EEG abnormalities since the former are less likely to have a family history of the disorder (12,15–17) than the latter. The spike and slow wave abnormalities are also not specific to BD (13,14), though they are reported more often in bipolar and schizoaffective than schizophrenic patients (14).

The obvious clinical question that follows from these observations is whether or not the presence of an EEG abnormality in a patient with BD has therapeutic implications. It has been shown that EEG abnormalities are a risk factor for treatment resistance in BD (20). A small observational study suggests that the presence of EEG abnormalities is predictive of a poor lithium response (21), while another supports this and suggests that EEG abnormalities do not mitigate against response to valproate (22). Further, one nonrandomized study in 115 patients who were predominantly lithium refractory found that, perhaps not surprisingly, patients with a history of seizures were more likely to respond to the anticonvulsant valproate. However, those patients who simply had some EEG abnormality but no previous seizures only "tended" to respond better to valproate (23). To date there are no randomized controlled trials of BD patients who exhibit EEG abnormalities. Until there are, it is difficult to be clear about the implications of any abnormalities detected in clinical practice, though they may reflect a subset of BD patients with no family history of the disorder and perhaps a different underlying pathology.

EFFECTS OF LITHIUM ON THE EEG

From a clinical perspective, the EEG currently has one main utility in BD: to identify lithium neurotoxicity. Lithium has a narrow therapeutic range with serious side effects seen at high plasma concentrations. Toxic symptoms become increasingly common above serum levels of 1.2 mmol/L and affect the gastrointestinal tract, cerebellum, and cerebrum. Early signs are mental lassitude sometimes occurring with agitation. This can progress to confusion with vomiting and diarrhea, coarse tremor, Parkinsonism, ataxia, and dysarthria. Finally, seizures, coma, and death can result. Of great concern is that long-term neurological deficits can result from acute lithium toxicity in about 10% of cases (24). The problem the clinician faces is that neurotoxicity can occur at therapeutic plasma concentrations of lithium (25). The most common presentation is of an encephalopathy, but (rarely) focal neurological disturbances, psychotic episodes, or cognitive deficits can occur (26). Lithium neurotoxicity appears to be more common in patients with pre-existing EEG abnormalities [particularly temporal lobe epilepsy (27)]. Prompt diagnosis and treatment of neurotoxicity may decrease the risk of long term sequelae. Essentially, the diagnosis is made on clinical grounds, with lithium plasma levels being of little value unless clearly raised. However, the EEG has a characteristic pattern in lithium neurotoxicity and so can aid diagnosis. There is an increase in theta and delta activity with diffuse slowing and background disorganization (28). While clinical improvement can be paralleled by improvements in the EEG, full resolution of all EEG abnormalities can be delayed by some months.

CONVENTIONAL ERP ABNORMALITIES IN BIPOLAR DISORDER

Perhaps the most studied conventional ERP paradigm in BD is the auditory evoked "P300" (or "P3"). This usually involves presentation of auditory tones to subjects, the majority of which are identical with random deviant tones occurring at a low frequency (perhaps 1 in 10 tones). When the ERPs related to the standard tones are compared with those related to deviant ones, the "deviant ERPs" show a large positive going deflection occurring around 300 ms following the onset of the tone. These "P300" deflections reflect multiple cognitive processes including attention, working memory, and higher auditory processing and originate from multiple cortical areas (29). In BD, as in schizophrenia (30) and depression (31), there are reports of a delay in the latency of the P300 component compared with that seen in healthy subjects (30,32,33). However, while in schizophrenia there are also reports of a reduction in P300 amplitude (30), this is a more equivocal finding in BD (30,33,34). The problem comes from interpreting these findings and considering their implications. It has been found that an increased P300 latency correlates with ratings of insomnia in patients (31) and may reflect impaired attentional processing (31,33). Alternatively or additionally, prolonged P300 latency may indicate possible abnormalities with auditory processing itself, though the case for this is stronger in schizophrenic patients. This is argued on the basis that abnormalities of earlier auditory ERP components (N1 and P2) are abnormal in some studies of schizophrenic but not bipolar patients (33). This may relate to psychosis per se rather than particular diagnoses since abnormalities in early auditory ERP components in bipolar patients have been reported to correlate with a longitudinal history of psychosis (35). It has also been suggested that the P300 may predict response to certain treatments, such that a "normal" P300 predicts rapid response to ECT in melancholic depression (36). However, there are few such studies in BD. Perhaps of greatest significance is a suggestion that P300 abnormalities may be a manifestation of some underlying trait abnormality since they are seen at an increased frequency in at least some families of patients with BD (32). It has subsequently been shown that P300 amplitude and latency cosegregates with the recently described DISC1 and DISC2 genes on chromosome 1 (37) that are risk factors for both schizophrenia and BD (38). It may be that these ERP findings can thus be used as a biological marker of genetic risk.

Another auditory ERP that has been extensively studied in psychiatric populations is the MMN paradigm. This ERP technique involves presenting subjects with auditory stimuli that follow some regular pattern. When there is some deviation to this standard pattern, a negative deflection is seen around 150 ms after the onset of the stimulus, even when subjects do not attend to the task. It is argued that this deflection reflects sound processing (39). Of particular importance, the MMN has been shown in animal and human studies to index *N*-methyl-*D*-aspartate (NMDA) functioning (40,41). This is of interest since the MMN is reliably found to be abnormal in schizophrenia (42–44), supporting hypotheses of a glutamate abnormality underlying this illness. It is reported that the MMN is normal in BD (44), though it would be of interest to explore this further examining for abnormalities in patients who have experienced or are experiencing psychotic phenomena.

FUTURE OPPORTUNITIES FOR EEG AND ERP STUDIES IN BIPOLAR DISORDER

With the exception of the diagnosis of lithium neurotoxicity, the EEG has little current clinical utility in the management of BD. Further, conventional EEG analysis and classical ERP studies have proved of limited value in the study of the etiopathogenesis of the disorder. However, to disregard EEG techniques in the study of BD is to discard a potentially valuable technique that supplements other research methodologies. The utility of the EEG in the study of any psychiatric illness is hampered by the largely unsolved problem of extracting the relevant information from the recording in a theory-guided manner. However, there are an increasing number of ways that this problem can be solved. This relates to new statistical methods of examining the EEG and coupling recordings with pharmacological and/or cognitive challenges. This will be illustrated below.

Evolving Quantitative EEG Techniques

As described above, in the past, analog EEG signals were recorded on pen and ink chart recorders. Now it is possible to digitize the signal for storage and analysis on computers. Quantitative analysis of the EEG (qEEG) using FFTs is able to identify alterations in the frequency components of the signal that would not be visible to the naked eye. Such techniques have extended the previous observations of abnormal EEGs in BD, for example showing increases in right temporal lobe theta and left occipital beta frequencies (45). However, perhaps the most exciting of the new methods of EEG analysis relate to methods for analyzing alterations in the synchronization of various frequency components. It has been argued that synchronous neuronal activity (precisely that recordable by EEG techniques), particularly at high frequencies (upper alpha, 10–13 Hz; beta, 13–30 Hz; and gamma, >30 Hz) is a mechanism by which there can be a functional integration of brain networks during cognitive tasks. For example, during a working memory task, EEG recordings demonstrate an increase in synchronization of beta activity across bilateral sites (46). By simultaneously recording EEG and fMRI it is possible to show that this change in beta synchronization is associated with linkage between cross-hemispheric regions (left angular gyrus and right superior parietal gyrus) and among anterior–posterior regions (right dorsolateral prefrontal cortex, putamen, and right superior temporal gyrus) (46). These are interesting observations not least because they illustrate that by combining EEG and fMRI techniques, the advantages of both (the ability to directly record neuronal electrical activity with high temporal resolution, and high spatial resolution, respectively) can be combined. Using this EEG methodology, it has been shown, for example, that synchronization decreases with age and correlates with decrease in executive function performance (47), and is lower in Alzheimer's disease (48) and schizophrenia (49). It has been argued that these findings reflect a breakdown of the synchrony of neural networks and that this underlies the cognitive impairment seen in these conditions. There have been relatively few similar studies in BD, but it has been reported that long-range synchrony is reduced (50) and that there is abnormal high frequency synchronization (51). The great importance of synchronization experimentally is that it can be studied in vivo and in vitro in animals. This has, for example, allowed the demonstration that nicotinic acetylcholine receptor alpha subunits (52) and $GABA_A$ $alpha_5$ receptors (53) play an important role in the regulation of gamma oscillations in hippocampus. It is hoped that as knowledge of the physiology and pharmacology of this brain oscillatory behavior becomes increasingly understood, the EEG will be able to act as an important tool to explore hypothesized abnormalities in various psychiatric conditions including BD and assess the effects of pharmacological challenges. In this regard, it is interesting to note that repetitive transcranial magnetic stimulation (rTMS) has been shown to influence gamma activity in prefrontal cortex (54) and that perhaps this plays a role in the therapeutic activity of this treatment.

A number of studies have specifically linked changes in synchronous EEG activity with various cognitive functions. The increase in synchronization of beta activity during working memory tasks has been described above (46), which is of interest given the findings of working memory deficits in BD (55). Another example that may also be relevant to BD are findings suggesting that the ratio between delta and beta frequency activity is a biological maker of motivation, particularly the subject's relative balance between reward- and punishment-driven behavior (56). Interestingly, this ratio of delta to beta activity is influenced by endogenous cortisol concentrations as a result, it has been argued, of cortisol-enhancing subcortical–cortical communication, particularly strengthening cortical control of subcortical drives (57). This provides a possible mechanism by which motivation and behavior is influenced by stress as well as in BD, given findings of abnormal hypothalamic-pituitary-adrenal (HPA) axis functioning (58). There are clearly a number of lines of investigation needed to further clarify these potential interactions.

Pharmaco-EEG Techniques

Pharmaco-EEG offers promise to the study of BD by following the example of MMN studies in schizophrenia (described above) in identifying EEG indices of neurotransmitter function. A number of possibilities exist. For example, it has been shown by a number of groups that the systemic administration of buspirone to healthy subjects leads to a negative shift of the awake EEG frequency spectrum (due to an increase in theta and decrease in alpha activity) following buspirone (59–62). Although buspirone is a relatively nonselective drug, the effects on the EEG frequency spectrum may be mediated by somatodendritic 5-HT_{1A} receptors. This is supported by a number of observations. Firstly, the shift in EEG frequency spectrum is mimicked by other more selective 5-HT_{1A} partial agonists in man (62,63) and is seen in animals following administration of buspirone as well as the highly selective full 5-HT_{1A} receptor agonist 8-OH-DPAT (64). Secondly, source localization of the effect of buspirone using low resolution electromagnetic tomographic analysis (LORETA) demonstrates a significant increase in theta EEG activity in the hippocampus as well as neighboring cortical areas (62). Hippocampal theta activity is well-known to be under ascending serotonergic control (65). Animal studies using local application of 8-OH-DPAT into the raphe have indicated that activation of somatodendritic 5-HT_{1A} receptors causes a decrease in hippocampal theta (66,67), an effect blocked by 5-HT_{1A} antagonists (68). Thirdly, a somatodendritic location for a 5-HT_{1A} receptor-mediated negative shift in the EEG frequency spectrum in man is also inferred from findings of acute administration of SSRIs causing such a shift (69) and that pindolol mimics, rather than blocks, the effect of buspirone on the awake EEG (70). Following acute administration, SSRIs increase concentrations of 5-HT in the raphe nuclei leading to somatodendritic 5-HT_{1A} activation (71); pindolol acts as a 5-HT_{1A} antagonist at postsynaptic receptors but a partial agonist at somatodendritic receptors (72). Taken together, these data indicate that the effect of buspirone on the EEG frequency spectrum is a potentially valid index of somatodendritic 5-HT_{1A} receptor function. Using this methodology, it has recently been reported that drug-free depressed patients have an increase in this autoreceptor function that would be predicted as leading to a decrease in the overall activity of the serotonergic system (73). It would be of great interest to extend these studies into BD patients.

Cognitive ERP Techniques

In the same way as identifying the pharmacological basis of various EEG or ERP measures can provide valuable information when studied in patient populations, so can linking cognitive function with EEG techniques. This is being increasingly done with measures of synchronization as described above. However, traditional ERP methods are also able to provide information regarding neuronal activity underlying cognitive function that is not available with other methodologies. A good example of this relates to the study of episodic memory retrieval.

When a past episodic memory is recollected, such as when a subject is presented with a "retrieval cue," a mental representation of the past event is generated involving a conscious reconstruction (74). It has been argued that there are "preretrieval" processes operating on the retrieval cue and "postretrieval" processes operating on the products of retrieval (75). This hypothesis, and its neural underpinning, can be examined by studying the effects of brain lesions. For example, bilateral damage to temporal lobes leads to an impairment of acquisition of new memories and recollection of premorbid events, while prefrontal cortex lesions have a limited effect unless highly elaborate encoding or retrieval is required (76). The problem with lesion studies is that it is impossible to separate effects on encoding and retrieval. To do this, functional imaging is required. fMRI studies have shown that episodic memory retrieval is associated with bilateral parietal lobe activation, reflecting activation of hippocampal memory stores (75). However, in addition, complex retrieval tasks also lead to activation of the right dorsolateral prefrontal cortex (75). This has been hypothesized to reflect the monitoring and evaluation of the products of retrieval, since the activity is greatest when the information being retrieved is more ambiguous (77). ERP studies using specific episodic memory retrieval paradigms have been used to explore the temporal relationship between these two areas of activity and confirms that the right prefrontal activation begins later after stimulus presentation compared with the bilateral parietal activity (78,79) consistent with the hypothesis. This illustrates the utility of EEG studies in parallel with other forms of functional imaging in the investigation of cognitive function.

This episodic memory retrieval ERP paradigm has been used to explore the effects of corticosteroids given theories of a casual link between hypercortisolemia and cognitive impairment in affective disorders (80). This has shown that cortisol not only impairs recollection performance but also alters frontal cortex activity (79), while the glucocorticoid functional antagonist dehydroepiandrosterone (DHEA) improves performance apparently mediated by effects on parietal lobe activity (81). Currently this ERP technique is being implemented in a cohort of BD patients undergoing treatment with a glucocorticoid receptor antagonist and the results are eagerly awaited.

CONCLUSIONS

EEG recordings have a role in the clinical management of patients with BD in aiding diagnosis of lithium neurotoxicity. They have also demonstrated that there is a significant minority of patients with a diagnosis of BD who exhibit EEG abnormalities. Intriguingly, these patients may have a rather different underlying pathophysiology as suggested by the lack of a family history of affective disorders. It is disappointing that this has not been followed up to any extent to date. Likewise, while there are hints that such patients may respond differently to medication, with a preferential response to anticonvulsants, there have been few if any attempts to develop specific EEG criteria to classify this subgroup of patients or randomized controlled trials (RCTs) performed to establish if there really is a management implication in this differentiation.

The future holds immense potential for modern EEG techniques to add a great deal of knowledge regarding the pathophysiology of BD. Quantitative analysis, for example examining synchronous neuronal activity, may offer a window into neuronal activity not available by any other experimental technique. Such studies, in parallel with in vitro pharmacological investigations as well as neurocognitive and molecular genetics, hold great promise. Similarly, the use of increasingly specific cognitive and pharmacological challenges in combination with EEG and ERP recordings offers the possibility of testing specific hypotheses regarding brain function in patients.

Nevertheless, EEG techniques have not, so far, delivered the number and importance of findings in BD (or psychiatric illness in general) that might have been hoped for when they first became available for use in humans over 65 years ago. The subsequent pessimism regarding EEGs that is found in many psychiatric researchers, however, ignores some of the important findings that have been made to date (which have often been inadequately followed up), but more importantly fails to recognize the complementary role EEGs can play along side other research techniques such as neuroimaging, cognitive neuroscience, neuropharmacology, and neuroendocrinology. These opportunities are coming about because of exciting developments in EEG methodology and analysis. In illustrating a number of potential future avenues for research in BD using EEGs it is hoped that the vital role these techniques can play will not be underestimated.

REFERENCES

1. Caton R. The electric currents of the brain. Brit Med J 1875; 2:278.
2. Berger H. Uber das elektrenkephalogramm des menschen. Arch Psychiat Nervenkr 1929; 87:527–570.
3. Harden CL. The co-morbidity of depression and epilepsy: epidemiology, etiology, and treatment. Neurology 2002; 59(6 suppl 4):S48–S55.
4. McAllister-Williams RH. Relapse prevention in bipolar disorder: a critical review of current guidelines. J Psychopharm 2006; 20(2 Suppl):12–16.
5. Harden CL, Goldstein MA. Mood disorders in patients with epilepsy: epidemiology and management. CNS Drugs 2002; 16(5):291–302.
6. Nishida T, Kudo T, Nakamura F, et al. Postictal mania associated with frontal lobe epilepsy. Epilepsy & Behavior 2005; 6(1):102–110.
7. Schmitz B. Depression and mania in patients with epilepsy. Epilepsia 2005; 46(suppl 4): 45–49.
8. Altshuler LL, Devinsky O, Post RM, et al. Depression, anxiety, and temporal lobe epilepsy. Laterality of focus and symptoms. Arch Neurology 1990; 47(3): 284–288.
9. Hesdorffer DC, Hauser WA, Annegers JF, et al. Major depression is a risk factor for seizures in older adults. Ann Neurology 2000; 47(2):246–249.
10. Post RM. Sensitization and kindling perspectives for the course of affective illness: toward a new treatment with the anticonvulsant carbamazepine. Pharmacopsychiat 1990; 23(1):3–17.
11. Goodwin FK, Jamison KR. Manic-depressive Illness. New York: Oxford University Press, 1990.
12. Hays P. Etiological factors in manic-depressive psychoses. Arch Gen Psychiatry 1976; 33(10):1187–1188.
13. Hughes JR. The EEG in psychiatry: an outline with summarized points and references. Clin Electroencephalogr 1995; 26(2):92–101.

14. Inui K, Motomura E, Okushima R, et al. Electroencephalographic findings in patients with DSM-IV mood disorder, schizophrenia, and other psychotic disorders. Biol Psychiatry 1998; 43(1):69–75.
15. Kadrmas A, Winokur G. Manic depressive illness and EEG abnormalities. J Clin Psychiatry 1979; 40(7):306–307.
16. Cook BL, Shukla S, Hoff AL. EEG abnormalities in bipolar affective disorder. J Affect Disord 1986; 11(2):147–149.
17. Small JG, Milstein V, Malloy FW, et al. Clinical and quantitative EEG studies of mania. J Affect Disord 1999; 53(3):217–224.
18. Bruder GE, Stewart JW, Towey JP, et al. Abnormal cerebral laterality in bipolar depression: convergence of behavioral and brain event-related potential findings. Biol Psychiatry 1992; 32(1):33–47.
19. Clementz BA, Sponheim SR, Iacono WG, et al. Resting EEG in first-episode schizophrenia patients, bipolar psychosis patients, and their first-degree relatives. Psychophys 1994; 31(5):486–494.
20. Cole AJ, Scott J, Ferrier IN, et al. Patterns of treatment resistance in bipolar affective disorder. Acta Psychiatr Scand 1993; 88(2):121–123.
21. Ikeda A, Kato N, Kato T. Possible relationship between electroencephalogram finding and lithium response in bipolar disorder. Prog Neuro-Psychopharm Biol Psychiatr 2002; 26(5):903–907.
22. Reeves RR, Struve FA, Patrick G. Does EEG predict response to valproate versus lithium in patients with mania? Ann Clin Psychiat 2001; 13(2):69–73.
23. Stoll AL, Banov M, Kolbrener M, et al. Neurologic factors predict a favorable valproate response in bipolar and schizoaffective disorders. J Clin Psychopharmacol 1994; 14(5): 311–313.
24. Schou M. Long-lasting neurological sequelae after lithium intoxication. Acta Psychiatr Scand 1984; 70(6):594–602.
25. Bell AJ, Cole A, Eccleston D, et al. Lithium neurotoxicity at normal therapeutic levels. Brit J Psychiatry 1993; 162:689–692.
26. Sheean GL. Lithium neurotoxicity. Clin Exp Neurology 1991; 28:112–127.
27. Ferrier IN, Tyrer SP, Bell AJ. Lithium therapy. Adv Psychiat Treat 1995; 1:102–110.
28. Bartha L, Marksteiner J, Bauer G, et al. Persistent cognitive deficits associated with lithium intoxication: a neuropsychological case description. Cortex 2002; 38(5):743–752.
29. Linden DE. The p300: where in the brain is it produced and what does it tell us? Neuroscientist 2005; 11(6):563–576.
30. Souza VB, Muir WJ, Walker MT, et al. Auditory P300 event-related potentials and neuropsychological performance in schizophrenia and bipolar affective disorder. Biol Psychiatry 1995; 37(5):300–310.
31. Bruder GE, Towey JP, Stewart JW, et al. Event-related potentials in depression: influence of task, stimulus hemifield and clinical features on P3 latency. Biol Psychiatry 1991; 30(3):233–246.
32. Blackwood DH, Sharp CW, Walker MT, et al. Implications of comorbidity for genetic studies of bipolar disorder: P300 and eye tracking as biological markers for illness. Brit J Psychiatry 1996; 168(suppl 30):85–92.
33. O'Donnell BF, Vohs JL, Hetrick WP, et al. Auditory event-related potential abnormalities in bipolar disorder and schizophrenia. Int J Psychophysiol 2004; 53(1):45–55.
34. Defrance JF, Ginsberg LD, Rosenberg BA, et al. Topographical analysis of adolescent affective disorders. Int J Neurosci 1996; 86(1–2):119–141.
35. Olincy A, Martin L. Diminished suppression of the P50 auditory evoked potential in bipolar disorder subjects with a history of psychosis. Am J Psychiatry 2005; 162(1):43–49.
36. Ancy J, Gangadhar BN, Janakiramaiah N. "Normal" P300 amplitude predicts rapid response to ECT in melancholia. J Affect Disord 1996; 41(3):211–215.
37. Blackwood DH, Visscher PM, Muir WJ. Genetic studies of bipolar affective disorder in large families. Brit J Psychiatry 2001; 178(suppl 41):s134–s136.
38. Devon RS, Anderson S, Teague PW, et al. Identification of polymorphisms within disrupted in schizophrenia 1 and disrupted in schizophrenia 2, and an investigation of their association with schizophrenia and bipolar affective disorder. Psychiatric Genetics 2001; 11(2):71–78.

39. Nyman G, Alho K, Laurinen P, et al. Mismatch negativity (MMN) for sequences of auditory and visual stimuli: evidence for a mechanism specific to the auditory modality. Electroencephalogr Clin Neurophys 1990; 77(6):436–444.
40. Javitt DC, Steinschneider M, Schroeder CE, et al. Role of cortical N-methyl-D-aspartate receptors in auditory sensory memory and mismatch negativity generation: implications for schizophrenia. Proc Natl Acad Sci USA 1996; 93(21):11962–11967.
41. Umbricht D, Koller R, Vollenweider FX, et al. Mismatch negativity predicts psychotic experiences induced by NMDA receptor antagonist in healthy volunteers. Biol Psychiatry 2002; 51(5):400–406.
42. Gene-Cos N, Ring HA, Pottinger RC, et al. Possible roles for mismatch negativity in neuropsychiatry. Neuropsychiatry Neuropsychol Behav Neurol 1999; 12(1):17–27.
43. Michie PT. What has MMN revealed about the auditory system in schizophrenia? Int J Psychophysiol 2001; 42(2):177–194.
44. Umbricht D, Koller R, Schmid L, et al. How specific are deficits in mismatch negativity generation to schizophrenia? Biol Psychiatry 2003; 53(12):1120–1131.
45. El Badri SM, Ashton CH, Moore PB, et al. Electrophysiological and cognitive function in young euthymic patients with bipolar affective disorder. Bipolar Disorders 2001; 3(2): 79–87.
46. Mizuhara H, Wang LQ, Kobayashi K, et al. Long-range EEG phase synchronization during an arithmetic task indexes a coherent cortical network simultaneously measured by fMRI. Neuroimage 2005; 27(3):553–563.
47. Paul RH, Clark CR, Lawrence J, et al. Age-dependent change in executive function and gamma 40 Hz phase synchrony. J Integrative Neurosci 2005; 4(1):63–76.
48. Stam CJ, Montez T, Jones BF, et al. Disturbed fluctuations of resting state EEG synchronization in Alzheimer's disease. Clin Neurophysiol 2005; 116(3):708–715.
49. Symond MP, Harris AW, Gordon E, et al. "Gamma synchrony" in first-episode schizophrenia: a disorder of temporal connectivity? Am J Psychiatry 2005; 162(3): 459–465.
50. Bhattacharya J. Reduced degree of long-range phase synchrony in pathological human brain. Acta Neurobiol Exp 2001; 61(4):309–318.
51. O'Donnell BF, Hetrick WP, Vohs JL, et al. Neural synchronization deficits to auditory stimulation in bipolar disorder. Neuroreport 2004; 15(8):1369–1372.
52. Song C, Murray TA, Kimura R, et al. Role of alpha7-nicotinic acetylcholine receptors in tetanic stimulation-induced gamma oscillations in rat hippocampal slices. Neuropharm 2005; 48(6):869–880.
53. Towers SK, Gloveli T, Traub RD, et al. Alpha 5 subunit-containing GABAA receptors affect the dynamic range of mouse hippocampal kainate-induced gamma frequency oscillations in vitro. J Physiol 2004; 559(Pt 3):721–728.
54. Schutter DJ, van Honk J, d'Alfonso AA, et al. High frequency repetitive transcranial magnetic over the medial cerebellum induces a shift in the prefrontal electroencephalography gamma spectrum: a pilot study in humans. Neurosci Lett 2003; 336(2):73–76.
55. Thompson JM, Gallagher P, Hughes JH, et al. Neurocognitive impairment in euthymic patients with bipolar affective disorder. Brit J Psychiatry 2005; 186:32–40.
56. Schutter DJ, van Honk J. Electrophysiological ratio markers for the balance between reward and punishment. Cogn Brain Res 2005; 24(3):685–690.
57. Schutter DJ, van Honk J. Salivary cortisol levels and the coupling of midfrontal delta-beta oscillations. Int J Psychophysiol 2005; 55(1):127–129.
58. Watson S, Gallagher P, Ritchie JC, et al. Hypothalamic-pituitary-adrenal axis function in patients with bipolar disorder. Brit J Psychiatry 2004; 184:496–502.
59. Murasaki M, Miura S, Ishigooka J, et al. Phase I study of a new antianxiety drug, buspirone. Prog Neuropsychopharmacol Biol Psychiatry 1989; 13(1–2):137–144.
60. Barbanoj MJ, Anderer P, Antonijoan RM, et al. Topographical pharmaco-EEG mapping of increasing doses of buspirone and its comparison with diazepam. Human Psychopharm 1994; 9:101–109.
61. Holland RL, Wesnes K, Dietrich B. Single dose human pharmacology of umespirone. Eur J Clin Pharmacol 1994; 46:461–468.
62. Anderer P, Saletu B, Pascual-Marqui RD. Effect of the 5-HT(1A) partial agonist buspirone on regional brain electrical activity in man: a functional neuroimaging study using low-resolution electromagnetic tomography (LORETA). Psychiatry Res 2000; 100(2):81–96.

63. Saito A, Kinoshita T, Okajima Y, et al. Quantitative pharmaco-EEG study of ipsapirone (BAY q 7821) in healthy volunteers. Jap J Neuropharm 1993; 15:359–373.
64. Bogdanov NN, Bogdanov MB. The role of 5-HT_{1A} serotonin and D_2 dopamine receptors in buspirone effects on cortical electrical activity in rats. Neurosci Lett 1994; 177(1–2):1–4.
65. Vertes RP. Brain stem generation of the hippocampal EEG. Prog Neurobiol 1982; 19(3):159–186.
66. Vertes RP, Kinney GG, Kocsis B, et al. Pharmacological suppression of the median raphe nucleus with $serotonin_{1A}$ agonists, 8-OH-DPAT and buspirone, produces hippocampal theta rhythm in the rat. Neurosci 1994; 60(2):441–451.
67. Nitz DA, McNaughton BL. Hippocampal EEG and unit activity responses to modulation of serotonergic median raphe neurons in the freely behaving rat. Learn Mem 1999; 6(2):153–167.
68. Marrosu F, Fornal CA, Metzler CW, et al. 5-HT1A agonists induce hippocampal theta activity in freely moving cats: role of presynaptic 5-HT1A receptors. Brain Res 1996; 739(1–2):192–200.
69. Saletu B, Grunberger J, Linzmayer L. On central effects of serotonin re-uptake inhibitors: quantitative EEG and psychometric studies with sertraline and zimelidine. J Neural Trans 1986; 67(3–4):241–266.
70. McAllister-Williams RH, Massey AE. EEG effects of buspirone and pindolol: a method of examining 5-HT(1A) receptor function in humans. Psychopharm 2003; 166(3):284–293.
71. Bel N, Artigas F. Fluvoxamine preferentially increases extracellular 5-hydroxytryptamine in the raphe nuclei: an in vivo microdialysis study. Eur J Pharmacol 1992; 229(1): 101–103.
72. Clifford EM, Gartside SE, Umbers V, et al. Electrophysiological and neurochemical evidence that pindolol has agonist properties at the 5-HT(1A) autoreceptor in vivo. Brit J Pharmacol 1998; 124(1):206–212.
73. McAllister-Williams RH, Marsh VR, Massey AE. Increased sensitivity of somatodendritic 5-HT_{1A} autoreceptors in depression. J Psychopharm 2004; 18(3 Suppl):A7.
74. Tulving E. Multiple memory systems and consciousness. Human Neurobiol 1987; 6(2):67–80.
75. Rugg MD, Otten LJ, Henson RN. The neural basis of episodic memory: evidence from functional neuroimaging. Philos Trans R Soc Lond B Biol Sci 2002; 357(1424):1097–1110.
76. Eskes GA, Szostak C, Stuss DT. Role of the frontal lobes in implicit and explicit retrieval tasks. Cortex 2003; 39(4–5):847–869.
77. Henson RN, Rugg MD, Shallice T, et al. Confidence in recognition memory for words: dissociating right prefrontal roles in episodic retrieval. J Cogn Neurosci 2000; 12(6): 913–923.
78. Allan K, Wilding EL, Rugg MD. Electrophysiological evidence for dissociable processes contributing to recollection. Acta Psychol (Amst) 1998; 98(2–3):231–252.
79. McAllister-Williams RH, Rugg MD. Effects of repeated cortisol administration on brain potential correlates of episodic memory retrieval. Psychopharm 2002; 160(1):74–83.
80. McAllister-Williams RH, Ferrier IN, Young AH. Mood and neuropsychological function in depression: the role of corticosteroids and serotonin. Psychol Med 1998; 28(3):573–584.
81. Alhaj HA, Massey AE, McAllister-Williams RH. Effects of DHEA administration on episodic memory, cortisol and mood in healthy young men: a double-blind, placebo-controlled study. Psychopharm 2006; 188(4):541–551.

13 Genetics of Bipolar Disorder

Nick Craddock
Department of Psychological Medicine, Wales School of Medicine, Cardiff University, Cardiff, U.K.

INTRODUCTION

Mental disorders in general, and mood (affective) disorders in particular, are leading causes of morbidity which affect human populations around the world (1). The term "affective disorder" includes a wide variety of conditions, from mild and common mood variations to some of the most severe episodes of psychotic illness seen in clinical practice. Co-occurrence of other clinical syndromes, such as anxiety or substance abuse, is common. Genetic factors are known to play an important role in influencing susceptibility to all these illnesses (2). Here, discussion will be restricted to bipolar spectrum illness.

DIAGNOSIS AND EPIDEMIOLOGY

Affective disorders are complex genetic disorders in which the core feature is a pathological disturbance of mood ranging from extreme elation or mania to severe depression. Other symptoms also found in these disorders include disturbances in thinking and behavior, which may include psychotic symptoms, such as delusions and hallucinations. Historically, affective disorders have been classified in a number of ways, with distinctions between endogenous and reactive episodes, psychotic and neurotic symptomatology, and affective disorders arising de novo (primary) and those episodes arising in the context of another disorder (secondary) (3,4). The main nosological division in modern classification systems such as ICD-10 (5) or DSM-IV (6) is between the unipolar and bipolar forms of the condition. The diagnosis of bipolar disorder (BD) (also known as manic depressive illness) requires that an individual has suffered one or more episodes of mania with or without episodes of depression at other times during the life history. This requirement for the occurrence of an episode of mania at some time during the course of illness distinguishes BD from unipolar disorder (also commonly known as unipolar major depression, or simply unipolar depression) in which individuals suffer one or more episodes of depression without ever experiencing episodes of pathologically elevated mood. Although bipolar and unipolar disorders are not completely distinct nosological entities, their separation for the purposes of diagnosis and research is supported by evidence from outcome, treatment, and genetic studies (4,7). Indeed, it was family-genetic studies that persuaded the field to move to classifications that separated bipolar and unipolar mood disorders (8,9). In DSM-IV, BD is sub-classified into bipolar I disorder, in which episodes of clear-cut mania occur, and bipolar II disorder, in which only milder forms of mania (so-called "hypomania") occur. Although there is evidence to support this distinction (10), the validity of this sub-classification awaits robust validation. The lifetime prevalence of narrowly defined BD is in the region of 0.5% to 1.5% with similar rates in males and females and a mean age of onset around the age of 21 years (11).

Unipolar disorder is substantially more common than bipolar illness, but measured prevalence rates differ markedly according to the diagnostic criteria, methodology, and sample employed. For example, the large U.S. multi-site Epidemiological Catchment Area (ECA) study reported a lifetime population prevalence for DSM-III major depression of approximately 4.4% (12), whereas the U.S. National Comorbidity Survey estimated the lifetime prevalence of DSM-IIIR major depression to be 17.1% with 10.3% of the population experiencing a major depressive episode in the preceding 12 months (13). In contrast to bipolar illness, the rate of unipolar disorder for women is about twice that for men: 21.3% and 12.7%, respectively, in the U.S. National Comorbidity Survey (13), and this gender difference is a consistent finding, at least in studies in the developed world. Affective disorders are associated with high levels of service utilisation and morbidity and often prove fatal, with up to 15% of patients eventually committing suicide (14). Reasonably effective treatments are available for both manic and depressive episodes (15), but current treatments have undesirable side effects, are not effective in all patients, and the pathogenesis of affective disorders remains poorly understood. These facts act as a major motivation for genetic investigation of affective illness with its promise of improved understanding of etiology and more effective treatments.

CLASSICAL GENETIC EPIDEMIOLOGY OF BIPOLAR DISORDER

Many classical genetic studies of mood disorders have been undertaken. As mentioned above, prior to the mid-1960s, family studies of mood disorder did not make the bipolar/unipolar distinction. However, these studies provided evidence for familial aggregation of the broad mood disorder phenotype (16). Subsequent family studies have provided persuasive evidence of familial aggregation of both bipolar and unipolar disorder; twin and adoption studies point to genes as an important cause of this familial resemblance (17–19). In all, there is a consistent and impressive body of evidence that supports the existence of mood disorder susceptibility genes. These studies also demonstrate a graduation in risk of mood disorder between various classes of relatives with monozygotic co-twin showing highest risk, through first degree relative to unrelated member of the general population showing the lowest risk. Because of differences in methodologies and diagnostic classifications, the absolute measures of estimated lifetime risk vary between studies. Table 1 shows a representative range of estimates for relative risks and heritabilities.

As can be seen from Table 1, the classical genetic epidemiology literature demonstrates that there is not a neat genetic separation between bipolar and unipolar disorders. It is not only bipolar illness that is found in the families of bipolar probands. Other affective disorders also occur at increased rates compared

TABLE 1 Genetic Epidemiology of Mood Disorders

	Bipolar disorder	Unipolar depression
Recurrence risk in sibling of a proband (λ_S)	5–10	2.5–3.5
Proband-wise MZ twin concordance	45–70%	40–50%
Heritability estimate	80–90%	33–42%

with the population. Indeed, the absolute risk of unipolar depression in first-degree relatives of bipolar probands is actually numerically greater than the absolute risk of bipolar illness. However, the background population prevalence is much higher for unipolar illness than for BD (of the order of 10% vs. 1%) and therefore the *relative* increase in risk is much lower, at approximately a doubling of risk (20). According to one estimate, two-thirds to three-quarters of cases of unipolar depression in the relatives of a bipolar individual can be considered to be "genetically bipolar," that is, they share a common genetic susceptibility with the bipolar form of affective illness (21).

Mode of Inheritance

Although rare, it is likely that some families exist in which single genes may play the major role in determining disease susceptibility. Early segregation analyses on large multiply affected pedigrees produced mixed results with some studies consistent with single gene models (22–25) and others unable to demonstrate major locus transmission (26–28). However, caution is required in interpreting these early positive results because of the limited power of the studies to distinguish between single gene and oligogenic models and because of the failure to take account of an important parameter, the recurrence risk in monozygotic (identical) cotwins of a bipolar proband (29). The observed very rapid decrease in recurrent risk from identical co-twins to first degree relatives and back to the general population (as shown in Table 1) is not consistent with single gene modes of inheritance (30). Rather, the substantial body of data from classical genetic epidemiological studies is consistent with models of inheritance that include multiple genes that interact with each other (often referred to in genetics as "epistasis") and environmental factors to confer susceptibility to illness (30). As a consequence, the default model used for most current genetic studies is that multiple genes interact with each other and with multiple environmental factors to influence susceptibility to illness. This is essentially the traditional multifactorial model. Other genetic/molecular mechanisms that are known to produce complex patterns of inheritance and have been suggested as possible contributors to BD include mitochondrial inheritance (31), and dynamic mutation (32).

X-linkage: Several large pedigrees were reported to show cosegregation of X-linked markers (e.g., color-blindness or glucose-6-phosphatase deficiency) and BD (33–35). These reports have been criticized on methodological grounds (36), and, despite over half a century of debate, the contribution of X-linked genes to the pathogenesis of BD remains uncertain. If X-linked genes do contribute, analyses suggest that they can only account for a modest proportion of cases (26,37).

CHROMOSOME STUDIES

A potential shortcut to identifying the genomic location of susceptibility genes for a disorder is recognition of the co-occurrence of the disorder and gross changes at the chromosomal level. Affective disorders have not been found to be consistently associated with chromosome abnormalities, although a number of such reports have appeared in the literature (38). One interesting observation is that individuals with trisomy 21 appear to be less susceptible to mania than are members of the general population (39). This is consistent with the existence of a bipolar susceptibility gene on chromosome 21, a possibility that finds support from some linkage studies. In contrast, chromosomal abnormalities have had an important influence on the molecular genetic investigation of schizophrenia. The velocardiofacial syndrome (VCFS), caused by deletions at chromosome 22q11, is associated with a high lifetime risk of psychosis. This has focused attention on the chromosome 22q11 region as likely to harbor one or more genes involved in susceptibility to psychosis (40). The chromosome 1q42 region similarly was implicated through its involvement in a chromosomal translocation in a large Scottish pedigree in which multiple members had both psychiatric illness and the translocation (41). Study of this pedigree resulted in identification of a gene that was disrupted by the translocation, and this was named Disrupted in Schizophrenia 1 (DISC-1) (42). Of relevance to the current chapter, and as will be discussed later, there is increasing evidence that the psychiatric phenotypes associated with the chromosomal abnormalities at 1q42 and 22q11 have a substantial mood component and may, therefore, contribute to bipolar spectrum illness.

MOLECULAR GENETIC STUDIES

Molecular genetic approaches have already achieved great success in discovering the mutations that lead to simple (mendelian) genetic diseases. The continuing challenge is to use the developing methodologies to uncover susceptibility genes for complex diseases, such as BD. Conceptually, molecular genetic studies can be divided into positional and candidate gene approaches. The positional approach assumes no knowledge of disease pathophysiology but determines the broad chromosomal locations of susceptibility genes, usually by linkage studies using multiply affected families. The candidate gene approach, however, involves the investigator making educated guesses at what genes may be involved in the pathophysiology of a condition and then testing the involvement of these genes, usually by association studies in which a set of cases is compared with a set of controls. The candidate gene approach is built on the assumption that the investigator has sufficient understanding of disease biology to be able to recognise suitable candidate genes. In practice, both positional and candidate approaches are often combined (43,44).

Linkage Studies in Bipolar Disorder

Linkage studies employing very large pedigrees and based on the assumption of a single major gene are appropriate for simple mendelian disorders, but when applied to complex disorders this approach can be problematic. In the late 1980s two high-profile claims for linkage appeared in the journal *Nature*: Baron et al. (35) reported linkage to X-chromosome markers in several Israeli pedigrees and Egeland et al. (45) reported linkage to markers on chromosome 11p15 in a large pedigree of the Old Order Amish community. Other workers were unable to replicate these findings and eventually in both cases the original groups published updated and extended analyses of their own data in which the significant evidence of linkage all but vanished (46,47). Reasons for this dramatic change in findings included: (*i*) family members originally diagnosed as unaffected became ill for the first time during follow-up, (*ii*) new family members were examined who did not show evidence for linkage, and (*iii*) additional DNA markers were examined which reduced the evidence for linkage. More detailed discussion of this issue and subsequent linkage findings in the 11p15 chromosomal region can be found elsewhere (48).

Such a major setback was of course a disappointment but the field moved forward with the development of methodologies more appropriate for the study of complex genetic traits including a trend towards the use of smaller families (particularly affected sibling pairs) and analyses less sensitive to diagnostic changes (44). Large, systematic molecular genetic linkage studies have been reported and are ongoing in several centers around the globe. A variety of types of sample set have been used, ranging from large densely affected pedigrees in genetic isolates to large numbers of affected sibling pairs. The pattern of findings is consistent with there being no gene of major effect to explain the majority of cases of BD and with that expected in the search for genes for a complex disorder (49–51). Chromosomal regions of interest are typically broad (often >20–30 cM). No finding replicates in all datasets and for individual positive findings levels of statistical significance and estimated effect sizes are usually modest. Table 1 shows chromosomal regions that have received support with genome-wide significance in at least one study. Several regions have been implicated repeatedly by individual studies, but usually at a lower level of significance and not sufficiently consistently to be highlighted by meta-analyses.

Two meta-analyses of BD genome scans have been conducted. Badner and Gershon (52) found the strongest evidence for susceptibility loci on 13q and 22q when examining seven published genome scans for BD. However, the more recent and detailed meta-analysis of Segurado et al. (53) conducted using the bin-ranking methodology did not find genome-wide significant evidence for linkage but provided a more modest level of support for regions on chromosomes 9p22.3–21.1, 10q11.21–22.1, 14q24.1–32.12, and regions of chromosome 18. This meta-analysis demonstrated lesser consistency in the findings from bipolar scans than from schizophrenia scans (54). Possible causes for this difference are discussed elsewhere (55).

Since publication of the meta-analyses, several further genome-wide scans in independent samples have been published, with several regions identified that meet genome-wide significant or suggestive evidence for linkage. Of particular note is the 6q21–q25 region which was not implicated in either meta-analysis, but which is supported by one genome-wide significant (56) and three genome-wide suggestive signals (57–59), making it one of the best-supported regions for BD. Indeed, in the recent combined collaborative analysis of eleven bipolar linkage scans, this region achieved genome wide significance (60).

Figure 1 shows chromosomal regions that have received genome-wide significant support in at least one scan of BD. Of particular note are the 6q21–q25 region mentioned above and the 12q23–q24 region, which has two genome scans reporting genome-wide significance (58,61) and is also supported by linkage analysis in unipolar disorder (62) and by linkage of this region in two pedigrees that show cosegregation bipolar spectrum illness and an autosomal dominant skin disease, Darier's disease (63).

Gene Studies of BD

To date there has not been unambiguous demonstration of a susceptibility gene identified for BD by positional cloning. Potentially interesting findings have come from the study of functional candidates and, most recently, investigation of

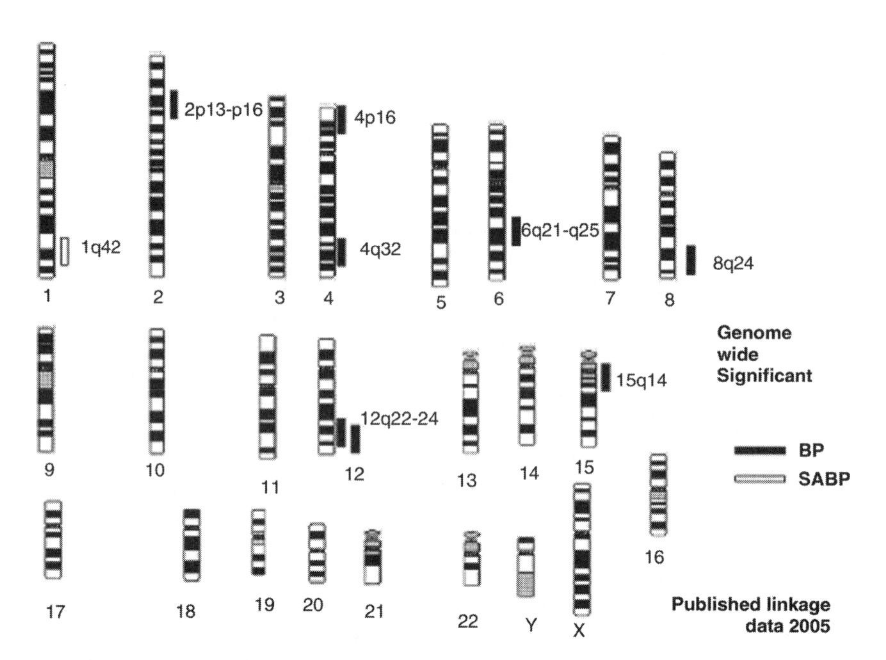

FIGURE 1 Chromosome ideograms showing locations of genome-wide significant linkages for bipolar spectrum phenotypes. The predominant phenotype used in the analysis is shown as: BP: bipolar disorder; SABP: schizoaffective disorder, bipolar type. Note that most genome scans of bipolar disorder have used a range of definitions of the bipolar phenotype from narrow (only bipolar I disorder) to broad (including also bipolar II disorder, schizoaffective disorder, and unipolar disorder).

genes first implicated in schizophrenia (some of which map in linkage regions of interest in BD). However, none of the findings yet achieve the level of support that dysbindin and neuregulin 1 (NRG1) have received in schizophrenia (55).

Functional Candidates

Most studies in the literature have focused on neurotransmitter systems influenced by medications employed in the management of the disorder, particularly the dopamine, serotonin, and noradrenaline systems (64). For most genes studied, the usual pattern has been for one or a few positive studies along with an even greater number of negative replications. However, polymorphisms of known functional relevance in three of the genes have been reported as significant at the $P < 0.05$ level in at least some published meta-analyses: *MAOA* (65), *COMT* (66), and *5HTT* (67,68), all with modest effect sizes (odds ratios, OR $\ll 2$). The findings for these functional polymorphisms remain to be tested in the necessary number of independent, large samples that will be required to determine unambiguously whether and to what extent variation within these genes contributes to susceptibility to BD, or to some intermediate clinical phenotype (69,70). It is of interest that *COMT* has also been implicated in schizophrenia (71) and has received support from the same group in a modestly sized study of BD (72) although an updated meta-analysis of the literature available in December 2005 found no significant evidence for association at *COMT* (70).

Most of the candidate gene reports in the literature describe studies in modestly sized samples (a few hundred individuals) that are likely to be underpowered for plausible effects sizes (69,70). As for all complex disorders, the trend in candidate gene studies of BD is for the use of larger samples, with increased power to detect modest or small effect sizes, and examination of candidate genes predicated on more sophisticated models of pathogenesis or directed by positional information from linkage studies. Recently, replicable positive findings have started to emerge from these approaches.

D-Amino Acid Oxidase Activator (G72)/G30 Locus

At least five independent datasets contribute evidence that variation at the *DAOA/G30* locus on chromosome 13q influences susceptibility to BD. This locus was implicated originally as being involved in susceptibility to schizophrenia (73). It was a novel locus with a designation of *G72*. The locus was renamed as D-amino acid oxidase activator (DAOA), because biological studies suggested that the gene product activated the enzyme, D-amino acid oxidase (DAO); genetic evidence was also found in this original study for association of alleles at DAO with susceptibility to schizophrenia. Subsequently linkage disequilibrium (LD) at the DAOA locus was also reported with BD in two U.S. family samples (74), and this was replicated in a further U.S. family sample (75), German case-control sample (76), and our own large U.K. case-control sample (77). In all studies, evidence for LD came from individual single nucleotide polymorphisms (SNPs) as well as multilocus haplotypes, although there is variation between studies in the SNPs and haplotypes showing LD. No pathologically relevant variant has yet been identified and the biological mechanism remains to be elucidated. It is of interest that DAO lies in the 12q23 region implicated in linkage studies of both bipolar and unipolar disorder (Fig. 1). DAO has been examined in only one study of BD (76), which found no evidence of LD. However, in view of the findings with DAOA, DAO warrants more thorough study in BD.

Brain Derived Neurotrophic Factor

A functional candidate gene that has attracted a great deal of recent interest is *Brain Derived Neurotrophic factor* (*BDNF*) (78). *BDNF* is a member of the neurotrophin superfamily. Neurotrophins are synthesized in neurons as proforms that can be cleaved intra- or extracellularly, and both their synthesis and secretion depends upon neuronal activity. *BDNF* plays an important role in promoting and modifying growth, development, and survival of neuronal populations and, in the mature nervous system, is involved in activity-dependent neuronal plasticity (79). These processes are prominent in the synaptic plasticity hypothesis of mood disorder, which focuses on the functional and structural changes induced by stress and antidepressants at the synaptic level. The *BDNF* gene lies on the reverse strand of chromosome 11p13 and encodes a precursor peptide (pro*BDNF*), which is cleaved proteolytically to form the mature protein (80). The 11p13 chromosomal location of *BDNF* has been implicated in some linkage studies of BD, but not in meta-analyses of linkage studies.

Consistent with the strong evolutionary conservation of the *BDNF* coding sequence across species, only one frequent, nonconservative polymorphism in the human *BDNF* gene has been identified, a SNP at nucleotide 196 within the 5′pro-*BDNF* sequence that causes an amino acid substitution of valine to methionine at codon 66 (Val66Met). There is cross-species conservation of the precursor portion of pro*BDNF* that is consistent with potential functional importance, and it is possible that the common Val66Met polymorphism could itself have a functionally relevant effect by modifying the processing and trafficking of *BDNF* (81).

There have been three positive reports using family based association studies of Caucasian BD samples of European-American origin and the Val66Met SNP: two were based on adult bipolar samples (82,83) and one was based on a small childhood onset sample (84). All have shown over-transmission of the common Val allele. Evidence with multilocus haplotypes was stronger in one study (83). There have been four case-control association studies [of European (85,86), Chinese (87), and Japanese origin (88)] to date, in which there is no evidence for an allelic or genotypic association. In our own Caucasian bipolar case-control sample ($N = 3062$) we found no overall evidence of allele or genotype association. However, we found significant association with disease status in the subset of cases that had experienced rapid cycling (four or more episodes per year) at some time, and a similar association on reanalysis of our previously reported family-based association sample (89). This suggests that variation at the Val66Met polymorphism of *BDNF* may not play a major role in influencing susceptibility to BD as a whole but, rather, may be associated with susceptibility to a specific aspect of the clinical bipolar phenotype. This receives support from reanalysis of one of the original family-based samples that shows the over-transmission of Val alleles is explained by the rapid cycling individuals in the sample (90). It should, however, be noted that the Val66Met polymorphism lies within a large haplotype block so it is difficult to determine which variant(s) within the block is (are) pathogenically relevant.

Substantial additional genetic and biological work will be required to confirm (or refute) the role of *BDNF* in influencing susceptibility to BD. Systematic study of variation across the whole gene is required with study in further independent samples.

Other Genes

We will briefly mention three other genes that have recently received attention as potential BD susceptibility genes. Two of these are in the 22q chromosome linkage region of interest. *G-protein receptor kinase 3* (*GRK3*) was implicated through positional follow-up of a linkage signal in a set of U.S. pedigrees and was supported also by expression data in a rodent model of mania (91). However, this has not yet received independent support. *XBP1*, a pivotal gene in the endoplasmic reticulum (ER) stress response, was reported to show association at a promoter polymorphism with BD susceptibility in two small association samples (92). Some degree of circumstantial biological support for a functional role for this polymorphism came from a cellular model of the action of mood-stabilizer medications. However, this report is highly likely to be a type I error because the putative functionally relevant variant was found to have no influence on susceptibility in independent family-based and case-control association samples six times larger than those in the initial report (93). This illustrates the caution that is required in interpreting weakly supported putative genetic effects even in the face of substantial biological plausibility (70). More promisingly, but as yet not widely tested, is the report that *P2X7*, in the 12q24 region of linkage interest, influences susceptibility to both BD and unipolar depression (94). It can be expected that several, probably many, other susceptibility genes will be identified over the coming years—no single study will be definitive and the true status of each candidate is unlikely to become clear until several thousand cases have been studied.

CONSIDERATION OF THE CLINICAL PHENOTYPE: CLINICAL COVARIATES AND SUBTYPES

As discussed elsewhere (17), mood disorder researchers have been taking an interest in a variety of clinical subtypes and covariates over recent years, as a way of testing subsets of cases with increased clinical (and hopefully genetic) homogeneity. Examples include rapid cycling (95), lithium responsiveness (96). bipolar affective puerperal psychosis (triggering of bipolar episodes in females by parturition) (97,98), early age at onset (99), and occurrence of psychotic features during illness (100–102).

Consideration of the occurrence of psychotic features in BD raises the interesting and biologically important issue of the overlap in genetic findings in BD and schizophrenia. This is discussed in the next section.

OVERLAP IN FINDINGS BETWEEN BIPOLAR DISORDER AND SCHIZOPHRENIA

Traditionally, psychiatric research in general, and the search for predisposing genes in particular, has proceeded under the assumption that schizophrenia and mood disorder are separate disease entities with separate underlying etiologies (and treatments): the so-called "Kraepelinian dichotomy." This distinction has pervaded Western psychiatry since Emil Kraepelin's influential nosological writings (103) and survives in current operational classification systems such as ICD-10 (5) and DMS-IV (6), although some workers, such as Crow, have argued for a continuum approach to psychosis (104). The clinical reality is that many individuals with severe psychiatric illness have features that fall between these two "extremes" and have *both* prominent mood and psychotic features (often classified as "schizoaffective disorder" or some similar atypical diagnosis). This suggests that there may not be a neat biological distinction between schizophrenia and BD. This possibility finds support in several observations from genetic research, including the following:

Family Studies

Although schizophrenia and BD may "breed true" (105–107), families are known where there are multiple cases of schizophrenia, BD, and cases with both psychosis and mood disorder (108). Further, some studies have shown statistically significant evidence that BD occurs at an increased rate in the relatives of probands with schizophrenia (109) and that BD occurs at an increased frequency in the relatives of bipolar probands (110). Schizoaffective disorder has been shown to occur at an increased rate in the families of probands with schizophrenia (111), and in the families of probands with BD (112). Both schizophrenia and BD have been shown to occur at increased rates in the families of probands with schizoaffective disorder (112). Together, these data suggest a more complex relationship between the psychoses than is reflected in the conventional dichotomous view.

Twin Studies

Only one twin study has used an analysis that was unconstrained by the diagnostic hierarchy inherent in current classification systems (i.e., the principle that schizophrenia "trumps" mood disorder in diagnosis). This study demonstrated a clear overlap in the genetic susceptibility to syndromally defined mania and schizophrenia (113). The findings suggested the existence of some susceptibility genes that are specific to schizophrenia, others that are specific to BD, and yet others that influence susceptibility to schizoaffective disorder, schizophrenia, and BD.

Linkage Studies

Genetic linkage studies have identified some chromosome regions that show convergent or overlapping regions of interest in BD and schizophrenia, including regions of 13q, 22q, 18 (52,114), and 6q. The hypothesis that loci exist that influence susceptibility across the schizophrenia-bipolar divide receives further support from the observation that a genome scan, using families ascertained on the basis of a proband with schizoaffective disorder (a form of illness with prominent features of both schizophrenia and BD), demonstrated genome-wide significance at 1q42 and suggestive linkage at 22q11, with linkage evidence being contributed equally from "schizophrenia families" (i.e., where other members had predominantly schizophrenia) and "bipolar families" (i.e., where other members had predominantly BD) (115).

Gene Studies

Most persuasively, several recent reports implicate variation at the same loci as influencing susceptibility to both schizophrenia and BD.

Currently the best-supported locus for BD is G72(DAOA)/G30 on chromosome 13q (74–77), which also has positive association reported in schizophrenia (73,76,116,117).

The *DISC-1* locus at 1q42 receives linkage support in schizophrenia (42,118,119), BD (120) and schizoaffective disorder (115) and, although it has been named *Disrupted in Schizophrenia 1*, the family in which the translocation was observed contained cases of both psychosis and mood disorder (42). Evidence for allelic association at polymorphisms at this locus has been reported for schizophrenia, BD, and schizoaffective disorder (121).

Neuregulin 1 is one of the best supported schizophrenia susceptibility genes with several studies showing evidence that a so-called Icelandic "core haplotype" is associated with increased risk in Icelandic, Scottish, and U.K. populations (122–124). We have found that this same haplotype is significantly associated with risk for BD and that it may exert a specific effect in the subset of functional psychosis that has both manic and mood-incongruent psychotic features (125).

The *COMT* gene lies at 22q11, a region implicated in BD and schizophrenia (and the region usually deleted in VCFS) (52). It is extremely likely that genetic variation in this region influences susceptibility across the psychosis spectrum (71,72), although it is not yet clear that *COMT* itself is the (or the major) susceptibility gene at this locus.

These gene findings provide strong evidence that, as suggested by the family and twin data, there are genetic loci that contribute susceptibility across the Kraepelinian divide to schizophrenia, BD, and schizoaffective disorders. These findings have important implications for classification of the major psychiatric disorders because they demonstrate an overlap in the biological basis of disorders that, over the last one hundred years, have been assumed to be distinct entities (126). Molecular genetic findings are likely to catalyze a reappraisal of psychiatric nosology as well as providing a path to understanding the pathophysiology that will facilitate development of improved treatments. Rather than classifying psychosis as a dichotomy, a more useful formulation may be to conceptualize a spectrum of clinical phenotype with susceptibility conferred by overlapping sets of genes (126).

CONCLUSION

Positive findings are emerging from molecular genetic studies of BD. Replication in large, well-characterized samples are required to determine their robustness and generalizability. It will be necessary to undertake detailed phenotype-genotype studies across the mood-psychosis spectrum as well as functional biological studies to determine how biological variation influences clinical phenotype. It can be expected that some current findings will prove to be false positives and there will be many more susceptibility or disease-modifying genes to be identified in future studies. New methodologies including whole genome association studies can be expected to complement existing approaches to facilitate progress.

In the past, psychiatric genetics has often attracted the pessimistic view that it is an area of endeavor that is so complex that advances were unlikely. However, promising findings are now emerging and the potential benefits for the practice of clinical psychiatry should not be underestimated (127). In addition to facilitating the development of treatments better targeted at the biochemical lesions involved in disease, it is also likely to lead to the development of a more rational etiologically-based classification system which will provide a much better guide to treatment and prognosis than current systems. Importantly, identifying susceptibility genes will facilitate the identification of environmental factors that alter risk. Once these environmental factors are characterized, it may prove possible to provide helpful occupational, social, and psychological advice to individuals at genetic risk of affective disorders. It is also likely that along this path we will learn much about the biological basis of normal affective responses.

In addition to the undoubted benefits, the potential costs must also be considered (127). Major advances raise major ethical issues. Many of these issues are no different from those that arise in the context of other complex familial disorders, but the combination of genetics and mental illness raises particular concerns and has justifiably received close scrutiny of ethical and psychosocial issues (128). It is important that we continue to address potential problems such as the availability of services, the right to information, and the testing of individuals below the age of consent. The challenge is to translate advances in understanding of the complex etiology of affective disorders into tangible improvements in clinical care.

SUMMARY

Key points are listed in Table 2. The enormous public health importance of mood disorders, when considered alongside their substantial heritabilities, has stimulated much work, predominantly in BD, aimed at identifying susceptibility genes using both positional and functional molecular genetic approaches. Several regions of interest have emerged in linkage studies and, recently, evidence implicating specific genes has been reported; the best supported include *BDNF* and *DAOA* but further replications are required and phenotypic relationships and biological mechanisms need investigation. The complexity of psychiatric phenotypes is demonstrated by the evidence accumulating for an overlap in genetic susceptibility across the

TABLE 2 Key Points

Family, twin, and adoption studies provide compelling evidence for the existence of susceptibility genes for bipolar spectrum mood disorders.
Estimates of heritability typically exceed 80% for narrowly defined bipolar disorder.
The pattern of inheritance is consistent with the action of multiple genes and environmental factors that interact to influence risk of illness.
Several chromosomal regions have been implicated repeatedly in linkage studies.
Evidence is accumulating to support the involvement of specific genes, including *DAOA(G72)* and *BDNF*, in susceptibility to, or modification of the course of, bipolar disorder.
No gene has yet received the level of systematic study in large samples that is likely to be necessary to characterize its contribution to bipolar illness.
Increasing genetic data (both classical and molecular) challenge the traditional dichotomous classification of functional psychoses and highlight the need for changes to psychiatric nosology.

traditional classification systems that divide disorders into schizophrenia and mood disorders. Although no clear examples of gene–environment interaction have yet been demonstrated in BD, it is to be expected that such mechanisms will contribute to the genetic complexity.

ACKNOWLEDGMENTS

The author is grateful for research support to the U.K. Medical Research Council and the Wellcome Trust and is indebted to all the individuals who have participated in the studies.

REFERENCES

1. Murray CJL, Lopez AD, eds. The Global Burden of Disease: A Comprehensive Assessment of Mortality, Injuries, and Risk Factors in 1990 and Projected to 2020. Cambridge, MA: WHO, 1996.
2. McGuffin P, Owen M, Gottesman II, eds. Oxford: Psychiatric Genetics and Genomics. Oxford: Oxford University Press, 2002.
3. Kendell RE. The classification of depression: a review of contemporary confusion. Br J Psychiatry 1976; 129:15–28.
4. Farmer A, McGuffin P. The classification of the depressions. contemporary confusion revisited. Br J Psychiatry 1989; 155:437–443.
5. World Health Organisation. The International Classification of Diseases 10 Classification of Mental and Behavioural Disorders. Diagnostic Criteria for Research. Geneva: WHO, 1993.
6. American Psychiatric Association. Diagnostic and Statistical Manual of Mental Disorders (Fourth Edition, Text Revision). Washington DC: APA, 2000.
7. Kendell RE. Diagnosis and classification of functional psychoses. Br Med Bull 1987; 43:499.
8. Perris C. A study of bipolar (manic-depressive) and unipolar recurrent depressive psychoses. I. Genetic investigation. Acta Psychiatr Scand Suppl 1966; 194:15–44.
9. Angst J. On the etiology and nosology of endogenous depressive psychoses. A genetic, sociologic and clinical study. Monogr Gesamtgeb Neurol Psychiatr 1966; 112:1–118.
10. Simpson SG, McMahon FJ, McInnis MG, et al. Diagnostic Reliability of Bipolar II Disorder. Arch Gen Psych 2002; 59:736–740.
11. Smith AL, Weissman MM. Epidemiology. In: Paykel ES, ed. Handbook of Affective Disorders. Edinburgh: Churchill Livingstone, 1992:111–129.

12. Weissman MM, Leaf PJ, Tischler GL, et al. Affective disorders in five United States communities. Psychol Med 1988; 18:141–153.
13. Kessler RC, McGonagle KA, Zhao S, et al. Lifetime and 12-month prevalence of DSM IIIR psychiatric disorders in the United States: results from the National Comorbidity Survey. Arch Gen Psych 1994; 51:8–19.
14. Guze SB, Robins E. Suicide and primary affective disorders. Br J Psychiatry 1970; 117:437–438.
15. Daly I. Mania. Lancet 1997; 349:1159–1160.
16. Tsuang MT, Faraone SV. The Genetics of Mood Disorders. Baltimore: The Johns Hopkins University Press, 1990.
17. Craddock N, Jones I. Genetics of Bipolar Disorder. J Med Genet 1999; 36:585–594.
18. Jones I, Kent L, Craddock N. Genetics of affective disorders. In: McGuffin P, Owen M, Gottesman II, eds. Psychiatric Genetics and Genomics. Oxford: Oxford University Press, 2002:211–245.
19. Sullivan PF, Neale MC, Kendler KS. Genetic epidemiology of major depression: review and meta-analysis. Am J Psychiatry 2000; 157:1552–1562.
20. McGuffin P, Katz R. The genetics of depression and manic-depressive disorder. Br J Psychiatry 1989; 155:294–304.
21. Blacker D, Tsuang MT. Unipolar relatives in bipolar pedigrees: are they bipolar? Psychiatr Genet 1993; 3:5–16.
22. Rice J, Reich T, Andreasen NC, et al. The familial transmission of bipolar illness. Arch Gen Psychiatry 1987; 44:441–447.
23. Pauls DL, Morton LA, Egeland JA. Risks of Affective illness among first-degree relatives of bipolar I old-order Amish Probands. Arch Gen Psychiatry 1992; 49:703–708.
24. Crowe RR, Smouse PE. The genetic implications of age-dependent penetrance in manic-depressive illness. J Psychiatr Res 1977; 13:273.
25. Spence MA, Flodman PL, Sadovnik AD, Bailey-Wilson JE, Ameli H, Remick RA. Bipolar disorder: evidence for a major locus. Am J Med Genet (Neuropsychiatr Genet) 1995; 60:370–376.
26. Bucher KD, Elston RC, Green R, et al. The transmission of manic depressive illness—II. Segregation analysis of three sets of family data. J Psychiatr Res 1981; 16:65–78.
27. Goldin LR, Gershon ES, Targum SD, Sparkes RS, McGinnis M. Segregation and linkage analyses in families of patients with bipolar, unipolar, and schizoaffective mood disorders. Am J Hum Genet 1983; 45:274–287.
28. Sham PC, Morton NE, Rice JP. Segregation analysis of the NIMH Collaborative study. Family data on bipolar disorder. Psychiatr Genet 1991; 2:175–184.
29. Craddock N, Van Eerdewegh P, Reich T. Single major locus models for bipolar disorder are implausible. Am J Med Genet 1997; 74:18.
30. Craddock N, Khodel V, Van Eerdewegh P, et al. Mathematical limits of multilocus models: the genetic transmission of bipolar disorder. Am J Hum Genet 1995; 57:690–702.
31. McMahon FJ, Stine OC, Meyers DA, Simpson SG, DePaulo JR. Patterns of maternal transmission in bipolar affective disorder. Am J Hum Genet 1995; 56:1277–1286.
32. O'Donovan M, Jones I, Craddock N. Anticipation and repeat expansion in bipolar disorder. Am J Med Genet C Semin Med Genet 2003; 123:10–17.
33. Reich T, Clayton P, Winokur G. Family history studies: V. The genetics of mania. Am J Psychiatry 1969; 125:1358–1368.
34. Mendlewicz J, Fleiss JL, Fieve RR. Evidence for X-linkage in the transmission of manic-depressive illness. J Am Med Assoc 1992; 222:1624.
35. Baron M, Risch N, Hamburger R, et al. Genetic linkage between X-chromosome markers and bipolar affective illness. Nature 1987; 326:289–292.
36. Hebebrand J. A critical appraisal of X-linked bipolar illness. Evidence for the assumed mode of inheritance is lacking. Br J Psychiatry 1992; 160:7–11.
37. Risch N, Baron M, Mendlewicz J. Assessing the role of X-linked inheritance in bipolar-related major affective disorder. J Psychiatr Res 1986; 20:275–288.
38. Craddock N, Owen M. Chromosomal aberrations and bipolar affective disorder. Br J Psychiatry 1994; 164:507–512.

39. Craddock N, Owen M. Is there an inverse relationship between Down's syndrome and bipolar affective disorder? Literature review and genetic implications. J Intellect Dis Res 1994; 38:613–620.
40. Murphy KC, Owen MJ. Velo-cardio-facial syndrome: a model for understanding the genetics and pathogenesis of schizophrenia. Br J Psychiatry 2001; 179:397–402.
41. St Clair D, Blackwood D, Muir W, et al. Association within a family of a balanced autosomal translocation with major mental illness. Lancet 1990; 336:13–16.
42. Millar JK, Wilson-Annan JC, Anderson S, et al. Disruption of two novel genes by a translocation co-segregating with schizophrenia. Hum Mol Genet 2000; 9:1415–1423.
43. Collins FS. Positional cloning moves from periditional to traditional. Nat Genet 1996; 9:347–350.
44. Craddock N, Owen MJ. Modern molecular genetic approaches to psychiatric disease. Br Med Bull 1996; 52:434–452.
45. Egeland JA, Gerhard DS, Pauls DL, et al. Bipolar affective disorders linked to DNA markers on chromosome 11. Nature 1987; 325:783–787.
46. Kelsoe JR, Ginns EI, Egeland JA, et al. Re-evaluation of the linkage relationship between chromosome 11p loci and the gene for bipolar affective disorder in the Old Order Amish. Nature 1989; 342:238–243.
47. Baron M, Freimer NF, Risch N, et al. Diminished support for linkage between manic-depressive illness and X-chromosome markers in three Israeli pedigrees. Nat Genet 1993; 3:49–55.
48. Craddock N, Lendon C. Chromosome Workshop: Chromosomes 11, 14, and 15. Am J Med Genet (Neuropsychiatr Genet) 1999; 88:244–254.
49. Suarez BK, Hample CL, Van Eerdewegh P. Problems of replicating linkage claims in psychiatry. Genetic approaches to mental disorders. Washington DC: American Psychiatric Press Inc., 1984:23–46.
50. Lander ES, Schork NJ. Genetic dissection of complex traits. Science 1994; 265:2037–2048.
51. Kruglyak L, Lander ES. High-resolution genetic mapping of complex traits. Am J Hum Genet 1995; 56:1212–1223.
52. Badner JA, Gershon ES. Meta-analysis of whole-genome linkage scans of bipolar disorder and schizophrenia. Mol Psychiatry 2002; 7:405–411.
53. Segurado R, Detera-Wadleigh SD, Levinson DF, et al. Genome scan meta-analysis of schizophrenia and bipolar disorder, part III: Bipolar disord. Am J Hum Genet 2003; 73(1):49–62.
54. Lewis CM, Levinson DF, Wise LH, et al. Genome scan meta-analysis of schizophrenia and bipolar disorder, part II: Schizophrenia. Am J Hum Genet 2003; 73(1):34–48.
55. Craddock N, O'Donovan MC, Owen MJ. The genetics of schizophrenia and bipolar disorder: dissecting psychosis. J Med Genet 2005; 42:193–204.
56. Middleton FA, Pato MT, Gentile KL, et al. Genomewide linkage analysis of bipolar disorder by use of a high-density single-nucleotide-polymorphism (SNP) genotyping assay: a comparison with microsatellite marker assays and finding of significant linkage to chromosome 6q22. Am J Hum Genet 2004; 74(5):886–897.
57. Dick DM, Foroud T, Flury L, et al. Genomewide linkage analyses of bipolar disorder: a new sample of 250 pedigrees from the National Institute of Mental Health Genetics Initiative. Am J Hum Genet 2003; 73(1):107–114. Erratum in: Am J Hum Genet 2003; 73(4):979.
58. Ewald H, Flint T, Kruse TA, et al. A genome-wide scan shows significant linkage between bipolar disorder and chromosome 12q24.3 and suggestive linkage to chromosomes 1p22–21, 4p16, 6q14–22,10q26 and 16p13.3. Mol Psychiatry 2002; 7(7):734–744.
59. Lambert D, Middle F, Hamshere ML, et al. Stage 2 of the Wellcome Trust UK-Irish bipolar affective disorder sibling-pair genome screen: evidence for linkage on chromosomes 6q16–q21, 4q12–q21, 9p21, 10p14–p12 and 18q22. Mol Psychiatry 2005; 10(9):831–841.
60. McQueen MB, Devlin B, Faraone SV, et al. Combined analysis from eleven linkage studies of bipolar disorder provides strong evidence of susceptibility Loci on chromosomes 6q and 8q. Am J Hum Genet 2005; 77(4):582–595.

61. Shink E, Morissette J, Sherrington R, et al. A genome-wide scan points to a susceptibility locus for bipolar disorder on chromosome 12. Mol Psychiatry 2005; 10:545–552.
62. Abkevich V, Camp NJ, Hensel CH, et al. Predisposition locus for major depression at chromosome 12q22–12q23.2. Am J Hum Genet 2003; 73:1271–1281.
63. Green E, Elvidge G, Jacobsen N, et al. Localization of bipolar susceptibility locus by molecular genetic analysis of the chromosome 12q23–24 region in two pedigrees with bipolar disorder and Darier's disease. Am J Psychiatry 2005; 162:35–42.
64. Craddock N, Dave S, Greening J. Association studies of bipolar disorder. Bipolar Disord 2001; 3:284–298.
65. Preisig M, Bellivier F, Fenton BT, et al. Association between bipolar disorder and monoamine oxidase A gene polymorphisms: results of a multi-center study. Am J Psychiatry 2000; 157:948–955.
66. Jones I, Craddock N. Candidate gene studies of Bipolar Disorder. Ann Med 2001; 33:248–256.
67. Anguelova M, Benkelfat C, Turecki G. A systematic review for association studies investigating genes coding for serotonin receptors and the serotonin transporter: I. Affective disorders. Mol Psychiatry 2003; 8:574–591.
68. Lasky-Su JA, Faraone SV, Glatt SJ, et al. Meta-analysis of the association between two polymorphisms in the serotonin transporter gene and affective disorders. Am J Med Genet 2005; 133:110–115.
69. Wang WY, Barratt BJ, Clayton DG, Todd JA. Genome-wide association studies: theoretical and practical concerns. Nat Rev Genet 2005; 6:109–118.
70. Craddock N, Owen MJ, O'Donovan MC. The catechol-O-methyltransferase (COMT) gene as a candidate for psychiatric phenotypes: evidence and lessons. Mol Psychiatry 2006; 11(5):446–458.
71. Shifman S, Bronstein M, Sternfeld M, et al. A highly significant association between a COMT haplotype and schizophrenia. Am J Hum Genet 2002; 71:1296–1302.
72. Shifman S, Bronstein M, Sternfeld M, et al. COMT: a common susceptibility gene in bipolar disorder and schizophrenia. Am J Med Genet 2004; 128B(1):61–64.
73. Chumakov I, Blumenfeld M, Guerassimenko O, et al. Genetic and physiological data implicating the new human gene G72 and the gene for D-amino acid oxidase in schizophrenia. Proc Natl Acad Sci USA 2002; 99:13675–13680.
74. Hattori E, Liu C, Badner JA, et al. Polymorphisms at the G72/G30 gene locus, on 13q33, are associated with bipolar disorder in two independent pedigree series. Am J Hum Genet 2003; 72:1131–1140.
75. Chen YS, Akula N, Detera-Wadleigh SD, et al. Findings in an independent sample support an association between bipolar affective disorder and the G72/G30 locus on chromosome 13q33. Mol Psychiatry 2004; 9:87–92.
76. Schumacher J, Jamra RA, Freudenberg J, et al. Examination of G72 and D-amino-acid oxidase as genetic risk factors for schizophrenia and bipolar affective disorder. Mol Psychiatry 2004; 9:203–207.
77. Williams NM, Green EK, Macgregror S, et al. Variation at the DAOA/G30 locus influences susceptibility to major mood episodes but not psychosis in schizophrenia and bipolar disorder. Arch Gen Psychiatry 2006; 63(4):366–373.
78. Green E, Craddock N. Brain-derived neurotrophic factor as a potential risk locus for bipolar disorder: evidence, limitations, and implications. Curr Psychiatry Rep 2003; 5:469–476.
79. Duman RS. The neurochemistry of mood disorders: preclinical studies. In: Charney DS, Nestler EJ, Bunney BS, eds. The Neurobiology of Mental Illness. New York: Oxford University Press, 1999:333–347.
80. Seidah NG, Benjannet S, Pareek S, Chretien M, Murphy RA. Cellular processing of the neurotrophin precursors of NT3 and BDNF by the mammalian proprotein convertases. FEBS Lett 1996; 379:247–250.
81. Egan MF, Kojima M, Callicott JH, et al. The BDNF Val66Met polymorphism affects activity-dependent secretion of BDNF and human memory and hippocampal function. Cell 2003; 112:257–269.

82. Sklar P, Gabriel SB, McInnis MG, et al. Family-based association study of 76 candidate genes in bipolar disorder: BDNF is a potential risk locus. Brain-derived neurotrophic factor. Mol Psychiatry 2002; 7:579–593.
83. Neves-Pereira M, Mundo E, Muglia P, et al. The brain-derived neurotrophic factor gene confers susceptibility to bipolar disorder: evidence from a family-based association study. Am J Hum Genet 2002; 71:651–655.
84. Geller B, Badner JA, Tillman R, et al. Linkage disequilibrium of the brain-derived neurotrophic factor Val66Met polymorphism in children with a prepubertal and early adolescent bipolar disorder phenotype. Am J Psychiatry 2004; 161(9):1698–1700.
85. Oswald P, Del-Favero J, Massat I, et al. Non-replication of the brain-derived neurotrophic factor (BDNF) association in bipolar affective disorder: a Belgian patient-control study. Am J Med Genet 2004; 129B(1):34–35.
86. Skibinska M, Hauser J, Czerski PM, et al. Association analysis of brain-derived neurotrophic factor (BDNF) gene Val66Met polymorphism in schizophrenia and bipolar affective disorder. World J Biol Psychiatry 2004; 5(4):215–220.
87. Hong CJ, Huo SJ, Yen FC, et al. Association study of a brain-derived neurotrophic-factor genetic polymorphism and mood disorders, age of onset and suicidal behaviour. Neuropsychobiology 2003; 48(4):186–189.
88. Nakata K, Ujike H, Sakai A, et al. Association study of brain-derived neurotrophic factor (BDNF) gene with bipolar disorder. Neurosci Lett 2003; 337:17–20.
89. Green E, Raybould R, McGregor S, et al. Genetic variation at brain-derived neurotrophic factor (BDNF) is associated with rapid cycling in a UK bipolar case-control sample of over 3000 individuals. Br J Psychiatry 2006; 188:21–25.
90. Müller DJ, De Luca V, Sicard T, et al. Brain-derived neurotrophic factor (BDNF) gene and rapid cycling bipolar disorder: family-based association study. Br J Psychiatry 2006; 189:317–323.
91. Barrett TB, Hauger RL, Kennedy JL, et al. Evidence that a single nucleotide polymorphism in the promoter of the G protein receptor kinase 3 gene is associated with bipolar disorder. Mol Psychiatry 2003; 8(5):546–557.
92. Kakiuchi C, Iwamoto K, Ishiwata M, Bundo M, et al. Impaired feedback regulation of XBP1 as a genetic risk factor for bipolar disorder. Nat Genet 2003; 35(2):171–175.
93. Cichon S, Buervenich S, Kirov G, et al. Lack of support for a genetic association of the XBP1 promoter polymorphism with bipolar disorder in probands of European origin. Nat Genet 2004; 36(8):783–784.
94. Barden N, Harvey M, Shink E, et al. Identification and characterisation of a gene predisposing to both bipolar and unipolar affective disorders (abstract). Am J Med Genet 2004; 130B(1):122.
95. Kirov G, Murphy KC, Arranz MJ, et al. Low activity allele of catechol-O-methyltransferase gene associated with rapid cycling bipolar disorder. Mol Psychiatry 1998; 3:342–345.
96. Turecki G, Grof P, Grof E, et al. Mapping susceptibility genes for bipolar disorder: a pharmacogenetic approach based on excellent response to lithium. Mol Psychiatry 2001; 6(5):570–578.
97. Jones I, Craddock N. Familiality of the puerperal trigger in bipolar disorder: results of a family study. Am J Psychiatry 2001; 158:913–917.
98. Coyle N, Jones I, Robertson E, et al. Variation at the serotonin transporter gene influences susceptibility to Bipolar affective puerperal psychosis. Lancet 2000; 356:1490–1491.
99. Faraone SV, Glatt SJ, Su J, et al. Three potential susceptibility loci shown by a genome-wide scan for regions influencing the age at onset of mania. Am J Psychiatry 2004; 161(4):625–630.
100. O'Mahony E, Corvin A, O'Connell R, et al. Sibling pairs with affective disorders: resemblance of demographic and clinical features. Psychol Med 2002; 32(1):55–61.
101. Craddock N, Jones I, Kirov G, et al. The bipolar affective disorder dimension scale (BADDS)—a dimensional scale for rating lifetime psychopathology in bipolar spectrum disorders. BMC Psychiatry 2004; 4:19.

102. Potash JB, Zandi PP, Willour VL, et al. Suggestive linkage to chromosomal regions 13q31 and 22q12 in families with psychotic bipolar disorder. Am J Psychiatry 2003; 160(4):680–686.
103. Kraepelin E. Manic-Depressive Insanity and Paranoia (trans. Barclay RM). Edinburgh: Livingstone, 1919.
104. Crow TJ. The continuum of psychosis and its genetic origins. The sixty-fifth Maudsley lecture. Br J Psychiatry 1990; 156:788–797.
105. Gershon ES, Hamovit J, Guroff JJ, et al. A family study of schizoaffective, bipolar I, bipolar II, unipolar, and normal control probands. Arch Gen Psychiatry 1982; 39:1157–1167.
106. Frangos E, Athanassenas G, Tsitourides S, et al. Prevalence of DSM III schizophrenia among the first-degree relatives of schizophrenic probands. Acta Psychiatr Scand 1985; 72:382–386.
107. Baron M, Gruen R, Asnis L, et al. Schizoaffective illness, schizophrenia and affective disorders: morbidity risk and genetic transmission. Acta Psychiatr Scand 1982; 65:253–262.
108. Pope HG Jr, Yurgelun-Todd D. Schizophrenic individuals with bipolar first-degree relatives: analysis of two pedigrees. J Clin Psychiatry 1990; 51:97–101.
109. Tsuang MT, Winokur G, Crowe RR. Morbidity risks of schizophrenia and affective disorders among first degree relatives of patients with schizophrenia, mania, depression and surgical conditions. Br J Psychiatry 1980; 137:497–504.
110. Valles V, Van Os J, Guillamat R, et al. Increased morbid risk for schizophrenia in families of in-patients with bipolar illness. Schizophr Res 2000; 42:83–90.
111. Kendler KS, Karkowski LM, Walsh D. The structure of psychosis: latent class analysis of probands from the Roscommon Family Study. Arch Gen Psychiatry 1998; 55:492–499.
112. Rice J, Reich T, Andreasen NC, et al. The familial transmission of bipolar illness. Arch Gen Psychiatry 1987; 44:441–447.
113. Cardno AG, Rijsdijk FV, Sham PC, et al. A twin study of genetic relationships between psychotic symptoms. Am J Psychiatry 2002; 159:539–545.
114. Berrettini W. Evidence for shared susceptibility in bipolar disorder and schizophrenia. Am J Med Genet 2003; 123C:59–64.
115. Hamshere ML, Bennet P, Williams N, et al. Genomewide linkage scan in schizoaffective disorder: significant evidence for linkage at 1q42 close to DISC1, and suggestive evidence at 22q11 and 19p13. Arch Gen Psych 2005; 62(10):1081–1088.
116. Wang X, He G, Gu N, et al. Association of G72/G30 with schizophrenia in the Chinese population. Biochem Biophys Res Commun 2004; 319:1281–1286.
117. Korostishevsky M, Kaganovich M, Cholostoy A, et al. Is the G72/G30 locus associated with schizophrenia? single nucleotide polymorphisms, haplotypes, and gene expression analysis. Biol Psychiatry 2004; 56(3):169–176.
118. Ekelund J, Hovatta I, Parker A, et al. Chromosome 1 loci in Finnish schizophrenia families. Hum Mol Genet 2001; 10:1611–1617.
119. Ekelund J, Hennah W, Hiekkalinna T, et al. Replication of 1q42 linkage in Finnish schizophrenia pedigrees. Mol Psychiatry 2004; 9(11):1037–1041.
120. Macgregor S, Visscher PM, Knott SA, et al. A genome scan and follow-up study identify a bipolar disorder susceptibility locus on chromosome 1q42. Mol Psychiatry 2004; 9(12):1083–1090.
121. Hodgkinson CA, Goldman D, Jaeger J, et al. Disrupted in schizophrenia 1 (DISC1): association with schizophrenia, schizoaffective disorder, and bipolar disorder. Am J Hum Genet 2004; 75(5):862–872.
122. Stefansson H, Sigurdsson E, Steinthorsdottir V, et al. Neuregulin 1 and susceptibility to schizophrenia. Am J Hum Genet 2002; 71:877–892.
123. Stefansson H, Sarginson J, Kong A, et al. Association of neuregulin 1 with schizophrenia confirmed in a Scottish population. Am J Hum Genet 2003; 72:83–87.
124. Williams NM, Preece A, Spurlock G, et al. Support for genetic variation in neuregulin 1 and susceptibility to schizophrenia. Mol Psychiatry 2003; 8:485–487.

125. Green EK, Raybould R, Macgregor S, et al. The schizophrenia susceptibility gene, neuregulin 1 (NRG1), operates across traditional diagnostic boundaries to increase risk for bipolar disorder. Arch of Gen Psych 2005; 62:642–628.
126. Craddock N, Owen MJ. The beginning of the end for the Kraepelinian dichotomy. Br J Psychiatry 2005; 186:364–366.
127. Jones I, Kent L, Craddock N. Clinical implications of psychiatric genetics in the new millennium—nightmare or nirvana? Psychiatric Bulletin 2001; 25:129–131.
128. Nuffield Council on Bioethics. Mental Disorders and Genetics: the Ethical Context. London: Nuffield Council on Bioethics, 1998.

14 Neurocognitive Findings in Bipolar Disorder

David C. Glahn
Department of Psychiatry and Research Imaging Center, University of Texas Health Science Center at San Antonio, San Antonio, Texas, U.S.A.

Carrie E. Bearden
Semel Institute for Neuroscience and Human Behavior, University of California, Los Angeles, California, U.S.A.

INTRODUCTION

The study of neuropsychological functioning in patients with bipolar disorder (BD) is expanding rapidly. Indeed, there was a threefold increase in the number of peer-reviewed publications on this topic from 2000 to 2005, as compared with the preceding five-year period (Fig. 1). Although significant increase in publication rate is expected in areas of new exploration, the study of neuropsychological functioning in BD is neither new nor dependent upon the development of novel technologies. Rather, this increase in scientific interest is driven by a recognition that poor neuropsychological functioning in patients with BD is at least partially independent of mood state (1–5), that cognitive problems may contribute significantly to lack of full functional recovery from affective episodes (6,7), and that neuropsychological deficits may provide clues into the neurophysiologic and neuroanatomic abnormalities implicated in the pathophysiology of the illness (8,9). While it is unclear at this time how common cognitive impairments are among individuals diagnosed with BD, a significant portion of patients with BD complain of cognitive difficulties (10,11). However, formal neuropsychological deficits have also been documented in asymptomatic patients who do not complain of cognitive difficulties (10,11), indicating that neuropsychological impairments may be more widespread than clinical experience would suggest.

In this chapter, we review the current literature on the cognition of BD. While earlier reviews of this literature have been conducted by our group and others (12–14), the exponential increase in recent publications warrants an updated review and synthesis of the literature. Our review is organized around a set of guiding questions:

- Is neuropsychological dysfunction limited to specific cognitive domains in BD?
- To what extent are impairments explained by current mood state?
- Does the course of illness impact neuropsychological functioning?
- Is neuropsychological dysfunction in BD explained by comorbid psychiatric illnesses (e.g., substance abuse, anxiety)?
- Is impairment secondary to the use of psychotropic medications?

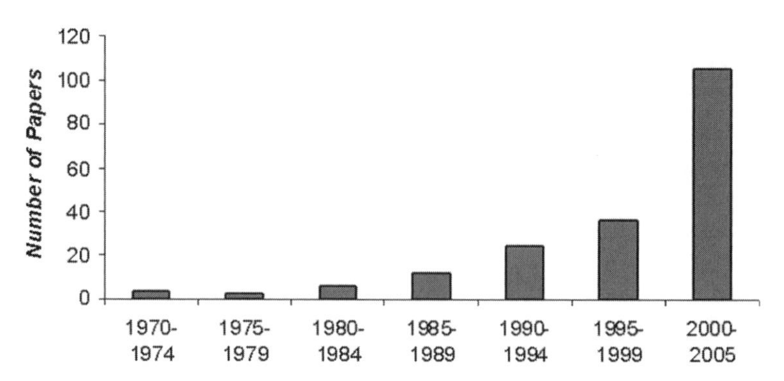

FIGURE 1 The number of peer-reviewed journals involving neurocognitive impairment in bipolar affective disorder.

TAXONOMY OF COGNITION

As more is learned about how information is processed in the brain, parcellation of cognitive processes into theoretically distinct domains becomes increasingly difficult. Current neuroscience models propose that most neuropsychological processes are supported by distributed large scale neural networks with numerous spatially distinct brain regions (15). In such networks, regional specialization is best discussed in terms of a continuum, where several regions may be responsible, to varying degrees, for similar types of information processing (16). Furthermore, there is increasing evidence that specific brain regions (17) and even specific neurons (18,19) can be engaged in putatively distinct cognitive processes.

Despite growing evidence that the organization of the central nervous system is complex and does not easily map on to a simple taxonomy (e.g., memory, attention, problem-solving), the organization of cognitive processes into specific domains or functions has been embraced in clinical and experimental neuropsychology and is a useful way to conceptualize an individual's functional impairment or disability (20). In this chapter we review evidence for neuropsychological dysfunction in BD using organizing principles developed for the measurement and treatment research to improve cognition in schizophrenia (MATRICS) initiative (21). As part of this initiative, a panel of experts was convened to determine the cognitive domains implicated in schizophrenia that are relatively independent of each other. Based on empirical evidence and expert opinion, the panel concluded that treatment research in schizophrenia should focus on the following cognitive domains: attention/vigilance, working memory, verbal learning and memory, visual learning and memory, reasoning and problem solving, and speed of processing (21,22). Although these domains were originally designed to assess cognitive change in schizophrenia, applying them to BD is a natural extension, given that measures assessing these same domains are also commonly applied in studies of cognition in BD patients. Since the intent of this panel was to develop a consensus cognitive performance battery to be used in clinical trials of promising pro-cognitive agents, general verbal ability or intelligence was excluded because of its marked resistance to change (20). As our goal is to review cognitive dysfunction in BD more generally, we will additionally include a discussion of general intellectual ability. Thus, we focus on seven putatively distinct cognitive domains.

ATTENTION/VIGILANCE

Attention is the ability to focus "awareness" to a specific stimulus in the environment and, often, to respond to that stimulus. Although a myriad of potential attentional subsystems have been proposed, one important distinction is that of top-down versus bottom-up processing (23). Top-down processing refers to the flow of information from "higher" to "lower" brain regions (24), conveying knowledge derived from previous experience rather than sensory stimulation in preparing and applying goal-directed selection for stimuli and responses. This system is also modulated by the detection of stimuli (25,26). A complex network of brain systems serves top-down attentional processing, including parts of the intraparietal cortex and superior frontal cortex (27,28). Bottom-up processing refers to information processing that proceeds in a single direction from sensory input, through perceptual analysis, towards motor output, without involving feedback information flowing backwards from "higher" to "lower" cortices (23). This system is specialized for the detection of behaviorally relevant stimuli, particularly when they are salient or unexpected (25). Bottom-up processing is associated with temporo-parietal and inferior frontal cortices (29,30), and is largely lateralized to the right hemisphere.

While attentional dysfunction can denigrate performance on a wide variety of cognitive measures, attention is most often measured with a continuous performance task (CPT) (31). During the CPT, subjects attend to a series of stimuli and respond in some way when a particular stimulus or groups of stimuli appear. CPT tasks involve several aspects of top-down attention including vigilance, rapid encoding of stimuli, response readiness, and stimulus-response mapping. Dependent measures of the CPT include target detection, or how often a subject responds when the appropriate stimulus is present, and false alarms, or how often a subject responds to inaccurate stimuli. Other (less commonly used) attentional measures include the span of apprehension task (SPAN), a sustained attention measure (32); the dichotic listening task (a test of auditory selective attention); and the stroop color and word tests (SCWT), which measure attentional interference (33).

Evidence for Impairment

At the level of clinical observation, patients with BD have difficulty concentrating for extended periods of time and are often quite distractible (34), suggesting that attentional dysfunction is associated with the illness. Consistent with these observations, clinically stable but symptomatic patients with BD are impaired on measures of target detection from the CPT (35–38). This impairment appears to be relatively independent of potential working memory confounds (39).

Visual backward masking measures are associated with bottom-up attentional processes and have been used to investigate early sensory processing in bipolar illness. The prototypical visual backward masking paradigm presents a target stimulus followed quickly (e.g., 20–50 milliseconds) by a masking stimulus designed to remove any sensory trace of the target. The object is to determine the extent to which a subject can process the target stimulus before the mask disrupts ongoing visual processing. Backward masking deficits have been reported in manic bipolar patients, and these impairments persist when mania resolves (40). MacQueen et al. (41) also reported that euthymic bipolar patients were significantly slower and had more errors than healthy controls on two complementary visual backward masking tasks: one requiring subjects to locate a target stimulus, the other requiring identification of a target stimulus.

Effect of Mood State

Several studies have documented an association between poor CPT performance and the severity of manic symptoms (36,42–47). Clark and Goodwin (43) report that while manic patients with BD have impaired target detection and increased rates of false alarms, euthymic patients only show target detection deficits, even after controlling for residual affective symptoms. Elsewhere, Clark et al. (35) report that after statistically controlling for residual symptoms, the only neuropsychological measure that distinguished euthymic patients with BD from matched healthy controls was target detection on the CPT. Given that two additional groups have replicated this finding (3,5), poor target detection may represent a trait marker for BD (13,43). However, currently there is little evidence that unaffected relatives of patients with bipolar illness suffer attentional deficits (48).

Impact of Clinical Course

While most investigators report that attentional processing is negatively correlated with duration or severity of illness (43,49,50), others find no relationship (37). Although hospitalized patients with bipolar illness improve from admission to discharge, their target detection remains significantly worse than the general population (36), suggesting that the extremes of depressive or manic symptoms exacerbate attentional dysfunction already present in these patients. Finally, visual backward masking performance may be sensitive to the total number of depressive episodes (41).

SPEED OF PROCESSING

Speed of processing is the time needed for one to complete a simple cognitive task (20), which often includes encoding some information, making a decision about that information, and formulating and executing a response. The cognitive demands of speed-of-processing tests are relatively simple, involving perceptual and motor components, and emphasize speed of performance. For example, neuropsychological tests of speed of processing assess the speed with which digit/symbol pairings can be completed (Digit Symbol), target symbols can be located (letter or symbol cancellation), number or number/letter sequences on a page can be identified and connected (Trails A/B), and colors can be named (stroop color-word interference test). Although verbal fluency is traditionally thought of as a language or semantic task, since the dependent measure is typically the number of words starting with a given letter that can be generated in a brief time period, it is also conceptualized as a speed-of-processing task (22).

While speed-of-processing measures can be influenced by a host of environmental or illness-related factors and by age-related declines (51), these measures are thought to index neuronal efficiency and to be sensitive to subtle brain dysfunction (52). Current cognitive neuroscience models of processing-speed tasks highlight the timely interplay of a prefrontal decision-making node and posterior heteromodal cortical regions, including parietal and temporal regions (53,54). These models emphasize the integration of information across spatially distinct brain regions, rather than the activity of a specific region, suggesting that cognitive slowing as indexed by processing rate is directly related to neuronal efficiency (52).

Evidence for Impairment

Depending on the specific measure employed, neuropsychological investigations of BD that include measures tapping this construct tend to find evidence for some degree of impairment (14); specifically, on a rapid visual-information processing task (35), the digit-symbol subtest of the Wechsler Adult Intelligence Scale version III (WAIS-III) (5), Trail-making Tests A and B (2,55,56), and verbal fluency measures (3). Moreover, performance on the digit-symbol subtest discriminated patients with BD with a family history of psychosis from bipolar patients without history of psychosis (57).

Slow processing speed can reduce neuropsychological performance on a wide variety of tests (58), particularly learning and memory measures (59,60). Speed-of-processing impairments have also been reported in unipolar depression, and may mediate memory deficits found in these patients (61). Kieseppa et al. found that reduced speed of processing is related to poor verbal learning and memory in euthymic patients with BD (62). However, others have not replicated this finding (63).

Effect of Mood State

Clinical observation would suggest that speed of processing increases in mania and decreases with depression. Depressed mood is related to impairment on speed-of-processing tests in both unipolar (61) and bipolar disorders (9,64). Contrary to clinical intuition, manic or hypomanic patients also show cognitive slowing on neuropsychological measures (9,64–66). This may be because manic patients have difficulty focusing on the task at hand, and thus perform more slowly. Euthymic bipolar patients are significantly impaired on the digit-symbol substitution test (5), a classic speed-of-processing measure. In contrast, Martinez-Aran et al. (3) reported that depressed bipolar patients generated fewer words than manic or euthymic bipolar patients, who did not differ from comparison subjects, suggesting that verbal fluency may be influenced by mood state. Together, these data suggest that speed of processing is impaired in BD; however, the specific neuropsychological measure employed is important, and there are no consistent findings across mood state.

Impact of Clinical Course

Denicoff and colleagues (49) report that with increasing illness duration, performance on a timed letter-cancellation task decreased significantly. Similarly, with increasing number of manic episodes, patients' performance on the letter fluency test is reduced (67). Others have reported poorer performance on processing speed measures to be associated with both overall illness duration (3,5) and more hospitalization episodes (56), suggesting that both duration and severity of illness may reduce speed of processing. While it is tempting to posit that history of psychotropic medication usage is associated with reduced processing speed, Clark and coworkers (35) found that current or past medication usage did not impact this measure. Furthermore, when this sample was limited to those patients not on medications ($n = 11$), patients were still significantly slower than healthy comparison subjects.

WORKING MEMORY

The information-processing capacity of the central nervous system is limited, necessitating a time-restricted approach to the processing of ambient sensory information (15,68,69). One conceptualization of working memory is that it provides the temporary storage facilities needed to prolong the neuronal response to a brief sensory event beyond the duration of the actual event or its iconic (or echoic) representation, and thus allows for 'higher' cognitive processes such as reasoning, planning, language, and other forms of abstract thought (69a,70). Baddeley proposed a tripartite organization of working memory consisting of a central executive component, which controls the manipulation of information held in memory and the distribution of finite processing resources (e.g., attention), and two slave systems, the articulatory loop and visuospatial scratch pad, which maintain mental representations of verbal and visuospatial information, respectively (71,72). Generally, the results of neuroimaging studies are consistent with the functional and anatomical segregation of the storage aspects of working memory into separate systems specialized for verbal and spatial information (73–77). These systems appear to follow the general pattern of hemispheric specialization, with verbal material maintained primarily in left hemisphere regions and spatial material maintained primarily in right hemisphere regions (73) and with segregated regions for passive storage versus active rehearsal within each domain (77). Executive or manipulation processes rely on prefrontal processing, specifically in the dorsolateral prefrontal cortex (74,77).

Evidence for Impairment

Neuropsychological measures of working memory include those designed to assess maintenance [e.g., delayed response task (78)] and manipulation [e.g., letter-number sequencing (79) or n-back paradigms (80)] with either verbal or visual spatial information. Several papers report that patients are not impaired on the forward condition of the digit span subtest of the WAIS-III (3,81,82), a measure associated with verbal memory capacity and attention. Using the CANTAB computerized neurocognitive test battery, Rubinsztein et al. (83) found that clinically remitted bipolar patients showed impairment on tests assessing recognition for geometric patterns and spatial locations (Table 1). Using the same tests in a larger sample of euthymic bipolar patients, Thompson et al. (5) observed impairment only on spatial (but not pattern) recognition memory, with additional deficits in paired associate learning, a measure assessing memory for location of specific visual patterns, and on a delayed (but not simultaneous). Matching to Sample task, suggesting difficulty with the memory component of this task. In addition, while Ferrier et al. (2) described significant impairment in delayed visuospatial memory in remitted bipolar patients, this measure was no longer significantly impaired after covarying for affective symptoms. While some reports suggest that bipolar patients are not impaired on spatial delayed match to sample tests (35,82,84,85), others find spatial maintenance impairments, particularly in the manic phase of the illness (9,86). Discrepant results may be due to heterogeneity across patient samples, or differences in the specific task demands of the working memory measures employed. Indeed, Glahn et al. found that all clinically stable but symptomatic patients with BD were impaired on a verbal manipulation task [backwards digit span subtest of the WAIS-III (The psychological Corporation 1997)]; however, only patients with a lifetime history of psychosis were impaired on a spatial delayed response task (81).

Many investigators report dysfunction on tasks requiring on-line manipulation of information in patients with BD (3,5,83,87,88); however this finding

TABLE 1 Visual Spatial Memory Tasks

Visual-spatial memory task	Description
Pattern recognition test (85a)	12 abstract colored visual patterns are presented sequentially. In the recognition phase, the same patterns (each paired with a novel pattern) are presented in the reverse order and subjects are asked to respond by touching the pattern that they have already seen.
Spatial recognition test (130)	Five squares are presented sequentially in different locations. In the recognition phase, each square is presented again, now paired with a novel location. Subjects must touch the correct location of the box they had been presented with earlier.
Rey-Osterreith complex figure test (124)	The subject copies a complex figure and then, after a delay, reproduces the figure from memory.
Simultaneous and delayed matching to sample task (130)	This task assesses subjects' ability to recognize complex visual designs after different delay intervals (0, 4, and 12 sec). Subjects are shown a target at screen center, and after the delay interval four surrounding stimuli are presented.
Pattern recognition memory task (130)	After a series of patterns is displayed, subjects are presented pairs of patterns, one shown previously and one being a novel pattern. They are asked to indicate the pattern they were shown previously.
Visual-spatial paired associate learning task (130)	Designs are presented in boxes on the screen at varying locations. The designs are then presented sequentially in the center of the screen and subjects are instructed to indicate the box in which each design was initially presented.
Spatial recognition memory task (130)	Five squares are presented in sequence at different locations on the screen, and then subjects are presented a pair of squares and asked to identify which is at a location where a square was previously presented.

is not universal (9,35). To date, few studies have employed the same measures of cognitive manipulation, making it difficult to interpret inconsistent results.

Effect of Mood State

Most reports of clinically symptomatic (82,84) and symptom-free (35) patients with BD do not show evidence for impaired short-term maintenance of information in BD. However, Sweeney et al. (9) found that, with increasing delay interval, both depressed and mixed/manic patients with BD were impaired, and that with a 12 second delay, the mixed/manic patients performed significantly worse than depressed patients with BD. As most reports of working memory maintenance did not include such long delay intervals, it is difficult to determine if these findings are at odds with other reports. In contrast, working memory tests that require the manipulation of remembered information appear to be impaired across phases of illness to a comparable degree (3,5,50), but see (9).

Impact of Clinical Course

Rubinsztein and associates (83) reported a significant association between the number of months hospitalized and visual working memory impairment. Thompson et al. (5)

documented significant relationships between total number of hospitalizations and spatial working memory. In contrast, Clark et al. (35) reported no significant relationships with the total number of hospital admissions, but found that the number of admissions specifically for depressive episodes was positively related to errors on a spatial working memory task. While Frangou and coworkers (143) report that increased duration of illness impacts executive functioning, others report that duration of illness and/or number of hospitalizations does not significantly affect working memory performance in bipolar patients (81,89). At issue may be the exact measures used, as Frangou and colleagues (143) only found an association between duration of illness and executive dysfunction on just one of several working memory measures employed. Current use of antipsychotic medications was associated with poorer manipulation of information held online in BD (50).

LEARNING AND MEMORY: VERBAL AND VISUAL

The explicit recall of previously learned information is known as declarative memory. In contrast to working memory, which involves the ability to maintain and/or manipulate information over a brief time period, this process relies on the ability to adequately encode, store, and retrieve information from long-term memory (90,91). Encoding refers to the process that converts a perceived event into a lasting neurophysiological trace (91), while retrieval involves reactivation of a stored representation, leading to an explicit 'memory' of the event. These cognitive functions (encoding, storage, and retrieval) are thought to be subserved by distinct brain regions. In particular, the temporal-hippocampal system has been demonstrated by both animal and human work to play a critical role in both encoding and retrieval of verbal information (92–94), while strategic or executive aspects of memory are heavily dependent upon prefrontal cortical function (95,96).

In studies of adult lesion patients, there appears to be a distinction between visual and verbal memory circuits that mirrors the hemispheric organization of the rest of the brain; that is, lesions to the right hippocampus or right temporal lobes have been shown to impair memory of visual and spatial information, including recall of complex geometric figures, paired-associate learning of nonsense figures, and performance on maze-learning, face-recognition, and spatial memory span tasks (97–101), while left-sided lesions appear to differentially impair memory for verbal information (99,100,102,103). Such findings have been validated by neuroimaging studies (104–106).

Evidence for Impairment: Verbal Learning and Memory

Verbal declarative memory impairments are among the most consistently reported cognitive difficulties in patients with BD (7,63,65,88,89,107–113). Given that these deficits have been observed in euthymia (2,3,5), have also been observed in unaffected relatives of patients with bipolar illness (48,114), and may lead to clinically significant impairment in everyday functioning (1,2,4,6), declarative memory dysfunction may represent a vulnerability marker for bipolarity.

Despite the potential importance of memory deficits for psychosocial and occupational outcome of patients with bipolar illness (6), the nature of these impairments is not well understood. There is some evidence that poor memory performance is secondary to strategic or organizational dysfunction rather than impaired memory processes per se (110); given that executive functions (i.e., the ability to plan and reason, and inhibit behaviors) are known to play a significant role in learning and memory performance (115), and are frequently shown to be impaired in bipolar patients (see discussion above), this seems unsurprising. However, few studies have directly investigated the role of learning strategies in memory performance in bipolar illness; while Deckersbach et al. (110) found that recall difficulties in euthymic bipolar patients were partially accounted for by poor semantic clustering strategies, our group did not replicate this finding (63). Rather, we find that verbal declarative memory deficits are more consistent with encoding impairments in adults and children with BD (63,116).

Effect of Mood State: Verbal Learning and Memory

Sweeney et al. (9) identified widely distributed cognitive deficits in mixed/manic bipolar patients, while depressed bipolar and unipolar patients demonstrated impairments only on an episodic memory test, suggesting a more selective dysfunction in mesial temporal lobe function during depressive episodes. Moreover, Henry et al. (117) found that bipolar patients in the manic state displayed poorer performance on a verbal learning task relative to their own performance when clinically remitted. However, they were not impaired on short-term free recall, regardless of mood state, suggesting that manic or depressive symptomatology may affect complex processing or memory functions, but not simpler cognitive processes (117).

Using the California verbal learning test (CVLT) (118), two studies of acutely ill inpatients in various mood states demonstrated significantly decreased recognition performance relative to healthy comparison subjects (65,108). Several studies using similar verbal learning tests have found that euthymic patients have normal performance on recognition measures (3,5,88,107,111,112), suggesting that verbal recognition deficits may be state-related, rather than a core deficit of BD (119). In contrast to recognition memory, free recall deficits appear to be relatively independent of mood state (3,63), suggesting that retrieval of verbal information from memory may reflect a trait-related impairment in patients with BD.

Impact of Clinical Course: Verbal Learning and Memory

Verbal memory performance may be adversely affected by severity of illness, as measured by number of hospitalizations, number and duration of manic and/or depressive episodes, and age at onset (35,56,109,120). Five out of six studies found at least one significant correlation between number of manic episodes and poorer verbal memory performance (3,35,109,110,113). In particular, four of these investigations found a relationship between past manic episodes and poorer delayed verbal memory (recall or recognition of a word list after a delay) (121). However, other studies have found no relationship between illness variables and memory performance (1,122).

Although one study found the number of depressive episodes to be inversely correlated with verbal memory performance (110), four others failed to find a relationship between number of depressive episodes and verbal memory performance (3,5,109,112). Martínez-Arán et al. (3) found a significant relationship between the total number of hospitalizations and poorer verbal memory performance. However, there is little evidence that age at illness onset bears relationship to verbal declarative memory performance, with several studies showing no relationship (3,5,119,123). Similarly, of the few existing studies that have examined length of time euthymic in relation to cognitive function (5,35,87), none have found an association between these variables.

Evidence for Impairment: Visual Learning and Memory

In general, there is less consistent evidence for impairment in nonverbal memory processes in patients with bipolar illness than exists for the verbal domain (12,14). However, this may be attributable to fact that relatively few studies have examined aspects of visual and spatial memory in bipolar patients, and those that have rarely employed the same measures, leading to difficulty synthesizing information across studies. Two studies found euthymic bipolar patients to be impaired on the Rey-Osterrieth complex figure test (ROCFT) (124), a nonverbal memory test that allows assessment of organizational strategies during learning (107,119). Furthermore, Deckersbach et al. (119) found that, compared with control participants, euthymic bipolar-I patients relied less on organizational strategies during encoding on this test. However, other studies have reported no impairment on this task (125,126). In addition, visuo-constructive ability, as measured by the WAIS Block Design task, appears unimpaired in euthymic bipolar patients (113,127).

Effect of Mood State: Visual Learning and Memory

There is increasing evidence that nonverbal memory impairment is not purely a function of mood state (107,119). However, visual and spatial memory impairments may be exacerbated during the mixed/manic phase of illness, and thus are not entirely state-independent (9).

Impact of Clinical Course: Visual Learning and Memory

One study reported a relationship between the number of both manic and depressive episodes and increasing impairment on a visual memory task, the ROCFT (110), although others found no significant relationships between manic episodes and performance on computerized measures of visuospatial memory (41,83). Thompson et al. (5) additionally noted poorer performance on the Paired Associates Learning task in those with longer duration of illness. Of studies that have instead examined the total number of mood episodes, regardless of polarity, El-Badri and colleagues (87) observed that number of episodes was negatively related to visual memory performance. However, other studies have found no relationship between number of lifetime mood episodes and both verbal and visual memory performance (88). Furthermore, Ferrier and co-workers categorized patients based on number of episodes (no more than two major mood episodes in the past five years, vs. three or more major mood episodes) and found no differences between the groups on any neuropsychological measure.

REASONING AND PROBLEM SOLVING

The MATRICS group notes that the label Reasoning and Problem Solving has the advantage of distinguishing this domain from the executive processes of working memory (128) and demonstrate that this domain is empirically dissociable (based on factor analyses) from working memory (22). That being said, many of the tasks included in the reasoning and problem-solving domain are often discussed as measures of executive functioning. The primary differences between 'executive' working memory and reasoning and problem-solving tasks is that working memory tasks tend to have explicit memory demands (e.g., n-back paradigms) or are designed to determine working memory capacity (e.g., letter-number sequencing).

Cognitive measures conceptualized as Reasoning and Problem Solving include sorting cards using an abstract principle that changes over time [e.g., Wisconsin Card Sorting Test (WCST) (129), ID/ED shift (130)], nonverbal reasoning to complete a sequence of visual patterns [e.g., Matrix Reasoning (131) or the Raven's Progressive Matrices (131a)] or to construct a visual pattern [e.g., Block Design (131)], moving round disks between pegs in the smallest number of steps to achieve a specific order [e.g., Tower of London (130)], and similar verbal and nonverbal problem-solving tasks (Table 2). Such higher-level cognitive processes often demand relatively intact lower-level processes, but also involve additional complex strategic planning and decision-making skills. While these tasks involve a myriad of cognitive processes, they tend to be highly correlated and are often described as 'frontal lobe' measures (132). Human functional neuroimaging studies demonstrate that reasoning and problem-solving tasks engage a wide range of brain regions, including prefrontal, temporal, parietal, occipital,

TABLE 2 Reasoning and Problem Solving Findings in Bipolar Patients

Article	Task	Finding
Altshuler et al., 2004	Wisconsin card sort	Clinically stable bipolar patients (n = 40) had more perseverative errors and fewer categories than healthy subjects (n = 22)
Frangou et al., 2005	Wisconsin card sort	Euthymic bipolar patients (n = 44) did not differ from healthy subjects (n = 44)
Martinez-Aran et al., 2004b	Wisconsin card sort	Depressed (n = 30), manic/hypomanic (n = 34), and euthymic (n = 44) bipolar patients had more perseverative errors than healthy subjects (n = 30)
Zubieta et al., 2001	Wisconsin card sort	Euthymic bipolar patients (n = 15) had more perseverative errors and fewer correct responses than healthy subjects (n = 15)
Clark et al., 2001	ID/ED shift	Manic bipolar patients (n = 15) had more reversal and extradimensional shifting errors than healthy subjects (n = 30)
Clark et al., 2002	ID/ED shift	Euthymic bipolar patients (n = 30) had more extradimensional shifting errors than healthy subjects (n = 30)
Olley et al., 2005	ID/ED shift	Euthymic bipolar patients (n = 15) did not differ from healthy subjects (n = 13)
Glahn et al., 2006a	Progressive matrices	Bipolar outpatients in various clinical states (n = 60) did not differ on a standard progressive matrices from healthy comparison subjects (n = 60)
Badcock et al., 2005	Tower of London/ stockings of Cambridge	Manic bipolar patients (n = 14) did not differ from healthy comparison subjects (n = 33)
Rubinsztein et al., 2000	Tower of London	Euthymic bipolar patients (n = 18) took longer to provide correct responses than healthy comparison subjects (n = 18)
Thompson et al., 2005	Tower of London	Euthymic bipolar patients (n = 63) were required more moves and longer times than healthy comparison subjects (n = 63)
Olley et al., 2005	Tower of London/ stockings of Cambridge	Euthymic bipolar patients (n = 15) did not differ from healthy subjects (n = 13)
Clark et al., 2002	Tower of London	Euthymic bipolar patients (n = 30) did not differ from healthy subjects (n = 30)

and cerebellar cortices (133–135). Although these data suggest that reasoning and problem solving tasks evoke large-scale cortical networks, Duncan and Owen (132) note that three prefrontal regions are consistently recruited: the mid-dorsolateral, mid-ventrolateral, and dorsal anterior cingulate regions.

Evidence for Impairment

Table 2 presents findings from four prototypical Reasoning and Problem Solving tests. Patients with BD appear to be impaired on specific reasoning and problem-solving measures (e.g., WCST), but not on other measures (e.g., progressive matrices). On other measures like the ID-ED shift task and the Tower of London (Stockings of Cambridge) task, findings are mixed, with some investigators reporting impairments and others not (Table 2). The inconsistency of findings suggests that the reasoning and problem-solving domain includes clearly dissociable cognitive processes, only some of which may be disrupted in BD. Patients with BD consistently make more perseverative errors than healthy comparison subjects on the WCST, suggesting difficulties with cognitive flexibility (3,107,123). However, findings on the ID-ED test, which is conceptually very similar to the WCST, are mixed, indicating that other cognitive factors (e.g., strategy or poor inhibition) may lead to at least some of the observed impairments on the WCST. Alternately, differences in the composition of the patient and comparison samples studied may lead to discrepant results.

While most investigators find that planning, as indexed by the Tower of London task, is impaired in BD (5,83), others find no evidence for impairment (35,136). Olley et al. (137) found that patients with BD trended towards impairment on this measure ($p = 0.068$).

Patients with BD do not appear to be impaired on progressive matrices. As these measures are closely related to general intellectual processing and given that there is little evidence that bipolar patients have reduced intellectual abilities (12,14), it is unsurprising that performance on progressive matrices is intact (138).

Effect of Mood State

While several investigators propose that cognitive flexibility (83,139) and planning (65) are markedly more impaired in mania, others report no differences between depressed, manic, or euthymic patients on these same measures (3). In one of the more detailed examinations of a reasoning and problem-solving test, Murphy and colleagues (64) measured decision making in manic and depressed bipolar patients and healthy comparison patients. To do so, they employed an experimental risk-taking task previously shown to engage inferior and orbital prefrontal cortex (140) where subjects were asked to win as many points as possible by choosing outcomes based on variably-weighted probabilities and by placing 'bets' on each decision. Depressed and manic patients were impaired on this task (64), as evidenced by slower deliberation times, a failure to accumulate as many points as controls, and suboptimal betting strategies. Interestingly, manic, but not depressed, patients made suboptimal decisions this measure. However, manic patients were not impaired on a conceptually similar gambling task (65).

Impact of Clinical Course

Given the variability of findings discussed above, it is difficult to draw clear conclusions concerning the effects of clinical course on performance on reasoning and problem-solving tests in BD. Clark et al. (35) report that Tower of London performance was negatively correlated with duration of illness in euthymic patients, potentially explaining some of the variable results reported in the literature on this measure. In addition, there is some evidence that rule learning and cognitive flexibility (as measured by the Wisconsin Card Sort) are impaired prior to illness onset and impairment on this measure in children at risk for BD predicts latter illness onset (141). Furthermore, performance on this measure worsens with increasing numbers of manic or depressive episodes (123) and longer duration of illness (3).

GENERAL INTELLECTUAL FUNCTIONING

Given that the frontal lobes occupy 30% to 40% of the neocortex (142), many theorists ascribe intellectual capabilities to frontal lobe function. To date, the cognitive neuroscience of general intellectual function is underdeveloped, largely because there is little evidence to suggest regional specificity in brain regions underlying these functions (132,134). Nevertheless, some researchers have proposed that working memory capacity, or the capability for executive attention, is the psychological core of the construct of general intelligence, or "g" (143). Because the dorsolateral prefrontal cortex (DLPFC) is critical to working memory capacity, normal individual differences in both working memory capacity and "g" may be mediated by individual differences in DLPFC function (132,142).

Evidence for Impairment

There is little evidence for impairment in global cognitive abilities in patients with bipolar illness, particularly during periods of euthymia (12,14). Although early studies reported relatively lower nonverbal or performance IQ (PIQ) than verbal IQ (VIQ) in bipolar patients (144,145), both verbal and performance IQ were well within the normal range. In addition, because the majority of these earlier studies did not describe the patients' clinical state at the time of testing, it is not clear what impact current clinical symptomatology may have had on IQ performance. Further, in contrast to patients with schizophrenia, epidemiologic studies offer little evidence for impairment in premorbid intellectual ability in patients with BD (146,147).

Effect of Mood State

In one of the few test-retest investigations of IQ in BD, Donnelly and colleagues found evidence that the same patients, when in a hypomanic or euthymic state, have higher IQs than when depressed (148). This finding was not attributable to practice effects, as the patients' IQ scores declined again when retested in the depressed state. However, this finding has yet to be replicated with patients meeting current diagnostic criteria for bipolar illness. Little is currently known about the effects of clinical course on IQ.

COMORBID PSYCHIATRIC ILLNESSES

While the effects of comorbid attention deficit hyperactivity disorder (ADHD) have been investigated in the small number of studies examining neurocognitive function in pediatric bipolar patients (149,159), cognitive differences as a function of psychiatric comorbidity have only been examined in a few studies of adult patients with BD. In particular, van Gorp et al. (113) found that bipolar patients both with and without alcohol dependence performed more poorly than controls on tests of verbal memory; however, bipolar subjects with a history of alcohol dependence had additional decrements in executive functions as compared with controls. While our group also found that current substance abuse was associated with poorer verbal learning and memory in bipolar patients, this association did not reach statistical significance after correcting for multiple comparisons (63). In addition, comorbid anxiety disorder did not affect verbal memory performance. Given that 30% to 50% of patients with BD have comorbid substance abuse (151), as well as very high rates of other comorbidities, particularly anxiety and personality disorders (152), the impact of comorbid disorders on cognitive function in bipolar illness clearly warrants further investigation.

While gender differences have rarely been investigated in studies of cognition in BD, there is some evidence that gender may modulate the severity of cognitive deficits in BD, with male patients demonstrating poorer neurocognitive performance (9). Other clinical factors, particularly history of psychosis, may adversely impact cognitive function (81), though systematic, large-scale investigations of the effect of psychosis on neuropsychological functioning in BD have not yet been published.

EFFECTS OF PSYCHOTROPIC MEDICATIONS

Although current and past use of psychotropic medications could impact neurocognitive functioning, systematic investigation of the cognitive impact of these agents in patients with BD has been limited. A prior qualitative review concluded that while lithium had a negative effect on memory and speed of information processing, patients were often unaware of these deficits (153). Although Kessing et al. (120) found both the number of episodes and length of lithium treatment to be inversely correlated with performance on two of five tests of global cognitive function, Engelsmann and coworkers (154) found that mean memory test scores remained quite stable over a six-year interval in lithium-treated bipolar patients. Further, there were no significant differences between patients with short- versus long-term lithium treatment on any measure, after controlling for age and initial memory scores, suggesting that long-term lithium usage is unlikely to cause progressive cognitive decline (154).

Some antidepressant medications have been shown to have adverse cognitive effects, particularly those with anticholinergic properties (155). Although evidence to date does not indicate cognitive side effects of selective serotonin reuptake inhibitors (e.g., paroxetine, sertraline), the long-term impact of these medications on cognition is not yet known (155). While few studies have examined neurocognitive performance in unmedicated bipolar patients, we previously found comparably impaired verbal memory in patients receiving psychotropic medication ($n = 32$) and those who were drug-free ($n = 17$) (63). Further, Strakowski and colleagues recently studied euthymic, unmedicated bipolar subjects with functional MRI, while performing a sustained attention task, the continuous performance task-identical pairs version (CPT-IP) (156,157). Despite comparable task performance to controls, the bipolar patients showed a different pattern of brain activation, involving overactivation of anterior limbic areas, with corresponding abnormal activation in visual associational cortical areas. Taken together, these findings suggest that cognitive deficits, and underlying abnormalities in neuronal activation, in patients with bipolar illness are not primarily attributable to the use of psychotropic medications. However, large-scale, longitudinal investigations of bipolar patients on different medication regimens are warranted to fully address this question.

CONCLUSIONS

Table 3 summarizes our findings of cognitive dysfunction in BD. In rather broad strokes, we find evidence for neuropsychological impairment in six of the seven cognitive domains reviewed.

However, this impairment is often subtle, or presents only specific measures of a cognitive construct and not others. Indeed, by closely examining the types of tests on which patients with BD show impairment, we conclude that the pattern of deficits suggests dysfunction in prefrontal and medial temporal lobe circuits. Evidence for frontal lobe dysfunction includes poor top-down attention as indexed by CPT tests, reduced speed-of-processing measures that include decision making (e.g., digit-symbol coding), impairment on working memory tasks requiring manipulation of information, and reduced cognitive flexibility and planning abilities. Evidence for temporal lobe dysfunction includes poor recall of verbal and nonverbal information. At this time, it is impossible to determine if these impairments are independent or interactive with one another or even if these impairments are seen in the same patients. Indeed, it is possible that patients with different pathologies are included in the overarching bipolar diagnostic category, and that neuropsychological measures may be useful in dissociating subgroups of patients. While the current review cannot address this issue, as sample sizes increase, such segregation analyses become possible. Such analyses could potentially delineate different groups of bipolar patients with distinct neuropsychological profiles, and potentially with different functional outcomes or treatment responses.

Relatively few of the cognitive deficits reported in BD are entirely explained by mood state at the time of assessment. In general, acute manic or depressive symptoms appear to intensify neuropsychological impairments found in euthymia. However, there are a number of notable exceptions to this generalization. For example, recognition of verbal memoranda is seemingly disrupted in manic and depressed patients but generally intact in euthymic patients. Other neuropsychological measures, particularly those involving verbal or semantic fluency, are reduced in bipolar patients when depressed, but not in manic or euthymic patients. Yet, deficits on most measures requiring cognitive flexibility or planning, vigilance, manipulation of information held on line, or recall of verbal or visual information appear to be present regardless of mood state. It should be noted that very few longitudinal neuropsychological studies have been conducted with bipolar patients and that within subject measurements are needed to definitively address this issue.

The impact of clinical course is difficult to estimate. Most current investigations correlate the number of manic or depressive episodes or total duration of illness or hospitalizations with cognitive measures. However these indices of clinical course are often dependent upon accurate patient report. For example, the number of manic or depressive episodes is difficult to determine in patients with mixed manic and depressive symptoms, and does not provide an indication of how long these symptoms persisted, or the intensity of these symptoms. Duration of illness is difficult to accurately determine, given that illness onset is often subtle

TABLE 3 Neuropsychological Findings in Bipolar Disorder

Cognitive domain	Impairment in bipolar disorder	Effect of mood state	Impact of clinical course
Attention	Consistent evidence for top-down and bottom-up impairments	Manic and depressive mood states exacerbate poor attentional processing found in euthymic patients	Poor clinical course is associated with increased impairment
Speed of processing	Evidence for consistent deficits on measures other than verbal fluency. Verbal fluency impairments reports in some studies. Impairments may be associated with memory difficulties.	Depressed and manic patients are impaired on numerous measures. Euthymic patients are impaired on the digit-symbol test	Poor clinical course is associated with increased impairment
Working memory	Little evidence for impaired maintenance of information; significant evidence for manipulation/ executive dysfunction.	Depressed, manic and euthymic patients are impaired on manipulation/ executive measures.	Reports are mixed whether duration of illness is associated with poor manipulation/ executive processing
Verbal learning and memory	Consistent evidence for impairment in recall of verbal information	Recognition impairments are linked to mood state, recall deficits are not	Number of manic episodes and number of hospitalizations are negatively correlated with memory performance
Visual learning and memory	Although there are relatively few studies, consistent evidence for impairment	Euthymic patients are impaired on measures of visual recall	Unclear
Reasoning and problem solving	Impairment on specific problem solving or planning measures, but not progressive matrices	Manic and depressive mood states may exacerbate poor problem solving performance found in euthymic patients	Duration of illness and number of manic episodes are negatively correlated with cognitive flexibility and planning performance
General intellectual functioning	Little evidence for impairment	Unknown	Unknown

and the measure does not provide an indication of clinically significant mood changes. Depending on health care availability, the number of hospitalizations could be biased by the socioeconomic status of the patient rather than his or her degree of psychiatric distress. Perhaps a better measure of clinical course would be the number of days well (or sick) in a given time period. However, this measure is subject to a number of the same concerns, particularly in cross-sectional studies. These difficulties point to the importance of obtaining collateral information (i.e., hospital records, informant reports) to more accurately quantify history of illness variables. Despite the difficulties with standard indices of clinical course, the literature to date generally indicates that patients with more severe clinical presentations tend to have poorer memory functioning and cognitive flexibility. It is unclear whether these patients truly represent a "severe" subgroup distinct from patients with less severe clinical course, or if the increased impairment in those patients with a more chronic, severe illness course are secondary to different medication regimens or other clinical or behavioral factors. In addition, while early studies often failed to characterize current mood state in cognitive studies of bipolar illness, researchers are increasingly employing rigorous definitions of euthymia (based not only on current symptom ratings, but length of time euthymic) to characterize patient samples. This practice should become the "gold standard" in all future studies of neurocognition in bipolar illness.

Patients with BD are often prescribed multiple psychotropic medications as well as other nonpsychiatric agents. This practice makes determining the effects of a single medication on cognitive processing difficult. There are a number of studies currently underway on medication-naïve or -free subjects and we should soon have a better understanding of the neuropsychological function of BD independent of psychotropic medications. Ultimately, the study of bipolar patients alone cannot determine whether neurocognitive deficits are the result of the underlying, pre-existing pathophysiology or the result of the illness itself and related factors (such as sub-syndromal symptoms, acute or chronic medication effects, or permanent structural changes wrought by prior episodes of acute illness). One potential way to address these issues is to study those without overt symptom expression, but with high genetic risk for BD. Although several manuscripts focusing on unaffected siblings and relatives of bipolar patients have been published in the last year (48,62,114,125,158–160), to date the majority of these studies include relatively small sample sizes, limiting potential conclusions. However, current findings do tend to corroborate findings in euthymic patients of subtle learning and memory deficits (62,114,125) and possibly impaired planning and attentional function as well (48,161,162). Further study of such at-risk individuals will elucidate neurocognitive and neurobiological vulnerability markers that can provide important clues into the underlying pathophysiology of the illness.

Cognitive impairment in patients with bipolar illness is increasingly understood as both persistent during symptom-free periods, and to be associated with functional outcome. Improving our understanding of the clinical implications of such neurocognitive dysfunction, and its effects on psychosocial and occupational functioning, is of primary importance in improving long-term outcome for this highly disabling mental illness.

REFERENCES

1. Atre-Vaidya N, Taylor MA, Seidenberg M, Reed R, Perrine A, Glick-Oberwise F. Cognitive deficits, psychopathology, and psychosocial functioning in bipolar mood disorder. Neuropsychiatry Neuropsychol Behav Neurol 1998; 11:120–126.
2. Ferrier IN, Stanton BR, Kelly TP, Scott J. Neuropsychological function in euthymic patients with bipolar disorder. Br J Psychiatry 1999; 175:246–251.

3. Martinez-Aran A, Vieta E, Reinares M, et al. Cognitive function across manic or hypomanic, depressed, and euthymic states in bipolar disorder. Am J Psychiatry 2004b; 161:262–270.
4. Scott J. Psychotherapy for bipolar disorder. Br J Psychiatry 1995; 167(5):581–588.
5. Thompson JM, Gallagher P, Hughes JH, et al. Neurocognitive impairment in euthymic patients with bipolar affective disorder. Br J Psychiatry 2005; 186:32–40.
6. Dickerson FB, Boronow JJ, Stallings CR, Origoni AE, Cole S, Yolken RH. Association between cognitive functioning and employment status of persons with bipolar disorder. Psychiatric Services 2003; 55:54–58.
7. Martinez-Aran A, Vieta E, Colom F, et al. Cognitive impairment in euthymic bipolar patients: implications for clinical and functional outcome. Bipolar Disord 2004a; 6:224–232.
8. Soares JC. Contributions from brain imaging to the elucidation of pathophysiology of bipolar disorder. Int J Neuropsychol (CINP) 2003; 6(2):171–180.
9. Sweeney JA, Kmiec JA, Kupfer DJ. Neuropsychologic impairments in bipolar and unipolar mood disorders on the CANTAB neurocognitive battery. Biological Psychiatry 2000; 48:674–685.
10. Burdick KE, Endick CJ, Goldberg JF. Assessing cognitive deficits in bipolar disorder: are self-reports valid? Psychiatry Res 2005; 136:43–50.
11. Martinez-Aran A, Vieta E, Colom F, et al. Do cognitive complaints in euthymic bipolar patients reflect objective cognitive impairment? Psychother Psychosom 2005; 74:295–302.
12. Bearden CE, Hoffman KM, Cannon TD. The neuropsychology and neuroanatomy of bipolar affective disorder: a critical review. Bipolar Disord 2001; 3:106–150; discussion 151–153.
13. Glahn DC, Bearden CE, Niendam TA, Escamilla MA. The feasibility of neuropsychological endophenotypes in the search for genes associated with bipolar affective disorder, 2004; Vol 63.
14. Quraishi S, Frangou S. Neuropsychology of bipolar disorder: a review. J Affect Disord 2002; 72(3):209–226.
15. Mesulam MM. From sensation to cognition. Brain 1998; 121:1013–1052.
16. O'Reilly RC, Braver TS, Cohen JD. A biologically based computational model of working memory. In Miyake A, Shah P, et al. (eds), Models of working memory: Mechanisms of active maintenance and executive control. New York, NY, USA: Cambridge University Press, 1999:375–411.
17. Price CJ, Friston KJ. Degeneracy and cognitive anatomy. Trends Cogn Sci 2002; 6: 416–421.
18. Miller EK. The prefrontal cortex: complex neural properties for complex behavior. Neuron 1999; 22:15–17.
19. Miller EK, Cohen JD. An integrative theory of prefrontal cortex function. Annu Rev Neurosci 2001; 24:167–202.
20. Lezak M. Neuropsychological Assessment, 3rd ed. New York: Oxford University Press, 1995.
21. Green MF, Nuechterlein KH, Gold JM, et al. Approaching a consensus cognitive battery for clinical trials in schizophrenia: the NIMH-MATRICS conference to select cognitive domains and test criteria. Biol Psychiatry 2004; 56:301–307.
22. Nuechterlein KH, Barch DM, Gold JM, Goldberg TE, Green MF, Heaton RK. Identification of separable cognitive factors in schizophrenia. Schizophr Res 2004; 72: 29–39.
23. Corbetta M, Shulman GL. Control of goal-directed and stimulus-driven attention in the brain. Nat Rev Neurosci 2002; 3:201–215.
24. Luria A. Higher cortical functions in man, 2nd ed. New York, NY: Plenum Publishing Corp., 1980.
25. Pashler H. The psychology of attention. Cambridge, MA: MIT press, 1999.
26. Shulman GL, Ollinger JM, Akbudak E, et al. Areas involved in encoding and applying directional expectations to moving objects. J Neurosci 1999; 19:9480–9496.
27. Kanwisher N. Neural events and perceptual awareness. Cognition 2001; 79:89–113.

28. Kanwisher N, Wojciulik E. Visual attention: insights from brain imaging. Nat Rev Neurosci 2000; 1:91–100.
29. Corbetta M, Shulman GL, Miezin FM, Petersen SE. Superior parietal cortex activation during spatial attention shifts and visual feature conjunction. Science 1995; 270: 802–805.
30. Gottlieb JP, Kusunoki M, Goldberg ME. The representation of visual salience in monkey parietal cortex. Nature 1998; 391:481–484.
31. Rosvold HE, Mirsky AF, Sarason I, Bransome JED, Beck LH. A continuous performance test of brain damage. J Consult Psychol 1956; 20:343–350.
32. Asarnow RF, MacCrimmon DJ. Span of apprehension deficits during the postpsychotic stages of schizophrenia. A replication and extension. Arch Gen Psychiatry 1981; 38: 1006–1011.
33. Stroop JR. Studies of interference in serial verbal reactions. J Exp Psychol 1935; 18: 643–661.
34. Goodwin FK, Jamison KR. Manic-Depressive Illness. New York, NY: Oxford University Press. 1990.
35. Clark L, Iversen SD, Goodwin GM. Sustained attention deficit in bipolar disorder. Br J Psychiatry 2002; 180:313–319.
36. Liu SK, Chiu CH, Chang CJ, Hwang TJ, Hwu HG, Chen WJ. Deficits in sustained attention in schizophrenia and affective disorders: stable versus state-dependent markers. Am J Psychiatry 2002; 159:975–982.
37. Najt P, Glahn D, Bearden CE, et al. Attention deficits in bipolar disorder: a comparison based on the Continuous Performance Test. Neurosci Lett 2005; 379:122–126.
38. Wilder-Willis KE, Sax KW, Rosenberg HL, Fleck DE, Shear PK, Strakowski SM. Persistent attentional dysfunction in remitted bipolar disorder. Bipolar Disord 2001; 3:58–62.
39. Harmer CJ, Clark L, Grayson L, Goodwin GM. Sustained attention deficit in bipolar disorder is not a working memory impairment in disguise. Neuropsychologia 2002; 40:1586–1590.
40. McClure RK. Backward masking in bipolar affective disorder. Prog Neuropsychopharmacol Biol Psychiatry 1999; 23:195–206.
41. MacQueen GM, Young LT, Galway TM, Joffe RT. Backward masking task performance in stable, euthymic out-patients with bipolar disorder. Psychol Med 2001; 31:1269–1277.
42. Addington J, Addington D. Attentional vulnerability indicators in schizophrenia and bipolar disorder. Schizophr Res 1997; 23:197–204.
43. Clark L, Goodwin GM. State- and trait-related deficits in sustained attention in bipolar disorder. Eur Arch Psychiatry Clin Neurosci 2004; 254:61–68.
44. Fleck DE, Sax KW, Strakowski SM. Reaction time measures of sustained attention differentiate bipolar disorder from schizophrenia. Schizophr Res 2001; 52:251–259.
45. Neuchterlein KH, Dawson ME, Ventura J, Miklowitz D, Konishi G. Information-processing anomalies in the early course of schizophrenia and bipolar disorder. Schizophr Res 1991; 5:195–196.
46. Sax KW, Strakowski SM, Keck PE Jr., McElroy SL, West SA, Stanton PA. Symptom correlates of attentional improvement following hospitalization for a first episode of affective psychosis. Biol Psychiatry 1998; 44:784–786.
47. Sax KW, Strakowski SM, Zimmerman ME, DelBello MP, Keck PE, Jr., Hawkins JM. Frontosubcortical neuroanatomy and the continuous performance test in mania. Am J Psychiatry 1999; 156:139–141.
48. Ferrier IN, Chowdhury R, Thompson JM, Watson S, Young AH. Neurocognitive function in unaffected first-degree relatives of patients with bipolar disorder: a preliminary report. Bipolar Disord 2004; 6:319–322.
49. Denicoff KD, Ali SO, Mirsky AF, et al. Relationship between prior course of illness and neuropsychological functioning in patients with bipolar disorder. J Affect Disord 1999; 56:67–73.
50. Frangou S, Donaldson S, Hadjulis M, Landau S, Goldstein LH. The Maudsley Bipolar Disorder Project: executive dysfunction in bipolar disorder I and its clinical correlates. Biol Psychiatry 2005a; 58:859–864.

51. Joy S, Kaplan E, Fein D. Speed and memory in the WAIS-III Digit Symbol–Coding subtest across the adult lifespan. Arch Clin Neuropsychol 2004; 19:759–767.
52. Rypma B, Berger JS, Genova HM, Rebbechi D, D'Esposito M. Dissociating age-related changes in cognitive strategy and neural efficiency using event-related fMRI. Cortex 2005; 41:582–594.
53. Fry AF, Hale S. Relationships among processing speed, working memory, and fluid intelligence in children. Biol Psychol 2000; 54:1–34.
54. Hansell NK, Wright MJ, Luciano M, Geffen GM, Geffen LB, Martin NG. Genetic Covariation Between Event-Related Potential (ERP) and Behavioral Non-ERP Measures of Working-Memory, Processing Speed, and IQ. Behav Genet 2005.
55. Hawkins KA, Hoffman RE, Quinlan DM, Rakfeldt J, Docherty NM, Sledge WH. Cognition, negative symptoms, and diagnosis: a comparison of schizophrenic, bipolar, and control samples. J Neuropsychiatry Clin Neurosci 1997; 9:81–89.
56. Tham A, Engelbrektson K, Mathe AA, Johnson L, Olsson E, Aberg-Wistedt A. Impaired neuropsychological performance in euthymic patients with recurring mood disorders. J Clin Psychiatry 1997; 58:26–29.
57. Tabares-Seisdedos R, Balanza-Martinez V, Salazar-Fraile J, et al. Specific executive/attentional deficits in patients with schizophrenia or bipolar disorder who have a positive family history of psychosis. J Psychiatr Res 2003; 37:479–486.
58. Salthouse TA. Aging and measures of processing speed. Biol Psychol 2000; 54: 35–54.
59. Bryan J, Luszcz MA. Speed of information processing as a mediator between age and free-recall performance. Psychol Aging 1996; 11:3–9.
60. Lindenberger U, Mayr U, Kliegl R. Speed and intelligence in old age. Psychol Aging 1993; 8:207–220.
61. Hart RP, Kwentus JA. Psychomotor slowing and subcortical-type dysfunction in depression. J Neurol Neurosurg Psychiatry 1987; 50:1263–1266.
62. Kieseppa T, Tuulio-Henriksson A, Haukka J, et al. Memory and verbal learning functions in twins with bipolar-I disorder, and the role of information-processing speed. Psychol Med 2005; 35:205–215.
63. Bearden CE, Glahn DC, Monkul ES, et al. Sources of declarative memory impairment in bipolar disorder: Mnemonic processes and clinical features. J Psychiatr Res 2005.
64. Murphy FC, Rubinsztein JS, Michael A, et al. Decision-making cognition in mania and depression. Psychol Med 2001; 31:679–693.
65. Clark L, Iversen SD, Goodwin GM. A neuropsychological investigation of prefrontal cortex involvement in acute mania. Am J Psychiatry 2001; 158:1605–1611.
66. Martinez-Aran A, Vieta E, Colom F, et al. Neuropsychological performance in depressed and euthymic bipolar patients. Neuropsychobiology 2002; 46 (suppl 1):16–21.
67. Lebowitz BK, Shear PK, Steed MA, Strakowski SM. Verbal fluency in mania: relationship to number of manic episodes. Neuropsychiatry Neuropsychol Behav Neurol 2001; 14:177–182.
68. Fuster JM. The prefrontal cortex: anatomy, physiology, and neuropsychology of the frontal lobe, 3rd ed. Philadelphia: Lippincott-Raven, 1997.
69. Goldman-Rakic PS. The prefrontal landscape: implications of functional architecture for understanding human mentation and the central executive. Philosophical Transactions of the Royal Society of London - Series B: Biological Sciences 1996; 351: 1445–1453.
69a. Fuster JM. Memory in the cerebral cortex: An empirical approach to neural networks in the human and nonhuman primate. Cambridge, MA, USA: The Mit Press, 1995.
70. Ericsson KA, Delaney PF. Long-term working memory as an alternative to capacity models of working memory in everyday skilled performance. In Miyake A, Shah P, et al. (eds), Models of working memory: Mechanisms of active maintenance and executive control. New York, NY, USA: Cambridge University Press, 1999; 257–297.
71. Baddeley A. Working memory. Science 1992; 255:556–559.
72. Baddeley A. The fractionation of working memory. Proc Int Conf Intell Syst Mol Biol 1996; 93:13468–13472.

73. D'Esposito M, Aguirre GK, Zarahn E, Ballard D, Shin RK, Lease J. Functional MRI studies of spatial and nonspatial working memory. Brain Research. Cognitive Brain Research 1998; 7:1–13.
74. Owen A, McMillan K, Laird A, Bullmore E. The n-back working memory paradigm: a meta-analysis of normative fMRI studies. Hum Brain Mapp 2005; 25: 46–59.
75. Ruchkin DS, Johnson R, Grafman J, Canoune H. Distinctions and similarities among working memory processes: An event-related potential study. Cognitive Brain Research 1992; 1:53–66.
76. Ruchkin DS, Johnson R, Jr., Grafman J, Canoune H. Multiple visuospatial working memory buffers: Evidence from spatiotemporal patterns of brain activity. Neuropsychologia 1997; 35:195–209.
77. Smith EE, Jonides J. Neuroimaging analyses of human working memory. Proc Int Conf Intell Syst Mol Biol 1998; 95:12061–12068.
78. Hunter WS. The delayed reaction in animals and children. Behavioral Monogr 1913; 2:1–86.
79. Gold JM, Carpenter C, Randolph C, Goldberg TE, Weinberger DR. Auditory working memory and Wisconsin Card Sorting Test performance in schizophrenia. Archives of General Psychiatry 1997; 54:159–165.
80. Gevins A, Cutillo B. Neuroelectric evidence for distributed processing in human working memory. Electroencephalogr Clin Neurophysiol 1993; 87:128 –143.
81. Glahn DC, Bearden CE, Cakir S, et al. Differential working memory impairment in bipolar disorder and schizophrenia: effects of lifetime history of psychosis. Bipolar Disord 2006b; 8:117–123.
82. Park S, Holzman PS. Schizophrenics show spatial working memory deficits. Arch Gen Psychiatry 1992; 49:975–982.
83. Rubinsztein JS, Michael A, Paykel ES, Sahakian BJ. Cognitive impairment in remission in bipolar affective disorder. Psychol Med 2000; 30:1025–1036.
84. Gooding DC, Tallent KA. The association between antisaccade task and working memory task performance in schizophrenia and bipolar disorder. J Nerv Ment Dis 2001; 189:8–16.
85. Park S, Holzman PS. Association of working memory deficit and eye tracking dysfunction in schizophrenia. Schizophr Res 1993; 11:55–61.
85a. Sahakian BJ, Morris RG, Evenden JL, et al. A comparative study of visuospatial memory and learning in Alzheimer-type dementia and Parkinson's disease. Brain 1988; 111 (Pt 3):695–718.
86. McGrath J, Chapple B, Wright M. Working memory in schizophrenia and mania: correlation with symptoms during the acute and subacute phases. Acta Psychiatrica Scandinavica 2001; 103:181–188.
87. El-Badri SM, Ashton CH, Moore PB, Marsh VR, Ferrier IN. Electrophysiological and cognitive function in young euthymic patients with bipolar affective disorder. Bipolar Disord 2001; 3:79–87.
88. Krabbendam L, Honig A, Wiersma J, et al. Cognitive dysfunctions and white matter lesions in patients with bipolar disorder in remission. Acta Psychiatr Scand 2000; 101:274–280.
89. Donaldson S, Goldstein LH, Landau S, Raymont V, Frangou S. The Maudsley Bipolar Disorder Project: the effect of medication, family history, and duration of illness on IQ and memory in bipolar I disorder. J Clin Psychiatry 2003; 64:86–93.
90. Gabrieli JD, Poldrack RA, Desmond JE. The role of left prefrontal cortex in language and memory. Proc Int Conf Intell Syst Mol Biol 1998; 95:906–913.
91. Kapur S, Tulving E, Cabeza R, McIntosh AR, Houle S, Craik FI. The neural correlates of intentional learning of verbal materials: a PET study in humans. Brain research. Cognitive brain research. 1996; 4(4):243–249.
92. Jones-Gotman M. Memory for designs: the hippocampal contribution. Neuropsychologia 1986a; 24:193–203.
93. Jones-Gotman M. Right hippocampal excision impairs learning and recall of a list of abstract designs. Neuropsychologia 1986b; 24:659–670.

94. Squire LR. The organization and neural substrates of human memory. Int J Neurol 1987; 21–22:218–222.
95. Cabeza R, Nyberg L. Imaging cognition II: An empirical review of 275 PET and fMRI studies. J Cogn Neurosci 2000; 12:1–47.
96. Lepage M, Habib R, Tulving E. Hippocampal PET activations of memory encoding and retrieval: the HIPER model. Hippocampus. 1998; 8(4).
97. Jones-Gotman M. Localization of lesions by neuropsychological testing. Epilepsia 1991; 32(Suppl 5):S41–52.
98. Jones-Gotman M, Milner B. Right temporal-lobe contribution to image-mediated verbal learning. Neuropsychologia 1978; 16:61–71.
99. Milner B. Disorders of learning and memory after temporal lobe lesions in man. Clin Neurosurg 1972; 19:421–446.
100. Petrides M. Deficits on conditional associative-learning tasks after frontal- and temporal-lobe lesions in man. Neuropsychologia 1985; 23:601–614.
101. Taylor MA, Greenspan B, Abrams R. Lateralized neuropsychological dysfunction in affective disorder and schizophrenia. Am J Psychiatry 1979; 136:1031–1034.
102. Petrides M, Milner B. Deficits on subject-ordered tasks after frontal- and temporal-lobe lesions in man. Neuropsychologia 1982; 20:249–262.
103. Samuels I, Butters N, Fedio P. Short term memory disorders following temporal lobe removals in humans. Cortex 1972; 8:283–298.
104. Gur RC, Ragland JD, Resnick SM, et al. Lateralized increases in cerebral blood flow during performance of verbal and spatial tasks: relationship with performance level. Brain Cogn 1994; 24:244–258.
105. Hannay HJ, Boyer CL. Sex differences in hemispheric asymmetry revisited. Percept Mot Skills 1978; 47:315–321
106. Smith EE, Jonides J, Koeppe RA. Dissociating verbal and spatial working memory using PET. Cereb Cortex 1996; 6:11–20.
107. Altshuler LL, Ventura J, van Gorp WG, Green MF, Theberge DC, Mintz J. Neurocognitive function in clinically stable men with bipolar I disorder or schizophrenia and normal control subjects. Biol Psychiatry. 2004; 56:560–9.
108. Basso MR, Lowery N, Neel J, Purdie R, Bornstein RA. Neuropsychological impairment among manic, depressed, and mixed-episode inpatients with bipolar disorder. Neuropsychology. 2002; 16:84–91.
109. Cavanagh JT, Van Beck M, Muir W, Blackwood DH. Case-control study of neurocognitive function in euthymic patients with bipolar disorder: an association with mania. Br J Psychiatry 2002; 180:320–326.
110. Deckersbach T, Savage CR, Reilly-Harrington N, Clark L, Sachs G, Rauch SL. Episodic memory impairment in bipolar disorder and obsessive-compulsive disorder: the role of memory strategies. Bipolar disorders 2004b; 6(3):233–244.
111. Fleck DE, Shear PK, Zimmerman ME, et al. Verbal memory in mania: effects of clinical state and task requirements. Bipolar Disord 2003; 5:375–380.
112. van Gorp WG, Altshuler L, Theberge DC, Mintz J. Declarative and procedural memory in bipolar disorder. Biological Psychiatry 1999; 46:525–531.
113. van Gorp WG, Altshuler L, Theberge DC, Wilkins J, Dixon W. Cognitive impairment in euthymic bipolar patients with and without prior alcohol dependence. A preliminary study. Arch Gen Psychiatry 1998; 55:41–46.
114. Keri S, Kelemen O, Benedek G, Janka Z. Different trait markers for schizophrenia and bipolar disorder: a neurocognitive approach. Psychol Med 2001; 31:915–922.
115. Duff K, Planel E. Untangling memory deficits. Nat Med 2005; 11:826–827.
116. Glahn DC, Bearden CE, Caetano S, et al. Declarative memory impairment in pediatric bipolar disorder. Bipolar Disord 2005; 7:546–554.
117. Henry GM, Weingartner H, Murphy DL. Influence of affective states and psychoactive drugs on verbal learning and memory. Am J Psychiatry 1973; 130:966–971.
118. Delis D, Kaplan E. Delis-Kaplan Executive Function System™ (D–KEFS™). San Antonio: Harcourt Assessment Company. 2001.
119. Deckersbach T, McMurrich S, Ogutha J, Savage CR, Sachs G, Rauch SL. Characteristics of non-verbal memory impairment in bipolar disorder: the role of encoding strategies. Psychological Medicine 2004a; 34:823–832.

120. Kessing LV. Cognitive impairment in the euthymic phase of affective disorder. Psychological medicine 1998; 28(5):1027–1038.
121. Robinson LJ, Ferrier IN. Evolution of cognitive impairment in bipolar disorder: a systematic review of cross-sectional evidence. Bipolar Disord 2006; 8:103–116.
122. Verdoux H, Liraud F. Neuropsychological function in subjects with psychotic and affective disorders. Relationship to diagnostic category and duration of illness. Eur Psychiatry 2000; 15:236–243.
123. Zubieta JK, Huguelet P, O'Neil RL, Giordani BJ. Cognitive function in euthymic bipolar I disorder. Psychiatry Res 2001; 102:9–20.
124. Rey A. L'Examen Clinique en Psychologie. Paris: Press Universitaire de France. 1964.
125. Gourovitch ML, Torrey EF, Gold JM, Randolph C, Weinberger DR, Goldberg TE. Neuropsychological performance of monozygotic twins discordant for bipolar disorder. Biol Psychiatry 1999; 45:639–646.
126. Jones BP, Duncan CC, Mirsky AF, Post RM, Theodore WH. Neuropsychological profiles in bipolar affective disorder and complex partial seizure disorder. Neuropsychology 1994; 8:55–64.
127. Sapin LR, Berrettini WH, Nurnberger JI, Jr, Rothblat LA. Mediational factors underlying cognitive changes and laterality in affective illness. Biological Psychiatry 1987; 22:979–986.
128. Baddeley A. The central executive: a concept and some misconceptions. J Int Neuropsychol Soc 1998; 4:523–526.
129. Heaton R, Cfhelune G, Talley J, Kay G, Curtiss G. Wisconsin Card Sorting Test Manual: Revised and expanded. Odessa, FL: Psychological Assessment Resources, Inc. 1993.
130. Robbins TW, James M, Owen AM, Sahakian BJ, McInnes L, Rabbitt P. Cambridge Neuropsychological Test Automated Battery (CANTAB): A factor analytic study of a large sample of normal elderly volunteers. Dementia 1994; 5:266–281.
131. The Psychological Corporation WAIS-III–WMS-III technical manual. San Antonio, TX: The Psychological Corporation. 1997.
131a. Carpenter PA, Just MA, Shell P. What one intelligence test measures: a theoretical account of the processing in the Raven Progressive Matrices Test. Psychological Review 1990; 97:404–431.
132. Duncan J, Owen AM. Common regions of the human frontal lobe recruited by diverse cognitive demands. Trends in neurosciences 2000; 23:475–483.
133. Baker SC, Rogers RD, Owen AM, et al. Neural systems engaged by planning: a PET study of the Tower of London task. Neuropsychologia 1996; 34:515–526.
134. Duncan J, Seitz RJ, Kolodny J, et al. A neural basis for general intelligence. Science 2000; 289:457–460.
135. Ragland JD, Glahn DC, Gur RC, et al. PET regional cerebral blood flow change during working and declarative memory: relationship with task performance. Neuropsychology 1997; 11:222–231.
136. Badcock JC, Michiel PT, Rock D. Spatial working memory and planning ability: contrasts between schizophrenia and bipolar I disorder. Cortex 2005; 41:753–763.
137. Olley AL, Malhi GS, Bachelor J, Cahill CM, Mitchell PB, Berk M. Executive functioning and theory of mind in euthymic bipolar disorder. Bipolar Disord 2005; 7(Suppl 5): 43–52.
138. Glahn DC, Bearden CE, Bowden CL, Soares JC. Reduced educational attainment in bipolar disorder. J Affect Disord 2006a; 92:309–312.
139. Dixon T, Kravariti E, Frith C, Murray RM, McGuire PK. Effect of symptoms on executive function in bipolar illness. Psychol Med 2004; 34:811–821.
140. Rogers RD, Owen AM, Middleton HC, et al. Choosing between Small, Likely Rewards and Large, Unlikely Rewards Activates Inferior and Orbital Prefrontal Cortex. J Neurosci 1999; 20:9029–9038.
141. Meyer SE, Carlson GA, Wiggs EA, et al. A prospective study of the association among impaired executive functioning, childhood attentional problems, and the development of bipolar disorder. Dev Psychopathol 2004; 16:461–476.
142. Kane MJ, Engle RW. The role of prefrontal cortex in working-memory capacity, executive attention, and general fluid intelligence: an individual-differences perspective. Psychon Bull Rev 2002; 9:637–671.

143. Engle RW, Tuholski SW, Laughlin JE, Conway AR. Working memory, short-term memory, and general fluid intelligence: a latent-variable approach. J Exp Psychol Gen 1999; 128:309–331.
144. Dalby J, Williams R. Preserved reading and spelling ability in psychotic disorders. Psychol Med 1986; 16:171–175.
145. Sackeim H, Decina P, Epstein D, Bruder G, Malitz S. Possible reversed affective lateralization in a case of bipolar disorder. Am J Psychiatry 1983; 140: 1191–1193.
146. Cannon M, Caspi A, Moffitt TE, et al. Evidence for early-childhood, pan-developmental impairment specific to schizophreniform disorder: results from a longitudinal birth cohort. Arch Gen Psychiatry 2002; 59:449–456.
147. Reichenberg A, Weiser M, Rabinowitz J, et al. A population-based cohort study of premorbid intellectual, language, and behavioral functioning in patients with schizophrenia, schizoaffective disorder, and nonpsychotic bipolar disorder. Am J Psychiatry 2002; 159:2027–2035.
148. Donnelly EF, Murphy DL, Goodwin FK, Waldman IN. Intellectual function in primary affective disorder. Br J Psychiatry 1982; 140:633–636.
149. Dickstein DP, Treland JE, Snow J, et al. Neuropsychological performance in pediatric bipolar disorder. Biol Psychiatry 2004; 55:32–39.
150. Doyle AE, Wilens TE, Kwon A, et al. Neuropsychological functioning in youth with bipolar disorder. Biol Psychiatry 2005; 58:540–548.
151. Regier DA, Farmer ME, Rae DS, et al. Comorbidity of mental disorders with alcohol and other drug abuse. Results from the Epidemiologic Catchment Area (ECA) Study. Jama 1990; 264:2511–2518.
152. Grant BF, Stinson FS, Hasin DS, et al. Prevalence, correlates, and comorbidity of bipolar I disorder and axis I and II disorders: results from the National Epidemiologic Survey on Alcohol and Related Conditions. J Clin Psychiatry 2005; 66:1205–1215.
153. Honig A, Arts BM, Ponds RW, Riedel WJ. Lithium induced cognitive side-effects in bipolar disorder: a qualitative analysis and implications for daily practice. Int Clin Psychopharmacol 1999; 14:167–171.
154. Engelsmann F, Katz J, Ghadirian AM, Schachter D. Lithium and memory: a long-term follow-up study. J Clin Psychopharmacol 1988; 8:207–212.
155. Amado-Boccara I, Gougoulis N, Poirier Littre MF, Galinowskı A, Loo H. Effects of antidepressants on cognitive functions: a review. Neurosci Biobehav Rev 1995; 19: 479–493.
156. Strakowski SM, Adler CM, Holland SK, Mills N, DelBello MP. A preliminary FMRI study of sustained attention in euthymic, unmedicated bipolar disorder. Neuropsychopharmacology 2004; 29:1734–1740.
157. Strakowski SM, Adler CM, Holland SK, Mills NP, DelBello MP, Eliassen JC. Abnormal FMRI brain activation in euthymic bipolar disorder patients during a counting Stroop interference task. Am J Psychiatry 2005; 162:1697–1705.
158. Clark L, Sarna A, Goodwin GM. Impairment of executive function but not memory in first-degree relatives of patients with bipolar I disorder and in euthymic patients with unipolar depression. Am J Psychiatry 2005; 162:1980–1982.
159. Frangou S, Haldane M, Roddy D, Kumari V. Evidence for deficit in tasks of ventral, but not dorsal, prefrontal executive function as an endophenotypic marker for bipolar disorder. Biol Psychiatry 2005b; 58:838–839.
160. McIntosh AM, Harrison LK, Forrester K, Lawrie SM, Johnstone EC. Neuropsychological impairments in people with schizophrenia or bipolar disorder and their unaffected relatives. Br J Psychiatry 2005; 186:378–385.
161. Sobczak S, Honig A, Nicolson NA, Riedel WJ. Effects of acute tryptophan depletion on mood and cortisol release in first-degree relatives of type I and type II bipolar patients and healthy matched controls. Neuropsychopharmacology 2002; 27:834–842.
162. Zalla T, Joyce C, Szoke A, et al. Executive dysfunctions as potential markers of familial vulnerability to bipolar disorder and schizophrenia. Psychiatry Res 2004; 121:207–217.

15 Biology vs. Environment: Stressors in the Pathophysiology of Bipolar Disorder

Morgen A. R. Kelly
Western Psychiatric Institute and Clinic, University of Pittsburgh School of Medicine, Pittsburgh, Pennsylvania, U.S.A.

Stefanie A. Hlastala
Children's Hospital and Regional Medical Center, Seattle, Washington, U.S.A.

Ellen Frank
Western Psychiatric Institute and Clinic, University of Pittsburgh School of Medicine, Pittsburgh, Pennsylvania, U.S.A.

INTRODUCTION

Fifty years after the discovery of lithium (1), bipolar disorder (BD) remains a major therapeutic challenge. Extensive research has demonstrated that people with BD typically suffer multiple recurrences of mania and/or depression throughout their lifetimes, often despite continuous maintenance pharmacotherapy (2–6). Markar and Mander (4), for example, followed a group of bipolar patients on lithium prophylaxis and found that only 30% to 40% remained well over a three-year period. Recently, clinical research has begun to examine the efficacy of adding psychosocial treatments to pharmacotherapy for BD (7). Outcomes in these studies are often measured both in terms of symptom reduction and in terms of the duration of well periods (8–10). The inclusion of time to relapse as an outcome measure in these studies presumes that most individuals with BD will relapse over time, often despite use of so-called maintenance medications. The occurrence of breakthrough episodes coupled with the poor work, family, and social functioning found over the long-term course of illness in patients with BD (11) makes it critical that we examine environmental as well as biological factors in an effort to better understand mechanisms underlying the pathophysiology of BD.

Examining the role that environmental stressors play in the onset, recovery, and recurrence of bipolar episodes provides a way to understand how psychosocial factors may influence the course of a disorder that is often thought of as purely biologically driven. Indeed, our own clinical experience would suggest that changes in social roles (e.g., getting a divorce, becoming a parent), changes in routines (e.g., travel across multiple time zones), and interpersonal losses (e.g., death of a loved one) are frequently associated with the onset of new episodes in bipolar patients. In this chapter, we review research that has examined the relationship between life stress and the long-term course of BD as well as the methodological inconsistencies that are found in this research. We also address some of the biological and behavioral processes that may modify this relationship. Additionally, the question of whether stress plays a greater role earlier as opposed to later in the course of illness is examined.

METHODOLOGICAL ISSUES

Before embarking on an examination of the relationship between environmental stressors and the long-term course of BD, the pervasive methodological inconsistencies and inadequacies found in the previous research on life stress and the course of BD need to be discussed. Two methods of life event measurement have dominated research on stress and BD: checklist measures and personal interview measures. Checklist approaches have several weaknesses. Often the items on these lists are ambiguous, which may lead to inaccurate interpretations and unreliable reports of basic information by respondents (12,13). Checklists often allow a wide range of severity in one category (e.g., a bad flu, terminal cancer, and chronic arthritis could all be reported under a "serious illness" category). Additionally, idiosyncratic interpretations influence checklist events such as "serious illness of close family member" by allowing the respondent to determine which illnesses are "serious" and which family members are "close" (14).

Perhaps one of the greatest problems that can arise when examining the role of psychosocial events in episode onset is that events may occur as a result rather than a cause of the disorder. Checklist measures are unable to sort out this confound because of the lack of specificity in the timing of events in relation to the onset of symptoms of a new episode. Studies using checklist methods are more likely to misdate distant events into a more recent time period (15,16). One-on-one interviewing, on the other hand, is more effective in obtaining accurate answers and dating (17), which is critical when trying to separate the events from early symptoms of psychopathology. The optimal strategy involves the use of reliable and accurate techniques for dating episode onset in relation to the occurrence of a life event. Such techniques are more likely to be associated with interview methods.

Many studies use retrospective designs that include prolonged recall periods and questionnaire methods that contribute to biased or incomplete reporting, errors in recall, inadequate sampling of important events, and inaccuracies in dating event onset and duration. One of the main problems with retrospective designs is the limitations in the long-term recall of stressful life events. Additionally, patients may experience some memory bias regarding life stress over time to fit their experience to their conceptualization of their illness (i.e., a "search after meaning").

Another prominent issue in stress disorder research is whether to use objective ratings (investigators' interpretations of threat level) or subjective ratings (respondents' interpretations of threat level) of stress. Selye (18), for example, noted the importance of differentiating between objective external stressors and perceived stress, stating "the stressor effects depend not so much upon what we do or what happens to us but on how we take it." However, research examining the role stress might play in the etiology and course of a disorder is complicated by the fact that it can be difficult to distinguish a person's *perception* of stress from the disorder itself or some prodromal sign of it. Moreover, psychological processes such as minimization, denial, and exaggeration may lead to a respondent being unable to report accurately on his or her response to an event. However, if one takes the respondent entirely out of the equation and merely looks at the event isolated from any personal context, the meaning may be entirely lost and the process does indeed become "arbitrary" as Selye noted.

In an effort to integrate all the aforementioned concerns, the Bedford College Life Events and Difficulties Schedule (LEDS) (19) uses a *contextual* interview-based approach to obtain an accurate portrayal of a subject's life events and difficulties. In the LEDS system, "an assessment of meaning or understanding on part of an investigator can take into account not only the immediate situation (a woman losing a job) but the wider context (she is unmarried, in debt, and living with her school-aged child)" (20). The contextual severity of threat and a variety of other dimensions are rated using the "event dictionary" that contains extensive rules, criteria, and more than 2000 case exemplars of life events and difficulties. Contextual ratings of threat are based on the notion of the likely response of an average person to an event occurring in the context of a particular set of biographical circumstances. The ratings reflect the threat associated with an event, taking into account the subject's particular set of circumstances while excluding both the respondent's emotional reaction and any psychiatric or physical symptoms that followed the event.

The LEDS system enables researchers to obtain information on life events in a way that minimizes self-report problems including denial, minimization, lack of spontaneous recall, and exaggeration of past life events. Additionally, the use of a semistructured interview format allows comprehensive assessment of events across broad domains. Interviewers use "anchors" (e.g., holidays, birthdays) as probes to date events accurately within the period of interest. Most important, the LEDS system has the capability of taking life circumstances into account so that the meaning of the event is not lost. This approach to the assessment of life events has yielded consistently strong associations between life stress and the onset or recurrence of a variety of physical and psychiatric illnesses (20). Of note, differing treatment outcome results were found in a comparison of checklist and LEDS methodologies (21). Results indicated that LEDS events predicted decreased likelihood of remission while the checklist method was not associated with differential outcome in the same sample. Because of the relative strength of interview and/or LEDS-based methodology, the following review places more emphasis on research using those methods.

STRESSORS IN DEPRESSION AND MANIA

A vast amount of evidence indicates a role for life events in the onset and course of unipolar depression (14,19). The effects of life events on BD, on the other hand, have not been as extensively studied. The lack of strong research in this area may be a result of the long-held assumption that biological factors play a more important role than psychosocial factors in the onset and timing of new episodes in BD. Despite this, a handful of relatively well-designed studies have recently been published regarding the relationship between life stress and the onset or recurrence of bipolar episodes. Although the existing evidence remains limited, the data have suggested that stressful life events influence the onset of both first and subsequent episodes of BD (22).

EPISODE ONSET

Among the first questions that may be answered using life event methodology is "are stressful life events related to the initial onset of BD?" Despite the fact that the study of disorder onset is exceedingly difficult to conduct, two studies have demonstrated an association between life stress and the initial onset of mood disorders in general or BD in specific. Hillegers et al. (23) adapted the LEDS for use with adolescents and administered the interview to 140 children of bipolar parents. They found that overall life event load was associated with about a 10% increase in the risk of mood disorder onset for these children. Unfortunately, this study was not able to determine the relationship between life events and onset of bipolar versus unipolar depression, given that this determination is impossible in cases where a depressive episode is the initial episode of a mood disorder. Using the Holmes-Rahe Social Readjustment scale, Glassner and Haldipur (24) found evidence that individuals with late onset BD (defined as onset after age 20) retrospectively reported more life events associated with both their first and their most recent mood episodes. These studies provide some evidence that life stress may be related to the onset of mood episodes, but are methodologically limited as well.

Perhaps because of the challenges of assessing the impact of stress on first episodes of mood disorder as well as the clinical importance of understanding relapse triggers, most extant research in this area has concentrated on the association between life stress and either recurrence or exacerbation of mood episodes. Although the balance of evidence suggests that stressful life events are associated with return or increase of mood symptoms, not all studies find such a relationship. Studies using *checklist* methodologies for assessing life events have generally failed to find a relationship between life events and bipolar episode onset (25,26) (see Table 1 for more specific methodological information). Yet each study had extensive design flaws that limit the strength of its results. Hall et al. (25) studied 38 bipolar I patients who were asked to complete a questionnaire consisting of 86 events over a 10-month period. They found that the 17 patients who relapsed into an episode of either depression or hypomania did not differ in frequency of life events from the 21 patients who did not relapse within the 10-month interval. However, the study may not have used an adequate observation period and did not control for the effects of medication compliance. Additionally, because depressive and manic episodes were defined as deviations from normal mood, the patients categorized as relapsers may have been relatively similar to those who did not relapse, thus resulting in no significant differences. Finally, the time studied was not limited to three months or less before episode onset, which may be a more accurate predictor of relapse. Mayo (26) also found no excess of stressful events in the six months before hospitalization in 28 bipolar, 7 unipolar, and 5 schizoaffective patients. However, this study also had methodological flaws (e.g., no analyses by diagnostic group; life event recall went back, on average, 18 years).

Most studies using more rigorous *interview-based* life event assessment methods have found a significant relationship between life events and episode onset (30–35). The only exceptions are Chung et al. (36) and McPherson et al. (37). Using the interview-based LEDS system, Chung et al. (36) assessed life event rates in the six months before episode onset and did not find a significant difference between hypomanic patients and surgical control subjects, schizophrenic patients, and schizophreniform patients. The hypomanic patients and control subjects did not exhibit a significant difference in the number of events experienced over six months. However, given the basic differences typically found between bipolar I and II patients [e.g., differences in episode severity and genetics (38,39)], it is questionable whether life event data on hypomanic episodes are generalizable to bipolar I patients. McPherson et al. (37) interviewed 61 bipolar I inpatients at three-month intervals over a two-year period. Life events were collected using

TABLE 1 Studies Finding No Significant Relationship Between Life Events and Episode Onset

Study	Sample	Comparison group	Life events measure	Life events time period	Definition of episode onset
Hall et al. (1977)	38 bipolar I, rapid cyclers excluded	17 relapsers vs. 21 nonrelapsers	SLE questionnaire	10 mo	Feighner criteria
McPherson et al. (1993)	58 bipolar I	Within subject control periods	RLE interview	3 mo	SADS
Mayo (1970)	28 bipolar I	6 mo before and after hospitalization	Interview based on checklist	6 mo	Structured interview
Chung et al. (1986)	14 hypomanic patients	Surgical control subjects	LEDS	6 mo	DSM-III
Reilly-Harrington et al. (1999)	49 with lifetime bipolar	97 with lifetime unipolar depression; 23 with no diagnosis	LES	1 mo between assessments	SADS

Abbreviations: LEDS, life events and difficulties schedule (19); LES, life experiences survey (27); SADS, schedule for affective disorders and schizophrenia (28); SLE, modified recent life events questionnaire (29).

the semistructured Interview for Recent Life Events (40). The events were also rated for threat and independence using modified LEDS criteria. Onset of relapse was taken from the time of the first clear symptom. Fourteen of 61 (23%) manic/hypomanic relapses and 9 of 32 (28%) depressive relapses were preceded by a moderate to severe independent event in the previous month. However, these rates were not significantly greater than the life event rates occurring during control periods.

Although Hunt et al. (Table 2) (34) used the identical measures and design as McPherson et al. (37) (Table 1), they did find empirical support for a relationship between life events and new episodes. In the patients who relapsed, 10 of 52 (19%) had at least one severe event in the month before relapse compared with 7 of 144 (5%) during control periods ($p < 0.01$). McPherson et al. suggest they may have failed to find a relationship because their sample was more seriously ill and at a later stage of illness and because life events may play a more important role in earlier episodes of the illness, as Post (46) has suggested. Additionally, their sample was from a more affluent area, in contrast to Hunt et al.'s inner-city sample, which had a greater proportion of ethnic minorities. This suggests that the threshold for relapse might be lowered by chronic psychosocial difficulties.

As seen in Table 2, many additional studies have found a relationship between life events and bipolar episodes. For example, Kennedy et al. (30) compared 20 manic inpatients with orthopedic outpatients and control subjects matched for age, sex, marital status, social class, and immigration status who were interviewed 6 to 21 months after hospital discharge. They found that rates of events with severe, marked, or moderate objective negative impact were elevated in the four months before psychiatric hospital admission when compared with post-discharge rates. Ambelas (47) found that 28% of manic patients experienced a stressful life event in the four weeks preceding their hospitalization, compared with only 6% of surgical control subjects. Joffe et al. (48), using the PERI-M (42), found that subjects with mania had significantly more unanticipated and uncontrollable events as compared with bipolar patients in an episode-free period.

The most methodologically strong studies have used interview-based assessments of life events. Using the LEDS interview-based method, Sclare and Creed (33) found that 11 of 24 (44%) bipolar I patients experienced a severe event in the 26 weeks before onset, compared with 5 of 24 (21%) experiencing a severe event in the 26 weeks after recovery. Beside its use of a well-validated method of life event measurement, another strength of this study was its use of symptom onset as the definition of a new episode. Because hospitalization often does not occur until several weeks after the onset of symptoms (49), one cannot be sure if an event precipitated onset or if it occurred as a result of the disorder when hospitalization is used as the onset date.

Two other interview-based studies also found an association between life events and episode onset. Using a modified version of the LEDS, Bebbington et al. (35) found higher rates of severe life events in the six months before onset than in a comparable period among normal control subjects. In a prospective design, Ellicott et al. (32) examined the impact of life stress on the course of bipolar illness over a 2-year period, while controlling for medication compliance. Using a LEDS-based method of life stress assessment of 61 bipolar I outpatients, they found that patients with a high level of stress were at 4.53 times higher risk of experiencing a new episode than patients experiencing no stress. Patients with low and average levels of stress did not have a greater risk of relapse than those without stress, indicating a threshold at which a patient becomes vulnerable to

TABLE 2 Studies Finding a Significant Relationship Between Life Events and Episode Onset

Study	Sample	Comparison group	Life events measure	Life events time period	Definition of episode onset
Ambelas (1979)	67 manic or hypomanic inpatients	60 surgical controls	Chart review	1 mo	Feighner criteria for chart review
Glassner et al. (1979)	25 bipolar I	25 community controls	Modified SRA	1 yr	Feighner criteria
Glassner and Haldipur (1983)	46 bipolar patients	None	Modified SRA	One time interview	DSM-III diagnosis by 2 independent psychiatrists
Kennedy et al. (1983)	20 manic patients	Orthopedic outpatients	RLE interview	4 mo	Renard diagnostic interview
Bidzinska (1984)	50 bipolar I and II	47 unipolar patients, 100 controls	Life event questionnaire	3 mo	Not reported
Miklowitz et al. (1988)	23 bipolar I	Relapsers vs. nonrelapsers	CFI-expressed emotion	9 mo	PSE-DSM-III
Joffe et al. (1989)	14 bipolar I in- and outpatients	Episode-free matched bipolar patients	PERI-M	1 yr	SADS-RDC

(Continued)

TABLE 2 Studies Finding a Significant Relationship Between Life Events and Episode Onset (*Continued*)

Study	Sample	Comparison group	Life events measure	Life events time period	Definition of episode onset
Ellicot et al. (1990)	61 bipolar I outpatients	None	RLE and LEDS-based interview	3 mo	DSM-III-R
Sclare and Creed (1990)	25 manic inpatients	Within-subject control periods	LEDS	6 mo	PSE
Hunt et al. (1992)	62 bipolar I patients	Within-subject control periods	RLE interview	3 mo	SADS
Bebbington et al. (1993)	31 manic patients	Psychiatric and non-psychiatric controls	Modified LEDS	6 mo	DSM-III
Johnson et al. (2000)	43 bipolar I inpatients	None	LEDS	Monthly	Increased BMRS or HRSD scores
Malkoff-Schwartz et al. (2000)	66 bipolar I	44 unipolar depressed	LEDS	1 yr	SCID interview
Johnson et al. (2000)	43 bipolar I inpatients	None	LEDS	Monthly	Increased BMRS or HRSD scores

Abbreviations: CFI, Camberwell family interview (41); LEDS, life events and difficulties schedule (19); LES, life experiences survey (27); PERI-M, psychiatric research interview modified life events scale (42); RDC, research diagnostic criteria (43); SADS, schedule for affective disorders and schizophrenia (28); SLE, modified recent life events questionnaire (29); SRA, Homes-Rahe social readjustment scale (44); BRMS, Bech-Rafaelgen Mania Scale (45); HRSD, Hamilton rating scale for depression (45).

the impact of threatening events. There were also no significant differences between the group that relapsed and the group that did not in terms of maintenance medication levels or medication compliance ratings. More recently, Cohen and colleagues (50), using Hammen's interview for episodic stress, found a relationship between level of stress and relapse rates over a one-year period.

As evidence suggesting a relationship between stressful events and mood disturbance in BD has accumulated, researchers have begun to turn their attention to the potentially differential impact of differing types of stress on recurrence of depressed versus manic symptoms as well as to interaction effects between life stress and cognitive or personality styles. In terms of type of episode triggered, Johnson and colleagues (51), using LEDS methods, found that events having to do with goal attainment were associated with increased manic symptoms even when baseline level of mania was statistically controlled. Malkoff-Schwartz and colleagues (52) reported that patients who became manic endorsed more severe life events in the 20 weeks prior to a recent episode compared with depressed bipolar patients. Similarly, Joffe et al. (48) showed that, although manic and nonmanic bipolar patients did not differ in terms of severity of events or total number of stressful events experienced, patients with recent mania reported more uncontrollable or unexpected events prior to the onset of mania. In contrast, Reilly-Harrington et al. (53) did not find a relationship between the number of events reported by bipolar and unipolar patients in terms of depressive symptom onset. Cohen et al. (50) found a marginal trend suggesting that life stress may play a larger role in the onset of depression than mania.

Other research has turned its attention to the impact of interactions between personality types or cognitive styles and stressful events on mood episodes among bipolar patients. Reilly-Harrington and colleagues (53) showed that negative cognitive style and the occurrence of life events interacted to predict depressive symptoms in both unipolar and bipolar patients. Further, Alloy and colleagues (54) found that negative attributional style interacted with life events to predict dysphoria and that positive attributional style interacted with positive events to predict hypomania among hypomanic or dysthymic undergraduates. However, these results were only found at the second assessment and not at baseline or third assessment, calling the robustness of these findings into question. Two other studies have examined the impact of events that are congruent with sociotropic or autonomous personality styles. Hammen et al. (55) found that events congruent with sociotropic or autonomous personality type were related to episode type for unipolar but not bipolar individuals. This lack of association for bipolar individuals was repeated in a later study from the same group (56).

Another area of related research has examined levels of familial expressed emotion (EE), which is often conceptualized as chronic interpersonal stress. For example, Miklowitz et al. (31) studied levels of familial EE and psychiatric relapse in 23 bipolar manic patients over nine months. Ratings of EE were based on the amount of criticism, hostility, and emotional overinvolvement found in the key relatives of patients. Manic or depressive relapse was 5.5 times more likely in patients from high EE homes than those from low EE homes. Priebe et al. (57) conducted a small study of EE in the key relatives of 21 patients with bipolar and schizoaffective disorder who had been well established on lithium therapy. Patients living with high EE relatives were found to be less stable, both in the three-year period before and in the nine-month period after the initial assessment. Similarly, Yan and colleagues (58) found that EE was marginally predictive of depressive relapse ($p < .06$) but not manic relapse among 47 medicated bipolar patients. Interestingly, Miklowitz and colleagues (59) found that although severity of criticism received from relatives was not associated with symptom outcome or proportion of well days at one year, the degree to which the patient was upset by criticism did predict symptom severity ($p < .02$) as well as well days ($p < .003$ when controlling for several other factors). However, this may speak more to the impact of individual differences in interpersonal sensitivity rather than EE per se. Regardless, high levels of familial EE (or patient negative reaction to EE) can easily be viewed as a chronic interpersonal stressor, and these results provide additional evidence suggestive of a significant association between stress and relapse.

STRESS AND RECOVERY FROM AN EPISODE

Based on the assumption that "reactive" episodes may be more responsive to treatment than "endogenous" episodes, several studies have set out to examine the connections between stress and recovery from affective episodes. In unipolar depressed patients, the findings have been mixed. Some studies reported that life stress occurring before treatment entry predicted a positive treatment response (60), whereas others found the opposite (61,62) or found no association at all (63). Unfortunately, only two studies to date have looked at stress and recovery from bipolar episodes. Johnson and Miller (64) found that bipolar I patients who experienced at least one severe negative life event during an episode took more than three times as long to achieve recovery as those without severe life events. However, a study by Hlastala and colleagues (65) did not replicate these results. In that sample of 96 bipolar I patients, the occurrence of acute and chronic stressors occurring during treatment was not significantly related to time to recovery from manic, depressed, or mixed/cycling episodes. Several differences between the studies may have contributed to these discrepant findings. Primary among these differences is that patients in the Hlastala et al. study (65) were participating in a randomized controlled study with strictly defined treatment protocols. In contrast, the Johnson and Miller study (64) was a naturalistic one with variable types, quantity, and quality of treatments. Therefore, in the Hlastala et al. study, it is possible that the "ceiling" effect of standardized treatment attenuated the impact of life events on time to remission. Obviously, more research needs to focus on the relationship between life stress and such important long-term course parameters as recovery and treatment response.

SUMMARY

Although the methodological inconsistencies of the previous research make it difficult to draw any strong conclusions, stressful life events appear to operate as contributing factors in a significant number of patients with BD. Most previous studies have shown evidence for this relationship, whereas the few studies that were unable to confirm the relationship were laden with methodological problems. Many researchers who failed to find a relationship between life events and episode onset, in fact, concluded that the null findings may reflect the possibility that life stress only plays a role in precipitating episodes early in the course of illness, as Post (46) has suggested (see subsequently and Chapter 16 of this volume).

BIPOLAR DISORDER-SPECIFIC MECHANISMS OF THE STRESS-EPISODE RELATIONSHIP

Kindling and Behavioral Sensitization

According to the kindling/behavioral sensitization model of recurrent affective disorders (46,66), the first episodes of BD are more likely to be associated with major psychosocial stressors than are episodes occurring later in the course of the illness. The kindling model relies on two principles of neuropsychological research: electrophysiological kindling (progressive vulnerability to seizures) and behavioral sensitization (progressive change in psychomotor stimulant response). Research on the development of amygdala-kindled seizures in response to electrical stimulation has shown that after a sufficient number of electrically induced seizures, spontaneous epilepsy will occur, even in the absence of electrical triggers. When used to understand recurrent affective disorders, the model is a *nonhomologous* paradigm that hypothesizes that "exogenous" stress (i.e., a precipitating "stimulant") will be less likely to be associated with the onset of new episodes as the disorder progresses, resulting in spontaneous episodes and/or rapid cycling. In other words, this model implies that bipolar patients become so sensitized to stress, as a consequence of experiencing multiple episodes, that they eventually require only the slightest stimulus to precipitate relapse.

Post (46) has described two types of sensitization: stressor sensitization and episode sensitization. Episode sensitization is the phenomenon of increased severity, increased symptom profile, and decreased time between episodes over the course of the disorder. Some research has shown that cycle length tends to shorten with each recurrence (67). However, this phenomenon is not found in all bipolar patients. Roy-Byrne et al. (68) noted after following 50 bipolar patients for over five years that approximately 52% showed a sensitization pattern of progressively shorter intervals during the course of the illness, whereas 48% showed no particular pattern with longer intervals randomly distributed throughout the course of the illness. Additionally, in those patients who experience a rapid-cycling pattern, most return to non–rapid-cycling episode patterns within a three-year period (69).

Stressor sensitization is said to occur when the provocation of new episodes, later in the course of the illness, requires lower and lower levels of stress. The role of psychosocial stressors in the initial onset of affective disorders was noted by early descriptive psychopathologists. Kraepelin (70) stated that new episodes of manic-depressive illness "begin not infrequently after the illness or death of a relative," but stress should not be viewed as necessary for the onset of recurrent episodes because they may be "to an astonishing degree independent of external influences." Additionally, Stern (71) postulated that subsequent episodes, following the activation of the "manic mechanism," require less in the way of stress.

Post (46), in his review of the evidence for the kindling model, based his conclusions primarily on studies examining recurrent unipolar depression. Only six studies have looked at the possibility of differential effects of stress as a function of the number of episodes experienced in bipolar I patients. Of those six studies, three lend support to the model, whereas three do not. Dunner et al. (72) examined the occurrence of stressful life events before the initial and subsequent episodes in 79 bipolar I patients. Fifty percent reported an event before the first episode, whereas only 15% reported an event preceding subsequent episodes. Ambelas (73) also found a significant difference in life event rates among bipolar patients with one episode versus those with repeated episodes. Sixty-six percent of patients in their first episode of mania experienced a stressful event in the four weeks preceding onset, whereas only 50% of individuals in repeat admissions appeared to have experienced an event.

Similarly, Glassner et al. (74) found a stronger association between stressful life events and onset in the first episode compared with subsequent episodes in 25 bipolar I patients. Seventy-five percent reported experiencing a life event in the year before their first episode, whereas 56% reported experiencing a life event in the year before their most recent episode. However, after increasing their sample size, they did not find the same pattern (24). The rates of stressful life events in bipolar patients who were classified as having either an early ($\leq$20 years old) or late onset were similar in the year before their first episode or latest episode.

Swendsen et al. (75) used a LEDS-based methodology to examine the occurrence of stressful life events in 45 bipolar I patients for one year after achieving clinical remission or their best clinical state. They found that patients who experienced moderate to severe levels of stress were significantly more likely to experience a recurrence than patients who experienced no or minimal stress, regardless of how many previous episodes they had experienced. High stress was also a significant predictor of recurrence among those with 12 or greater previous episodes. The author concluded that these results, obtained using a more rigorous measure of life stress than previous studies, are inconsistent with the kindling model.

In a more direct test of the kindling model, Hammen and Gitlin (76) followed 52 bipolar I patients for two years, while interviewing every three months for major life events using a LEDS-based methodology. Over the two-year follow-up period, 36 patients experienced either a relapse or recurrence. The relapse/recurrence group was more likely than the episode-free group to have experienced a severe life event within the six months before the relapse/recurrence. Forty percent of the subjects with eight or less previous episodes experienced a major life event in the six months preceding relapse/recurrence, whereas an even greater proportion (76%) of those subjects with nine or more previous episodes experienced a major life event in the six months preceding relapse/recurrence. Interestingly, backward survival analyses revealed that those with nine or more episodes relapsed more rapidly after a major event than the group with fewer episodes. These results seem to provide contradictory evidence when applied to the kindling model. They suggest that patients with more episodes may be more reactive to stress because they relapsed more quickly after a life event, which supports a model of stressor sensitization. However, the data also suggest that patients with a greater number of episodes may still need high levels of stress preceding onset, which contradicts what has traditionally been viewed as support for the kindling model.

Our own work testing Post's hypotheses (77) has not demonstrated compelling evidence in support of the kindling model. Sixty-four bipolar I patients, who were participants in the Pittsburgh Study of Maintenance Therapies in Bipolar Disorder, were interviewed about a year before onset of their index episode. The Bedford College LEDS (19) was used to determine severe events (events with threat level rated as the highest two points on a four-point scale) and nonsevere events (events with threat level rated as the lowest two points on a four-point scale) occurring during the three months before episode onset and during a three month episode-free control period. Based on these ratings, each patient was categorized as having experienced either "high," "moderate," or "low" stress during the three month observation periods. Cumulative logit analyses were used to examine the relationship between number of previous episodes and the type of stress (i.e., high, moderate, and low) experienced in both the preonset and control period. The analyses did not support the hypotheses that the number of episodes experienced would predict stress level in the preonset and control periods. However, age was found to be a significant predictor of stress level in the preonset period ($p = 0.03$), with older subjects evidencing lower levels of stress before episode onset than younger subjects. These results suggest that a more complex relationship may exist among age, stress, and onset of new episodes than can be adequately explained by the kindling model. Additionally, previous research supporting a "kindling" process in bipolar patients may be an artifact of ignoring the effects of age when examining a longitudinal process.

Summary

Although Post's model has been accepted by many researchers and clinicians, the evidence for kindling/behavioral sensitization in bipolar patients is equivocal at best. The studies that appear to support the model have extensive methodological weaknesses, including reliance on chart review (73) and self-report checklist questionnaires of stressful life events (72). Additionally, one study (74) expected the subjects to recall as far back as 10 years or more to obtain information on life events preceding their first episodes. Although the studies that are inconsistent with the kindling model (75–77) are methodologically stronger (i.e., interview- based LEDS methodology and prospective design), further research is needed before any strong conclusions can be made.

Finally, Post's kindling model is a very difficult model to test adequately because of its internal inconsistencies. Implicit in this model are two distinct phenomena that are tied together—kindling and sensitization. The kindling aspect of the model presupposes that the illness becomes autonomous after experiencing a certain number of episodes and stressors. The sensitization aspect of the model is inherently different, however. Implicit in a sensitization model is that patients with recurrent affective disorder become increasingly sensitized to stress as a function of how many affective episodes they have experienced. Thus, very minor amounts of stress should play an increasingly greater role in the onset of new episodes.

This inconsistency has led to multiple interpretations of past research, such that depending on how the data are approached, the kindling/sensitization models (which are really two distinct although conceptually linked models) may or may not be "supported." For example, Hammen and Gitlin (76), as described previously, found that patients with a greater number of episodes relapsed more quickly than patients with fewer episodes after a stressful life event. They concluded that the data do not support the kindling model because these patients did not have "autonomous" episodes (i.e., stress had an even greater effect on episode onset among those who had experienced many episodes). However, another conclusion from this data could be that those patients with more episodes were "sensitized" to stress, thus they relapsed more quickly after a stressful event. Although Post (46) concedes that "the processes involved in sensitization may be more directly analogous to those occurring in the affective disorders because of the behavioral rather than convulsive endpoints observed" (p.1001), the model continues to be interpreted in various ways by researchers and clinicians. For an excellent review of these issues as they pertain to depression, see Monroe and Harkness' 2005 review (78).

STRESS, HPA AXIS DYSFUNCTION, AND BIPOLAR EPISODES

A large body of research has shown a strong relationship between stress and increased hypothalamic-pituitary-adrenal (HPA) axis activity. Elevations of epinephrine, norepinephrine, and cortisol have repeatedly been found among persons experiencing chronic and acutely stressful events (79). Thus, early studies demonstrating elevated cortisol levels in depressed patients were generally regarded as a normal adaptational response to a stressed state (i.e., depression) (80). However, some investigators viewed such elevations as being the result of an abnormality in the HPA system (81).

Extensive research on unipolar depressed patients partially supports this hypothesis. Elevated levels of glucocorticoids in cerebrospinal fluid, plasma, saliva, and/or urine, which is commonly used as evidence for HPA axis dysfunction, have been found in more than one half of hospitalized depressed patients (82,83). In addition, a large proportion of depressed patients show evidence of impaired HPA feedback inhibition (84,85) and abnormal circadian regulation of the HPA axis (86,87).

Unfortunately, the findings on the relationship between HPA axis dysfunction and bipolar episodes are not as clear. In general, the previous research has found more evidence for increased HPA activity in bipolar depression than in mania. Some studies have found normal HPA activity in patients experiencing pure manic episodes (88–90), although some studies have not (84,91). Some work has shown that DST response is more exaggerated in bipolar depression than in unipolar depression, even when in recovery from a major depressive episode (92). In a critical review of the literature on dexamethasone suppression test (DST) in bipolar episodes, Goodwin and Jamison (67) concluded that DST nonsuppression occurs more frequently in the depressive and mixed phases than pure manic phases of the illness. One interesting study (93) found that among three rapid-cycling patients, DST results were abnormal in and before depressed episodes but normal during manic episodes. However, contradictory evidence can be found in the work of Cervantes and colleagues (94). In their small sample of patients with bipolar I disorder, abnormal cortisol levels were found among patients in both depressed ($n = 5$) and manic ($n = 5$) states. Many studies agree that HPA abnormality usually resolves itself after recovery from an episode (95,96), indicating that DST nonsuppression is a state marker rather than an underlying biological abnormality. However, not all studies reach the same conclusion. For example, Cervantes et al. (94) found elevated cortisol levels among even euthymic bipolar patients ($n = 8$) when compared with non-psychiatric controls.

There has been some evidence that HPA dysregulation may have utility in predicting episode onset. Goodyer et al. (97) found that HPA abnormality prospectively predicted the occurrence of major depressive episodes among adolescents at high risk for depression. Further, Vieta and colleagues (98) found evidence that remitted bipolar patients with elevated corticotrophin levels had higher incidents of mania during the subsequent six months. One possibility is that HPA dysfunction is caused by acute or chronic stress in vulnerable individuals, promoting mood episodes. Ellenbogen et al. (99) found that the non-disordered children of bipolar parents showed abnormally elevated cortisol levels across the day. They suggest that prolonged exposure to stress (due to the difficulties associated with living with a psychiatrically ill parent) may have promoted HPA changes in these children. Brown and colleagues further suggest that prolonged exposure to elevated cortisol levels may alter the structure of the hippocampus, both promoting mild cognitive abnormalities and increasing vulnerability to stress (100). They suggest that this feedback loop may be promoted by initial mood episodes and may facilitate future episodes.

In addition to a more biological mechanism, hypercortisolemia as a result of HPA axis dysfunction may play a role in the onset of bipolar episodes through its ability to affect sleep rhythms. It has been well established that most patients experiencing an episode of either mania or depression display specific disturbances in their biological rhythms (101,102), particularly in their sleep-wake cycle (101,103). Corticosteroids have an ability to disrupt sleep by causing increased awakenings and decreased slow wave sleep (104). Thus, a person may experience one or more "sleepness nights" not only because of the psychological sequelae, but because of the hormonal effects of acute or chronic stress resulting in a bipolar episode. This sleep disruption mechanism parallels another hypothesized pathway to the onset of manic and depressive episodes in bipolar patients—social rhythm disruption (105,106).

SOCIAL RHYTHM DISRUPTION AND BIPOLAR EPISODES

Another promising area of research connecting environmental stressors to the onset of new episodes has been through their ability to disrupt social and circadian rhythms (105,106). Goodwin and Jamison (67) have integrated large volumes of research on the pathophysiology of bipolar episodes, postulating that "instability is the fundamental dysfunction in manic depressive illness" (p.594). One particularly salient contributor to this instability may be a disruption in a patient's social routines. The social zeitgeber and biological rhythm theory (105,106) articulates this pathway, which is both biological and behavioral in nature, through which life events may precipitate and/or exacerbate affective episodes.

This theory suggests that stressful life events can often act as "social agents of circadian disruption" by causing alterations in social routines that ordinarily act to synchronize circadian rhythms, leading to the onset of affective episodes in vulnerable persons. For example, when an individual loses a typical "nine to five" job, he or she may not only experience certain psychological aspects of this loss but also lose a significant social zeitgeber (i.e., social cues and demands that act to entrain the biological clock.). Most likely that person will alter his or her regular sleeping times, meal times, and times of activity and rest. Such disruptions in social routines could act to disrupt circadian rhythms that have been implicated in the pathogenesis of both depression and mania (102,107–110).

To date, only one study has directly examined the role of social rhythm disruption (SRD) in the onset of depressive and manic episodes in bipolar I patients. Our research group (111) rated life event descriptions obtained using LEDS (20) for degree of SRD during eight-week preonset and control periods. Life events that were classified as disrupting social routines (i.e., "SRD events") were associated with the onset of manic, but not depressive, episodes. Because social rhythm disruption may have more gradual effects on depressive compared with manic onsets, the eight-week preonset window may have been of insufficient duration to observe an association between SRD events and bipolar depressive episodes. Alternatively, SRD events may be important in the onset of manic but not depressive episodes.

Further evidence of the relationship between social rhythm disruption and bipolar episodes comes from a treatment outcome study conducted in our clinic (112). This study compared Interpersonal and Social Rhythm Therapy (IPSRT) with intensive clinical management (ICM) for bipolar I disorder. The IPSRT treatment emphasized increased schedule regularity (in addition to psychotherapy and medication management) as one component of the treatment while the ICM arm consisted of education about BD, sleep hygiene, non-specific support, and medication management. Results indicated that participants who received IPSRT in the active phase of treatment had longer survival time before a bipolar recurrence and were less likely to have a recurrence in the two year maintenance phase. Further, increased social rhythm regularity during the acute treatment phase was associated with reduced odds of recurrence in the maintenance phase. This work provides empirical support for the theory that social rhythms play an important role in the course of BD.

CONCLUSIONS

External stressors appear to play a substantial role in the course of BD. Stressful life events appear to influence not only the timing of initial and recurrent episodes but also the recovery from what is often a difficult, even excruciating, struggle to recover from such episodes. Further elucidation of the mechanisms through which stress may affect bipolar episodes is clearly needed. Unfortunately, not enough is known about exactly how stress might precipitate bipolar episodes. Biological mechanisms, such as neuroendocrine abnormalities, and behavioral mechanisms, such as SRD, appear to yield interesting findings. Researchers and clinicians alike would benefit greatly from a better understanding of different biological, behavioral, and psychological pathways that relate external stressors to the onset and maintenance of affective episodes.

Beside the specific mediators of stress that were discussed earlier, moderators of the stress-episode relationship need to be examined. For example, in our own work, it remains to be seen whether increasing the regularity of daily routines mediates the SRD–onset relationship. Analysis of this relationship will be undertaken shortly. Other psychosocial variables such as personality factors, social support, and self-esteem may play a large role in how a patient with BD reacts to both chronic and acute stressors. Research on unipolar depressed patients suggests that life stress may have more potent effects if it is related to a patient's particular personality or cognitive vulnerabilities (113,114). However, this area of research has not received strong support in the literature so far. Although Hammen et al. (56) were able to find support for this phenomenon in unipolar depressed patients, they did not in bipolar patients (55). Interestingly, events that were rated as interpersonal in nature predicted higher symptom severity scores in all patients, regardless of their personality type. Perhaps bipolar patients are especially vulnerable to problems that arise in the context of their relationships with others.

The effects of patient-generated stressors are also commonly ignored. It is likely that patients with bipolar illness tend to generate moderately stressful events as a result of their illness and that these events in turn serve to delay or prevent remission or to provoke new symptomatic exacerbations. This idea is similar to that proposed by Hammen (115), who concluded that women with recurrent unipolar depression tend to generate stressful conditions by their symptoms, behaviors, characteristics, and the social context in which they find themselves. Both interpersonal and practical (e.g., lack of employment, housing, etc.) problems that occur as a consequence of a manic or depressive episode could contribute to a chronic or worsening course of illness over the long term.

Further emphasis on the role of stressors in the pathophysiology of BD is a promising way to integrate biological and psychosocial perspectives on the etiology and long-term course of BD. We must keep in mind that even the most "biologically based" illnesses exist in a psychosocial context; thus, a more complete understanding of psychosocial influences could provide useful information for the treatment and prevention of future episodes in patients suffering from this often-disabling illness.

REFERENCES

1. Cade JFJ. Lithium salts in the treatment of psychotic excitement. Med J Aust 1949; 36:349–352.
2. Gelenberg AJ, Kane JM, Keller MB, et al. Comparison of standard and low serum levels of lithium for maintenance treatment of bipolar disorder. N Eng J Med 1989; 321:1489–1493.
3. Goldberg JF, Harrow M, Grossman LS. Course and outcome in bipolar affective disorder: a longitudinal follow up study. Am J Psychiatry 1995; 152:379–384.
4. Markar HR, Mander AJ. Efficacy of lithium prophylaxis in clinical practice. Br J Psychiatry 1989; 155:496–500.
5. Prien RF, Kupfer DJ, Mansky PA, et al. Drug therapy in the prevention of recurrences in unipolar and bipolar affective disorder: report of the NIHM collaborative study group comparing lithium carbonate, imipramine, and a lithium carbonate-imipramine combination. Arch Gen Psychiatry 1984; 41:1096–1104.
6. Shapiro DR, Quitkin FM, Fleiss JL. Response to maintenance therapy in bipolar illness: effect of index episode. Arch Gen Psychiatry 1989; 46:401–405.
7. Swartz HA, Frank E, Kupfer JD. Psychotherapy of bipolar disorder. In: American Psychiatric Publishing Textbook of Mood Disorders. Stein DJ, Schwartzberg AF, eds. Washington, DC: American Psychiatric Press Publishing, 2006. In press.
8. Perry A, Tarrier N, Morriss R, McCarthy E, Limb K. Randomised controlled trial of efficacy of teaching patients with bipolar disorder to identify early symptoms of relapse and obtain treatment. BMJ (Clinical research ed.) 1999; 318:149–153.
9. Colom F, Vieta E, Martinez-Arán A, et al. A randomized trial on the efficacy of group psychoeducation in the prophylaxis of recurrences in bipolar patients whose disease is in remission. Arch Gen Psychiatry 2003; 60:402–407.
10. Lam DH, Watkins ER, Hayward P, et al. A randomized controlled study of cognitive therapy for relapse prevention for bipolar affective disorder: outcome of the first year. Arch Gen Psychiatry 2003; 60:145–152.
11. Gitlin MJ, Swendsen J, Heller TL, Hammen C. Relapse and impairment in bipolar disorder. Am J Psychiatry 1995; 152:1635–1640.
12. Brown G. Life events, psychiatric disorder, and physical illness. J Psychosom Res 1981; 25:461–447.
13. Dohrenwend BP, Link BG, Kern R, Shrout PE, Markowitz J. Measuring life events: the problem of variability within event categories. In: Cooper B, ed. Psychiatric Epidemiology: Progress and Prospects. London: Croom Helm, 1987:103–119.
14. Dohrenwend BS, Dohrenwend BP. Overview and prospects for research on stressful life events. In: Dohrenwend BS, Dohrenwend BP, eds. Stressful Life Events: Their Nature and Effects. New York: Wiley and Sons, 1974:313–331.
15. Raphael KG, Cloitre M, Dohrenwend BP. Problems with recall and misclassifications with checklist methods of measuring stressful life events. Health Psychol 1991; 10:62–74.

16. McQuaid JR, Monroe SM, Roberts JR, et al. Toward the standardization of life stress assessments: definitional discrepancies and inconsistencies in methods. Stress Med 1992; 8:47–56.
17. Cannell CF, Miller PV, Oksenberg L. Research on interviewing techniques. In: Leinhardt S, ed. Sociological Metholodology. San Francisco: Jossey-Bass 1981:389–436.
18. Selye H. The Stress of Life. New York: McGraw-Hill, 1956.
19. Brown GW, Harris TO. Social Origins of Depression: A Study of Psychiatric Disorder in Women. New York: Free Press, 1978.
20. Brown GW, Harris TO. Life Events and Illness. New York: Guilford Press, 1989.
21. McQuaid JR, Monroe S, Roberts JER, Kupfer DJ, Frank E. A comparison of two life stress assessment approaches: Prospective prediction of treatment outcome in recurrent depression. J Abnorm Psychol, 2000; 109(4):787–791.
22. Johnson SL, Roberts JE. Life events and bipolar disorder: implications from biological theories. Psychol Bull 1995; 17:434–449.
23. Hillegers MHJ, Burger H, Wals M, et al. Impact of stressful life events, familial loading and their interaction of the onset of mood disorders: study in a high-risk cohort of adolescent offspring of parents with bipolar disorder. Br J Psychiatry 2004; 185:97–101.
24. Glassner B, Haldipur CV. Life events and early and last onset of bipolar disorder. Am J Psychiatry 1983; 140:215–217.
25. Hall KS, Dunner DL, Zeller G, Fieve RR. Bipolar illness: a prospective study of life events. Comp Psychiatry 1977; 18:497–505.
26. Mayo JA. Psychosocial profiles of patients on lithium treatment. Int Pharmacopsychiatry 1970; 5:190–202.
27. Sarason IG, Johnson JH, Siegel JM. Assessing the impact of life changes: development of the life experiences survey. J Consult Clin Psychology 1978; 46:932–946.
28. Endicott J, Spitzer RL. A diagnostic interview: the schedule for affective disorders and schizophrenia. Archives of General Psychiatry 1978; 35:837–844.
29. Paykel ES. The inventory for recent life events. Psychological Medicine 1997; 27:301–310.
30. Kennedy S, Thompson R, Stancer HC, Roy A, Persad E. Life events precipitating mania. Br J Psychiatry 1983; 142:398–403.
31. Miklowitz DJ, Goldstein MJ, Neuchterlein KH, Snyder KS, Mintz J. Family factors and the course of bipolar affective disorder. Arch Gen Psychiatry 1988; 45: 225–231.
32. Ellicott A, Hammen C, Gitlin M, Brown G, Jamison K. Life events and the course of bipolar disorder. Am J Psychiatry 1990; 147:1194–1198.
33. Sclare P, Creed F. Life events and the onset of mania. Br J Psychiatry 1990; 156:508–514.
34. Hunt N, Bruce-Jones W, Silverstone T. Life events and relapse in bipolar affective disorder. J Affect Disord 1992; 25:13–20.
35. Bebbington P, Wilkins S, Jones P, et al. Life events and psychosis: Initial results from the Camberwell Collaborative Psychosis Study. Br J Psychiatry 1993; 162:72–79.
36. Chung RK, Langeluddecke P, Tennant C. Threatening life events in the onset of schizophrenia, schizophreniform psychosis and hypomania. Br J Psychiatry 1986; 148:680–685.
37. McPherson H, Herbison P, Romans S. Life events and relapse in established bipolar affective disorder. Br J Psychiatry 1993; 163:381–385.
38. Coryell W, Endicott J, Reich T, Andreasen N, Keller M. A family study of bipolar II disorder. Br J Psychiatry 1984; 145:49–54.
39. Endicott J, Nee J, Andreasen N, Clayton P, Keller M, Coryell W. Bipolar II: combine or keep separate? J Affect Disord 1985; 8:17–28.
40. Paykel ES. Interview for recent life events. London: Department of Psychiatry, St. George's Hospital Medical School, 1980, Unpublished manuscript.
41. Vaughn CE, Leff JP. The measurement of expressed emotion in the families of psychiatric patients. British Journal of Social and Clinical Psychology 1976; 15:157–165.
42. Dohrenwend BS, Krasnoff L, Askenasy AR, Dohrenwend BP. Exemplification of a method for scaling life events: the PERI Life Events Scale. J Health Social Behav 1978; 19:205–229.

43. Spitzer RL, Endicott J, Robins E. Research diagnostic criteria: rationale and reliability. Archives of General Psychiatry 1978; 35:773–782.
44. Holmes TH, Rahe RH. The social readjustment rating scale. J Psychosom Res 1967; 11:213–218.
45. Bech P, Bolwig TG, Kramp P, Rafaelsen OJ. The Bech-Rafaelsen Mania Scale and the Hamilton Depression Scale: Evaluation of homogeneity and inter-observer reliability. Acta Psychiatrica Scandinavica 1979; 59:420–430.
46. Post RM. Transduction of psychosocial stress into the neurobiology of recurrent affective disorder. Am J Psychiatry 1992; 149:999–1010.
47. Ambelas A. Psychologically stressful events in the precipitation of manic episodes. Br J Psychiatry 1979; 135:15–21.
48. Joffe RT, MacDonald C, Kutcher SP. Life events and mania: A case-controlled study. Psychiatry Res 1989; 30:213–216.
49. Francis A, Gasparo P. Interval between symptom onset and hospitalization in mania. J Affect Disord 1994; 31:179–185.
50. Cohen AN, Hammen C, Henry RM, Daley SE. Effects of stress and social support on recurrence in bipolar disorder. J Affect Disord 2004; 82:143–147.
51. Johnson SL, Sandrow D, Meyer B, et al. Increases in manic symptoms after life events involving goal attainment. J Abnorm Psychol 2000, 109(4):721–727.
52. Malkoff-Schwartz S, Frank E, Anderson BP, et al. Social rhythm disruption and stressful life events in the onset of bipolar and unipolar episodes. Psychol Med 2000; 30:1005–1016.
53. Reilly-Harrington NA, Alloy LB, Fresco DM, Whitehouse WG. Cognitive styles and life events interact to predict bipolar and unipolar symptomatology. J Abnorm Psychol 1999; 108(4):567–578.
54. Alloy LB, Reilly-Harrington N, Fresco DM, Whitehouse WG, Zechmeister JS. Cognitive styles and life events in subsyndromal unipolar and bipolar disorders: stability and prospective prediction of depressive and hypomanic mood swings. J Cog Psychother 1999; 13(1):21–40.
55. Hammen C, Ellicott A, Gitlin M, Jamison KR. Sociotropy/autonomy and vulnerability to specific life events in patients with unipolar depression and bipolar disorders. J Abnorm Psychol 1989; 98:154–160.
56. Hammen C, Ellicott A, Gitlin M. Stressors and sociotropy/autonomy: a longitudinal study of their relationship to the course of bipolar disorder. Cogn Ther Res 1992; 16:409–418.
57. Priebe S, Wildgrube C, Muller-Oerlinghausen B. Lithium prophylaxis and expressed emotion. Br J Psychiatry 1989; 154:396–399.
58. Yan LJ, Hammen C, Cohen AN, Daley SE, Henry RM. Expressed emotion versus relationship quality variables in the prediction of recurrence in bipolar patients. J Affect Disord 2004; 83:199–206.
59. Miklowitz DJ, Wisniewski SR, Miyahara S, Otto MW, Sachs GS. Perceived criticism from family members as a predictor of the one-year course of bipolar disorder. Psychiatry Res 2005; 136:101–111.
60. Monroe SM, Depue RA. Life stress and depression In: Becker J, Kleinman A, eds. Psychosocial Aspects of Depression. Hillsdale, NJ: Erlbaum, 1991:101–130.
61. Monroe SM, Kupfer DJ, Frank E. Life stress and treatment course of recurrent depression 1. Response during index episode. J Consul Clin Psychol 1992; 60:718–724.
62. Zimmerman M, Pfohl B, Coryell WS, Stangl D. The prognostic validity of DSM –III Axis IV in depressed inpatients. Am J Psychiatry 1987; 144:102–106.
63. Lloyd C, Zisook S, Click MJ, Jaffe KE. Life events and response to antidepressants. J Hum Stress 1981; 7:2–15.
64. Johnson SL, Miller I. Negative life events and time to recovery from episodes of bipolar disorder. J Abnorm Psychol 1997; 106:449–457.
65. Hlastala SA, Frank E, Rucci P, et al. The influence of psychosocial stressors and psychotherapy on treatment response in bipolar I disorder. Bipolar Disorders 2001, 3(suppl):42.

66. Post RM, Rubinow DR, Ballender JC. Conditioning and sensitization in the longitudinal course of affective illness. Br J Psychiatry 1986; 149:191–201.
67. Goodwin FK, Jamison KR. Manic Depressive Illness. New York: Oxford University Press, 1990.
68. Roy-Byrne P, Post RM, Uhde TW, Porcu T, Davis D. The longitudinal course of recurrent affective illness: life chart data from research patients at the NIMH. Acta Psychiatr Scand 1985; 71(suppl 317):1–34.
69. Coryell W, Endicott J, Keller M. Rapidly cycling affective disorder: demographics, diagnosis, family history, and course. Arch Gen Psychiatry 1992; 49:126–131.
70. Kraepelin E. Manic-Depressive Insanity and Paranoia. Edinburgh, Scotland: E.S. Livingston, 1921.
71. Stern ES. The psychopathology of manic-depressive disorder and involutional melancholia. Br J Med Psychol 1944; 20:20–32.
72. Dunner DL, Patrick V, Fieve RR. Life events at the onset of bipolar affective illness. Am J Psychiatry 1979; 136:508–511.
73. Ambelas A. Life events and mania: a special relationship. Br J Psychiatry 1987; 150:235–240.
74. Glassner B, Haldipur CV, Dessauersmith J. Role loss and working-class manic depression. J Nerv Ment Dis 1979; 167:530–541.
75. Swendsen J, Hammen C, Heller T, Gitlin M. Correlates of stress reactivity in patients with bipolar disorder. Am J Psychiatry 1995; 152:795–797.
76. Hammen C, Gitlin M. Stress reactivity in bipolar patients and its relation to prior history of disorder. Am J Psychiatry 1997; 154:856–857.
77. Hlastala SA, Frank E, Kowalski J, et al. Stressful life events, bipolar disorder and the "kindling model." J Abnorm Psychol 2000; 109(4):777–786.
78. Monroe SM, Harkness KL. Life stress, the "Kindling" hypothesis, and the recurrence of depression: Considerations from a life stress perspective. Psychol Rev 2005; 112(2):417–445.
79. Baum A, Grunberg N, Singer J. The use of psychological and neuroendocrinological measurements in the study of stress. Health Psychol 1982; 1:217–236.
80. Thase ME, Howland RH. Biological processes in depression. In: Beckham EE, Leber WR, eds. Handbook of Depression, 2nd ed. New York: Guilford Press, 1995; 213–279.
81. Carrol BJ, Curtis GC, Mendels J. Neuroendocrine regulation in depression. Limbic system-adrenocortical dysfunction. Arch Gen Psychiatry 1976; 33:1039–1044.
82. Haskett RF. The HPA axis and depressive disorders. In: Mann JJ, Kupfer DJ, eds. Biology of Depressive Disorders. Part A. A Systems Perspective. New York: Plenum Press, 1993:171–188.
83. Holsboer F. The hypothalamic-pituitary-adrenocortical system. In: Paykel ES, ed. Handbook of Affective Disorders. New York: Guilford Press, 1992:267–287.
84. Arana GW, Baldessarini RJ, Ornsteen M. The dexamethasone suppression test for diagnosis and prognosis in psychiatry: commentary and review. Arch Gen Psychiatry 1985; 42:1193–1204.
85. Stokes PE, Sikes DR. The hypothalamic-pituitary-adrenocortical axis in major depression. Endocrinol Metab Clin North Am 1988; 17:1–19.
86. Jarrett DB, Miewald JM, Fedorka IB, Coble P, Kupfer DJ, Greenhouse JB. Prolactin secretion during sleep: a comparison between depressed patients and healthy control subjects. Biol Psychiatry 1987; 22:1216–1226.
87. Sherman B, Pfohl B, Winokur G. Circadian analysis of plasma cortisol levels before and after dexamethasone administration in depressed patients. Arch Gen Psychiatry 1984; 41:271–275.
88. Swann AC, Stokes PE, Casper R, et al. Hypothalamic-pituitary-adrenocortical function in mixed and pure mania. Acta Psychiatr Scand 1992; 85:270–274.
89. Evans DL, Nemeroff CB. The dexamethasone suppression test in bipolar disorder. Am J Psychiatry 1983; 104:615–617.
90. Schlesser MA, Winokur G, Sherman BM. Hypothalamic-pituitary-adrenal axis activity in depressive illness. Arch Gen Psychiatry 1980; 37:737–743.

91. Graham PM, Booth J, Boranga G, et al. The dexamethasone suppression test in mania. J Affect Disorder 1982; 4:201–211.
92. Rybakowski JK, Twardowska K. The dexamethaxone/corticotropin releasing hormone test in depression in bipolar and unipolar affective illness. J Psychiatric Res 1999; 33:363–370.
93. Greden JF, DeVigne JP, Albala AA, et al. Serial dexamethasone suppression tests among rapidly cycling bipolar patients. Biol Psychiatry 1982; 17:455–462.
94. Cervantes P, Gelber S, Ng Ying Kin FNK, et al. Circadian secretion of cortisol in bipolar disorder. J Psychiatry Neurosci 2001; 26(5):411–416.
95. Carrol BJ. The dexamethasone suppression test for melancholia. Br J Psychiatry 1982; 140:292–304.
96. Joyce PR, Paykel ES. Predictors of drug response in depression. Arch Gen Psychiatry 1989; 46:89–99.
97. Goodyer IM, Herbert J, Tamplin A, Altham PME. First-episode major depression in adolescents: affective, cognitive and endocrine characteristics of risk status and predictors of onset. Br J Psychiatry 2000; 176:142–149.
98. Vieta E, Martinez-de-Osaba MJ, Colom F, et al. Enhanced corticotropin response to corticotropin-releasing hormone as a predictor of mania in euthymic bipolar patients. Psychological Medicine 1999, 29(4):971–978.
99. Ellenbogen MA, Hodgins S, Walker CD. High levels of cortisol among adolescent offspring of parents with bipolar disorder: a pilot study. Psychoneuroendocrinology 2004; 29:99–106.
100. Brown ES, Rush AJ, McEwen BS. Hippocampal remodeling and damage by corticosteroids: Implications for mood disorders. Neuropsychopharmacology 1999, 21(4):474–484.
101. Wehr TA, Sack DA, Rosenthal NE. Sleep reduction as a final common pathway in the genesis of mania. Am J Psychiatry 1987; 144:201–204.
102. Wehr TA, Wirz-Justice A. Internal coincidence model for sleep deprivation and depression. In: Koella WP, ed. Sleep. Basel, Switzerland: Karger 1980:26–33.
103. Kupfer DJ, Foster FG. Interval between onset of sleep and rapid-eye-movement sleep as an indicator of depression. Lancet 1972; 2:684–686.
104. Born J, DeKloet ER, Wenz H, Kern W, Fehm HL. Gluco- and antimineralocorticoid effects on human sleep: a role of central corticosteroid receptors. Am J Physiol 1991; 22:E183–188.
105. Ehlers CL, Frank E, Kupfer DJ. Social zeitgebers and biological rhythms: a unified approach to understanding the etiology of depression. Arch Gen Psychiatry 1988; 45:948–952.
106. Ehlers CL, Kupfer DJ, Frank E, Monk TH. Biological rhythms and depression: the role of Zeitgebers and Zeitstorers. Depression 1993; 1:285–293.
107. Healy D, Williams JMG. Dysrhythmia, dysphoria, and depression: the interaction of learned helplessness and circadian dysrhythmia in the pathogenesis of depression. Psychol Bull 1988; 103:163–178.
108. Healy D, Williams JMG. Moods, misattributions and mania: an interaction of biological and psychological factors in the pathogenesis of mania. Psychiatr Dev 1989; 1:49–70.
109. Leibenluft E, Frank E. Circadian rhythms in affective disorders. In: Takahashi JS, Turek F, Moore RY, eds. Handbook of Behavioral Neurobiology: Circadian Clocks. New York: Plenum Publishing, 2001:625–644.
110. Wehr TA. Effects of wakefulness and sleep on depression and mania. In: Montplaisir J, Godbout R, eds. Sleep and Biological Rhythms: Basic Mechanisms and Applications to Psychiatry. New York: Oxford University Press, 1990:42–86.
111. Malkoff-Schwartz S, Frank E, Anderson B, et al. Stressful life events and social rhythm disruption in the onset of manic and depressive bipolar episodes: a preliminary investigation. Arch Gen Psychiatry 1998; 55:702–707.
112. Frank E, Kupfer DJ, Thase ME, et al. Two-year outcomes for interpersonal and social rhythm therapy in individuals with bipolar I disorder. Arch Gen Psychiatry 2005; 62:996–1004.
113. Abramson L, Seligman M, Teasdale J. Learned helplessness in humans: critique and reformulation. J Abnorm Psychol 1978; 87:49–74.

114. Oatley K, Bolton W. A social-cognitive theory of depression in reaction to life events. Psychol Rev 1985; 92:372–388.
115. Hammen C. Generation of stress in the course of unipolar depression. J Abnorm Psychol 1990; 100:555–561.
116. Miller IW, Bishop S, Norman WH, Maddever H. The Modified Hamilton Rating Scale for depression: reliability and validity. Psychiatric Research 1985; 14:131–142.

16

The Kindling/Sensitization Model: Implications for the Pathophysiology of Bipolar Disorder

Robert M. Post

Mood and Anxiety Disorders Program, National Institute of Mental Health, Department of Health and Human Services, National Institutes of Health, Bethesda, Maryland, U.S.A.

HISTORICAL OVERVIEW

It is remarkable that almost one hundred years ago Dr. Emil Kraepelin (1) described the essential process of affective illness progression, based on his careful charting of episodes of affective illness and the potential precipitating circumstances with which they were associated. While he spoke of the inherent variability of the illness course both within and between subjects, he noted an overall tendency for the illness to increase in frequency over time, with a decreasing duration of well intervals between successive episodes. At the same time, he noted that the first episodes of mania or depression were often precipitated by psychosocial stressors, but that with the appearance of enough episodes, they would begin to recur in a highly similar form "quite without external occasion" (p. 181) (1). Thus, he captured the ideas of both increasing automaticity and vulnerability to recurrence as a function of number of prior episodes. He attributed this progression to inherent genetic mechanisms interacting with recurrences of stressors and episodes themselves.

In dissecting the inherent course of bipolar illness and its response to treatment, a careful Kraepelinian-like graphic depiction of episodes appeared fundamental to further understanding pathophysiology and treatment responsivity. The new system, termed the National Institute of Mental Health—Life Chart Methodology (NIMH-LCM™) increased the detail of Kraepelin's schema for documenting manic and depressive episodes (2–4). It also allowed for different levels of manic and depressive episode severity (from mild to low and high moderate to severe) based on the degree of dysfunction with which the manic and depressive symptoms were associated. A self-rated version of the NIMH-LCM is also available and includes a rating of mood from 0 to 100, with zero representing "most depressed ever," 50 representing "balanced or euthymic," and anything over 50 toward 100 would be "activated, energized, or more speeded up than usual."

On the patient's rating form, not only was this subjective mood assessment included, but number of hours of sleep, positive and negative life events and stressors, side effects, and comorbid symptoms could also be entered on a daily basis. These ratings and the severity of mania and depression are charted retrospectively on a *monthly* basis along with the medications the patient may or may not have taken, and prospectively on a *daily* basis including dose and numbers of pills taken.

In this way, the prior course of illness could be delineated in a retrospective fashion and continued to be monitored in much greater detail prospectively in order to assess eventual treatment response. A series of specific techniques were used for enhancing recall from patients, family members, and previous physicians, and hospital notes and records were examined to maximize the accuracy of the delineation of the prior illness course. The prospective LCM form has been validated against more conventional cross-sectional elements and, in addition, has inherent face validity and the benefit of a continuous rating system, not one that is intermittently measured with the potential liability of missing major fluctuations in mood and behavior in the interval (5–7).

Using retrospective life chart data, we delineated the overall Kraepelinian tendency for illness progression in our select group of treatment-refractory unipolar and bipolar affectively-ill patients. There was also a small subgroup who cycled very rapidly even from the onset of their illness. Examining the course in those not responding adequately to treatment is a critical point to be emphasized. It is more likely to mirror the Kraepelinian delineation of the course of illness prior to the advent of major psychopharmacological interventions. It is also an area subject to a great deal of misinterpretation, as some critics have suggested that somehow the sensitization/kindling models employed to conceptualize mechanisms underlying illness progression mean that this process is invariably relentless and not subject to treatment intervention. On the contrary, episode recurrence, cycle acceleration, and illness progression can most often be interrupted with adequate treatment. However, intervention later rather than earlier in the course of illness may be associated with greater difficulty in achieving remission and require more complex treatment regimens (8).

Stress and Episode Sensitization: Homologous Models for Affective Illness Prognosis

The fundamentals of the kindling and sensitization model and the predictions derived from them have been spelled out in detail with appropriate supporting references (9–15). Briefly, the sensitization component of the model derives from observations that animals exposed to repeated doses of psychomotor stimulants, in the same behavioral context, show increasing motor activity and stereotypic disorganization to the same dose over time. The psychomotor stimulants in their early phase of use in humans are excellent models for hypomania and mania, and later in the course of use are associated with dysphoric mania and paranoid psychosis. Thus, the sensitization model in animals provides potential mechanisms for considering how recurrent manic episodes could increase severity and invulnerability to recurrence.

In addition, a second component to the sensitization model is important in relationship to predictive validity. Stimulant-induced behavioral sensitization shows cross-reactivity to a variety of stressors (16–18). Following the elucidation of the cross-sensitization between stimulants and stressors, stress-sensitization paradigms began to be studied in their own right. Depending on the quality, intermittency, intensity, and phase of development, initial stressors could lead to long-term increases in reactivity to subsequent stressors.

These preclinical data support evidence for long-lasting modification of reactivity in adults from early life experiences in newborns and infants (19,20). There is also evidence for cross-reactivity between stressors and subsequent adoption of cocaine self-administration habits (Table 1) (21–26), indicating that in some instances there are bi-directional cross sensitivities between stressors and stimulants. These data are particularly noteworthy from the perspective that there is a

TABLE 1 Relationship of Early and Concurrent Stressors to Acquisition, Maintenance, and Reinstatement of Cocaine Self Administration (in Rodents)

Stressor	Age	Cocaine effect	Authors
Social stress	Adult	Acquisition of cocaine self-administration	Haney et al., 1995 (21)
Uncontrollable foot-shock	Adult	Cocaine self-administration (only if corticosteroids elevated)	Goeders and Guerin, 1994 (22)
Social defeat × 4 (sensitization to stress)	Adult	Cocaine amount (binge size) Motor response to stimulants (i.e., cross-sensitization)	Covington and Miczek, 2001 (23)
Stressors or cocaine cues	Adult	Cocaine lever pressing Reinstatement of extinguished response	Goeders and Clampitt, 2002 (24)
Three hour maternal deprivation stress X 7 (neonate)	Stress (neonate) drug intake (adult)	Alcohol and cocaine self administration (reversal with antidepressant treatment)	Huot et al., 2001 (25); Meaney et al., 2002 (26)

markedly increased incidence of substance abuse in those with recurrent unipolar and bipolar affective disorders compared with the general population (27,28).

Kindling as a Nonhomologous Model for Illness Progression

The kindling paradigm provides another, altogether different, model for understanding increased physiological and behavioral reactivity to repeated stimulation of the brain. The most commonly studied type is amygdala kindling, where animals are stimulated electrically once a day for one second at intensities below the threshold for afterdischarges (ADs) (29). The repeated stimulation both lowers the AD threshold, and the ADs emerge with increasing duration and complexity in the amygdala, and subsequently spread throughout the brain. This AD spread is associated with progressive increases in the severity of behavioral response from stage 1 (behavior immobility with whisker twitching) to stage 3 (unilateral forepaw convulsions) to generalized stage 5 (bilateral forepaw convulsions with rearing and falling) (30,31). There is also a parallel spatiotemporal induction of neurotrophic factors, immediate early genes, and late effector genes that likely mediate both the primary long-lasting kindled "memory trace" and short-lived endogenous anticonvulsant adaptations (32). The relative balance of these primary pathological versus secondary adaptive changes in gene expression could explain the intermittent emergence (or not) of seizures and affective episodes.

Goddard et al. (29), who first described the kindling phenomenon, considered it an interesting model for neuronal learning and memory because the effects appeared to be long-lasting, if not permanent. Intermittent stimulation was important, because animals stimulated continuously or once every five minutes never kindled, whereas more periodic stimulation, optimally at once every 24 hours, led to kindling progression (29).

If fully kindled animals are stimulated repeatedly, many will go on to have spontaneous seizures, that is, identical appearing stage 4 and 5 behavioral convulsions to those observed before, but now occurring without amygdala stimulation or even prior to the animal's being picked up or handled. Although there is some homology of affective disorder behaviors and inducing factors in the preclinical models of stress and psychomotor stimulant sensitization noted above, there is obviously no behavioral homology in the kindling paradigm where seizures (as opposed to affective episodes) are the endpoints of interest (33).

Nonetheless, the kindling paradigm is pertinent not only for considering increased reactivity to the same stimulation over time, but also how this process progresses and emerges into a new phenomenon of spontaneity or automaticity in which seizures occur in the absence of the inducing stimuli. Understanding how animals progress from triggered amygdala-kindled seizures to spontaneous ones may give hints to similar transitions occurring in the affective disorders, especially since initial episodes are often triggered by psychosocial stressors and later episodes can occur more spontaneously (10,14). However, a caveat is that the precise neuroanatomical and neurotransmitter systems underlying seizure progression are likely quite different from those underlying affective episode progression.

EVIDENCE FROM ANIMAL STUDIES

The potential mechanisms underlying behavioral sensitization to psychomotor stimulants and amygdala-kindling seizure evolution are reviewed elsewhere (34–42). Here we focus on selected newer data from animal studies which are particularly revealing, and potentially relevant to the clinical observations of long-lasting increased vulnerability to stressors and the precipitation of pathological affective behaviors.

Persisting Effects of Stress on Neurobiology and Behavior

One paradigm well-studied by Levine et al. (43,44) involves separation of rat pups from their mothers for a period of 24 hours (while body temperature is maintained). This results in a long-lasting increase in cortisol secretion and anxiety-like behaviors into adulthood. New data suggest that this one-day stressor is associated with increases in preprogrammed cell death (apoptosis) (45) and in decreases in brain-derived neurotrophic factor (BDNF) (45,46) and calcium calmodulin kinase-II (CaMK-II) (47). These latter two changes are of considerable interest in relation to the critical roles of BDNF and CaMK-II in long-term learning and memory. Animal strains where either of these substances is knocked out fail to remember how to navigate a previously learned water maze and show deficits in long-term potentiation (LTP) (48–50).

Animals subjected to repeated neonatal separation stress show long-lasting decrements in not only hippocampal BDNF (51), but also BDNF in prefrontal cortex (52). Again, these findings are particularly noteworthy, in that decrements in plasma and brain BDNF have been noted in some studies of patients with affective disorders and in brain autopsy specimens (especially when these subjects were not treated with antidepressants) (53–57). Decrements in prefrontal cortical CaMK-II have also been observed selectively in BD, but not in patients with schizophrenia or controls (58).

Interestingly, repeated stressors in neonatal rodents lead to deficits in hippocampal volume and associative learning. In contrast, the amygdala increases in size compared with litter-mate controls and is associated with increased anxiety-like emotional hyper-reactivity (53,54). The elegant studies of Plotsky and Meany deserve special focus because of their potential relevance to long-lasting changes in neurochemistry, endocrinology, and behavior based on early life experience in the rodent. Neonatal rat pups subjected to ten days of 15-minute periods of maternal separation not only show few adverse consequences, but have increased hippocampal volume and preserved learning and memory performance into old age.

In contrast, those neonatal rat pups subjected to repeated maternal separations for three hours (rather than 15 minutes) show lifelong increases in the rat equivalent of human cortisol (corticosterone) and in anxiety-like behaviors (59–62). These stressors have now been associated with increased levels of corticotropin-releasing hormone (CRH) both in hypothalamus and amygdala (59). These and related studies indicate that depending on the quality, duration, and timing of early stressors, the central nervous system (CNS) can have a lifelong change in its set point for peptide, endocrine, and anxiety-related behaviors.

Interestingly, the mechanisms underlying these long-term changes greatly depend on reactions in the infant/mother dyad. In the 15-minute separation paradigm, the rat pup upon reunion is greeted with increased licking and grooming. In contrast, in the three-hour separation paradigm, the frantic mother rodent no longer recognizes the returned rat pup as her own and in her agitation ignores and sometimes even tramples over the returned pup. If a substitute rat pup is put in the litter during the separation period (i.e., the mother is not aware that one of her pups is missing), she remains calm and the rat pup returning from the three-hour separation is dealt with normally and never develops the hypercorticosteronemia or anxiety-related behaviors.

Naturally occurring high-licking maternal behavior is associated with offspring who are low in anxiety and corticosterone. Conversely, low-licking mothers tend to have high anxious, high corticosterone offspring. This apparently familial trait is not genetic because cross-fostering studies indicate that, again, it is maternal behavior and not genetic inheritance that drives these long-lasting changes. Moreover, if a rat pup born of a low-licking mother is fostered by a high-licking mother and that pup has offspring as an adult, those pups retain the high-licking, low-anxiety and low-corticosterone signature, indicating that there can be transgenerational transmission of what may have previously been considered a genetic trait (63).

More recent studies of Meaney et al. have implicated changes in DNA methylation in such transgenerational cross-fostering experiences (64). Moreover, the elegant studies of Insel and associates have also revealed that many traits in animals are based on either: (*i*) in utero environmental experiences, or (*ii*) those occurring in early infancy, or both. These data were gleaned from pregnant females who were cross-implanted or rat pups who were cross-fostered, or both (65).

We cite these studies in the current context to emphasize the environmentally-based malleability to many different types of behavior and endocrine reactivity that may have previously been thought to be transmitted on a genetic basis. In parallel, these studies offer alternative insights for considering how pathological changes in neurochemistry may occur in patients with BD, not only on the basis of well-documented genetic vulnerability, but likely through environmental-experiential routes as well.

Bi-Directional Cross Sensitization Between Stressors and Substances of Abuse

The stressor models noted previously show that maternal stress predisposes animals to the adoption of substance self-administration (25,26), which is particularly pertinent to the recurrent affective disorders. Patients with BD are at markedly increased risk for the adoption of alcohol and substance abuse problems. Not only is manic behavior associated with greater degrees of indiscretion and poor judgment (66), but the stress and episode sensitization mechanisms occurring concurrently with the affective episodes themselves may yield increased vulnerability to the adoption of substance abuse.

Obviously, the cooccurrence of two chronic, difficult-to-treat illnesses (BD and a substance abuse disorder) could markedly complicate treatment and overall illness outcome (67). This result would also be consistent with the neurochemistry data indicating that many of the changes in gene expression induced by chronic stimulant administration appear to mirror (and potentially exacerbate) some of the underlying pathophysiological processes occurring in BD (Fig. 1) (58,68–94).

However, this convergence and overlap of potential pathophysiological mechanisms raises the possibility of treatment simplification and a double improvement effect from a single treatment intervention. This clinically optimistic perspective so far is not in line with the available data indicating that those with comorbid substance abuse have a more difficult course of bipolar illness than those without (95,96). Moreover, recent analyses indicate that even those with a prior history of alcohol abuse (and not current problems) continue to have a more difficult course of BD than those without such a prior history of such difficulties (97). In support of the therapeutic viewpoint, however, are the data that many patients who experience treatment-related stabilization of their bipolar illness are able to abstain from alcohol and substance abuse (98). Moreover, a number of

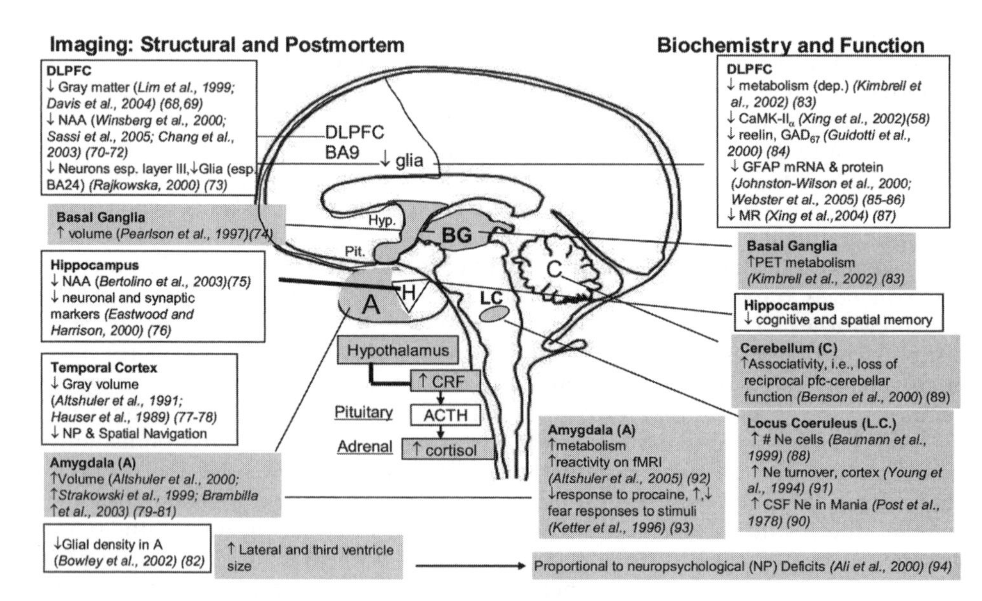

FIGURE 1 Convergence of structural, biochemical, and functional abnormalities in bipolar illness.

medications have been found in controlled studies to be useful in primary substance abuse problems, whether or not these agents are also mood stabilizers (99,100).

EVIDENCE FROM THE COURSE OF ILLNESS AND NEUROIMAGING STUDIES

The data of Kessing et al. (101) based on the Danish case registry, and the more recent replication of this data in Switzerland (102), provide very strong evidence for the episode sensitization phenomenon. These investigators find that the best predictor of increased incidence and shorter latency to a relapse is the number of prior hospitalizations. It is also important and potentially disturbing to note that this relapse occurred despite naturalistic treatment in the community, and one looks forward to assessing whether optimal prophylactic intervention would have prevented this general pattern in those who received it.

The stress sensitization component of recurrent affective disorders is most elegantly dissected in the twin studies of Kendler et al. (103,104) in recurrent unipolar depression. These investigators found that stressors were more likely to occur in close relationship to the precipitation of the initial nine episodes of depression, but became less relevant thereafter. They also examined how this sensitization/ kindling phenomenon interacted with genetic vulnerability. They noted that those with higher risks related to the presence of a strong family history of affective illness appeared to be prekindled, that is, those patients were found to have lesser degrees of stress at the onset of even their initial episodes compared with those without this genetic familial vulnerability.

These data are the precursors to elegant studies of environmental/genetic interactions based on the delineation of common variants in single nucleotide polymorphisms (SNPs) that may be associated with illness onset or treatment response. The data of Caspi et al. (105) provide a remarkable demonstration of such an interaction. They examined a large group of individuals who suffered early traumas in childhood (typically physical or sexual abuse) who were then re-exposed to a stressor in adulthood. The serotonin (5-HT) transporter (5-HT T) to which the serotonin-selective reuptake inhibitor (SSRI) antidepressants bind and are thought, in part, to exert their antidepressant actions, has two common variants, or SNPs. The long (l) variant is a more efficient transporter (5-HT-Tll) compared with the short (s) variant (5-HT-Tss). In the study of Caspi et al. those patients at high risk for depression because of previous life experience who had the long variant (5-HT Tll) appeared relatively immune from stress-induced precipitation of a depressive episode during adulthood. In contrast, those patients with the 5-HT Tss form, and to a lesser extent, those with the sl SNP, showed a significantly higher incidence of depression following these stressors in adulthood.

These findings were partially replicated by Kendler et al. (106), and the paradigm of potent environmental–SNP interactions was also revealed to be of critical importance in another study by Caspi et al. (107). In this instance, marijuana smoking during adolescence yielded an increased incidence of psychosis and schizophrenia only in the subgroup of patients with increased genetic vulnerability because of the catechol-O-methyltransferase polymorphism (COMT val158val allele), which is associated with lesser degrees of prefrontal cortical dopamine levels.

One looks forward to the dissection of these types of more complex gene/environmental interactions in bipolar illness, in which a variety of environmental stressors and substance abuse comorbidities are known to be particularly prevalent. Uncovering these dual vulnerability mechanisms at the level of SNPs and the environment could also eventually assist in the earlier initiation of prophylactic treatment, and in the choice of the most appropriate one.

Evidence of Stressor Involvement in the Onset and Course of Bipolar Disorder

We have observed that childhood adversity (a self-reported history of physical or sexual abuse) in bipolar illness is associated with an earlier illness onset, a more difficult retrospective course of illness, and more Axis I, II, and III comorbidities compared with bipolar patients without such a history (108). These retrospective reports were validated prospectively by clinicians' ratings of patients during naturalistic treatment, and those patients with a history of early adversity had an increased severity of depression and more time depressed in prospective follow-up compared with those without (108,109).

Interestingly, those with a history of early adversity had an increased incidence of negative life events (but not positive events) at both the onset of their illness and in relationship to the most recent episode compared with those without this early adversity. These data raise the issue of not only stress vulnerability, but also of increased proclivity for exposure to stressful life events in those with these early histories of adversity. The sensitization model postulates increased stressor reactivity as a function of stressor recurrences rather than decreased reactivity, so that such an association of increased negative life events at illness onset and at the most recent episode does not contradict the sensitization prediction. Nonetheless, in the relative absence of negative life events, the prediction derived from the sensitization/kindling model would be that those patients with a greater number of prior episodes would be more vulnerable to recurrence than those with fewer episodes.

Neurochemistry and Neuroimaging Studies

We have previously reviewed some of the emerging evidence for functional and neurochemical alterations in patients with affective disorder based on brain imaging studies, and on autopsy specimens from those who died with this illness, compared with other psychiatric illnesses and controls (110–113). As illustrated in Figure 1, there is converging evidence for frontal deficits in both neuronal and glial systems associated with decreased functioning of the prefrontal cortex, especially during depression.

At the same time, there is also evidence for amygdala and ventral striatal overactivity in adults with bipolar illness. One study by Altshuler et al. (79), but not several others (74,114,115), indicated that increased amygdala volume was directly related to the number of prior hospitalizations for mania. A series of neuroimaging studies have now been performed in children and adolescents with bipolar illness, and the findings show that these youngsters had decreased volume of the amygdala compared with controls (114,116–118), in contrast to the preponderance of evidence for increases in amygdala volume in the adults (80,81,119).

These changing alterations as a function of age and course of illness deserve further attention and study. If replicated, they have very interesting clinical and theoretical implications. Does the initially smaller volume of the amygdala in adolescents relate to some of the neurocognitive and emotional facial recognition deficits (likely trait deficits) that even children with early onset bipolar illness seem to experience, and then with increasing age, course of illness, or other mechanisms this deficit changes to relative amygdala hypertrophy compared with controls later in life? The rodent studies of McEwen (120) suggest that the occurrence of neonatal stressors is associated in adulthood with relative increases in amygdala size and concordant prefrontal and hippocampal volume deficits. Thus, it would be important to examine the stress versus episode sensitization mechanisms (vs. genetic mechanisms) that could account for these developmental changes, based on comparing those patients with and without a history of early adversity for various indices of amygdala volume and function.

The transition from a small to large amygdala as a function of age and development could also be related to use-dependent neuroplasticity, similar to that observed with taxi drivers in London who appear to have increased hippocampal volumes compared with those whose work relies less on spatial navigation (121). Of course, it is possible that the hippocampal differences predated engaging in the taxi-driving occupation, and those who decided to become taxi drivers were self-selected based on their spatial navigational skills (and attendant increased hippocampal volume).

Nonetheless, experience-dependent changes in neuronal volume and plasticity have been demonstrated directly in a variety of other animal and human paradigms, raising the possibility that the experience of many bouts of pathological affect could be related to this change in trajectory of amygdala volume in those with bipolar illness. Clearly, further work remains to assess these findings and examine underlying pathophysiological mechanisms for an initially small amygdala volume in childhood bipolar illness changing to a larger volume in adulthood compared with controls.

Based on the sensitization/kindling hypothesis one would postulate neurochemical abnormalities might emerge as a function of number, severity, and duration of prior affective episodes and illness. A variety of neurobiological findings potentially compatible with such a hypothesis of illness progression have now been reported (13,112,122) although causal mechanisms remain to be demonstrated (Table 2) (13,79,83,123–133). It could be that those with increased brain pathology from the outset are more vulnerable to an adverse course of illness, rather than the long duration or greater recurrence of illness driving these neurological abnormalities.

Nonetheless, such course of illness and neurobiological correlates are worthy of consideration and at least raise the possibility that more active illness intervention and episode prevention could prevent some of these neurobiological changes (134). One highly suspicious marker in this regard is the finding of Kessing et al. (130) that the occurrence of one or two episodes of unipolar or bipolar depression is associated with a normal risk of experiencing dementia in late life compared with the general population. However, the occurrence of four or more episodes approximately doubles the risk for such late life cognitive impairment. Their more recent analysis further suggests that with every new episode of illness there is an increasing 6% to 13% likelihood of such cognitive difficulties.

TABLE 2 Neurobiological Correlates of Number of Episodes, Hospitalizations, or Duration of Recurrent Affective Illness

Finding	UP/BP	Correlates	Authors
Hippocampus			
Volume decreased (MRI)	UP	Increased number of previous episodes, longer illness duration	MacQueen et al., 2003 (123)
Volume decreased (MRI)	UP	Longer duration of illness	Sheline et al., 1999 (124)
Vermal brain regions			
Vermal subregion V2 volume decreased (MRI)	BP	Increased number of previous episodes	Mills et al., 2005 (125)
Amygdala			
Volume increased (MRI)	BP	Increased number of previous manic episodes	Altshuler et al., 2000 (79)
Left subgenual and rostral anterior cingulate			
Decreased cerebral glucose metabolism	UP	Increased number of previous episodes	Kimbrell et al., 2002 (83)
Neurocognitive			
Increased dysfunction and disability	BP	Increased number of previous episodes, longer illness duration	Denicoff et al., 1999 (126)
	BP	Increased number of hospitalizations, longer illness duration	Thompson et al., 2005 (127)
	BP	Increased number of hospitalization episodes	Tham et al., 1997 (128)
	BP	Increased number of previous months hospitalized	Rubinsztein et al., 2000 (129)
Risk of dementia	UP/BP	Increased number of previous episodes	Kessing et al., 2004 (130)
Endocrine			
Increased dexamethasone-CRH response	UP/BP	Increased number of previous episodes	Kunzel et al., 2003 (131)
	UP/BP	Increased number of previous episodes	Hatzinger et al., 2002 (132)
Decreased pharmacological response to			
Lithium		Increased number of previous episodes or hospitalizations	Nine studies (Post et al., 2003) (13)
Lamotrigine		Increased number of previous episodes	Obrocea et al., 2002 (133)
Gabapentin		Increased number of previous episodes	Obrocea et al., 2002 (133)

Abbreviations: BP, bipolar; CRH, corticotropin-releasing hormone; MRI, magnetic resonance imaging; UP, unipolar.

Regardless of whether or not adequate pharmacological prophylaxis would prevent the increased incidence of late-life dementia syndromes associated with four or more depressive episodes, from the clinical perspective there would still be much merit in attempting to intervene in this fashion. Effective prophylactic treatment would have at least prevented these several episodes of clinical depression, which would have considerable benefits whether or not the fundamental course or any of its neurobiological correlates (Table 2) were altered in a positive fashion as well. In other words, there is little to lose and much to gain from more aggressive attempts at pharmacological prophylaxis.

Early Age of Onset and Greater Number of Episodes: Poor Prognosis Factors

The number of episodes of prior depression or the occurrence of rapid or faster cycling may also be associated with a poor prognosis for affective illness recurrence and eventual treatment response. Both adult- and childhood-onset illness is under-diagnosed and under-recognized in the community and in clinical populations (135–137). To the extent that episode sensitization is changing the long-term course of illness in an adverse direction, altered strategies for early intervention are strongly indicated.

A strategy for early intervention would be particularly important in childhood- and adolescent-onset bipolar illness, which is associated with serious long-term consequences and poorer treatment response in adulthood (108,109,138,139). However, childhood-onset illness (prior to age 13) several decades ago was associated with an average 16-year delay prior to the first pharmacological treatment for mania or depression, and adolescent-onset (ages 13–19) was also associated with a disturbingly long average 12-year delay between the onset of first symptoms of the illness and first treatment. In contrast, the delays are much shorter for early and late adult-onset bipolar illness (139).

These findings, together with the observation that early onset bipolar illness is associated with an extremely adverse outcome (even in adulthood at average age 40) when patients are treated naturalistically by experts, strongly point out the need for earlier recognition, diagnosis, and treatment of these young children. Again, it would be hoped that introducing such effective treatment would help ameliorate the poor prognosis in adulthood, or potentially even prevent the development of more full-blown illness altogether.

What is now remarkable, but not widely recognized in the community, are the findings that lithium, valproate, and the antidepressants can all increase brain-derived neutrotrophic factor (BDNF) and neurogenesis (54,140–143), potentially ameliorating some of the neurobiological deficits listed in Table 2. Thus, these agents have clinical and theoretical support in not only helping to prevent episodes, but also potentially either reversing or preventing the progression of biological alterations in the illness (Table 3). Such a possible dual benefit of treatment might help some patients overcome their reluctance to start or sustain long-term prophylactic treatment.

IMPLICATIONS FOR PSYCHOPHARMACOLOGICAL INTERVENTIONS

We have already outlined the strong clinical and theoretical rationale—based on the kindling and stress sensitization models—for earlier and more effective

TABLE 3 Potential Neurotrophic and Neuroprotective Effects of Lithium, Valproate, and the Antidepressants: Opposite Effects of Illness

	Stress	Glucocorticoids	Bipolar illness	Lithium	Valproate	Antidepressants
Neurotrophic factors						
BDNF	↓	↓		↑		↑ (and block stress effects)
Bcl-2				↑	↑	
Cell death factors						
Bax, p53	↑	↑		↓	↓	
Stroke model						
Severity	↑	↑	↑, ?	↓	↓ (?)	
Calcium signaling						
Ca_i (intracellular calcium)	↑		↑ Ca_i, white blood cells	(↓)		↑
CaMK-II			↓ Prefrontal cortex CaMK-II			↑
GR			↓ GR			↑
MR			↓ MR			↑
New neurons and glia						
Neurogenesis	↓	↓	(↓ neurons)	↑		↑
Gliogenesis			(↓ glia)	↑		
Neuronal integrity						
N-acetylaspartate (prefrontal cortex > hippocampus)			↓ NAA	↑ cortex		↑ hippocampus
Gray matter on MRI			↓ in prefrontal cortex, anterior cingulate	↑		
Clinical suicide	↑	(↑)	↑ ↑ incidence	↓	?	↓ ?
Excess medical mortality if depressed		↑	↑ incidence	↓	?	?

Abbreviations: BDNF, brain-derived neurotrophic factor; CaMK-II, calcium calmodulin kinase II; GR, glucocorticoid receptors; MR, mineralocorticoid receptors; MRI, magnetic resonance imaging; ↑, increase; ↓, decrease; ?, unknown or not definite.

prophylactic treatment intervention in bipolar illness, for the possibility of not only preventing episodes, but also preventing their associated episode sensitization, stress sensitization, cross-sensitization to comorbid substance abuse, and the neurobiological alterations that may evolve and progress with each of these difficulties. In addition to emphasizing the importance of early intervention, the kindling and sensitization models help to conceptualize other psychopharmacological approaches and treatment predictions noted below.

Effective Treatment May Differ as a Function of Stage of Illness Intervention

Both the kindling and sensitization paradigms reveal that treatment effectiveness may differ as a function of stage of illness intervention. Different psychopharmacological interventions prevent the development of behavioral sensitization as opposed to its later expression. Similarly, the data are extremely clear that pharmacological interventions that prevent the development of kindling are not necessarily the same that prevent the full-blown expression of completed amygdala-kindled seizures (10,11,14).

While this might be expected, it is even more remarkable that once amygdala-kindled animals have made the transition from triggered to spontaneous seizures, there appears to be another pharmacological divergence. Most striking is the observation that diazepam, which is effective in the initial development and completed middle phase of amygdala-kindled seizures, becomes ineffective in preventing the late spontaneous seizures (144). Conversely, in a double-dissociation, phenytoin, which is not effective in preventing kindling development and is ambiguous on the completed amygdala-kindled seizure phase, becomes highly effective in prophylaxis of the spontaneous variety.

These striking differences in pharmacological effectiveness of interventions as a stage of syndrome evolution are consistent with the observations that the neurochemical and neuroanatomical substrates involved in both sensitization and in kindling likewise evolve over time. The initial effects on physiology, immediate early genes, and peptides are largely confined to the amygdala in the earliest stages of kindling evolution, but then progressively become more widely distributed throughout the brain, involving unilateral and then bi-lateral hippocampal and diverse cortical areas as full-blown kindled seizures are engendered (10,14). Likewise, the mechanisms underlying behavioral sensitization to the psychomotor stimulants appear to evolve from those that are local in the ventral-tegmental area (VTA) to those in the amygdala and nucleus accumbens (145,146), depending on the context-specific nature of the sensitization, and then eventually into a variety of cortical areas.

Thus, to the extent that bipolar illness also evolves from well-state vulnerability, to symptomatic illness, to fully-developed affective episodes initially triggered but eventually occurring spontaneously with faster cycling and ultradian switching of mood phases, one would also expect differences in pharmacological responsivity as a function of these stages of illness evolution. This obviously becomes critically important in assessing which very early interventions at first symptoms (secondary prevention), or primary prevention strategies in those at highest risk, might be effective in preventing initial phases of illness development, as opposed to the drug response data in adults that are acquired after the illness has become fully manifest.

A person's age and developmental phase would also combine to produce differences in neurochemistry. Drug responses in childhood might not be identical with those observed in adulthood. For example, we know that the GABA-B agonist baclofen is an effective anticonvulsant on amygdala-kindled seizures in very young rodents (147), but not in adults. Similar questions arise in the therapeutics of BD as to whether or not effective interventions in adults will apply in parallel to those in the youngest children and adolescents. Clearly, more empirical data are needed in this realm.

Combination Therapy May Be Required

In adults, a modicum of data support changes in illness responsivity as a function of number of prior episodes and/or rapidity of cycling. With the exception of isolated studies (148), most data suggest that prophylactic treatments are less effective in those with a high number of prior episodes or rapid cycling compared with those with fewer episodes or nonrapid cycling. This reduced responsiveness in those with greater number of prior episodes not only includes response to lithium, but also emerging studies on a poorer response to valproate, carbamazepine, lamotrigine, and even more recently, the atypical antipsychotics (149–153).

Partially convergent with this perspective are the recent findings that the majority of patients in a variety of clinical and academic treatment settings require complex pharmacotherapy in order to bring the illness into partial or full remission. The data of Frye et al. (8) are interesting in this regard because they illustrate the need over the past several decades for increasingly complex treatment (or polypharmacy) in order to discharge bipolar patients from a clinical research hospital and achieve approximately the same degree of success of about 80% improvement. In the 1970s and 1980s, the average number of drugs required to achieve this acute improvement at discharge were one and two, respectively, and this increased to three in the 1990s, and in the most recently discharged cohorts, to an average of 3.5 to 4 drugs.

This increased need for complex combination therapy coincided with the more recent patients having an earlier onset of their illness, greater amounts of time depressed prior to coming to NIH, and increased rapid cycling, each of which has been associated in some studies with greater degrees of treatment resistance. Whether these population characteristics change over time relates to an inherent selection bias toward more treatment-refractory patients in more recent cohorts, or whether there is an increasing severity and difficulty in achieving response to treatment in the general community, remains for further examination.

Many other clinics have observed this trend for the requirement of an average of three or four medications during naturalistic treatment of bipolar illness, including the Pittsburgh group (154,155), and the outpatient collaborative network as described by Kupka et al. (156). In the Kupka et al. study, those patients with rapid cycling required an average of 4.5 classes of psychopharmacological medications over a year of treatment in contrast to 3.5 in those without a history of rapid cycling. Gitlin et al. (157) also report a high incidence of complex combination treatment in their setting, as well as others (158,159).

This trend toward a greater need for combination treatment and polypharmacy appears to extend to childhood onset bipolar illness cohorts as well. The majority of patients in the study of Kowatch et al. (160) required combination treatment and the more recent studies of Findling et al. indicated a very high failure rate on lithium or valproate monotherapy (161). A report from the American Psychiatric Institute for Research and Education described children with bipolar illness as having the most complex pharmacological regimens compared with all other children with psychiatric diagnoses (162). Birmaher et al. (163) also described the need for complex regimens in order to bring youngsters with bipolar illness into remission; this group not only required multiple medications, but 11.5 months on the average until remission for bipolar I, 9.2 months for bipolar II, and most distressingly, more than two years (34 months) for those with bipolar NOS.

In these instances and in the studies of Geller et al. (164), only a relatively short period of time was required for what was considered a remission. Both of these prospective cohorts found a very high incidence of relapse even in those who had previously been "stabilized." Thus, the nature and complexity of combination treatment optimally required to achieve and maintain remission in childhood-onset bipolar illness requires much more concerted psychopharmacological research attention than does the adult bipolar variety.

Despite having a large number of classes of psychopharmacological interventions and a number of agents in each class, there is little systematic guidance for clinicians and patients to choose among these agents, especially after the first or second treatment intervention. While a large variety of factors appear to contribute to the increasing recognition of the generally poor prognosis of even adult-onset bipolar illness in many academic settings, the failure to have a systematic database from which to make informed clinical treatment decisions is likely an important contributing factor. In the kindling model, combination treatment slows tolerance development more effectively than monotherapy (14,15,165).

Difficulties in Achieving and Maintaining Remission

While many psychopharmacological studies describe clinically successful endpoints of 50% improvement in manic or depressive symptomatology, this endpoint is often inadequate from a patient or clinician's perspective. Here, one is striving for complete remission of bipolar symptomatology and return to one's usual employment, educational, or social role with a minimum of illness-related interference. Not only is this state highly sought after, but preliminary evidence suggests that minor mood fluctuations may be precursors to more major ones and full-blown episode breakthrough (135,166,167). Thus, there is increasing movement toward the goal of achieving complete remission in both the unipolar and bipolar recurrent affective disorders. This will likely also contribute to the need for increasingly complex psychopharmacological regimens, because such "miracle" monotherapy treatment responses are less frequently observed than one would desire (168). In the kindling paradigm, minor breakthrough episodes predict more complete loss of drug effectiveness via tolerance (14,15).

Remission is not an endpoint with a high incidence rate in many psychopharmacological studies, even in those in which patients are selected for lack of comorbidities and symptom extremes such as suicidality. Since the majority of bipolar patients experience considerable illness comorbidity and some 25% to 50% of bipolar patients make a medically serious suicide attempt prior to average age 40 (169,170), the presence of these characteristics in representative patients often requires highly individualized additional psychopharmacological approaches.

Fortunately, a variety of interventions have been demonstrated to be effective in patients with the primary comorbidity of substance abuse (i.e., those with alcohol or cocaine abuse in the absence of bipolar illness) based on double-blind controlled clinical trials, but, unfortunately, relatively few of these options have been specifically tested in patients with these comorbidities in the context of BD (99,100). Thus, in this area as well, the patient and clinician are mostly building treatment regimens based on indirect inferences and best guesses from other data sets, rather than a strong body of controlled evidence. To the extent that inadequate definition of appropriate treatment paradigms both for the primary illness and its comorbidities contribute to illness progression and, eventually, treatment resistance, this tremendous lack of systematic data requires major remedies. In the kindling tolerance paradigm, increasing illness drive via increased intensity of stimulation facilitates and speeds the loss of efficacy via tolerance. In bipolar illness, comorbid conditions may be conceptualized in formal ways of increasing illness drive that require specific treatment.

Two Types of Acquired Treatment Resistance

A further complicating factor is that once a period of illness remission is achieved in a given patient, there is still no guarantee that it can be maintained over the long term. We and others have described two different types of acquired development of treatment resistance, that is, the development of lack of responsivity to a pharmacological agent after an initial extended period of good response. These two types of treatment-resistance include (*i*) tolerance and (*ii*) treatment-discontinuation-related refractoriness.

Tolerance

In a cohort of treatment-refractory bipolar patients, we have observed a considerable incidence of the development of gradual loss of efficacy via a tolerance-like process. We call it 'tolerance-like' because there is often a lack of incontrovertible data that patients remained highly compliant and took the appropriate doses of medication, achieving the same blood levels to which they had previously responded. However, based on patient report this is often the case, and despite excellent adherence, episodes begin to break through a previously effective regimen with increasing frequency, severity, or duration. We have observed this with lithium, valproate, and carbamazepine and potentially with lamotrigine and gabapentin as well (10). Others have seen this phenomenon with lithium, such as in the experience of Maj et al., where patients who were completely remitted on lithium for a period of two years were then followed for a total of five years, and a substantial subgroup of patients began to show illness re-emergence (171).

As described in detail elsewhere, tolerance to most anticonvulsant interventions develops readily in the amygdala-kindled seizure model and provides a basis for exploring treatment approaches that may sustain anticonvulsant efficacy for a longer period of time or indefinitely (15). Based on these preclinical studies we have observed that: (*i*) some drugs such as valproate are much less susceptible to tolerance development than others, such as carbamazepine, lamotrigine, or levetiracetam; (*ii*) higher or consistent doses appear less likely to be associated with tolerance than minimally effective doses that are gradually escalated; (*iii*) combinations of agents with different mechanisms of action may be more effective than either agent alone; (*iv*) intervening earlier in the development of full-blown amygdala-kindled seizures, that is, shortly after these have begun to be observed, also appears associated with a reduced likelihood of developing tolerance compared with the same dose of drug given to animals late in their course after many dozens of kindled seizures.

Once tolerance has developed, the preclinical tolerance model suggests both the utility of adding drugs with a novel mechanism of action, or returning to an originally effective drug after a brief period of time off that agent (11,165). None of these predictions from the preclinical model have been directly tested in patients with recurrent BD, although they clearly need to be.

One caveat that requires attention is that a considerably longer period of time well (in remission) compared with that observed during previous well intervals between episodes should be required prior to considering a putative psychopharmacological agent likely effective in prophylaxis. It is only after this demonstration of sustained efficacy that one can begin to consider a true tolerance process. If only a short well interval is observed on a given agent, it is not clear whether the drug was effective, or ineffective and merely accompanied by the expected course-of-illness variation between episodes.

Lithium Discontinuation-Induced Refractoriness

A second type of acquired treatment resistance is that exemplified by lithium-discontinuation-induced treatment refractoriness (172,173). In this instance, a sustained and long period of illness remission has been observed on lithium (often six to eight years of wellness) compared with much shorter well intervals observed between episodes (on the order of months) prior to treatment. The patient and/or physician then decides on this basis of sustained wellness that lithium could or should be discontinued, and even after a very slow taper period, episodes of illness re-emerge. However, the patient is no longer responsive to the reinitiation of lithium treatment at equal or higher doses and blood levels than those that had previously been effective (174).

This phenomenon is not common, but may occur in about 10–15% of patients, and adds further indirect evidence for illness progression with the occurrence of new episodes potentially heightening "illness drive" and the vulnerability to successive recurrences. Also potentially complicating this process is the withdrawal of a drug such as lithium which considerable evidence suggests may be neuroprotective or neurotrophic, which could further enhance the development of treatment refractoriness. Moreover, an additional seizure episode in the kindling model may propel the patient to a new stage of illness evolution that is no longer responsive to the previously effective agent (15,165).

Neuroprotective Effects of the Mood Stabilizers

Lithium not only increases cell survival factors such as BDNF and Bcl-2, but it inhibits cell death factors such as Bax and p53 (175,176). Lithium also facilitates the rate of neurogenesis and has facilitatory effects on glial activity, growth, and survival as well. Thus, new episodes occurring in the absence of lithium may be occurring in the absence of lithium's neurotrophic and neuroprotective effects, thus incurring a greater degree of pathophysiological change than might have occurred if the patient had continued lithium or some other maintenance treatment with similar properties.

Whatever the mechanisms involved, we know a number of patients who have experienced such a phenomenon, and each of these individuals greatly regrets ever having stopped their previously effective lithium treatment. Whether similar phenomena of discontinuation-induced refractoriness occur in a small percentage of patients after discontinuation of other effective stabilizers remains to be delineated. However, in recurrent unipolar illness, we have also observed a number of individuals who have repeatedly responded acutely to traditional antidepressants or monoamine oxidase inhibitors, but following repeated drug discontinuations and depressive episode recurrences then begin to fail to respond to these agents in subsequent episodes. Even if this pattern occurs in only a small percentage of patients who repeatedly discontinue effective antidepressant treatment for recurrent unipolar depression, it could have a substantial public health impact. These preliminary observations are convergent with the data of Keller et al. (166) that every episode of recurrent depression carries about a 10% increased risk of the development of treatment-nonresponsive illness.

Developing clinical and biological markers to which patients may be vulnerable to these types of acquired treatment resistance would have considerable benefit for clinical therapeutics. Similarly, there is only a modicum of data to suggest which clinical factors may help increase the chance of a given patient's responding to a given mood stabilizer or atypical antipsychotic.

Prediction of Individual Treatment Response

Given the large number of agents potentially available, it would appear wise to consider the suggestions of David Cox of Perlagen Sciences, Inc., for using a profile of some 25 to 30 SNPs, not to define new pathophysiological mechanisms and treatment approaches, but in helping (more immediately) to delineate which patients are likely to be responders to specific existing treatment or be subject to their rare extreme adverse events (177). Use of multiple SNP illness vulnerability factors could also help assess the risk of bipolar illness onset and, at the same time, assist in the assessment of the risk-to-benefit ratio for early treatment initiation.

In this fashion, Cox argues that the expected benefits of the molecular biology revolution could be quite readily and rapidly applied for clinical therapeutic benefit a great many years before any of these factors are used to develop new illness targets and therapeutic strategies aimed at such single dysfunctions. Gene-driven new therapies appear to be an inordinately difficult problem even in Huntington's disease where the single gene and protein defect are known. Treatment advances based on this information have yet to be realized. In contrast, the process of gene-driven therapeutics would appear extraordinarily more difficult for most psychiatric illnesses in which there are likely multiple genes of small effect. In this case, even directly altering one or more such small pathophysiological defects may or may not be therapeutically successful.

If using such SNP profiles in concert with a series of other clinical and biological predictors of response, one might hope to be able to increase the percentage a good initial treatment response from the approximately 50% range that is consistently observed in acute studies in mania (for lithium, the mood stabilizing anticonvulsants, and the atypical antipsychotics) to an 80% or 90% likelihood of response. In this fashion one might be able to intervene both earlier and more effectively with the appropriate agent(s) and help prevent more recurrent episodes, but also their potential adverse illness course consequences predicted by the kindling and sensitization models.

SUMMARY OF FINDINGS

Bipolar disorder has emerged in the last several decades as an illness that is more difficult to treat than previously surmised. If the goal is long-term sustained remission, about two-thirds of the bipolar patients in academic outpatient settings would be considered treatment-resistant. Several different studies suggest that on average, patients with bipolar illness are ill about half the time despite treatment in the community or by experts in academia, and that time depressed exceeds time manic by a factor of approximately three. Correlates of these more adverse and poor prognosis outcomes include earlier age of onset of illness and more episodes experienced prior to study entry.

Both the preclinical models of sensitization and kindling and the empirical data on course of bipolar illness itself suggest that processes of illness progression occur in the absence of effective treatment. While the basic tenets of the sensitization and kindling model have been generally well-validated in the clinical literature (i.e., stress sensitization and episode sensitization), further precision is required to delineate the neurobiological mechanisms underlying stress, episode, and substance abuse sensitization, and how they may interact with each other and with genetic vulnerability.

Even in the absence of compelling evidence for the course of illness modifications induced by stressors and episodes as postulated in the sensitization and kindling models, the utility of the clinical perspective derived from this longitudinal view of the illness would appear to have merit in its own right. Helping to ameliorate the impact of stressors and engage in the effective prevention of manic and depressive episodes would, in any case, be a critical goal and help avoid much suffering. To the extent that this model provides an additional theoretical rationale for early, effective, sustained prophylactic treatment for the recurrent affective disorders, it also may help support more active therapeutic and public health measures directed toward these potentially disabling illnesses. Most of the longitudinal course of illness predictions of the model are ultimately directly testable, but some may be ethically difficult and unjustified, such as randomizing patients to early intervention versus delayed or no treatment. Indirect inferences may nonetheless be gleaned from naturalistic treatment comparisons, and many direct randomized controlled studies of the relative effectiveness of two different early treatment approaches (178) are feasible and very much needed.

REFERENCES

1. Kraepelin E. Manic-depressive insanity and paranoia. Edinburgh: E.S. Livingstone, 1921.
2. Post RM, Roy-Byrne PP, Uhde TW. Graphic representation of the life course of illness in patients with affective disorder. Am J Psychiatry 1988; 145(7):844–848.
3. Leverich GS, Post RM. Life charting the course of bipolar disorder. Current review of mood and anxiety disorders 1996; 1(4):48–61.
4. Leverich GS, Post RM. Life charting of affective disorders. CNS Spectrums 1998; 3(5):21–37.
5. Denicoff KD, Leverich GS, Nolen WA, et al. Validation of the prospective NIMH-Life-Chart Method (NIMH-LCM-p) for longitudinal assessment of bipolar illness. Psychol Med 2000; 30(6):1391–1397.

6. Denicoff KD, Ali SO, Sollinger AB, et al. Utility of the daily prospective National Institute of Mental Health Life-Chart Method (NIMH-LCM-p) ratings in clinical trials of bipolar disorder. Depress Anxiety 2002; 15(1):1–9.
7. Denicoff KD, Smith-Jackson EE, Disney ER, et al. Preliminary evidence of the reliability and validity of the prospective life-chart methodology (LCM-p). J Psychiatr Res 1997; 31(5):593–603.
8. Frye MA, Ketter TA, Leverich GS, et al. The increasing use of polypharmacotherapy for refractory mood disorders: 22 years of study. J Clin Psychiatry 2000; 61(1):9–15.
9. Post RM. Transduction of psychosocial stress into the neurobiology of recurrent affective disorder. Am J Psychiatry 1992; 149(8):999–1010.
10. Post RM, Ketter TA, Speer AM, et al. Predictive validity of the sensitization and kindling hypotheses. In: Soares JC, Gershon S, eds. Bipolar Disorders: Basic Mechanisms and Therapeutic Implications. New York: Marcel Dekker Inc., 2000:387–432.
11. Post RM, Weiss SRB, Leverich GS, et al. Sensitization and kindling-like phenomena in bipolar disorder: implications for psychopharmacology. Clin Neurosci Res 2001; 1: 69–81.
12. Post RM. Kindling. In: Craighead WE, Nemeroff CB, eds. The Corsini Encyclopedia of Psychology and Behavioral Science. New York: John Wiley & Sons, 2001:833–835.
13. Post RM, Leverich GS, Weiss SR, et al. Psychosocial stressors as predisposing factors to affective illness and PTSD: potential neurobiological mechanisms and theoretical implications. In: Cicchetti D, Walker E, eds. Neurodevelopmental Mechanisms in Psychopathology. New York: Cambridge University Press, 2003:491–525.
14. Post RM. The status of the sensitization/kindling hypothesis of bipolar disorder. Current Psychosis and Therapeutics Reports 2004; 2(4):135–141.
15. Post RM, Zhang ZJ, Weiss SRB, et al. Contingent tolerance and cross tolerance to anticonvulsant effects in amygdala-kindled seizures: Mechanistic and clinical implications. In: Corcoran ME, Moshe SL, eds. Kindling VI. New York: Springer, 2005:305–314.
16. Antelman SM, Eichler AJ, Black CA, et al. Interchangeability of stress and amphetamine in sensitization. Science 1980; 207(4428):329–331.
17. Kalivas PW, Richardson-Carlson R, Van Orden G. Cross-sensitization between foot shock stress and enkephalin-induced motor activity. Biol Psychiatry 1986; 21:939–950.
18. Kalivas PW, Stewart J. Dopamine transmission in the initiation and expression of drug- and stress-induced sensitization of motor activity. Brain Research Reviews 1991; 16:223–244.
19. Green WH, Campbell M, David R. Psychosocial dwarfism: a critical review of the evidence. J Am Acad Child Psychiatry 1984; 23(1):39–48.
20. Field TM, Schanberg SM, Scafidi F, et al. Tactile/kinesthetic stimulation effects on preterm neonates. Pediatrics 1986; 77(5):654–658.
21. Haney M, Maccari S, Le Moal M, et al. Social stress increases the acquisition of cocaine self-administration in male and female rats. Brain Res 1995; 698(1–2):46–52.
22. Goeders NE, Guerin GF. Non-contingent electric footshock facilitates the acquisition of intravenous cocaine self-administration in rats. Psychopharmacol (Berl) 1994; 114: 63–70.
23. Covington HE III, Miczek KA. Repeated social-defeat stress, cocaine or morphine. Effects on behavioral sensitization and intravenous cocaine self-administration "binges." Psychopharmacol (Berl) 2001; 158(4):388–398.
24. Goeders NE, Clampitt DM. Potential role for the hypothalamo-pituitary-adrenal axis in the conditioned reinforcer-induced reinstatement of extinguished cocaine seeking in rats. Psychopharmacology (Berl) 2002; 161(3):222–232.
25. Huot RL, Thrivikraman KV, Meaney MJ, et al. Development of adult ethanol preference and anxiety as a consequence of neonatal maternal separation in Long Evans rats and reversal with antidepressant treatment. Psychopharmacol (Berl) 2001; 158(4):366–373.
26. Meaney MJ, Brake W, Gratton A. Environmental regulation of the development of mesolimbic dopamine systems: a neurobiological mechanism for vulnerability to drug abuse? Psychoneuroendocrinology 2002; 27(1–2):127–138.

27. Sbrana A, Bizzarri JV, Rucci P, et al. The spectrum of substance use in mood and anxiety disorders. Compr Psychiatry 2005; 46(1):6–13.
28. Chengappa KN, Levine J, Gershon S, et al. Lifetime prevalence of substance or alcohol abuse and dependence among subjects with bipolar I and II disorders in a voluntary registry. Bipolar Disord 2000; 2(3 Pt 1):191–195.
29. Goddard GV, McIntyre DC, Leech CK. A permanent change in brain function resulting from daily electrical stimulation. Exp Neurol 1969; 25(3):295–330.
30. Racine RJ. Modification of seizure activity by electrical stimulation. II. Motor seizure. Electroencephalogr Clin Neurophysiol 1972; 32:281–294.
31. Racine RJ. Modification of seizure activity by electrical stimulation. I. After-discharge threshold. Electroencephalogr Clin Neurophysiol 1972; 32:269–279.
32. Post RM, Weiss SRB. A speculative model of affective illness cyclicity based on patterns of drug tolerance observed in amygdala-kindled seizures. Mol Neurobiol 1996; 13(1):33–60.
33. Weiss SR, Post RM. Caveats in the use of the kindling model of affective disorders. Toxicol Ind Health 1994; 10(4–5):421–447.
34. Post RM, Weiss SRB, Pert A, et al. Chronic cocaine administration: sensitization and kindling effects. In: Raskin A, Fisher S, eds. Cocaine: Clinical and Biobehavioral Aspects. New York: Oxford University Press, 1987:109–173.
35. Post RM, Rubinow DR, Ballenger JC. Conditioning, sensitization, and kindling: implications for the course of affective illness. In: Post RM, Ballenger JC, eds. Neurobiology of Mood Disorders. Baltimore: Williams and Wilkins, 1984:432–466.
36. Sato M, Racine RJ, McIntyre DC. Kindling: basic mechanisms and clinical validity. Electroencephalogr Clin Neurophysiol 1990; 76:459–472.
37. Goddard GV, Douglas RM. Does the engram of kindling model the engram of normal long term memory? Can J Neurol Sci 1975; 2(4):385–394.
38. Post RM, Weiss SR, Pert A. Implications of behavioral sensitization and kindling for stress-induced behavioral change. Adv Exp Med Biol 1988; 245:441–463.
39. Post RM, Weiss SR. Psychomotor stimulant vs. local anesthetic effects of cocaine: role of behavioral sensitization and kindling. NIDA Res Monogr 1988; 88:217–238.
40. Post RM, Weiss SR, Pert A. Cocaine-induced behavioral sensitization and kindling: implications for the emergence of psychopathology and seizures. Ann NY Acad Sci 1988; 537:292–308.
41. Post RM, Weiss SRB, Pert A. Sensitization and kindling effects of chronic cocaine administration. In: Lakoski JM, Galloway MP, White FJ, eds. Cocaine: Pharmacology, Physiology, and Clinical Strategies. New Jersey: Telford Press, 1992:115–161.
42. Post RM, Weiss SR, Fontana D, et al. Conditioned sensitization to the psychomotor stimulant cocaine. Ann NY Acad Sci 1992; 654:386–399.
43. Stanton ME, Gutierrez YR, Levine S. Maternal deprivation potentiates pituitary-adrenal stress responses in infant rats. Behav Neurosci 1988; 102(5):692–700.
44. Levine S, Huchton DM, Wiener SG, et al. Time course of the effect of maternal deprivation on the hypothalamic- pituitary-adrenal axis in the infant rat. Dev Psychobiol 1991; 24(8):547–558.
45. Zhang LX, Levine S, Dent G, et al. Maternal deprivation increases cell death in the infant rat brain. Brain Res Dev Brain Res 2002; 133(1):1–11.
46. Roceri M, Hendriks W, Racagni G, et al. Early maternal deprivation reduces the expression of BDNF and NMDA receptor subunits in rat hippocampus. Mol Psychiatry 2002; 7(6):609–616.
47. Xing GQ, Smith MA, Levine S, Yang ST, Post RM, Zhang LX. Suppression of CaMKII and nitric oxide synthase by maternal deprivation in the brain of rat pups. Society for Neuro Abst 1998; 24 [Abstract 176.9], 452.
48. Korte M, Carroll P, Wolf E, et al. Hippocampal long-term potentiation is impaired in mice lacking brain-derived neurotrophic factor. Proc Natl Acad Sci USA 1995; 92(19):8856–8860.
49. Silva AJ, Stevens CF, Tonegawa S, et al. Deficient hippocampal long-term potentiation in alpha-calcium-calmodulin kinase II mutant mice. Science 1992; 257(5067):201–206.

50. Linnarsson S, Bjorklund A, Ernfors P. Learning deficit in BDNF mutant mice. Eur J Neurosci 1997; 9(12):2581–2587.
51. Kuma H, Miki T, Matsumoto Y, et al. Early maternal deprivation induces alterations in brain-derived neurotrophic factor expression in the developing rat hippocampus. Neurosci Lett 2004; 372(1–2):68–73.
52. Roceri M, Cirulli F, Pessina C, et al. Postnatal repeated maternal deprivation produces age-dependent changes of brain-derived neurotrophic factor expression in selected rat brain regions. Biol Psychiatry 2004; 55(7):708–714.
53. Chen B, Dowlatshahi D, MacQueen GM, et al. Increased hippocampal BDNF immunoreactivity in subjects treated with antidepressant medication. Biol Psychiatry 2001; 50(4):260–265.
54. Shimizu E, Hashimoto K, Okamura N, et al. Alterations of serum levels of brain-derived neurotrophic factor (BDNF) in depressed patients with or without antidepressants. Biol Psychiatry 2003; 54(1):70–75.
55. Karege F, Perret G, Bondolfi G, et al. Decreased serum brain-derived neurotrophic factor levels in major depressed patients. Psychiatry Res 2002; 109(2):143–148.
56. Aydemir O, Deveci A, Taneli F. The effect of chronic antidepressant treatment on serum brain-derived neurotrophic factor levels in depressed patients: a preliminary study. Prog Neuropsychopharmacol Biol Psychiatry 2005; 29(2):261–265.
57. Gonul AS, Akdeniz F, Taneli F, et al. Effect of treatment on serum brain-derived neurotrophic factor levels in depressed patients. Eur Arch Psychiatry Clin Neurosci 2005; 255(6):381–386.
58. Xing GQ, Russell S, Hough C, et al. Decreased prefrontal CaMKII α mRNA in bipolar illness. NeuroReport 2002; 13(4):501–505.
59. Plotsky PM, Thrivikraman KV, Nemeroff CB, et al. Long-term consequences of neonatal rearing on central corticotropin-releasing factor systems in adult male rat offspring. Neuropsychopharmacology 2005; 30(12):2192–2204.
60. Ladd CO, Huot RL, Thrivikraman KV, et al. Long-term behavioral and neuroendocrine adaptations to adverse early experience. Prog Brain Res 2000; 122:81–103.
61. Meaney MJ, Aitken DH, Van Berkel C, et al. Effect of neonatal handling on age-related impairments associated with the hippocampus. Science 1988; 239(4841 Pt 1): 766–768.
62. Plotsky PM, Meaney MJ. Early, postnatal experience alters hypothalamic corticotropin-releasing factor (CRF) mRNA, median eminence CRF content and stress-induced release in adult rats. Mol Brain Res 1993; 18:195–200.
63. Caldji C, Diorio J, Meaney MJ. Variations in maternal care in infancy regulate the development of stress reactivity. Biol Psychiatry 2000; 48(12):1164–1174.
64. Meaney MJ, Szyf M. Environmental programming of stress responses through DNA methylation: life at the interface between a dynamic environment and a fixed genome. Dialogues Clin Neurosci 2005; 7(2):103–123.
65. Francis DD, Szegda K, Campbell G, et al. Epigenetic sources of behavioral differences in mice. Nat Neurosci 2003; 6(5):445–446.
66. Bentall RR, Kinderman P, Manson K. Self-discrepancies in bipolar disorder: comparison of manic, depressed, remitted and normal participants. Br J Clin Psychology 2005; 44:457–473.
67. Strakowski SM, DelBello MP, Fleck DE, et al. The impact of substance abuse on the course of bipolar disorder. Biol Psychiatry 2000; 48(6):477–485.
68. Davis KA, Kwon A, Cardenas VA, et al. Decreased cortical gray and cerebral white matter in male patients with familial bipolar I disorder. J Affect Disord 2004; 82(3):475–485.
69. Lim KO, Rosenbloom MJ, Faustman WO, et al. Cortical gray matter deficit in patients with bipolar disorder. Schizophr Res 1999; 40(3):219–227.
70. Winsberg ME, Sachs N, Tate DL, et al. Decreased dorsolateral prefrontal N-acetyl aspartate in bipolar disorder. Biol Psychiatry 2000; 47(6):475–481.
71. Sassi RB, Stanley JA, Axelson D, et al. Reduced NAA levels in the dorsolateral prefrontal cortex of young bipolar patients. Am J Psychiatry 2005; 162(11):2109–2115.
72. Chang K, Adleman N, Dienes K, et al. Decreased N-acetylaspartate in children with familial bipolar disorder. Biol Psychiatry 2003; 53(11):1059–1065.

73. Rajkowska G. Postmortem studies in mood disorders indicate altered numbers of neurons and glial cells. Biol Psychiatry 2000; 48(8):766–777.
74. Pearlson GD, Barta PE, Powers RE, et al. Ziskind-Somerfeld Research Award 1996. Medial and superior temporal gyral volumes and cerebral asymmetry in schizophrenia versus bipolar disorder. Biol Psychiatry 1997; 41:1–14.
75. Bertolino A, Frye M, Callicott JH, et al. Neuronal pathology in the hippocampal area of patients with bipolar disorder: a study with proton magnetic resonance spectroscopic imaging. Biol Psychiatry 2003; 53(10):906–913.
76. Eastwood SL, Harrison PJ. Hippocampal synaptic pathology in schizophrenia, bipolar disorder and major depression: a study of complexin mRNAs. Mol Psychiatry 2000; 5(4):425–432.
77. Hauser P, Altshuler LL, Berrettini W, et al. Temporal lobe measurement in primary affective disorder by magnetic resonance imaging. J Neuropsychiatry Clin Neurosci 1989; 1:128–134.
78. Altshuler LL, Conrad A, Hauser P, et al. Reduction of temporal lobe volume in bipolar disorder: A preliminary report of magnetic resonance imaging [letter]. Arch Gen Psychiatry 1991; 48:482–483.
79. Altshuler LL, Bartzokis G, Grieder T, et al. An MRI study of temporal lobe structures in men with bipolar disorder or schizophrenia. Biol Psychiatry 2000; 48(2):147–162.
80. Strakowski SM, DelBello MP, Sax KW, et al. Brain magnetic resonance imaging of structural abnormalities in bipolar disorder. Arch Gen Psychiatry 1999; 56(3): 254–260.
81. Brambilla P, Harenski K, Nicoletti M, et al. MRI investigation of temporal lobe structures in bipolar patients. J Psychiatr Res 2003; 37(4):287–295.
82. Bowley M, Drevets W, Ongur D, et al. Low glial numbers in the amygdala in major depressive disorder. Biol Psychiatry 2002; 52(5):404.
83. Kimbrell TA, Ketter TA, George MS, et al. Regional cerebral glucose utilization in patients with a range of severities of unipolar depression. Biol Psychiatry 2002; 51(3): 237–252.
84. Guidotti A, Auta J, Davis JM, et al. Decrease in reelin and glutamic acid decarboxylase67 (GAD67) expression in schizophrenia and bipolar disorder: a postmortem brain study. Arch Gen Psychiatry 2000; 57(11):1061–1069.
85. Johnston-Wilson NL, Sims CD, Hofmann JP, et al. Disease-specific alterations in frontal cortex brain proteins in schizophrenia, bipolar disorder, and major depressive disorder. The Stanley Neuropathology Consortium. Mol Psychiatry 2000; 5(2): 142–149.
86. Webster MJ, O'Grady J, Kleinman JE, et al. Glial fibrillary acidic protein mRNA levels in the cingulate cortex of individuals with depression, bipolar disorder and schizophrenia. Neuroscience 2005; 133(2):453–461.
87. Xing GQ, Russell S, Webster MJ, et al. Decreased expression of mineralocorticoid receptor mRNA in the prefrontal cortex in schizophrenia and bipolar disorder. Int J Neuropsychopharmacol 2004; 7(2):143–153.
88. Baumann B, Danos P, Krell D, et al. Unipolar-bipolar dichotomy of mood disorders is supported by noradrenergic brainstem system morphology. J Affect Disord 1999; 54(1-2):217–224.
89. Benson BE, Willis MW, Ketter TA, et al. Altered relationships in rCMRglu associativity in bipolar and unipolar illness. Biol Psychiatry 2000; 47(8S):108S.
90. Post RM, Lake CR, Jimerson DC, et al. Cerebrospinal fluid norepinephrine in affective illness. Am J Psychiatry 1978; 135(8):907–912.
91. Young LT, Warsh JJ, Kish SJ, et al. Reduced brain 5-HT and elevated NE turnover and metabolites in bipolar affective disorder. Biol Psychiatry 1994; 35(2):121–127.
92. Altshuler L, Bookheimer S, Proenza MA, et al. Increased amygdala activation during mania: a functional magnetic resonance imaging study. Am J Psychiatry 2005; 162(6): 1211–1213.
93. Ketter TA, Andreason PJ, George MS, et al. Anterior paralimbic mediation of procaine-induced emotional and psychosensory experiences. Arch Gen Psychiatry 1996; 53(1): 59–69.

94. Ali SO, Denicoff KD, Altshuler LL, et al. A preliminary study of the relation of neuropsychological performance to neuroanatomic structures in bipolar disorder. Neuropsychiatry Neuropsychol Behav Neurol 2000; 13(1):20–28.
95. Sonne SC, Brady KT, Morton WA. Substance abuse and bipolar affective disorder. J Nerv Ment Dis 1994; 182:349–352.
96. Weiss RD, Ostacher MJ, Otto MW, et al. Does recovery from substance use disorder matter in patients with bipolar disorder? J Clin Psychiatry 2005; 66(6): 730–735.
97. Frye MA, Altshuler LL, McElroy SL, et al. Gender differences in prevalence, risk, and clinical correlates of alcoholism comorbidity in bipolar disorder. Am J Psychiatry 2003; 160(5):883–889.
98. Drake RE, Xie H, McHugo GJ, et al. Three-year outcomes of long-term patients with co-occurring bipolar and substance use disorders. Biol Psychiatry 2004; 56(10):749–756.
99. Post RM. Differing psychotropic profiles of the anticonvulsants in bipolar and other psychiatric disorders. Clin Neuosci Res 2004; 4(1–2):9–30.
100. Post RM. Adjunctive strategies in the treatment of refractory bipolar depression: clinician options in the absence of a systematic database. Expert Opin Pharmacother 2005; 6(4):531–546.
101. Kessing LV, Andersen PK, Mortensen PB, et al. Recurrence in affective disorder. I. Case register study. Br J Psychiatry 1998; 172:23–28.
102. Kessing LV, Hansen MG, Andersen PK, et al. The predictive effect of episodes on the risk of recurrence in depressive and bipolar disorders—a life-long perspective. Acta Psychiatr Scand 2004; 109(5):339–344.
103. Kendler KS, Thornton LM, Gardner CO. Stressful life events and previous episodes in the etiology of major depression in women: an evaluation of the "kindling" hypothesis. Am J Psychiatry 2000; 157(8):1243–1251.
104. Kendler KS, Thornton LM, Gardner CO. Genetic risk, number of previous depressive episodes, and stressful life events in predicting onset of major depression. Am J Psychiatry 2001; 158(4):582–586.
105. Caspi A, Sugden K, Moffitt TE, et al. Influence of life stress on depression: moderation by a polymorphism in the 5-HTT gene. Science 2003; 301(5631):386–389.
106. Kendler KS, Kuhn JW, Vittum J, et al. The interaction of stressful life events and a serotonin transporter polymorphism in the prediction of episodes of major depression: a replication. Arch Gen Psychiatry 2005; 62(5):529–535.
107. Caspi A, Moffitt TE, Cannon M, et al. Moderation of the effect of adolescent-onset cannabis use on adult psychosis by a functional polymorphism in the catechol-O-methyltransferase gene: longitudinal evidence of a gene X environment interaction. Biol Psychiatry 2005; 57(10):1117–1127.
108. Leverich GS, McElroy SL, Suppes T, et al. Early physical and sexual abuse associated with an adverse course of bipolar illness. Biol Psychiatry 2002; 51:288–297.
109. Leverich GS, Post RM. Earlier age of onset and more severe course of bipolar illness in those with a history of childhood trauma: Clinical and theoretical implications (commentary). Lancet 2006; in press.
110. Post RM, Speer AM, Hough CJ, et al. Neurobiology of bipolar illness: implications for future study and therapeutics. Ann Clin Psychiatry 2003; 15(2):85–94.
111. Post RM, Post SLW. Molecular and cellular developmental vulnerabilities to the onset of affective disorders in children and adolescents: Some implications for therapeutics. In: Steiner H, ed. Handbook of Mental Health Interventions in Children and Adolescents. San Francisco: Jossey-Bass, 2004:140–192.
112. Post RM. Neurobiology of seizures and behavioral abnormalities. Epilepsia 2004; 45(suppl 2):5–14.
113. Post RM. Neural substrates of psychiatric syndromes. In: Mesulam MM, ed. Principles of Behavioral and Cognitive Neurology, 2nd ed. New York: Oxford University Press, 2000:406–438.
114. Blumberg HP, Kaufman J, Martin A, et al. Amygdala and hippocampal volumes in adolescents and adults with bipolar disorder. Arch Gen Psychiatry 2003; 60(12):1201–1208.

115. Swayze VW, Andreasen NC, Alliger RJ, et al. Subcortical and temporal structures in affective disorder and schizophrenia: a magnetic resonance imaging study. Biol Psychiatry 1992; 31(3):221–240.
116. DelBello MP, Zimmerman ME, Mills NP, et al. Magnetic resonance imaging analysis of amygdala and other subcortical brain regions in adolescents with bipolar disorder. Bipolar Disord 2004; 6(1):43–52.
117. Dickstein DP, Milham MP, Nugent AC, et al. Frontotemporal alterations in pediatric bipolar disorder: results of a voxel-based morphometry study. Arch Gen Psychiatry 2005; 62(7):734–741.
118. Chang K, Karchemskiy A, Barnea-Goraly N, et al. Reduced amygdalar gray matter volume in familial pediatric bipolar disorder. J Am Acad Child Adolesc Psychiatry 2005; 44(6):565–573.
119. Altshuler LL, Bartzokis G, Grieder T, et al. Amygdala enlargement in bipolar disorder and hippocampal reduction in schizophrenia: an MRI study demonstrating neuroanatomic specificity [letter]. Arch Gen Psychiatry 1998; 55(7):663–664.
120. McEwen BS. Glucocorticoids, depression, and mood disorders: structural remodeling in the brain. Metabolism 2005; 54(5 suppl 1):20–23.
121. Maguire EA, Gadian DG, Johnsrude IS, et al. Navigation-related structural change in the hippocampi of taxi drivers. Proc Natl Acad Sci USA 2000; 97(8):4398–4403.
122. Post RM. Do the epilepsies, pain syndromes, and affective disorders share common kindling-like mechanisms? Epilepsy Res 2002; 50(1–2):203–219.
123. MacQueen GM, Campbell S, McEwen BS, et al. Course of illness, hippocampal function, and hippocampal volume in major depression. Proc Natl Acad Sci USA 2003; 100(3):1387–1392.
124. Sheline YI, Sanghavi M, Mintun MA, et al. Depression duration but not age predicts hippocampal volume loss in medically healthy women with recurrent major depression. J Neurosci 1999; 19(12):5034–5043.
125. Mills NP, DelBello MP, Adler CM, et al. MRI analysis of cerebellar vermal abnormalities in bipolar disorder. Am J Psychiatry 2005; 162(8):1530–1532.
126. Denicoff KD, Ali SO, Mirsky AF, et al. Relationship between prior course of illness and neuropsychological functioning in patients with bipolar disorder. J Affect Disord 1999; 56(1):67–73.
127. Thompson JM, Gallagher P, Hughes JH, et al. Neurocognitive impairment in euthymic patients with bipolar affective disorder. Br J Psychiatry 2005; 186:32–40.
128. Tham A, Engelbrektson K, Mathe AA, et al. Impaired neuropsychological performance in euthymic patients with recurring mood disorders. J Clin Psychiatry 1997; 58(1): 26–29.
129. Rubinsztein JS, Michael A, Paykel ES, et al. Cognitive impairment in remission in bipolar affective disorder. Psychol Med 2000; 30(5):1025–1036.
130. Kessing LV, Andersen PK. Does the risk of developing dementia increase with the number of episodes in patients with depressive disorder and in patients with bipolar disorder? J Neurol Neurosurg Psychiatry 2004; 75(12):1662–1666.
131. Kunzel HE, Binder EB, Nickel T, et al. Pharmacological and nonpharmacological factors influencing hypothalamic-pituitary-adrenocortical axis reactivity in acutely depressed psychiatric in-patients, measured by the Dex-CRH test. Neuropsychopharmacology 2003; 28(12):2169–2178.
132. Hatzinger M, Hemmeter UM, Baumann K, et al. The combined DEX-CRH test in treatment course and long-term outcome of major depression. J Psychiatr Res 2002; 36(5):287–297.
133. Obrocea GV, Dunn RM, Frye MA, et al. Clinical predictors of response to lamotrigine and gabapentin monotherapy in refractory affective disorders. Biol Psychiatry 2002; 51(3):253–260.
134. Post RM, Leverich GS, Xing G, et al. Developmental vulnerabilities to the onset and course of bipolar disorder. Dev Psychopathol 2001; 13(3):581–598.
135. Hirschfeld RM, Calabrese JR, Weissman MM, et al. Screening for bipolar disorder in the community. J Clin Psychiatry 2003; 64(1):53–59.

136. Das AK, Olfson M, Gameroff MJ, et al. Screening for bipolar disorder in a primary care practice. JAMA 2005; 293(8):956–963.
137. Hunt JI, Dyl J, Armstrong L, et al. Frequency of manic symptoms and bipolar disorder in psychiatrically hospitalized adolescents using the K-SADS mania rating scale. J Child Adolesc Psychopharmacol 2005; 15(6):918–930.
138. Ernst CL, Goldberg JF. Clinical features related to age at onset in bipolar disorder. J Affect Disord 2004; 82(1):21–27.
139. Perlis RH, Miyahara S, Marangell LB, et al. Long-term implications of early onset in bipolar disorder: data from the first 1000 participants in the systematic treatment enhancement program for bipolar disorder (STEP-BD). Biol Psychiatry 2004; 55(9): 875–881.
140. Fukumoto T, Morinobu S, Okamoto Y, et al. Chronic lithium treatment increases the expression of brain-derived neurotrophic factor in the rat brain. Psychopharmacol (Berl) 2001; 158(1):100–106.
141. Chen G, Rajkowska G, Du F, et al. Enhancement of hippocampal neurogenesis by lithium. J Neurochem 2000; 75(4):1729–1734.
142. Manji HK, Moore GJ, Chen G. Clinical and preclinical evidence for the neurotrophic effects of mood stabilizers: implications for the pathophysiology and treatment of manic-depressive illness. Biol Psychiatry 2000; 48(8):740–754.
143. Malberg JE, Eisch AJ, Nestler EJ, et al. Chronic antidepressant treatment increases neurogenesis in adult rat hippocampus. J Neurosci 2000; 20(24):9104–9110.
144. Pinel JPJ. Effects of diazepam and diphenylhydantoin on elicited and spontaneous seizures in kindled rats: a double dissociation. Pharmacol Biochem Behav 1983; 18: 61–63.
145. Cador M, Bjijou Y, Stinus L. Evidence of a complete independence of the neurobiological substrates for the induction and expression of behavioral sensitization to amphetamine. Neuroscience 1995; 65:385–395.
146. Kalivas PW, Alesdatter JE. Involvement of N-methyl-D-aspartate receptor stimulation in the ventral tegmental area and amygdala in behavioral sensitization to cocaine. J Pharmacol Exp Ther 1993; 267(1):486–495.
147. Wurpel JN, Sperber EF, Moshe SL. Baclofen inhibits amygdala kindling in immature rats. Epilepsy Res 1990; 5:1–7.
148. Baldessarini RJ, Tondo L, Floris G, et al. Effects of rapid cycling on response to lithium maintenance treatment in 360 bipolar I and II disorder patients. J Affect Disord 2000; 61(1–2):13–22.
149. Suppes T, Brown E, Schuh LM, et al. Rapid versus non-rapid cycling as a predictor of response to olanzapine and divalproex sodium for bipolar mania and maintenance of remission: Post hoc analyses of 47-week data. J Affect Disord 2005; 89(1–3):69–77.
150. Vieta E, Calabrese JR, Hennen J, et al. Comparison of rapid-cycling and non-rapid-cycling bipolar I manic patients during treatment with olanzapine: analysis of pooled data. J Clin Psychiatry 2004; 65(10):1420–1428.
151. Kupka RW, Luckenbaugh DA, Post RM, et al. Rapid and non-rapid cycling in bipolar disorder: A meta-analysis of clinical studies. J Clin Psychiatry 2003; 64(12):1483–1494.
152. Tondo L, Hennen J, Baldessarini RJ. Rapid-cycling bipolar disorder: effects of long-term treatments. Acta Psychiatr Scand 2003; 108(1):4–14.
153. Maj M, Pirozzi R, Magliano L, et al. Long-term outcome of lithium prophylaxis in bipolar disorder: a 5-year prospective study of 402 patients at a lithium clinic. Am J Psychiatry 1998; 155(1):30–35.
154. Levine J, Chengappa KN, Brar JS, et al. Psychotropic drug prescription patterns among patients with bipolar I disorder. Bipolar Disord 2000; 2(2):120–130.
155. Kupfer DJ, Frank E, Grochocinski VJ, et al. Demographic and clinical characteristics of individuals in a bipolar disorder case registry. J Clin Psychiatry 2002; 63(2):120–125.
156. Kupka RW, Luckenbaugh DA, Post RM, et al. Comparison of rapid-cycling and non-rapid-cycling bipolar disorder based on prospective mood ratings in 539 outpatients. Am J Psychiatry 2005; 162(7):1273–1280.
157. Gitlin MJ, Swendsen J, Heller TL, et al. Relapse and impairment in bipolar disorder. Am J Psychiatry 1995; 152(11):1635–1640.

158. Frangou S, Raymont V, Bettany D. The Maudsley bipolar disorder project. A survey of psychotropic prescribing patterns in bipolar I disorder. Bipolar Disord 2002; 4(6):378–385.
159. Goldberg JF, Harrow M, Sands JR. Lithium and the longitudinal course of bipolar illness. Psychiatric Ann 1996; 26(10):651–658.
160. Kowatch RA, Sethuraman G, Hume JH, et al. Combination pharmacotherapy in children and adolescents with bipolar disorder. Biol Psychiatry 2003; 53(11):978–984.
161. Findling RL, McNamara NK, Youngstrom EA, et al. Double-blind 18-month trial of lithium versus divalproex maintenance treatment in pediatric bipolar disorder. J Am Acad Child Adolesc Psychiatry 2005; 44(5):409–417.
162. Duffy FF, Narrow WE, Rae DS, et al. Concomitant pharmacotherapy among youths treated in routine psychiatric practice. J Child Adolesc Psychopharmacol 2005; 15(1): 12–25.
163. Birmaher B, Axelson D, Strober M, et al. Clinical course of children and adolescents with bipolar spectrum disorders. Arch Gen Psychiatry 2006; 63(2):175–183.
164. Geller B, Tillman R, Craney JL, et al. Four-year prospective outcome and natural history of mania in children with a prepubertal and early adolescent bipolar disorder phenotype. Arch Gen Psychiatry 2004; 61(5):459–467.
165. Weiss SR, Clark M, Rosen JB, et al. Contingent tolerance to the anticonvulsant effects of carbamazepine: relationship to loss of endogenous adaptive mechanisms. Brain Res Brain Res Rev 1995; 20(3):305–325.
166. Keller MB, Boland RJ. Implications of failing to achieve successful long-term maintenance treatment of recurrent unipolar major depression. Biol Psychiatry 1998; 44(5): 348–360.
167. Perlis RH, Ostacher MJ, Patel JK, et al. Predictors of recurrence in bipolar disorder: primary outcomes from the systematic treatment enhancement program for bipolar disorder (STEP-BD). Am J Psychiatry 2006; 163(2):217–224.
168. Post RM. What is an ideal mood stabilizer? Clinical Approaches in Bipolar Disorders 2006; in press.
169. Valtonen H, Suominen K, Mantere O, et al. Suicidal ideation and attempts in bipolar I and II disorders. J Clin Psychiatry 2005; 66(11):1456–1462.
170. Leverich GS, Altshuler LL, Frye MA, et al. Factors associated with suicide attempts in 648 patients with bipolar disorder in the Stanley Foundation Bipolar Network. J Clin Psychiatry 2003; 64(5):506–515.
171. Maj M, Pirozzi R, Kemali D. Long-term outcome of lithium prophylaxis in patients initially classified as complete responders. Psychopharmacology (Berl) 1989; 98(4):535–538.
172. Post RM, Leverich GS, Altshuler L, et al. Lithium-discontinuation-induced refractoriness: preliminary observations. Am J Psychiatry 1992; 149(12):1727–1729.
173. Post RM, Leverich GS, Pazzaglia PJ, et al. Lithium tolerance and discontinuation as pathways to refractoriness. In: Birch NJ, Padgham C, Hughes MS, eds. Lithium in Medicine and Biology. Lancashire, UK: Marius Press, 1993:71–84.
174. Maj M, Pirozzi R, Magliano L. Nonresponse to reinstituted lithium prophylaxis in previously responsive bipolar patients: prevalence and predictors. Am J Psychiatry 1995; 152:1810–1811.
175. Chen G, Zeng WZ, Yuan PX, et al. The mood-stabilizing agents lithium and valproate robustly increase the levels of the neuroprotective protein bcl-2 in the CNS. J Neurochem 1999; 72(2):879–882.
176. Chen RW, Chuang DM. Long term lithium treatment suppresses p53 and Bax expression but increases Bcl-2 expression. A prominent role in neuroprotection against excitotoxicity. J Biol Chem 1999; 274(10):6039–6042.
177. Cox D. Human genetic variation and common disease: A short-term approach for improving human health. NIH Director's Lecture, October 13th, 2004.
178. Post RM, Kowatch RA. The heath care crisis of childhood onset bipolar illness: Some recommendations for its amelioration. J Clin Psychiatry 2006; 67(1):115–125.

17 Biologic Factors in Different Bipolar Disorder Subtypes

Michael A. Cerullo and Stephen M. Strakowski
Division of Bipolar Disorders Research, Department of Psychiatry, University of Cincinnati College of Medicine, Cincinnati, Ohio, U.S.A.

INTRODUCTION

Modern psychiatric nosology, and the classification of mood disorders, did not really take shape until the early twentieth century with the work of Emil Kraepelin. Kraepelin was one of the pioneers in differentiating psychiatric conditions. However, he combined most forms of mood disorder, including unipolar and bipolar depression, into the large category of manic-depressive insanity (1). It wasn't until the late 1950s and 1960s that unipolar and bipolar depression were separated into unique diagnostic syndromes [see Akiskal (2) for review]. The key element that separates bipolar disorder (BD) and its subtypes from unipolar disorder is the occurrence of mania or hypomania. Over the last three decades there have been varying suggestions of how to subdivide BD (2). In this chapter, we focus on what we feel are the most well-established subgroups within and separate from BD type I, namely, BD type II, cyclothymic disorder, rapid cycling, and psychotic mania. Type II BD and cyclothymia are specific diagnoses in the DSM-IV (3), while rapid cycling and psychotic mania are DSM-IV specifiers that can be applied to either type I or II BDs (Table 1).

This chapter extends an earlier comprehensive review chapter on the subtypes of BD from the previous edition of this book (4). For each of the four major subtypes covered, we will discuss epidemiology, etiology, and treatment of the condition. The DSM-IV also includes two additional subtypes of BD: BD not otherwise specified (NOS), and mood disorder due to a general medical condition. There is very limited research on Bipolar Disorder NOS, hence this is not discussed in the present review. Secondary mania is an important and large topic in itself but is not discussed in this chapter. For a comprehensive review of secondary mania see Strakowski et al. (5).

Before discussing these specific subtypes of BD in detail, a few general remarks are warranted. Although much has been written on BD as a spectrum illness (6), the scarcity of empirical evidence prevents meaningful speculation in this regard. The major problem with the spectrum concept, and any other nosology with multiple subtypes of BD, is that so little is known about the etiology of even the most established (i.e., bipolar I) forms of the disorder. Hence multiple subtypes of the disorder become difficult to substantiate. Only when the molecular and neuroanatomic mechanisms of major subtypes (e.g., bipolar type I or II) are further understood will it be possible to determine whether other variants of the disorder are minor dysfunctions of the same systems or have separate and unique etiologies. For example, even the biologic differences between unipolar and bipolar depression

TABLE 1 Subtypes of Bipolar Disorder and Their DSM-IV Descriptors

Subtype	DSM-IV description
Bipolar I	At least one manic episode must occur; no requirement for a depressive episode
Bipolar II	At least one episode of hypomania and one depressive episode
Cyclothymic disorder	Two years of numerous periods of hypomanic symptoms and depressive symptoms, with no period of two or more months without symptoms
Rapid cycling	Applied to bipolar I disorder with four or more manic, mixed, or depressive episodes in one year; or bipolar type II disorder with four or more hypomanic or depressive episodes in one year
Psychotic mania	Applied to bipolar I patients during a manic or mixed episode with psychosis

remain to be elucidated (7,8). There is only limited research that compares and contrasts possible etiological mechanisms for the different subtypes of BD. Yet the rapid advances in our knowledge of BD should leave us optimistic about future understanding of the subtypes of this condition. As more is learned about the etiology of bipolar I, there will be molecular mechanisms and neuroanatomic circuits to compare when studying the variant subtypes.

BIPOLAR DISORDER, TYPE II

Bipolar disorder, type II is characterized in the DSM-IV (3) by the occurrence of at least one clear episode of hypomania (with no history of ever having a full manic episode) and the occurrence of at least one major depressive episode.

Epidemiology

Although first described by Dunner et al. in 1976 (9), BD type II was not recognized as a separate diagnostic entity until the publication of the DSM-IV in 1994. There is still considerable controversy surrounding estimates of the prevalence of bipolar II disorder. Although a recent review cites a prevalence of 3% to 5% (10), there is little consensus in the literature. Using the definition in DSM-III (11), the U.S. National Epidemiologic Catchment Area Study (12) found a lifetime prevalence of only 0.5% for hypomanic episodes. Angst (13) reviewed eight other studies examining the epidemiology of bipolar II and found prevalence ranges between 0.3% and 2%.

Judd and Akiskal (14) reanalyzed the U.S. National Epidemiologic Catchment Area data using less restrictive criteria for hypomania and found a prevalence of 6.4% for the bipolar spectrum. Similar results were found in the Zurich study, another important study of the epidemiology of bipolar II that followed 4547 subjects from the canton of Zurich from 1979 to 1993 (13). They found a prevalence rate of 5.5% for hypomania. Together, then, these estimates suggest that type II BD is relatively common, probably occurring in 2% to 4% of the population.

Biology

Genetic studies provide the strongest evidence to distinguish type I and II BDs. Maier (15) compared the lifetime prevalence of affective disorders in relatives of bipolar I and II patients. The risk of having bipolar I disorder was equivalent in relatives of bipolar I and II patients. However, the risk of having bipolar II was greater in relatives of patients with bipolar II (6.1%) compared with relatives of patients with bipolar I (1.8%). Gershon et al. (16) also found a higher prevalence of bipolar II disorder in relatives of patients with bipolar II and similar rates of bipolar I illness in relatives of patients with bipolar I and II. However, these conditions did not "breed true," suggesting considerable overlap in familial risk.

There have been few neuroimaging studies that compare bipolar I and II patients. The only structural MRI study comparing the two subtypes was by Sassi et al. (17) in which they examined the pituitary volume in bipolar versus unipolar patients. They separated the bipolar patients into bipolar I ($n = 18$) and bipolar II ($n = 5$), and did not find any differences in pituitary volume between these two groups. The small sample provided limited power, however, so that these results are not particularly informative. Functional imaging studies have also looked at differences between bipolar I and II disorder. A PET study examining cerebral glucose metabolism during a continuous performance test by Ketter et al. (18) compared 14 bipolar I and 29 bipolar II patients. The direct comparison of the two groups showed no differences in global metabolism, but bipolar I patients had increased metabolism in the supragenual anterior cingulate, right middle frontal gyrus, and right inferior parietal lobule compared with bipolar II patients. An fMRI study by Malhi et al. (19) examined 10 hypomanic patients during an emotional task and found activation in the caudate and thalamus compared with the control group. However, this study was limited in that the hypomanic patients were not further diagnosed as having either bipolar I or II illness.

A small number of magnetic resonance spectroscopy studies have compared patients with bipolar I and II disorder. Winsberg et al. (20) looked at ratios of NAA, choline, and myo-inositol to creatine-phosphocreatine (Cr-PCr) in the dorsolateral prefrontal cortex in 10 bipolar I and 10 bipolar II patients during euthymia. Compared with controls, bipolar I patients had lower NAA/Cr-PCr ratios in both hemispheres, while bipolar II patients had lower NAA/Cr-PCr ratios compared with controls only in the left hemisphere. When both groups were compared directly, bipolar II patients had higher NAA/Cr-PCr ratios in both hemispheres. Another study by Kato et al. (21) looked at the Cho/Cr-PCr ratio in the left subcortical region (which included the basal ganglia) in 10 bipolar I patients and 9 bipolar II patients during euthymia. The Cho/Cr-Pcr ratio was higher in the bipolar II patients compared with the bipolar I patients.

Taken together, these results provide some evidence that there may be biological differences between bipolar I and II disorder and provide incentive for further comparative studies. The genetic studies are consistent with several possible relationships between the two disorders, including the possibility that bipolar II is a less severe variant of bipolar I and the possibility that the two disorders have separate but overlapping etiology. Neuroimaging research suggests that there may be functional brain differences between the two groups in frontal brain networks, but this finding is based only a single study and clearly needs replication. The magnetic resonance spectroscopy studies provide evidence for metabolic difference in the dorsolateral prefrontal cortex and subcortical regions.

Treatment

Li

An early prospective study of lithium in bipolar type II patients suggested that lithium was effective in preventing depressive relapse (22), but the small sample size ($n = 18$) and lack of statistical analysis limited the study. An early double-blind placebo-controlled study looking at 22 bipolar II patients by Kane et al. (23) showed that lithium was effective in preventing relapse. A retrospective study (24) of 102 bipolar II patients showed that lithium was effective in preventing the relapse of depression. A more recent retrospective study of lithium therapy in 188 bipolar I and 129 bipolar II patients who were followed for an average of 8.4 years (25) found lithium to be more effective in bipolar II than bipolar I patients with decreased disease relapse, less percent of time spent ill, and fewer illness episodes per year.

Anticonvulsants

A double-blind, placebo-controlled study of lamotrigine as maintenance prophylaxis in rapid cycling BDs showed that in the subgroup of 52 bipolar II patients (with rapid cycling) compared with placebo there was a significantly higher percentage of patients who were stable without relapse at six months (26). Divalproex sodium monotherapy was shown to reduce Hamilton Depression Scale ratings by 50% in 19 depressed bipolar II subjects in a 12-week open label study (27).

Antidepressants

Amsterdam and Brunswick (28) recently reviewed the use of antidepressants in bipolar II depression. They concluded that the data supported the use of antidepressant therapy in bipolar II depression and that this use is associated with a low risk of inducing switching to mania or hypomania. Amsterdam et al. published several recent studies of the effectiveness of antidepressant treatment in bipolar II patients. In one double-blind study, 89 bipolar II patients in remission from a recent depressive episode were given either fluoxetine or placebo for 52 weeks (29). These subjects were compared with 89 unipolar patients matched for age and gender and 661 unmatched unipolar patients. Relapse prevention rates were equivalent to those in the unipolar patient groups. The rate of switching to mania or hypomania was equivalent to the matched unipolar controls but increased compared with the unmatched controls (3.8% vs. 0.3%). In another study, venlafaxine was compared in 17 bipolar II patients and 31 matched unipolar patients, with both groups in a current depressive episode (29). There was no difference in response in the two groups, and there were no episodes of switching in the bipolar II group. In a recent study by Amsterdam et al. (30), 37 bipolar II patients were given fluoxetine monotherapy for depression in an open label trial. Eleven of 23 patients who completed the eight-week trial showed a reduction in Hamilton Depression Rating scores greater than 50% and three patients (7.3%) had symptoms suggestive of hypomania.

RAPID CYCLING

Rapid cycling is a specifier in the DSM-IV (3) that can be applied to either bipolar I or bipolar II disorder when patients have multiple mood episodes in one year. In bipolar I disorder, rapid-cycling patients must have at least four manic, mixed, or depressive episodes in one year. Bipolar type II patients are required to have at least four hypomanic or depressive episodes in one year.

Epidemiology

Although psychiatrists have long been aware of rapid mood shifts in some patients, not until 1974 did Dunner and Fieve (31) introduce the current nosology of rapid-cycling BD in a paper studying factors correlated with poor lithium response in bipolar patients. In their paper, Dunner and Fieve arbitrarily set the criteria to be a minimum of four mood swings in a one year period. This nosology for rapid cycling was then included as a specifier in the DSM-IV for both bipolar I and bipolar II disorders.

Kupka et al. (32) recently completed an important meta-analysis examining every study of rapid-cycling BD published since 1974. Among 2054 bipolar I and II patients in eight studies, 335 (16.3%) had rapid cycling. These studies included both inpatient and outpatient subjects, and the prevalence of rapid cycling ranged from 12% to 24% in the individual studies. Women and those with bipolar II disorders showed higher rates of rapid cycling (although the overall size of this effect was small) than men and those with bipolar I respectively. Since this meta-analysis, there have been two other large studies looking at rapid-cycling BD. A study by Schneck et al. (33) looked at the data for the first 483 bipolar I or II patients enrolled in the Systematic Treatment Enhancement Program (STEP) for one year after their diagnosis. They found an overall rate of rapid cycling of 20%. There was a higher rate of rapid cycling among women (23% vs. 16%), and rapid-cycling patients had an earlier age of onset of illness, were more depressed at study entry, and had poorer global functioning in the year before study entry. There was no significant difference between the rate of rapid cycling in bipolar I versus bipolar II disorder. The next major study was by Kupka et al. (34). They followed 539 outpatients with bipolar I and II disorder for one year and found that 38.2% had rapid cycling. They found no difference in rate of rapid cycling between men and women. There was a significantly higher rate of rapid cycling in bipolar I (41.3%) versus bipolar II disorder (27.9%). The rapid-cycling patients spent more time in both manic (and hypomanic) and depressive states over the course of a year than those without rapid cycling. They also found that rapid cycling was associated with a greater number of previous mood episodes, previous rapid cycling, a history of childhood physical and/or sexual abuse, and lifetime drug abuse. The authors considered different cut-off ranges for number of episodes in a one-year period. Increasing the cut-off to eight episodes in a year, only 13.5% of the patients would meet criteria for rapid cycling. However, none of the cut-off ranges showed any evidence of nonlinearity that would suggest a natural cut-off point to separate rapid cyclers from nonrapid cyclers.

In addition to the prevalence of rapid cycling, it is also important to understand when it occurs during the course of bipolar illness and whether it is a temporary condition or a more long lasting subtype. Strakowski et al. (35) followed 144 subjects for up to five years after their first hospitalization for mania. At the end of the follow-up period, only 10% of patients had ever met the criteria for rapid cycling. These results suggest that rapid cycling may develop with illness progression over time. Koukopoulos et al. (36) followed the course of illness in 109 rapid-cycling bipolar patients for up to 36 years (with a minimum period of two years). They found that rapid cycling emerged in 96 patients (88%) after antidepressant and other medication treatment. The mean total duration of rapid cycling was eight years while the mean total duration of affective episodes was 22 years. Baldessarini et al. (37) found that among patients who met the criteria for rapid cycling at one time, these patients did not consistently maintain rapid-cycling status and had only moderately greater lifetime rates of rapid cycling compared with other bipolar patients. These last two results support the concept of rapid cycling as a temporary condition during BD.

Biology

Unfortunately, there have been few studies directly examining the structural and functional neuroanatomic networks or molecular mechanisms of bipolar patients with rapid cycling. One area which has been studied involves thyroid function in rapid-cycling patients. In their prior review, Gary et al. (4) discussed thyroid axis dysfunction in rapid-cycling patients. Some, but not all, of the studies reviewed suggested a higher prevalence of thyroid disorders in rapid-cycling versus non–rapid-cycling bipolar patients. Two of the studies showed increased hypothyroidism in rapid-cycling patients after treatment with lithium carbonate (38,39). A recent meta-analysis by Kupka et al. (32) also looked at studies of rapid-cycling that included measures of thyroid function, and only two of the seven studies (39,40) found significantly higher cases of hypothyroidism among rapid-cycling patients.

In their discussion, Gary et al. (4) suggested that rapid-cycling patients may be more severely ill and therefore more likely to be exposed to lithium, which in turn leads to the greater proportion of patients with thyroid dysfunction. Two recent papers looking at thyroid dysfunction in rapid-cycling patients address this concern. Kupka et al. (41) found increased rates of thyroperoxidase antibodies in bipolar patients, but this was not associated with either lithium treatment or rapid-cycling BD, suggesting there are independent risk factors in bipolar patients for developing hypothyroidism. Gyulai et al. (42) compared 20 rapid-cycling patients with controls on a lithium challenge. Rapid-cycling bipolar patients had a significantly higher change in maximum thyroid-stimulating hormone (TSH) after the lithium challenge than controls, supporting the idea that rapid-cycling bipolar patients are more vulnerable to lithium-induced thyroid dysfunction. Clearly, more studies are needed to help clarify these contradicting results.

Treatment

Li

The nosology of rapid-cycling arose from Dunner and Fieve's (31) study looking at bipolar patients unresponsive to lithium. They found that 82% of rapid cyclers failed lithium compared with 41% of nonrapid cyclers. This result was then replicated by Koukopoulos et al. (43) who found a poor response to lithium in 72% of rapid cyclers. However, several recent studies have challenged this finding. The meta-analysis by Kupka et al. (32) discussed previously also examined response to lithium. In patients who were treated with lithium prophylaxis, there were no differences in recurrence rates between rapid and non–rapid-cycling bipolar patients (47% vs. 34%, respectively, which was nonsignificant).

The most recent and methodologically sound study looking at lithium treatment in rapid cyclers was that by Baldessarini et al. (37). The authors followed 360 bipolar I and II patients over 13 years. Their results indicated that rapid-cycling BD was not associated with greater morbidity during lithium maintenance treatment. There were no significant differences in several critical outcome measures between rapid and non–rapid-cycling bipolar patients, including: months remaining stable on treatment, proportion of time spent in mania and depression, and psychiatric hospitalizations.

Valproate

Calabrese et al. (44) examined the effectiveness of valproate in 78 rapid-cycling bipolar patients in a 16-month open label trial. Valproate showed a good acute response in 54% of patients in a manic state, 72% of those in mixed states, and only 19% of those in a depressive state. A good prophylactic response was seen in 72% of manic patients, 94% of mixed patients, and 33% of depressive patients. Another open-label trial by Calabrese and Delucchi (45) examined 55 rapid-cycling patients over eight months on valproate. The results were similar to the Calabrese et al. study (44) but showed greater efficacy in depression (47% in acute depression and 76% in prophylactic treatment). Calabrese et al. (46) recently completed a large randomized double-blind study of valproate versus lithium in rapid-cycling bipolar patients. The chance of relapsing into either a manic or depressive episode was not statistically different for lithium (56%) versus valproate (50%).

Lamotrigine

Bowden et al. (47) looked at the efficacy of lamotrigine in 41 rapid-cycling and 34 non–rapid-cycling patients with BD in a 48-week open label prospective study. Patients with depressive and mild-to-moderate symptoms improved from baseline on lamotrigine in both patient groups (with no significant difference between the groups). However, rapid-cycling patients with severe mania showed little improvement in their symptoms.

A large double-blind, placebo-controlled study of lamotrigine as maintenance prophylaxis in 324 patients with rapid-cycling bipolar I and II disorders (26) showed a statistically significant greater percent of the treatment group was stable without relapse at six months compared with placebo (41% vs. 26%). Another smaller double-blind placebo-controlled study of lamotrigine showed an improved antidepressant response compared with placebo in rapid-cycling patients (48).

Carbamazepine

Okuma (49) looked at 215 bipolar patients treated with lithium versus carbamazepine in a retrospective study and found that rapid cycling (at the time of study or in the past) predicted nonresponse to both medicines. Denicoff et al. (50) completed a three-year double-blind crossover study of lithium, carbamazepine, and the combination of both medications in 52 outpatients with BD. Rapid-cycling patients responded poorly to both lithium (28% response) and carbamazepine (19% response), but did better on the combination (56.3% response).

Olanzapine

Gonzalez-Pinto et al. (51) studied olanzapine as an add-on in an open-label trial in 13 bipolar I patients during a mixed mood episode. All patients had a history of rapid cycling within the last year. Ten of the 13 patients responded to olanzapine (in which response was defined as a decrease of 50% score on the Young Mania Rating Scale and the Hamilton Rating Scale for Depression).

A recent meta-analysis by Vieta et al. (52) pooled the data from two double-blind placebo-controlled trials of olanzapine to include 90 patients with rapid cycling. The results indicated that olanzapine was more effective in the early treatment of mania in rapid-cycling versus non–rapid-cycling bipolar patients. However, olanzapine was less effective in the long-term treatment of rapid cyclers who were significantly more likely to experience recurrence (especially depression), to have rehospitalizations, and to have suicide attempts compared with nonrapid cyclers.

Comparisons of Treatments

A recent meta-analysis of long-term treatment of rapid-cycling bipolar patients by Tondo and Baldessarini (53) looked at treatments with carbamazepine, lamotrigine, lithium, topiramate, and valproate. They found lower effectiveness of all treatments in rapid-cycling versus non–rapid-cycling patients. They concluded that there were few studies providing direct comparisons of the different medications and there was no evidence of any medicine being superior.

CYCLOTHYMIC DISORDER

Cyclothymic disorder appears as one of four subtypes of BD in the DSM-IV. The diagnostic criteria require a period of hypomanic symptoms and depressive symptoms (which do not meet the criteria for a major depressive episode) that continue for at least two years with no period of two or more months without symptoms. The diagnosis also requires that no major depressive, manic, or mixed episode be present during the first two years of the disorder. If these episodes occur after two years of the start of the cyclothymic symptoms, then the appropriate mood disorder can be diagnosed concurrently with cyclothymic disorder.

Epidemiology

Kraepelin considered cyclothymia as a possible constitutional disposition that could lead to manic-depressive illness (54). In the DSM-II (55), cyclothymia was considered a part of the affective personality disorders category. In the DSM-III (11), cyclothymia was included in the mood disorders and the diagnosis required numerous hypomanic and depressive symptoms to be present continuously for at least two years without symptom-free periods of two months or greater. This nosology persisted in the DSM-IV (3).

Unfortunately, there are no epidemiological studies looking at cyclothymic disorder as defined by the DSM-III or IV. Weissman and Myers (56) interviewed 1095 households and found a lifetime rate of cyclothymic personality of 0.4% as defined by the Schedule for Affective Disorders and Schizophrenia Diagnostic Research Criteria. Placidi et al. (57) examined 1010 students between 14 and 26 years old and found a prevalence of 6.3% for cyclothymic temperament.

Biology

To our knowledge there are no studies that have attempted to examine the underlying molecular or neuroanatomic etiology of cyclothymic disorder.

Treatment

There are no double-blind placebo-controlled studies examining treatments for cyclothymic disorder. A retrospective study by Peselow et al. (24) looked at cyclothymic patients taking lithium over a two-year period. The probability of remaining free of a depressive episode was only 26% to 36% (compared with 42–55% for bipolar II patients and 31–42% for unipolar patients). The probability of suffering a depressive episode severe enough to require hospitalization during the two years was 69% in the cyclothymic patients (compared with 51% and 64% for bipolar II and unipolar patients, respectively).

A prospective study by Jacobsen (58) looked at valproate in cyclothymia. Twenty-six patients (15 cyclothymic, 11 bipolar II) out of 33 started on daily valproate doses between 125 and 250 mg and titrated up reported partial or complete stabilization of their mood. The cyclothymic patients required lower doses and blood levels to achieve stabilization compared with the bipolar II patients.

PSYCHOTIC MANIA

Psychosis is a specifier in DSM-IV that can be added to bipolar I disorder during a manic, mixed, or depressed episode. If psychosis is present the mood episode becomes "severe with psychotic features." In this chapter we are concerned with psychosis that occurs during a manic or mixed episode. Whether the psychosis is mood congruent or incongruent is also relevant, and most studies of psychotic mania classify the manic episode as psychotic only when the psychosis is mood incongruent.

Epidemiology

Distinguishing mood from psychotic disorders has been one of the major challenges of psychiatric nosology and was one of the major accomplishments of Kraepelin (1). Yet, many questions remain about the overlap of these two major groups of illness. The DSM-III (11) included psychosis as a specifier for a manic or depressive episode. In the DSM-IV (3), this specifier can also be applied to mixed episodes. A factor analytic study of 576 manic patients by Sato et al. (59) supports the concept of psychotic mania as a subtype of acute mania. The subgroup of psychotic mania was significantly different from three other subgroups of mania (pure mania, aggressive mania, and mixed mania) on measures of suicidality at admission, global assessment of functioning score at discharge, number of residual symptoms at discharge, and gender.

In their textbook on BD, Goodwin and Jamison (60) reviewed prior studies of BD and found a lifetime prevalence of 60% for having at least one psychotic symptom. More recent epidemiological figures are hard to come by, but Conus and McGorry (61) estimated the prevalence of psychotic symptoms in the range of 63–88% in the first episode of mania based on their review of first episode bipolar studies. In another study (62), 90 out of 139 (65%) manic patients had psychotic symptoms. Strakowski et al. (35) found that psychosis during mania was more common in bipolar I patients who had alcohol use disorders one year prior to the episode than patients with no history of alcohol use disorders. Two studies (63,64) have found that psychotic symptoms are associated with an earlier age of onset of illness in BD. Finally, a recent study of hallucinations in mood disorders and schizophrenia found that only 11.2% of manic patients had hallucinations (65).

Biology

Toni et al. (66) studied the family history of mental illness in 155 manic inpatients, 86 with mood-incongruent psychosis. Those with mood-incongruent psychosis had a family history of schizophrenia of 4% compared with 0% in those without mood-incongruent psychosis. However, this study included those with schizoaffective disorder as well as bipolar I, which limited the conclusions that could be drawn regarding the genetics of BD. Potash et al. (67) looked at 65 families of probands with bipolar I disorder. They found that psychotic symptoms during affective episodes occurred more often in family members of bipolar I subjects with psychosis than family members of bipolar I subjects without psychosis, suggesting that psychotic symptoms cluster in certain bipolar pedigrees. Potash et al. (68) replicated these results using 69 new BD pedigrees. Potash et al. (69) then performed genetic linkage analysis on the ten families with the highest number of psychotic mood disorders. They found linkage to chromosomal regions 13q31 and 22q12.

Strasser et al. (70) performed structural MRI scans on 23 bipolar subjects with psychosis (21 with bipolar I and 2 with bipolar II), 15 bipolar subjects without psychosis (nine with bipolar I and six with bipolar II), 33 schizophrenic subjects, and 44 healthy controls. Psychosis was defined as the occurrence of hallucinations or delusions during one affective episode (mania, mixed, or depressive). The bipolar subjects with psychosis and the schizophrenic subjects had enlarged lateral and third ventricles compared with healthy controls and bipolar patients without psychosis. There was no difference in hippocampal volume when the two groups of bipolar subjects were compared with each other or with healthy controls. In contrast, schizophrenic subjects showed reduced left hippocampal volume compared with healthy controls. Two PET studies (71,72) have also found similarities between schizophrenia and bipolar subjects with psychosis. Pearlson et al. (71) found increased basal ganglia D_2 dopamine receptor density in psychotic bipolar and schizophrenic subjects compared with healthy controls and bipolar subjects without psychosis. Wong et al. (72) found increased caudate D_2 dopamine receptor density in psychotic bipolar subjects compared with bipolar subjects without psychosis and healthy controls, and increased caudate D_2 dopamine receptor density in schizophrenic subjects compared with healthy controls. These neuroimaging studies suggest that there may be similar etiological mechanisms in BD with psychosis and schizophrenia, although it is possible the changes are merely correlative with psychosis.

Another major research focus in psychotic mania has been examining whether there is a worse prognosis and increased severity of symptoms in psychotic mania versus mania without psychosis. Although there have been several studies of this question, there is no clear consensus. In a retrospective study, Kessing (73) examined 149 manic patients without psychosis and 202 patients with psychosis. The patients with psychotic mania had longer admissions compared those without psychosis, but no difference was found in the risk of relapse between the two groups.

Strakowski et al. (74) followed 50 manic patients for eight months after their first psychiatric hospitalization for BD. Those patients with mood incongruent psychosis at hospitalization had significantly more weeks during follow-up of both mood-incongruent and mood-congruent psychosis as well as poorer overall functioning during the outcome interval of eight months. MacQueen et al. (75) studied 62 outpatients with bipolar I disorder (16 with psychosis) in a prospective study. Although psychotic patients were more symptomatic during the acute manic episode there were no differences in ratings of function and well-being when both groups were euthymic. In a prospective study, Miklowitz (76) examined 23 hospitalized manic patients, 11 with psychosis. Although the two groups did not differ in rates of relapse, those with psychotic symptoms had poorer social adjustment and were less medically compliant. Swann et al. (77) looked at the role of psychosis in the severity of symptoms and treatment response in a randomized double-blind treatment study of lithium, divalproex sodium, and placebo in 179 hospitalized manic patients. Those patients with psychosis had lower Global Assessment Scale scores but similar response to treatment.

Treatment

Li

In the prospective study of 139 bipolar patients (90 with psychotic mania) mentioned earlier, Coryell et al. (62) did not find any difference in response to lithium in psychotic mania versus mania without psychosis. As mentioned above, Swann et al. (77) did not find any differences in efficacy between lithium and divalproex in the treatment of psychotic mania. In a prospective study of 30 patients with acute mania (24 with psychotic symptoms), Zemlan et al. (78) found that lithium was more effective in those with psychotic symptoms. Rosenthal et al. (79) studied 66 bipolar I patients, 44 whom had psychosis at some point during their illness (30 during mania, 5 during depression, and 9 during both mania and depression). They found that psychosis during mania was a predictor of good response to lithium. However, in a retrospective study of 145 bipolar patients (94 with psychotic mania), Yazici et al. (80) found a first episode of mania that included psychosis was one of four variables that predicted a poor response to lithium.

Clozapine

Green et al. (81) studied clozapine in the treatment of refractory psychotic mania. In an open-label trial, 22 psychotic manic patients were given clozapine for 12 weeks. Clozapine proved effective in the treatment of these patients, showing a mean improvement of 56.7%, 56.6%, and 39.1% on the Brief Psychiatric Rating Scale (BPRS), Young Mania Rating Scale (YMRS), and Clinical Global Impressions (CGI), respectively.

Olanzapine

In an 18-month double-blind maintenance study of the prevention of relapse of bipolar I disorder, Tohen et al. (82) looked at olanzapine versus placebo added to valproate or lithium and found a significant reduction of symptomatic relapse with the addition of olanzapine. The presence of psychotic features had no effect on the outcome. In a double-blind study of olanzapine versus divalproex in acute mania (83), the subgroup of bipolar patients without psychosis showed decreased YMRS scores with olanzapine compared with divalproex. The presence of psychotic features had no effect on outcome. Tohen et al. (84) looked at olanzapine versus haldol in a 12-week double-blind study of the treatment of acute mania. Patients receiving olanzapine had significant improvement in manic symptoms (lower YMRS scores) in the subgroup of patients without psychosis, but not in the subgroup of patients with psychosis. Chengappa et al. (85) combined the results from two prior randomized double-blind studies of olanzapine versus placebo in acute bipolar mania and found that olanzapine was significantly better than placebo at reducing manic symptoms (lower YMRS scores) in patients with psychosis.

Risperidone

Sachs et al. (86) examined risperidone versus haldol or placebo as an add-on to lithium or divalproex in a double-blind study of acute mania. Risperidone was effective in lowering YMRS scores in patients without psychosis (mean change −13.3) and those with psychosis (mean change −15.4) compared with the placebo group. Hirschfeld et al. (87) studied the effectiveness of risperidone versus placebo for acute mania in a three-week double-blind trial. Risperidone was effective in lowering YMRS scores in patients with and without psychosis. In a double-blind study of risperidone compared with haldol and placebo in acute mania, Smulevich et al. (88) showed that risperidone was effective in acute mania (lower YMRS scores) compared with placebo at three weeks in patients with and without psychosis.

Quetiapine

McIntyre et al. (89) examined quetiapine versus haldol monotherapy in a 12-week double-blind placebo-controlled study of acute mania. In patients without psychosis, both quetiapine and haldol showed improvements in YMRS scores compared with placebo on days 21 and 84. However, in the subgroup of patients with psychosis, quetiapine did not show improvement in YMRS scores compared with placebo on either day 21 or 84. Sachs et al. (90) examined quetiapine versus placebo added to lithium or divalproex in a double-blind study of acute mania. Quetiapine was superior to placebo in improving YMRS scores, and the presence of psychotic symptoms did not show an interaction with the treatment effect, indicating quetiapine was effective in bipolar patients with psychosis. Yatham et al. (91) also looked at quetiapine versus placebo added to lithium or divalproex in a double-blind study of acute mania. They also found that the presence of psychosis did not alter the effectiveness of quetiapine.

SUMMARY

In this paper we reviewed the epidemiology, biology, and treatment of four subtypes of BD: bipolar II, rapid cycling, cyclothymia, and psychotic mania. Among the four disorders, bipolar II disorder has the strongest support for being a distinct diagnostic entity. There is evidence suggesting bipolar II patients have different genetic risk factors compared with bipolar I patients. The neuroimaging studies reviewed are beginning to find differences that if replicated could lead to elucidation of different etiological mechanisms and risk factors between bipolar I and II. Studies of BD with rapid cycling have focused on thyroid abnormalities and different response to treatments. The evidence regarding thyroid abnormalities is inconsistent and more studies are needed before any conclusions can be drawn. The studies of treatment response are also inconsistent, but the newer studies suggest that lithium can be effective in bipolar patients with rapid cycling. In addition, recent outcome studies suggest that rapid cycling may be a temporary condition during the course of BD.

Regarding psychotic mania, genetic studies suggest it may cluster in certain families, and biological studies have shown similar brain deficits to those found in schizophrenic patients. Recent studies of psychotic mania have shown that atypical antipsychotics are an effective treatment while studies regarding prognosis remain mixed. Almost no research has been done on cyclothymia and even the most basic epidemiological data remain uncertain in this disorder.

PERSPECTIVE

Clearly more research is needed to better understand the subtypes of BD. With so little evidence it is premature to consider BD a spectrum illness. When the molecular and neuroanatomic aspects of bipolar I disorder are better understood this should enable the comparison of the proposed subtypes to bipolar I disorder. These comparative studies may lead to the elucidation of the etiology of each subtype as well as the etiology of the entire spectrum of BDs.

In terms of treatment, we should also be wary of assuming that medications that are effective for bipolar I will be effective for other possible subtypes of the disorder. Only rigorous double-blind studies can assure us of the effectiveness of medications, and more such studies are needed for many of the subtypes of BD.

While psychiatric nosology has come a long way since Kraepelin's formulation of manic-depressive insanity, there is still a long road ahead. The most fundamental questions about the etiology of the mood disorders remain unanswered. Only an understanding of the etiology of the illnesses we treat can bring us to the next level of nosology beyond Kraepelin and the DSM-IV.

ACKNOWLEDGMENTS

Supported in part by the Stanley Medical Research Institute and NIMH awards MH066626, 068801, 071931 (Strakowski).

REFERENCES

1. Kraepelin E. Manic-Depressive Insanity and Paranoia. Edinburgh: Livingstone, 1921.
2. Akiskal H. Classification, diagnosis and boundaries of bipolar disorder: a review. In: Maj M, Akiskal H, Lopez-Ibor J, Sartorius N, eds. Bipolar Disorder. New York: John Wiley & Sons, 2002; 1–52.
3. American Psychiatric Association. Diagnostic and Statistical Manual of Mental Disorders, 4th ed. Washington D.C.: American Psychiatric Association, 1994.
4. Gary K, Zeph R, Winokur A, Leighton H. Biological factors in different bipolar disorder subtypes. In: Soares J, Gershon S, eds. Bipolar Disorders: Basic Mechanisms and Therapeutic Implications. New York: Marcel Decker Inc., 2000; 433–458.
5. Strakowski S, Sax K. Secondary mania: a model of the pathophysiology of bipolar disorder? In: Soares J, Gershon E, eds. Bipolar Disorders: Basic Mechanisms and Therapeutic Implications. New York: Marcel Dekker, 2000; 13–30.
6. Akiskal H. The prevalent clinical spectrum of bipolar disorders: beyond DSM-IV. J Clin Psychopharmacol 1996; 16(2 suppl 1):4S–14S.
7. Swann A. Is bipolar depression a specific biological entity? In: Young T, Joffe R, eds. Bipolar Disorder: Biological Models and Their Clinical Applications. New York: Marcel Dekker, 1997; 255–286.
8. Strakowski S. Differential brain mechanisms in bipolar and unipolar disorders considerations from brain imaging. In: Soares J, ed. Brain Imaging in Affective Disorders. New York: Marcel Dekker, 2003; 337–362.
9. Dunner D, Gershon E, Goodwin E. Heritable factors in the severity of affective illness. Biol Psychiatry 1976; 11:31–42.
10. Berk D. Bipolar II disorder: a review. Bipolar Disord 2005; 7:11–21.

11. American Psychiatric Association. Diagnostic and Statistical Manual of Mental Disorders, 3rd ed. Washington D.C.: American Psychiatric Association, 1980.
12. Regier D, Farmer M, Rae D, et al. Comorbidity of mental disorders with alcohol and other drug abuse. Results from the Epidemiologic Catchment Area (ECA) Study. JAMA 1990; 264(19):2511–2518.
13. Angst J. The emerging epidemiology of hypomania and bipolar II disorder. J Affect Disord 1998; 50(2–3):143–151.
14. Judd L, Akiskal H. The prevalence and disability of bipolar spectrum disorders in the U.S. population: re-analysis of the ECA database taking into account subthreshold cases. J Affect Disord 2003; 73:123–131.
15. Maier W. The distinction of bipolar II disorder from bipolar I and recurrent unipolar depression: results of a controlled family study. Acta Psychiatr Scand 1993; 87(4): 279–284.
16. Gershon E, Hamovit J, Guroff J, et al. A family study of schizoaffective, bipolar I, bipolar II, unipolar, and normal control probands. Arch Gen Psychiatry 1982; 39(10):1157–1167.
17. Sassi R, Nicoletti M, Brambilla P, et al. Decreased pituitary volume in patients with bipolar disorder. Society of Biol Psychiatry 2001; 50:271–280.
18. Ketter T, Kimbrell T, George M, et al. Effects of mood and subtype on cerebral glucose metabolism in treatment-resistant bipolar disorder. Society of Biol Psychiatry 2001; 49:97–109.
19. Malhi G, Lagopoulos J, Sachdev P, Mitchell P, Ivanovski B, Parker G. Cognitive generation of affect in hypomania: an fMRI study. Bipolar Disord 2004; 6:271–285.
20. Winsberg E, Sachs N, Tate D, Adalsteinsson E, Spielman D, Ketter T. Decreased dorsolateral prefrontal N-acetyl aspartate in bipolar disorder. Society of Biol Psychiatry 2000; 47:475–481.
21. Kato T, Hamakawa H, Shioiri T, et al. Choline-containing compounds detected by proton magnetic resonance spectroscopy in the basal ganglia in bipolar disorder. J Psychiatry Neurosci 1996; 21(4):248–254.
22. Fieve R, Kumbaraci T, Dunner D. Lithium prophylaxis of depression in bipolar I, bipolar II, and unipolar patients. Am J Psychiatry 1976; 133:925–929.
23. Kane J, Quitkin F, Rifkin A, Ramos-Lorenzi J, Nayak D, Howard A. Lithium carbonate and imipramine in the prophylaxis of unipolar and bipolar II illness: a prospective, placebo-controlled comparison. Arch Gen Psychiatry 1982; 39(9):1065–1069.
24. Peselow E, Dunner D, Fieve R, Lautin A. Lithium prophylaxis of depression in unipolar, bipolar II, and cyclothymic patients. Am J Psychiatry 1982; 139(6):747–752.
25. Tondo L, Baldessarini R, Hennen J, Floris G. Lithium maintenance treatment of depression and mania in bipolar I and bipolar II disorders. Am J Psychiatry 1998; 155:638–645.
26. Calabrese J, Suppes T, Bowden C, et al. A double-blind, placebo-controlled, prophylaxis study of lamotrigine in rapid-cycling bipolar disorder. J Clin Psychiatry 2000; 61: 841–850.
27. Winsberg M, DeGolia S, Strong C, Ketter T. Divalproex therapy in medication-naïve and mood-stabilizer-naïve bipolar II depression. J Affect Disord 2001; 67:213–219.
28. Amsterdam J, Brunswick D. Antidepressant monotherapy for bipolar type II major depression. Bipolar Disord 2003; 5:388–395.
29. Amsterdam J. Efficacy and safety of fluoxetine in treating bipolar II major depressive episode. J Clin Psychopharmacol 1998; 18(6):435–440.
30. Amsterdam J. Short-term fluoxetine monotherapy for bipolar type II or bipolar NOS major depression—low manic switch rate. Bipolar Disord 2004; 6:75–81.
31. Dunner D, Fieve R. Clinical factors in lithium carbonate prophylaxis failure. Arch Gen Psychiatry 1974; 30:229–233.
32. Kupka R, Luckenbaugh D, Post R, Leverich G, Nolen WA. Rapid and non-rapid cycling bipolar disorder: a meta-analysis of clinical studies. J Clin Psychiatry 2003; 64(12):1483–1494.
33. Schneck C, Miklowitz D, Calabrese J, et al. Phenomenology of rapid-cycling bipolar disorder: data from the first 500 participants in the systematic treatment enhancement program. Am J Psychiatry 2004; 161:1902–1908.

34. Kupka R, Luckenbaugh D, Post R, et al. Comparison of rapid-cycling and non-rapid-cycling bipolar disorder based on prospective mood ratings in 539 outpatients. Am J Psychiatry 2005; 162:1273–1280.
35. Strakowski S, DelBello M, Fleck D, et al. Effects of co-occurring alcohol abuse on the course of bipolar disorder following a first hospitalization for mania. Arch Gen Psychiatry 2005; 62:851–858.
36. Koukopoulos A, Sani G, Kukopulos AE, et al. Duration and stability of the rapid-cycling course: a long-term personal follow-up of 109 patients. J Affect Disord 2002; 73:75–85.
37. Baldessarini R, Tondo L, Floris G, Hennen J. Effects of rapid cycling on response to lithium maintenance treatment in 360 bipolar I and II disorder patients. J Affect Disord 2000; 61:13–22.
38. Cho J, Bone S, Dunner D, Colt E, Fieve R. The effect of lithium treatment on thyroid functions in patients with primary affective disorder. Am J Psychiatry 1979; 136:115–116.
39. Cowdry R, Wehr T, Zis A, Goodwin F. Thyroid abnormalities associated with rapid-cycling bipolar illness. Arch Gen Psychiatry 1983; 40:414–420.
40. Kusalic M. Grade II and grade III hypothyroidism in rapid-cycling bipolar disorder. Neuropsychobiol 1992; 25:177–181.
41. Kupka R, Nolen W, Post R, et al. High rate of autoimmune thyroiditis in bipolar disorder: Lack of association with lithium exposure. Society of Biol Psychiatry 2002; 51:305–311.
42. Gyulai L, Bauer M, Bauer M, Garcia-Espana F, Cnaan A, Whybrow P. Thyroid hypofunction in patients with rapid-cycling bipolar disorder after lithium challenge. Society of Biol Psychiatry 2003; 53:899–905.
43. Koukopoulos A, Reginaldi D, Laddomada P, Floris G, Serra G, Tondo L. Course of the manic-depressive cycle and changes caused by treatment. Pharmakopsychiatr Neuropsychopharmakol 1980; 13(4):156–167.
44. Calabrese J, Markovitz P, Kimmel S, Wagner S. Spectrum of efficacy of valproate in 78 rapid-cycling bipolar patients. J Clin Psychopharmacol 1992; 12(1 suppl):53S–56S.
45. Calabrese J, Delucchi G. Spectrum of efficacy of valproate in 55 patients with rapid-cycling bipolar disorder. Am J Psychiatry 1990; 147:431–434.
46. Calabrese J, Shelton M, Rapport D, et al. A 20-month, double-blind, maintenance trial of lithium versus divalproex in rapid-cycling bipolar disorder. Am J Psychiatry 2005; 162:2152–2161.
47. Bowden C, Calabrese J, McElroy S, et al. The efficacy of lamotrigine in rapid cycling and non-rapid cycling patients with bipolar disorder. Society of Biol Psychiatry 1999; 45: 953–958.
48. Frye M, Ketter T, Kimbrel LT, et al. A placebo-controlled study of lamotrigine and gabapentin monotherapy in refractory mood disorders. J Clin Psychopharmacol 2000; 20(6):607–614.
49. Okuma T. Effects of carbamazepine and lithium on affective disorders. Neuropsychobiol 1993; 27(3):138–145.
50. Denicoff K, Smith-Jackson E, Disney E, Ali S, Leverich G, Post R. Comparative prophylactic efficacy of lithium, carbamazepine, and the combination in bipolar disorder. J Clin Psychiatry 1997; 58(11):470–478.
51. Gonzalez-Pinto A, Tohen M, Lalaguna B, et al. Treatment of bipolar i rapid cycling patients during dysphoric mania with olanzapine. J Clin Psychopharmacol 2002; 22(5):450–454.
52. Vieta E, Calabrese J, Hennen J, et al. Comparison of rapid-cycling and non-rapid-cycling bipolar I manic patients during treatment with olanzapine: analysis of pooled data. J Clin Psychiatry 2004; 65(10):1420–1428.
53. Tondo L, Hennen J, Baldessarini R. Rapid-cycling bipolar disorder: effects of long-term treatments. Acta Psychiatr Scand 2003; 108:4–14.
54. Akiskal H. Dysthymia and cyclothymia in psychiatric practice a century after Kraepelin. J Affect Disord 2001; 62:17–31.
55. American Psychiatric Association. Diagnostic and Statistical Manual of Mental Disorders. 2nd ed. Washington D.C.: American Psychiatric Association, 1968.

56. Weissman M, Myers J. Affective disorders in a U.S. urban community: the use of research diagnostic criteria in an epidemiological survey. Arch Gen Psychiatry 1978; 35(11):1304–1311.
57. Placidi G, Signoretta S, Liguori A, Gervasi R, Maremmani I, Akiskal H. The semi-structured affective temperament interview (TEMPS-I) reliability and psychometric properties in 1010 14–26-year-old students. J Affect Disord 1998; 47:1–10.
58. Jacobsen F. Low-dose valproate: a new treatment for cylothymia, mild rapid cycling disorders, and premenstrual syndrome. J Clin Psychiatry 1993; 54(6):229–234.
59. Sato T, Bottlender R, Kleindienst N, Möller H. Syndromes and phenomenological subtypes underlying acute mania: a factor analytic study of 576 manic patients. Am J Psychiatry 2002; 159:968–974.
60. Goodwin F, Jamison K. Manic-Depressive Illness. New York: Oxford University Press, 1990.
61. Conus P, McGorry P. First-episode mania: a neglected priority for early intervention. Aust NZ J Psychiatry 2002; 36:158–172.
62. Coryell W, Leon A, Turvey C, Akiskal H, Mueller T, Endicott J. The significance of psychotic features in manic episodes: a report from the NIMH collaborative study. J Affect Disord 2001; 67:79–88.
63. Black D, Winokur G, Nasrallah A, Brewin A. Psychotic symptoms and age of onset in affective disorders. Psychopathology 1992; 25(1):19–22.
64. Rosen L, Rosenthal N, Van Dusen P, Dunner R, Fieve R. Age at onset and number of psychotic symptoms in bipolar I and schizoaffective disorder. Am J Psychiatry 1983; 140(11):1523–1524.
65. Baethge C, Baldessarini R, Freudenthal K, Streeruwitz A, Bauer M, Bschor T. Hallucinations in bipolar disorder: characteristics and comparison to unipolar depression and schizophrenia. Bipolar Disord 2005; 7(2):136–145.
66. Toni C, Perugi G, Mata B, Madaro D, Maremmani I, Akiskal H. Is mood-incongruent manic psychosis a distinct subtype? Eur Arch Psychiatry Clin Neurosci 2001; 251(1):12–17.
67. Potash J, Willour V, Chiu Y, et al. The familial aggregation of psychotic symptoms in bipolar disorder pedigrees. Am J Psychiatry 2001; 158:1258–1264.
68. Potash J, Chiu Y, MacKinnon D, et al. Familial aggregation of psychotic symptoms in a replication set of 69 bipolar disorder pedigrees. Am J Med Gen 2003; 116:B90–B97.
69. Potash J, Zandi P, Willour V, et al. Suggestive linkage to chromosomal regions 13q31 and 22q12 in families with psychotic bipolar disorder. Am J Psychiatry 2003; 160:680–686.
70. Strasser H, Lilyestrom J, Ashby E, et al. Hippocampal and ventricular volumes in psychotic and nonpsychotic bipolar patients compared with schizophrenia patients and community control subjects: a pilot study. Biol Psychiatry 2005; 57:633–639.
71. Pearlson G, Wong D, Tune L, et al. In vivo D_2 dopamine receptor density in psychotic and nonpsychotic patients with bipolar disorder. Arch Gen Psychiatry 1995; 52(6): 471–477.
72. Wong D, Pearlson G, Tune L, et al. Quantification of neuroreceptors in the living human brain: IV. Effect of aging and elevations of D_2-like receptors in schizophrenia and bipolar illness. J Cerebral Blood Flow & Metabolism 1997; 17:331–342.
73. Kessing L. Subtypes of manic episodes according to ICD-10—prediction of time to remission and risk of relapse. J Affect Disord 2004; 81:279–285.
74. Strakowski S, Williams J, Sax K, Fleck D, DelBello M, Bourne M. Is impaired outcome following a first manic episode due to mood-incongruent psychosis? J Affect Disord 2000; 61:87–94.
75. MacQueen G, Young T, Robb J, Cooke R, Joffe R. Levels of functioning and well-being in recovered psychotic versus nonpsychotic mania. J Affect Disord 1997; 46:69–72.
76. Miklowitz DJ. Longitudinal outcome and medication noncompliance among manic patients with and without mood-incongruent psychotic features. J Nerv Ment Dis 1992; 180(11):703–711.
77. Swann A, Daniel D, Kochan L, Wozniak P, Calabrese J. Psychosis in mania: specificity of its role in severity and treatment response. J Clin Psychiatry 2004; 65(6):825–829.

78. Zemlan F, Hirschowitz J, Garver D. Mood-Incongruent versus mood-congruent psychosis: differential antipsychotic response to lithium therapy. Psychiatry Res 1984; 11:317–328.
79. Rosenthal N, Rosenthal L, Stallone F, Fleiss J, Dunner D, Fieve R. Psychosis as a predictor of response to lithium maintenance treatment in bipolar affective disorder. J Affect Disord 1979; 1(4):237–245.
80. Yazici O, Kora K, Üçok A, Tunali D, Turan N. Predictors of lithium prophylaxis in bipolar patients. J Affect Disord 1999; 55:133–142.
81. Green A, Tohen M, Patel J, et al. Clozapine in the treatment of refractory psychotic mania. Am J Psychiatry 2000; 157:982–986.
82. Tohen M, Chengappa R, Suppes T, et al. Relapse prevention in bipolar I disorder: 18-month comparison of olanzapine plus mood stabilizer v. mood stabilizer alone. Br J Psychiatry 2004; 184:337–345.
83. Tohen M, Baker R, Altshuler L, et al. Olanzapine versus divalproex in the treatment of acute mania. Am J Psychiatry 2002; 159:1011–1017.
84. Tohen M, Goldberg J, Arrillaga A, et al. A 12-week, double-blind comparison of olanzapine vs haloperidol in the treatment of acute mania. Arch Gen Psychiatry 2003; 60: 1218–1226.
85. Chengappa K, Baker R, Shao L, et al. Rates of response, euthymia and remission in two placebo-controlled olanzapine trials for bipolar mania. Bipolar Disord 2003; 5:1–5.
86. Sachs G, Grossman F, Ghaemi S, Okamoto A, Bowden C. Combination of a mood stabalizer with risperidone or haloperidol for treatment of acute mania: a double-blind, placebo-controlled comparison of efficacy and safety. Am J Psychiatry 2002; 159: 1146–1154.
87. Hirschfeld R, Keck P, Kramer M, et al. Rapid antimanic effect of risperidone monotherapy: a 3-week multicenter, double-blind, placebo-controlled trial. Am J Psychiatry 2004; 161:1057–1065.
88. Smulevich A, Khanna S, Eerdekens M, Karcher K, Kramer M, Grossman F. Acute and continuation risperidone monotherapy in bipolar mania: a 3-week placebo-controlled trial followed by a 9-week double-blind trial of risperidone and haloperidol. Eur Neuropsychopharmacol 2004; 15(1):75–84.
89. McIntyre R, Brecher M, Paulsson B, Huizar K, Jones M. Quetiapine or haloperidol as monotherapy for bipolar mania—a 12-week, double-blind, randomized, parallel-group, placebo-controlled trial. Eur Neuropsychopharmacol 2005; 15:573–585.
90. Sachs G, Chengappa K, Suppes T, et al. Quetiapine with lithium or divalproex for the treatment of bipolar mania: a randomized, double-blind, placebo-controlled study. Bipolar Disord 2004; 6:213–223.
91. Yatham L, Paulsson B, Mullen J, Vagero M. Quetiapine versus placebo in combination with lithium or divalproex for the treatment of bipolar mania. J Clin Psychopharmacol 2004; 24(6):599–606.

18 Biological Factors in Bipolar Disorder in Childhood and Adolescence

Melissa A. Brotman, Daniel P. Dickstein, Brendan A. Rich, and Ellen Leibenluft
Mood and Anxiety Disorders Program, National Institute of Mental Health, Department of Health and Human Services, National Institutes of Health, Bethesda, Maryland, U.S.A.

INTRODUCTION

From its earliest descriptions, clinicians, and researchers have observed bipolar disorder (BD) in children and adolescents. Pediatric manifestations of BD were recognized in antiquity by Aretaeus and documented in case reports in the 19th century by Esquirol (1). However, despite this long history, the exact clinical presentation of juvenile BD remains highly controversial. Clinical, neurocognitive, and pathophysiological comparisons to adult BD require further study. In the past decade, a growing number of studies on pediatric BD indicate unparalleled academic interest in understanding the pathophysiology of the disorder and its developmental trajectory (2).

DIAGNOSING BIPOLAR DISORDER IN CHILDREN

Pathophysiological studies in psychiatry typically involve intensive biological assessments of a relatively small sample of patients. In such studies, investigators try to recruit a clinically homogeneous sample, since clinical homogeneity should be associated with decreased biological variability between patients, and thus with an increased probability of finding significant differences between patients and controls in the pathophysiological variable of interest. In the case of pediatric BD, considerable controversy has surrounded the diagnosis, complicating investigators' ability to recruit homogeneous patient samples and to compare results across sites. Therefore, to understand the current state of research on the biological underpinnings of BD in children, one must appreciate the nature of this diagnostic controversy, and how the diagnosis is applied differently in different research settings.

In diagnosing pediatric BD, researchers disagree as to whether distinct mood episodes, and/or euphoria, should be required to make the diagnosis. Diagnostic and Statistical Manual-Fourth Edition-Text Revision (DSM-IV-TR) clearly requires episodes of mania or hypomania for the diagnosis; indeed, technically, one diagnoses a manic episode, not mania per se. Criterion A of manic and hypomanic episodes requires "a *distinct period* of elevated, expansive, or irritable mood" (italics added) lasting at least seven days in the case of mania, or four days in the case of hypomania (3). However, based in part on earlier case reports (4), researchers suggested that children with BD differ from adults in having less distinct episodes, instead presenting with persistent and severe irritability accompanied by the "B" symptoms of mania, such as agitation, distractibility, and pressured speech (5–7). These investigators suggested that the clinical presentation of children with BD is usually that of a chronic mixed mania (5,8).

Researchers also debate whether euphoria should be required for the diagnosis of mania in children. DSM-IV-TR states that a manic episode can be characterized by either irritability or elated mood. However, the issue is contentious in pediatric psychiatry because irritability is very common in childhood psychopathology (9). Indeed, if one does not require an episodic course for the diagnosis of BD in children, then the boundary between BD and attention deficit hyperactivity disorder (ADHD) accompanied by severe irritability, is unclear (10–12).

In response to the controversies described above, Leibenluft et al. (13) proposed a phenotyping system for BD in children. This phenotyping system was designed to facilitate research by creating homogeneous subgroups among the various populations of children receiving the diagnosis. In this system, one subgroup of children, the narrow phenotype (NP-BD), has the least equivocal presentation; these children have had at least one distinct episode clearly meeting DSM-IV-TR criteria for mania or hypomania, with a duration of at least four days for hypomania and seven days for mania (Table 1). In fact, the criteria for narrow phenotype illness in Leibenluft et al. are narrower than the DSM-IV-TR criteria for a manic episode, since Leibenluft et al., like Geller et al. (14,15), require elation or grandiosity to be the episode's predominant mood state, whereas DSM-IV-TR allows irritability.

Two intermediate phenotypes were designed to address the disagreements concerning the required duration of an episode, and the question of whether euphoria should be required to make the diagnosis of mania (13). One intermediate phenotype includes children whose episodes meet full duration criteria for hypomania or mania, but exhibit irritability rather than euphoria. The second

TABLE 1 Narrow Phenotype Diagnostic Criteria

Mania: (hypo)mania, with full-duration episodes and hallmark symptoms
Modification to the DSM-IV-TR criteria for manic episode
The child must exhibit either elevated/expansive mood or grandiosity, while also meeting other DSM-IV-TR criteria for a (hypo)manic episode.
Guidelines for applying the DSM-IV-TR criteria
Episodes must meet the full duration criteria (i.e., ≥7 days for mania and ≥4 days for hypomania) and be demarcated by switches from other mood states (depression, mixed state, euthymia).
Episodes are characterized by a change from baseline in the patient's mood (DSM-IV-TR criterion A) and simultaneously, by the presence of the associated symptoms (DSM-IV-TR criterion B). For example, the distractibility of a child with ADHD would count toward a diagnosis of (hypo)mania only if his/her distractibility worsened at the same time that he/she experienced mood elevation.
Decreased need for sleep should be distinguished from insomnia (i.e., nonspecific difficulty sleeping, which is associated with fatigue).
Poor judgment per se is not a diagnostic criterion for (hypo)mania; the poor judgment must occur in the context of "increased goal-directed activity" or "excessive involvement in pleasurable activities that have a high potential for painful consequences."

Abbreviations: ADHD, attention deficit hyperactivity disorder; DSM-IV-TR, Diagnostic and Statistical Manual-Fourth Edition-Text Revision.

intermediate phenotype includes children who exhibit distinct euphoric episodes that are shorter than the four days required by DSM-IV-TR.

Finally, children meeting criteria for the "broad phenotype" exhibit chronic impairing irritability accompanied by hyperarousal symptoms; the latter are similar to those seen in children with ADHD. These youth have neither distinct episodes nor euphoria, and their receipt of a bipolar diagnosis has been perhaps the major source of controversy in the literature. While acknowledging that children with this clinical presentation have been called the "broad phenotype" of BD (16), Leibenluft et al. (13) label them as having "severe mood and behavioral dysregulation (SMD)," to emphasize the fact that the question of whether they are indeed suffering from a phenotype of BD is open to empirical investigation. These SMD criteria of Leibenluft et al. (Table 2) were designed to: (*i*) operationalize clearly the criteria for irritability (i.e., increased reactivity to negative emotional stimuli at least three times a week); (*ii*) ensure that the irritability was severe (specifically, severely impairing in one setting and at least mildly impairing in a second); (*iii*) ascertain a sample who, in addition to irritable outbursts, exhibited abnormal baseline mood, in order to exclude children with "tantrums" who are otherwise well-functioning; (*iv*) capture the population of children who exhibit those "B" criteria of mania that overlap with symptoms of ADHD (i.e., exhibit three or more of the following: insomnia, agitation, distractibility, pressured speech, flight of ideas or racing thoughts, intrusiveness); and (*v*) exclude children with an episodic illness or any of the symptoms specific to mania (i.e., euphoria, grandiosity, decreased need for sleep). In designing these criteria, a major goal was to clearly define a population of children who could be compared on a number of biological variables to children meeting narrow phenotype criteria for BD.

TABLE 2 Severe Mood Dysregulation (SMD) Diagnostic Criteria

Inclusion criteria
Age 7–17 years, with symptom onset before age 12.
Abnormal mood (specifically, anger or sadness) present at least 1/2 of the day most days and of sufficient severity to be noticeable by others (e.g., parents, teachers, or peers).
Hyperarousal, as defined by ≥3: insomnia, agitation, distractibility, racing thoughts or flight of ideas, pressured speech, or intrusiveness.
Compared with peers, the child exhibits markedly increased reactivity to negative emotional stimuli manifest verbally or behaviorally—temper tantrums out of proportion to the inciting event and/or child's developmental stage—occurring >3 times/week during past four weeks.
Symptoms are present for ≥12 months without ≥2 months symptom-free.
Symptoms are severe in at least 1 setting and at least mild in a second setting—e.g., school, home, peers.
Exclusion criteria
Child has any "cardinal" BD symptoms: elevated/expansive mood, grandiosity, or episodically decreased need for sleep.
Distinct episodes ≥4 days.
Individual meets diagnostic criteria for schizophrenia, schizophreniform disorder, schizoaffective illness, pervasive developmental disorder, or post-traumatic stress disorder.
Individual meets criteria for substance use disorder in the past three months.
IQ < 70.
Symptoms are direct physiological effect of drug abuse or general medical/neurological condition.

Abbreviations: BD, bipolar disorder; IQ, intelligent quotient.

PREVALENCE ESTIMATES OF PEDIATRIC BIPOLAR DISORDER

Despite the controversy surrounding the diagnostic boundaries of pediatric BD, estimates across epidemiologic studies reveal a relatively consistent prevalence. This consistency may be due to the fact that the epidemiologic studies have all used DSM conventions in making the diagnosis in children, requiring clear episodes with the requisite duration (i.e., these studies exclude children with the "broad phenotype" of the illness). Using the kiddie schedule for affective disorders and schizophrenia for school-age children (K-SADS) (17) to examine DSM-III-R BD in adolescents aged 14 to 18 years, Lewinsohn et al. (18) found a lifetime prevalence of approximately 1%. Not surprisingly, the prevalence of youth with subsyndromal BD (i.e., a distinct period of abnormally elevated, expansive, or irritable mood for less than four days, or a hypomanic episode with no history of a depressive episode) is higher (5.7%) than the prevalence of youth with BD (19,20). Similarly, Carlson and Kashani (21) examined the rates of BD in 150 nonreferred 14- to 16-year-old and found that approximately 1.5% met research diagnostic criteria (RDC) (22) criteria for BD-II or cyclothymia. No child met DSM-III RDC criteria for BD-I, because the degree of impairment was insufficient. In another study examining 717 adolescents 10 to 20 years old (mean age 15 years), Johnson and colleagues (23) found that 14 youth (1.9% of the sample) met criteria for BD according to the diagnostic interview schedule for children (DISC) (24). An additional 10 adolescents (1.4% of the sample) experienced subthreshold BD, as the number of their hypo/manic symptoms was one symptom below threshold. In a younger sample, data from the great smoky mountains study (GSMS) (25) found a 0.1% prevalence of hypomania and no cases of mania in children aged 9 to 13 years old. Applying Leibenluft et al.'s (13) phenotype criteria to the GSMS dataset, Brotman et al. (26) found a 3.3% prevalence of SMD, or the broad phenotype of BD. Taken together, epidemiologic data suggest that the prevalence of DSM-based BD is quite low; not surprisingly, when criteria are relaxed, that is, not requiring clear episodes, or allowing shorter episodes and/or fewer criteria "B" symptoms, population frequency estimates increase.

COGNITIVE AND EMOTIONAL PROCESSING

Classical cognitive paradigms assessing memory and attention can help to elucidate the neuropsychological deficits apparent in youth and adults with BD. Specifically, the study of cognitive impairments can clarify the neurobiological substrates of the disorder, such as deficits in cortical functioning, and potentially demonstrate the relationship between neural correlates and clinical symptoms (27,28). It is particularly important to differentiate nonspecific cognitive deficits from those associated with BD. In addition, it is essential to parse deficits that persist during euthymia from those that occur only during acute mood episodes in order to determine the extent to which BD patients do not achieve functional and full cognitive recovery during remission, and to identify possible endophenotypes of the illness, that is, behavioral markers associated with risk (29). Neuropsychological studies in BD adults have demonstrated state- and trait-related deficits in set shifting, planning, attention, and memory (27,30–36). However, as detailed below, fewer studies have explored neuropsychological impairment in pediatric BD. In addition, as also discussed below, more recent research has assessed patients' processing of emotional stimuli in an attempt to explore the emotional impairments of BD youth (27).

Memory, Learning, and Attention

In BD adults, several studies have demonstrated impairments in memory and verbal learning (35,37–39). Building on these studies, Dickstein et al. (40) used the Cambridge neuropsychological test automated battery (CANTAB) (Cambridge Cognition Ltd., Cambridge, U.K.) to assess visual/visuospatial memory and other neuropsychological parameters in NP-BD children. Consistent with adult BD findings (41,42), NP-BD youth exhibited impairments in pattern recognition memory (PRM). Moreover, NP-BD youth demonstrated decreased spatial memory span relative to control children, again similar to impairments seen in adult BD patients (42,43). Because such memory deficits have been observed in adults with unipolar depression (43) and adults with ADHD (44,45), it remains unclear if these impairments are specific to BD.

Also consistent with studies in adult BD (37,39), McClure et al. (46) found subtle memory deficits in NP-BD youth on the California verbal learning test for children (CVLT-C) (47). The impairments were particularly strong in NP-BD children with comorbid ADHD. In a related study controlling for ADHD comorbidity, Doyle et al. (28) found deficits in sustained attention, working memory, and processing speed in BD youth. However, it is important to note that the criteria for BD in these two studies differed. Whereas the former (46) employed Leibenluft et al.'s (13) narrow phenotype BD criteria, the criteria for the latter study (28) included youth with intense irritability, without euphoric/elated mood, and a nonepisodic illness.

In contrast to the developmental continuity of memory deficits found in adult and pediatric BD on the CANTAB and CVLT, studies using the continuous performance test (CPT) to study sustained attention have revealed more discrepant results. That is, while numerous adult BD studies have demonstrated attentional deficits with this task (34,48–50), results from pediatric BD studies have been less conclusive. In a pilot study, DelBello et al. (51) found no differences in sustained attention between adolescents with BD and controls. However, McClure et al. (52) found that NP-BD subjects did not make more commission errors than did controls, but did demonstrate lower discriminability (i.e., hit rate minus false alarms). These results are not dissimilar from what one would expect in a sample of bipolar adults (34).

Response Flexibility and Motor Inhibition

Response flexibility, or the ability to modify one's behavior based on changing contingencies, is germane to the study of BD for several reasons. First, response flexibility is mediated by the ventral prefrontal cortex (VPFC), which has been implicated in the pathophysiology of BD. Nonhuman primate and adult control studies have shown that behavioral responses to changing contingencies engage a neural circuit encompassing the amygdala, VPFC, and striatum (53). Both lesion (54) and functional magnetic resonance imaging (fMRI) studies (55–61) demonstrate functional abnormalities in these regions in patients with BD (see below for neuroimaging in pediatric BD). Integrating neural findings with clinical observation, our research group is testing the hypothesis that BD patients demonstrate inflexibility in their responses to changes in emotional stimuli. Clinically, depressed patients respond similarly to positive and negative stimuli, in that neither are rewarding. On the other hand, manic patients cannot respond with appropriate flexibility to negative stimuli (e.g., they either deny the existence of such stimuli, or become irritable in response to them). As detailed below, we have demonstrated that children with BD, regardless of mood state, have deficits in adapting to changes in emotional contingencies. We hypothesize that such response inflexibility, when present in euthymia, is a *forme fruste* of a more marked deficit that is present during mania or depression and is relevant to the clinical symptoms of these aberrant mood states. Thus, the examination of impairments in response flexibility may elucidate the pathophysiology of pediatric BD.

Several studies, using different behavioral paradigms, have demonstrated that NP-BD children and adolescents have impaired response flexibility. Two studies have used reversal learning, or the ability to reverse previously acquired stimulus/reward associations, to measure cognitive and response flexibility in NP-BD subjects. In the first, Dickstein et al. (40) showed that NP-BD subjects, compared with typically developing youth, have impaired attentional set shifting, specifically on the simple reversal stage, of the CANTAB's intradimensional/extradimensional (ID/ED) shift task (42). Similarly, Gorrindo et al. (62) found that euthymic NP-BD subjects made more errors during probabilistic reversal and took longer to learn the new rewarded object than did control subjects. In sum, two out of two studies that have examined reversal learning in pediatric BD have shown that NP-BD subjects have impaired response flexibility.

Motor inhibition is a specific type of response flexibility, because inhibiting a prepotent response, or substituting an alternate response for an inhibited one, requires cognitive and motor flexibility. From a clinical perspective, deficits in motor inhibition may contribute to the impulsivity and affective aggression characteristic of BD youth, as such youth may have difficulty inhibiting physical and emotional responses to environmental triggers. To investigate motor control in children with BD, McClure et al. (52) administered the stop and change paradigms (63). Again demonstrating response inflexibility, compared with controls, NP-BD youth had a significantly longer reaction time when substituting an alternate response for an inhibited behavior on the "change" task, and demonstrated a trend toward a delay in their ability to inhibit a prepotent response on the "stop" task.

Attention-Emotion Interactions and Emotion Recognition

To elucidate the pathophysiology of the emotional dysregulation seen in BD children, researchers have begun to explore the influence of emotional stimuli on attention allocation. Adverse attention–emotion interactions may be important in BD because the development of effective emotion regulation requires attentional control; that is, children's ability to deliberately control attention is essential to moderating internal affective states (64,65). For example, directing attention away from frustrating stimuli is a common method for modulating irritability and anger.

Although the disrupting effects of emotional stimuli on attentional allocation has received extensive study in patients with anxiety disorders (66) or depression (67,68), only recently have investigators begun to explore impaired attention–emotion interactions in pediatric BD. Rich et al. (65) used an affective Posner task (69) in which subjects completed the same attentional task three times: first at baseline, then with contingencies (i.e., monetary reward and punishment), and finally in the presence of frustration (i.e., rigged feedback). The BD patients, once again, demonstrated response inflexibility, in that controls, but not BD youth, responded to the introduction of contingencies and frustration with decreased response time. Furthermore, there was evidence that the patients' response inflexibility might be secondary to adverse attention–emotion interactions. Specifically, in controls, parietal P300 evoked response potentials (ERPs) showed increased amplitude in the frustration task compared with the other two tasks, whereas the P300 amplitude in patients was unchanged across the three tasks. Given that P300 amplitude is thought to be a measure of allocation of attention, these data indicate that controls marshaled increased attentional resources in the presence of a heightened emotional context, but BD youth were unable to do so (Fig. 1). These data further suggest that the BD patients' response inflexibility may be associated with deficits in executive attention arising only in emotional, but not in nonemotional, contexts. Speculatively, it is possible that the BD children were unable to focus their attention on the task while frustrated because they were distracted by the emotion that they were experiencing.

Other studies have assessed processing deficits and the influence of emotion on attention by examining children's ability to categorize facial displays of emotion. Clinically, BD youth demonstrate social problems and experience high rates of peer rejection. In addition, adults with BD display impaired recognition of facial emotions, particularly during acute mood states (70,71). To examine such potential impairments in BD youth, McClure et al. (52,72) administered the diagnostic analysis of nonverbal accuracy scale (DANVA) (73) and found that NP-BD youth performed worse than anxious adolescents and psychiatrically healthy control subjects on a facial emotional recognition task. Yet, it remains unclear if these impairments are specific to certain facial emotions, such as anger, or represent a more general and extensive emotion recognition deficit.

Further data by Rich et al. (74) suggest that BD children may mislabel facial emotions because their affective response to the emotional expression impedes the attention allocation needed for emotion categorization. Specifically, BD children and controls completed emotional (e.g., "how afraid are you?") and nonemotional (i.e., "How wide is the nose?") ratings of neutral faces. They found that compared

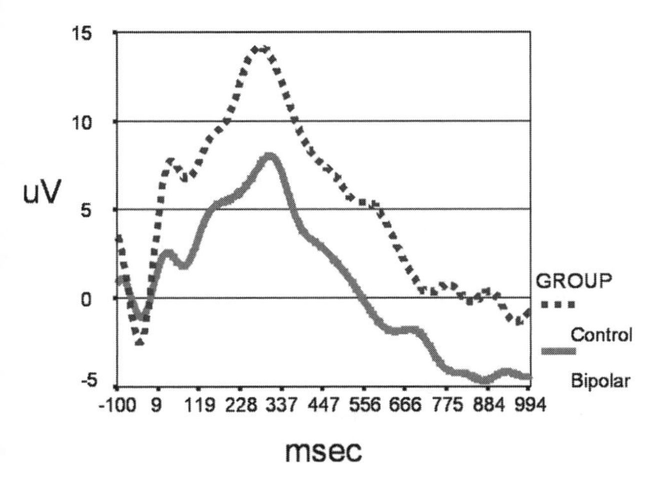

FIGURE 1 P3 ERP wave at parietal sites in the setting of frustration.

with control youth, NP-BD children reported more fear when viewing these relatively innocuous stimuli. However, NP-BD and control children did not differ in nonemotional rating of the faces, again suggesting impaired attention–emotion interactions in BD youth.

Once behavioral deficits are identified in NP-BD youth, compared with controls or other psychiatric populations, fMRI can be used to explore the brain mechanisms mediating these between-group differences. MRI and other neuroimaging techniques are capable of elucidating aberrant biological mechanisms in pediatric BD, including structural abnormalities and dysfunctional circuitry.

NEUROIMAGING IN PEDIATRIC BIPOLAR DISORDER

Unlike in adults, researchers cannot use positron emission tomography (PET) to study BD youth, due to concerns about radiation exposure. However, numerous groups have utilized magnetic resonance imaging (MRI) technology, which uses no radiation and is noninvasive, to evaluate structural, chemical, and functional alterations in pediatric psychopathology. Taken as a whole, these studies demonstrate that fronto-amygdala-striatal abnormalities differentiate BD children and adolescents from typically developing control youth (75).

STRUCTURAL NEUROIMAGING IN PEDIATRIC BIPOLAR DISORDER

The vast majority of published pediatric BD neuroimaging studies are structural MRI studies, that is, studies that use MRI to determine volumetric differences in brain structures between BD youth and controls. While size does not correlate directly with function, structural MRI studies allow the identification of regions-of-interest (ROI) that may mediate the manifestations of BD.

The primary focus of structural MRI studies in pediatric BD has been on limbic structures, most notably the amygdala. Studies have demonstrated that the amygdala mediates numerous functions central to emotion regulation in general, as well as to BD specifically. Such functions include modulating somatic signs of negatively-valenced emotions, such as fear, anxiety, and depression (76,77). The amygdala also shapes behavior in response to rewards and incentives (78–80), which is germane to BD given the DSM-IV-TR clinical criteria of excessive goal-directed activity and pleasure-seeking during mania, and of anhedonia during depression. Of the six published studies that have evaluated amygdala volume in pediatric BD, four showed that BD children and adolescents had decreased amygdala volume, especially on the left, compared with typically developing youth [including one (81) comparing NP-BD youth, in particular, to controls] (81–84). In the fifth study, a similar finding emerged as a trend (85). Only one study failed to show decreased amygdala volume in pediatric BD subjects (86), possibly due to the young average age of both BD and control subjects (mean age 11.3 ± 2.7 BD and 11.0 ± 2.6 controls) or due to BD subjects having a significantly lower IQ than controls (mean verbal IQ 99.2 ± 14.2 BD and 114.9 ± 11.4 controls). Moreover, Blumberg et al. (87) recently published a two-year follow-up study demonstrating that decreased amygdala volume in pediatric BD subjects remains a stable trait feature compared with healthy controls. The consistency of these structural MRI studies demonstrating decreased amygdala volume in pediatric BD subjects stands in marked contrast to neuroimaging studies of BD adults that show either increased or unchanged amygdala volume compared with controls (88–92). Such a consistent finding in pediatric BD suggests that there may be a neurobiological difference between prepubertal- and adult-onset BD.

Structural MRI studies have also evaluated the hippocampus, another limbic structure implicated in memory for emotional events. At present, two studies have found decreased hippocampal volume in pediatric BD compared with controls (82,86), while three have not (81,84,85). Frazier et al. (86), in the more recent of the two studies demonstrating decreased hippocampal volume in BD adolescents, found that this effect was driven by the female BD subjects. Again, structural MRI studies of hippocampal volume in BD adults have found inconsistent results; one study found decreased volume (93) while several others found no difference (89–92,94). These negative studies should be interpreted with caution, however, due to the possibility of type II errors.

Only recently have neuroimaging studies moved beyond the focus on limbic regions to evaluate other regions, such as the frontal cortex. These studies find that, compared with controls, NP-BD youth have decreased volume of the left dorsolateral prefrontal cortex (DLPFC) (Brodmann's area 9—Fig. 2) (81), an area that exerts top-down attentional control in response to emotionally-evocative stimuli (95). Similar decreases in DLPFC volume and density have been found in BD adults

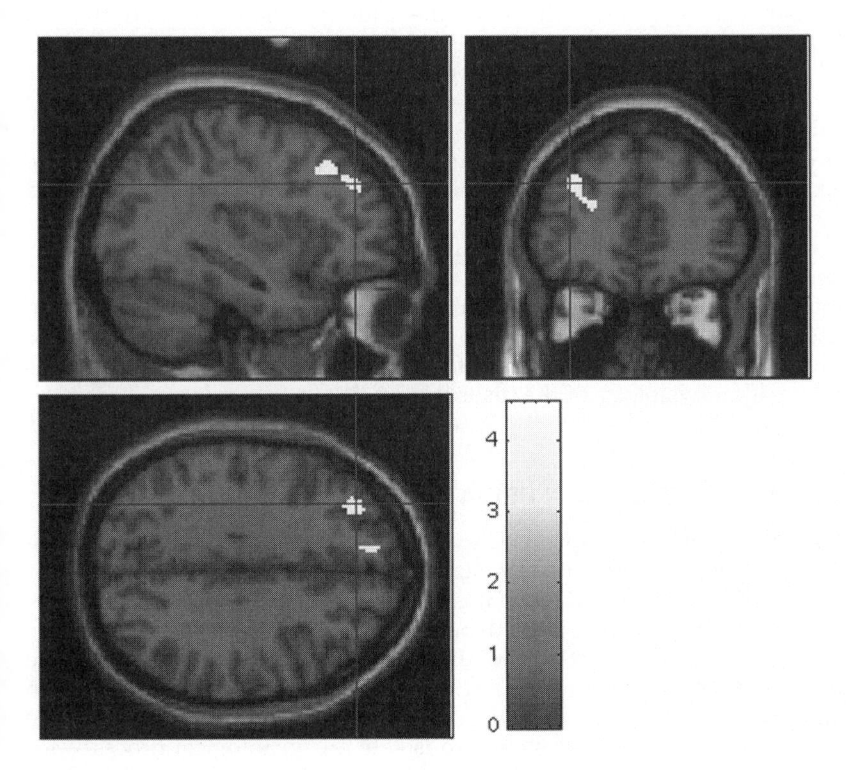

FIGURE 2 Pediatric BD subjects (N = 20) show volume reduction of left dorsolateral prefrontal cortex in comparison to age and gender matched controls (N = 20). 1.5 Tesla magnetic resonance imaging scan axial three-dimensional spoiled gradient recalled echo in the steady state (FSPGR) analyzed with voxel-based morphometry. *Note*: Scale (*right*) represents T-scores, increasing from black to white. ($x = -32$, $y = 42$, $z = 32$; $t = 4.52$, $Z = 4.01$; $p_{corrected} = 0.04$).

(96,97). Other structural MRI studies indicate that BD children and adolescents have decreased volume in the anterior cingulate cortex (ACC) (98), an area that mediates conflict monitoring, but BD youth do not appear to have decreased subgenual prefrontal cortex (PFC) volume (99), a region implicated in adult mood disorders (100). Given the frontal lobe's complex neuroanatomy and equally complex role in psychopathology, additional work is required to elucidate the frontal neuromorphometric alterations associated with pediatric BD.

The striatum, a neural region encompassing basal ganglia structures including the caudate, putamen, and accumbens area, is another area implicated by recent studies in the pathophysiology of pediatric BD. The striatum plays an important role in regulating attention and motor responses (101,102), as well as in adjusting behavior in response to rewarding or pleasurable stimuli (103–107). Thus far, four studies have evaluated striatal volume in pediatric BD, yielding inconsistent results. Specifically, compared with typically developing controls, one study showed that BD children and adolescents have greater basal ganglia volume (108). Another study found that pediatric BD subjects had greater putamen, but not caudate, volume (83). A third study did not find between-group differences in either caudate or putamen volume in BD versus control youth; however, results indicated that in the BD group there was an inverse relationship between age and volume of the caudate bilaterally and the left putamen (109). The fourth study to evaluate striatal volume in pediatric BD utilized voxel-based morphometry (VBM), an automated technique allowing volumetric evaluations of small ROIs, such as the accumbens area, and found decreased volume in the left accumbens area in NP-BD youth compared with controls (81). To the best of our knowledge, no other study has measured accumbens volume in BD subjects, whether pediatric or adult, largely due to the difficulty associated with hand-tracing this region reliably. Given the role of the ventral striatum—that is, accumbens area—in reward-processing, further work is necessary to determine the neurodevelopmental trajectory of striatal volumes in BD.

In sum, the most consistent structural MRI alteration found in BD children and adolescents is decreased amygdala volume. However, in total, structural neuroimaging research in pediatric BD implicates a fronto-amygdala-striatal circuit as mediating the pathophysiology of the disorder.

FUNCTIONAL NEUROIMAGING IN PEDIATRIC BIPOLAR DISORDER

The second neuroimaging technique used to examine the underlying pathophysiology of pediatric BD is fMRI. This neuroimaging technique relies upon the fact that the MR signal of blood is different when it is carrying oxygenated versus deoxygenated blood. This blood-oxygen level dependent (BOLD) signal change can be recorded from a subject's brain across time; therefore, one can compare the BOLD signal, and corresponding neural activity, occurring in specific brain regions during different event types. For example, subjects may be scanned while performing a computerized task designed to isolate a psychological process—for example, recognizing angry or neutral faces. Although there are numerous fMRI studies of BD adults, there are only two published fMRI studies of BD children and adolescents.

First, Blumberg et al. (56) demonstrated that, compared with controls, BD adolescents had increased activation in the left putamen and thalamus during incongruent versus congruent trials on a Stroop color-naming task. In comparison, previous fMRI studies of BD adults showed increased activation of the ventral PFC during similar tasks requiring conflict monitoring (55,110). In the second fMRI study in pediatric BD, Chang et al. (111) reported findings from two tasks: a visuospatial working memory task and a task with positively and negatively valenced pictures. Pediatric BD subjects had increased activation on the two-back working memory task in the bilateral ACC and left-sided putamen, thalamus, and DLPFC compared with controls. In addition, pediatric BD subjects exhibited greater activation than controls in the bilateral DLPFC, inferior frontal gyrus, and right insula while viewing negatively valenced nonface images. Finally, BD subjects had greater activation in the bilateral caudate and thalamus, left middle/superior frontal gyrus, and left ACC when viewing positively valenced pictures (111).

These fMRI studies are consistent with the structural MRI studies previously discussed. In total, they support the hypothesis that pediatric BD subjects have both structural and functional alterations of fronto-amygdala-striatal structures. Future fMRI studies will likely address the extent to which these are trait abnormalities versus mood-state dependent changes. Moreover, as high rates of comorbidity (particularly, ADHD and anxiety disorders) have been reported in BD youth (112), additional work is needed to assess the role of comorbid diagnoses on the neural abnormalities seen in BD youth. Finally, further research is necessary to determine whether these patterns of altered neural activity are specific to pediatric BD or if they are nonspecific markers of more generalized psychopathology.

The Neurochemistry of Pediatric Bipolar Disorder

The same MRI machines used for structural and functional imaging can also determine neurochemical alterations associated with psychopathology. This third form of neuroimaging, known as magnetic resonance spectroscopy (MRS), relies on the principle that the electron cloud surrounding all molecules is differentially and characteristically shifted by the applied magnetic field of the MR scanner. Instead of recording spatial information (structural MRI) or BOLD signal change (functional MRI), MRS records these characteristic chemical shifts in order to determine the concentration of known neurochemicals.

Spectroscopy studies of adults have advanced what is known about both the pathophysiology of BD and the mechanism of action of commonly used anti-manic medications, most notably lithium. Two chemicals in particular have been major foci of research in BD: (*i*) *N*-acetyl asparatate [(NAA), an intraneuronal marker whose levels are decreased in neuropathology, such as brain tumors—increased levels are thought to be markers of neuronal health] and (*ii*) myo-inositol (a component of the intracellular second messenger system). Several MRS studies have demonstrated that BD adults have decreased NAA in comparison to healthy controls in frontal ROIs (113), including the DLPFC (114). In addition, lithium increases frontal NAA in both BD adults and controls (115), a finding which suggests that lithium may have neuroprotective effects (116). Moreover, according to the "myo-inositol depletion hypothesis of mania," mania is due to an increase of second messengers, such as myo-inositol, causing an aberrant and excessive increase in cell signaling; in turn, anti-manic agents, such as lithium, deplete this excessive supply of myo-inositol, restoring balance to the cells and the patient (117). MRS studies indicate that lithium and valproate decrease frontal myo-inositol concentrations when taken by adults at therapeutic doses (118,119). However, although lithium can reduce myo-inositol to "normal" concentrations after only one week of treatment, and these levels remain decreased with maintenance lithium treatment, there is no significant correlation between reduction in mania and reduction in myo-inositol in adults (118).

As with all neuroimaging, far fewer MRS studies have been conducted in BD children and adolescents in comparison to the number done in BD adults. In fact, only three MRS studies have been published in pediatric BD. The first showed that pediatric patients with familial BD had significantly decreased NAA in the DLPFC compared with controls (120). The second found that acute lithium treatment (e.g., seven days) significantly decreased myo-inositol in the ACC and that this decrease was more prominent in lithium responders than in nonresponders (121). The third MRS study, conducted by the same group, demonstrated that pediatric BD subjects had significantly increased ACC myo-inositol compared with both controls and to youth with intermittent explosive disorder (122). Together, these MRS studies confirm that pediatric BD is characterized by specific neurochemical alterations that change in response to psychopharmacological treatment. Moreover, unlike in BD adults, these acute changes seem to correlate with treatment response in BD children and adolescents.

Future MRS studies will facilitate greater understanding of the neurochemical basis of BD in children and adolescents. Moreover, several research groups are working to enable reliable quantification of neurochemicals more directly linked to BD and other mood and anxiety disorders, such as gamma-aminobutyric acid (GABA). Additionally, MRS studies will allow the exploration of state versus trait issues and the therapeutic effect of psychotropic medications. In turn, MRS will enable the design of targeted interventions, including novel medications, to treat or even prevent morbidity and mortality from BD.

FUTURE DIRECTIONS

What will be the future direction of research into the biological mechanisms of pediatric BD? Forthcoming studies will likely expand upon those of the present by evaluating children with different phenotypes of pediatric BD. For example, future work may compare neuropsychological and imaging findings in NP-BD and SMD youth. Studies determining pathophysiological similarities and differences between youth with NP-BD and those with ADHD or MDD will enable researchers to ascertain specific neural and cognitive correlates of pediatric BD. Moreover, researchers must continue to examine the influence of medications, comorbidity status, and mood state while interpreting neuropsychological and fMRI results. In addition, researchers can build on recent neuropsychological and MRI work elucidating brain development (123) to determine how the brains of children and adolescents with BD grow, develop, and prune differently than typically developing youth.

Finally, a consensus is emerging that the search for risk-related genes in BD and other complex neuropsychiatric illnesses would be facilitated by the identification of endophenotypes, which are behavioral deficits (e.g., impaired attention–emotion interactions) or biological findings (e.g., abnormalities in the fronto-amygdala-striatal circuit) that are familial and associated with risk for an illness (124,125). The identification of such endophenotypes would allow the field to move beyond DSM-based phenotypic characterization of affected individuals and those at risk. Specifically, if non-symptomatic children with a first-degree relative with BD demonstrate similar deficits to those observed in NP-BD patients (e.g., deficits in response reversal or emotion identification), such impairments can be conceptualized as endophenotypes, or trait-markers of BD. Employing this strategy, researchers have begun to explore the psychiatric phenomenology of BD offspring (126) and to perform structural and functional neuroimaging in such youth (84,111). These initial studies suggest structural and functional differences in children at risk for BD compared with control youth. Studies are also beginning to explore neurocognitive function in first-degree relatives of BD patients (127,128) and children temperamentally "at risk" for BD (129). Future studies should continue to explore potential neuropsychological deficits in these at risk youth. Ultimately, it is likely that research will begin to integrate at-risk studies with behavioral and neuroimaging work, examining the relationship between endophenotypic markers, such as neuropsychological deficits, genes, and brain development.

REFERENCES

1. Faedda GL, Baldessarini RJ, Suppes T, Tondo L, Becker I, Lipschitz DS. Pediatric-onset bipolar disorder: a neglected clinicial and public health problem. Harv Rev Psychiatry 1995; 3:171–195.
2. Pavuluri MN, Birmaher B, Naylor MW. Pediatric bipolar disorder: a review of the past 10 years. J Am Acad Child adolesc Psychiatry 2005; 44:846–871.
3. Diagnostic and Statistical Manual of Mental Disorders 4th Edition Text Revision (DSM-IV-TR). Washington, D.C: American Psychiatric Association, 2000.
4. Weller E, Weller RA, Dogin JW. A rose is a rose is a rose. J Affect Disord 1998; 51:189–195.
5. Biederman J. Resolved. Mania is mistaken for ADHD in prepubertal children (affirmative). J Am Acad Child adolesc Psychiatry 1998; 37:1091–1093.
6. Spencer TJ, Biederman J, Wozniak J, Faraone SV, Wilens TE, Mick E. Parsing pediatric bipolar disorder from its associated comorbidity with the disruptive behavior disorders. Biol Psychiatry 2001; 49:1062–1070.
7. Wozniak J, Biederman J, Kiely K, et al. Mania-like symptoms suggestive of childhood-onset bipolar disorder in clinically referred children. J Am Acad Child adolesc Psychiatry 1995; 34:867–876.
8. Biederman J, Mick E, Faraone SV, Spencer T, Wilens TE, Wozniak J. Pediatric mania: a developmental subtype of bipolar disorder? Biol Psychiatry 2000; 48:458–466.
9. Leibenluft E, Blair RJR, Charney DS, Pine DS. Irritability in pediatric mania and other childhood psychopathology. Ann NY Acad Sci 2003; 1008:201–218.
10. Carlson GA. Mania and ADHD: comorbidity or confusion. J Affect Disord 1998; 51:177–187.
11. Carlson GA, Weintraub S. Childhood behavioral problems and bipolar disorder-relationship or coincidence? J Affect Disord 1993; 28:143–153.
12. Mick E, Spencer T, Wozniak J, Biederman J. Heterogeneity of irritability in attention-deficit/hyperactivity disorder subjects with and without mood disorders. Biol Psychiatry 2005; 58:576–582.
13. Leibenluft E, Charney DS, Towbin KE, Bhangoo RK, Pine DS. Defining clinical phenotypes of juvenile mania. Am J Psychiatry 2003; 160:430–437.
14. Geller B, Zimerman B, Williams M, et al. DSM-IV mania symptoms in a prepubertal and early adolescent bipolar disorder phenotype compared with attention-deficit hyperactive and normal controls. J Child Adolesc Psychopharmacol 2002; 12:11–25.
15. Geller B, Zimerman B, Williams M, DelBello MP, Frazier J, Berginger L. Phenomenology of prepubertal and early adolescent bipolar disorder: Examples of elated mood, grandiose behaviors, decreased need for sleep, racing thoughts and hypersexuality. J Child Adolesc Psychopharmacol 2002; 12:3–9.

16. Nottelmann E, Biederman J, Birmaher B, Carlson GA, Chang KD, Fenton WS. National Institute of Mental Health research roundtable on prepubertal bipolar disorder. J Am Acad Child adolesc Psychiatry 2001; 40:871–878.
17. Kaufman J, Birmaher B, Brent D, et al. Schedule for affective disorders and schizophrenia for school-age children-present and lifetime version (k-SADS-PL): initial reliability and validity data. J Am Acad Child Adolesc Psychiatry 1997; 36:980–988.
18. Lewinsohn PM, Klein DN, Seeley JR. Bipolar Disord in a community sample of older adolescents: prevalence, phenomenology, comorbidity, and course. J Am Acad Child adolesc Psychiatry 1995; 34:454–463.
19. Lewinsohn PM, Klein DN, Seeley JR. Bipolar disorder during adolescence and young adulthood in a community sample. Bipolar Disord 2000; 2:281–293.
20. Lewinsohn PM, Seeley JR, Klein DN. Bipolar disorder during adolescence. Acta Psychiatr Scand 2003; 108:47–50.
21. Carlson GA, Kashani JH. Manic symptoms in a non-referred adolescent population. J Affect Disord 1988; 15:219–226.
22. Spitzer RL, Endicott J, Robins E. Research Diagnostic Criteria (RDC). New York, New York State Psychiatric Institute, 1975.
23. Johnson JG, Cohen P, Brook JS. Associations between bipolar disorder and other psychiatric disorders during adolescence and early adulthood: a community-based longitudinal investigation. Am J Psychiatry 2000; 157:1679–1681.
24. Costello EJ, Edelbrock CS, Duncan MK, Kalas R. Testing the NIMH Diagnostic Interview Schedule for Children (DISC) in a clinical population: Final report to the Center for Epidemiological Studies NIMH. Pittsburgh, 1984.
25. Costello EJ, Angold A, Burns BJ, et al. The Great Smoky Mountains Study of Youth: goals, design, methods, and the prevalence of DSM-III-R disorders. Arch Gen Psychiatry1996; 53:1129–1136.
26. Brotman MA, Schmajuk M, Rich BA, et al. Prevalence, clinical correlates, and longitudinal course of severe mood dysregulation in children. Biol Psychiatry 2006; 60(9): 991–997.
27. Murphy FC, Sahakian BJ. Neuropsychology of bipolar disorder. Br J Psychiatry 2001; 178:s120–s127.
28. Doyle AE, Wilens TE, Kwon A, et al. Neuropsychological functioning in youth with bipolar disorder. Biol Psychiatry 2005; 58:540–548.
29. Tavares JVT, Drevets WC, Sahakian BJ. Editorial: cognition in mania and depression. Psychol Med 2003; 33:959–967.
30. Basso MR, Lowery N, Neel J, Purdie R, Bornstein RA. Neuropsychological impairment among manic, depressed, and mixed-episode inpatients with bipolar disorder. Neuropsychology 2002; 16:84–91.
31. Wilder-Willis KE, Sax KW, Rosenberg HL, Fleck DE, Shear PK, Strakowski SM. Persistent attentional dysfunction in remitted bipolar disorder. Bipolar Disord 2001; 3:58–62.
32. Bearden CE, Hoffman KM, Cannon TD. The neuropsychology and neuroanatomy of bipolar affective disorder: a critical review. Bipolar Disord 2001; 3:106–150.
33. Clark L, Goodwin GM. State- and trait-related deficits in sustained attention in bipolar disorder. Eur Arch Psychiatry Clin Neurosci 2004; 254:61–68.
34. Sax KW, Strakowski SM, McElroy SL, Keck PE Jr, Wcst SA. Attention and formal thought disorder in mixed and and pure mania. Biol Psychiatry 1995; 37: 420–423.
35. Martinez-Aran A, Vieta E, Reinares M, et al. Cognitive function across manic or hypomanic, depressed, and euthymic states in bipolar disorder. Am J Psychiatry 2004; 161:262–270.
36. Rubinsztein JS, Michael A, Paykel ES, Sahakian BJ. Cognitive impairment in remission in bipolar affective disorder. Psychol Med 2000; 30:1025–1036.
37. Cavanagh JT, Van Beck M, Muir W, Blackwood DH. Case-control study of neurocognitive function in euthymic patients with bipolar disorder: an association with mania. Br J Psychiatry 2002; 180:320–326.

38. Clark L, Iversen SD, Goodwin GM. A neuropsychological investigation of prefrontal cortex involvement in acute mania. Am J Psychiatry 2001; 158:1605–1611.
39. Fleck DE, Shear PK, Zimmerman ME, et al. Verbal memory in mania: effects of clinical state and task requirements. Bipolar Disord 2005; 5:375–380.
40. Dickstein DP, Treland JE, Snow J, et al. Neuropsychological performance in pediatric bipolar disorder. Biol Psychiatry 2004; 55:32–39.
41. Murphy FC, Sahakian BJ, Rubinsztein JS, et al. Emotional bias and inhibitory control processes in mania and depression. Psychol Med 1999; 29:1307–1321.
42. Sweeney JA, Kmiec JA, Kupfer DJ. Neuropsychologic impairments in bipolar and unipolar mood disorders on the CANTAB neurocognitive battery. Biol Psychiatry 2000; 48:674–684.
43. Kempton S, Vance A, Maruff P, Luk E, Costin J, Pantelis C. Executive function and attention deficit hyperactivity disorder: stimulant medication and better executive function performance in children. Psychol Med 1999; 29:527–538.
44. McLean A, Dowson J, Toone B, et al. Characteristic neurocognitive profile associated with adult attention-deficit/hyperactivity disorder. Psychol Med 2004; 34:681–692.
45. Dowson JH, McLean A, Bazanis E, et al. Impaired spatial working memory in adults with attention-deficit/hyperactivity disorder: comparisons with performance in adults with borderline personality disorder and in control subjects. Acta Psychiatr Scand 2004; 110:45–54.
46. McClure EB, Treland JE, Snow J, et al. Memory and learning in pediatric bipolar disorder. J Am Acad Child Adolesc Psychiatry 2005; 44:461–469.
47. Delis DC, Kramer JH, Kaplan E, Ober B. California Verbal learning Test—Children's Version (CVLT-C) Manual. San Antonio, Psychological Corporation, 1994.
48. Tham A, Engelbrekston K, Mathe AA, Johnson L, Olsson E, Aberg-Wistedt A. Impaired neuropsychological performance in euthymic patients with recurring mood disorders. J Clin Psychiatry 1997; 58:26–28.
49. van Gorp WG, Altshuler L, Theberge DC, Wilkins J, Dixon W. Cognitive impariemtns in euthymic bipolar patients with and without prior alcohol dependence: a preliminary study. Arch Gen Psychiatry 1998; 55:41–46.
50. Najt P, Glahn D, Bearden CE, et al. Attention deficits in bipolar disorder: a comparison based on the continuous performance test. Neurosci Lett 2005; 379: 122–126.
51. DelBello MP, Adler CM, Amicone J, et al. Parametric neurocognitive task design: a pilot study of sustained attention in adolescents with bipolar disorder. J Affect Disord 2004; 82S:S79–S88.
52. McClure EB, Treland JE, Snow J, et al. Deficits in social cognition and response flexibility in pediatric bipolar disorder. Am J Psychiatry 2005; 162:1644–1651.
53. Clark L, Cools R, Robbins TW. The neuropsychology of ventral prefrontal cortex: Decision-making and reversal learning. Brain Cogn 2004; 55:41–53.
54. Starkstein SE, Robinson RG. Mechanism of disinhibition after brain lesions. J Nerv Ment Dis 1997; 185:108–114.
55. Blumberg HP, Leung H, Skudlarski P, et al. A functional magnetic resonance imaging study of bipolar disorder. Arch Gen Psychiatry 2003; 60:601–609.
56. Blumberg HP, Martin A, Kaufman J, et al. Frontostriatal abnormalities in adolescents with bipolar disorder: preliminary observations from functional MRI. Am J Psychiatry 2003; 160:1345–1347.
57. Chang K, Adleman NE, Dienes K, Simeonova DI, Menon V, Reiss A. Anomalous prefrontal-subcortical activation in familial pediatric bipolar disorder. Arch Gen Psychiatry 2004; 61:781–792.
58. Blumberg HP, Kaufman J, Martin A, Charney DS, Krystal JH, Peterson BS. Significance of adolescent neurodevelopment for the neural circuitry of bipolar disorder. Ann NY Acad Sci 2004; 1021:376–383.
59. Yurgelun-Todd DA, Gruber SA, Kanayama G, Killgore WD, Baird AA, Young AD. fMRI during affect discrimination in bipolar affective disorder. Bipolar Disord 2000; 2:237–248.
60. Blumberg HP, Stern E, Ricketts S, et al. Rostral and orbital prefrontal cortex dysfunction in the manic state of bipolar disorder. Am J Psychiatry1999; 156:1986–1988.

61. Elliot R, Ogilvie A, Rubinsztein JS, Calderon R, Dolan RJ, Sahakian BJ. Abnormal ventral frontal response during performance of an affective go/no go task in patients with mania. Biol Psychiatry 2004; 55:1163–1170.
62. Gorrindo T, Blair RJR, Budhani S, Dickstein DP, Pine DS, Leibenluft E. Deficits on a probabilistic response-reversal task in patients with pediatric bipolar disorder. Am J Psychiatry 2005; 162:1975–1977.
63. Logan GD. On the ability to inhibit thought and action: a users' guide to the stop signal paradigm. In: Dagenbach D, Carr TH, eds. Inhibitory Processes in Attention, Memory, and Language. San Diego: Academic Press, 1994:189–239.
64. Kopp CB. Commentary: the codevelopments of attention and emotion regulation. Infancy 2002; 3:199–208.
65. Rich BA, Schmajuk M, Perez-Edgar K, Pine DS, Fox NA, Leibenluft E. The impact of reward, punishment, and frustration on attention in pediatric bipolar disorder. Biol Psychiatry 2005; 58:532–539.
66. Mogg K, Bradley BP. A cognitive-motivational analysis of anxiety. Behav Res Ther 1998; 36:809–848.
67. Lloyd GG, Lishman WA. Effect of depression on the speed of recall of pleasant and unpleasant experiences. Psychol Med 1975; 5:173–180.
68. Bradley BP, Mogg K, Millar N. Implicit memory bias in clinical and non-clinical depression. Behav Res Ther 1996; 34:865–879.
69. Perez-Edgar K, Fox N. A behavioral and electrophysiological study of children's selective attention under neutral and affective conditions. J Cogn Dev 2005; 6:89–116.
70. Harmer CJ, Grayson L, Goodwin GM. Enhanced recognition of disgust in bipolar illness. Biol Psychiatry 2002; 51:298–304.
71. Lembke A, Ketter TA. Impaired recognition of facial emotion in mania. Am J Psychiatry 2002; 159:302–304.
72. McClure EB, Pope K, Hoberman AJ, Pine DS, Leibenluft E. Facial expression recognition in adolescents with mood and anxiety disorders. Am J Psychiatry 2003; 160:1–3.
73. Nowicki S, Duke MP. Individual differences in the nonverbal communication of affect: the diagnostic analysis of nonverbal accuracy scale. J Nonverbal Behav 1994; 18:9–35.
74. Rich BA, Vinton DT, Roberson-Nay R, et al. Limbic hyperactivation during processing of neutral facial expressions in children with bipolar disorder. Proceedings of the National Academy of Sciences of the United States of America, 2006; 103:8900–8905.
75. Blumberg HP, Charney DS, Krystal JH. Frontotemporal neural systems in bipolar disorder. Semin Clin Neuropsychiatry 2002; 7:243–254.
76. LeDoux JE, Iwata J, Cicchetti P, Reis DJ. Different projections of the central amygdaloid nucleus mediate autonomic and behavioral correlates of conditioned fear. J Neurosci 1988; 8:2517–2529.
77. Davis M, Whalen PJ. The amygdala: vigilance and emotion. Mol Psychiatry 2001; 6: 13–34.
78. Baxter MG, Murray EA. The amygdala and reward. Nat Rev Neurosci 2002; 3:563–573.
79. Gottfried JA, O'Doherty J, Dolan RJ. Encoding predictive reward value in human amygdala and orbitofrontal cortex. Science 2003; 301:1104–1107.
80. Izquierdo A, Murray EA. Combined unilateral lesions of the amygdala and orbital prefrontal cortex impair affective processing in rhesus monkeys. J Neurophysiol 2004; 91:2023–2039.
81. Dickstein DP, Milham MP, Nugent AC, et al. Frontotemporal alterations in pediatric bipolar disorder: results of a voxel-based morphometry study. Arch Gen Psychiatry 2005; 62:734–741.
82. Blumberg HP, Kaufman J, Martin A, et al. Amygdala and hippocampal volumes in adolescents and adults with bipolar disorder. Arch Gen Psychiatry 2003; 60:1201–1208.
83. DelBello MP, Zimmerman ME, Mills NP, Getz GE, Strakowski SM. Magnetic resonance imaging analysis of amygdala and other subcortical brain regions in adolescents with bipolar disorder. Bipolar Disord 2004; 6:43–52.
84. Chang K, Karchemskiy A, Barnea-Goraly N, Garrett A, Simeonova DI, Reiss A. Reduced amygdalar gray matter volume in familial pediatric bipolar disorder. J Am Acad Child adolesc Psychiatry 2005; 44:565–573.

85. Chen BK, Sassi RB, Axelson D, et al. Cross-sectional study of abnormal amygdala development in adolescents and young adults with bipolar disorder. Biol Psychiatry 2004; 56:399–405.
86. Frazier JA, Chiu S, Breeze JL, et al. Structural brain magnetic resonance imaging of limbic and thalamic volumes in pediatric bipolar disorder. Am J Psychiatry 2005; 162:1256–1265.
87. Blumberg HP, Fredericks CA, Wang F, et al. Preliminary evidence for persistent abnormalities in amygdala volumes i—n adolescents and young adults with bipolar disorder. Bipolar Disord 2005; 7:570–576.
88. Pearlson GD, Barta PE, Powers RE, et al. Medial and superior temporal gyral volumes and cerebral asymmetry in schizophrenia versus bipolar disorder. Biol Psychiatry 1997; 41:1–14.
89. Altshuler LL, Bartzokis G, Grieder T, Curran J, Mintz J. Amygdala enlargement in bipolar disorder and hippocampal reduction in schizophrenia: an MRI study demonstrating neuroanatomic specificity. Arch Gen Psychiatry 1998; 55:663–664.
90. Strakowski SM, DelBello MP, Sax KW, et al. Brain magnetic resonance imaging of structural abnormalities in bipolar disorder. Arch Gen Psychiatry 1999; 56:254–260.
91. Altshuler LL, Bartzokis G, Grieder T, et al. An MRI study of temporal lobe structures in men with bipolar disorder or schizophrenia. Biol Psychiatry 2000; 48:147–162.
92. Brambilla P, Harenski K, Nicoletti M, et al. MRI investigation of temporal lobe structures in bipolar patients. J Psychiatr Res 2003; 37:287–295.
93. Altshuler LL, Conrad A, Hauser P, et al. Reduction of temporal lobe volume in bipolar disorder: A preliminary report of magnetic resonance imaging. Arch Gen Psychiatry 1991; 48:482–483.
94. Hauser P, Matochik J, Altshuler LL, et al. MRI-based measurements of temporal lobe and ventricular structures in patients with bipolar I and bipolar II disorders. J Affect Disord 2000; 60:25–32.
95. Posner MI, Rothbart MK. Attention, self-regulation and consciousness. Philosophical transactions of the Royal Society of London. Series B. Biol Sci 1998; 353:1915–1927.
96. Sax KW, Strakowski SM, Zimmerman ME, DelBello MP, Keck PE Jr, Hawkins JM. Frontosubcortical neuroanatomy and the continuous performance test in mania. Am J Psychiatry 1999; 156:139–141.
97. Lyoo IK, Kim MJ, Stoll AL, et al. Frontal lobe gray matter density decreases in bipolar I disorder. Biol Psychiatry 2004; 55:648–651.
98. Kaur S, Sassi RB, Axelson D, et al. Cingulate cortex anatomical abnormalities in children and adolescents with bipolar disorder. Am J Psychiatry 2005; 162:1637–1643.
99. Sanches M, Sassi RB, Axelson D, et al. Subgenual prefrontal cortex of child and adolescent bipolar patients: a morphometric magnetic resonance imaging study. Psychiatry Res 2005; 138:43–49.
100. Drevets WC, Ongur D, Price JL. Neuroimaging abnormalities in the subgenual prefrontal cortex: implications for the pathophysiology of familial mood disorders. Mol Psychiatry 1998; 3:220–221.
101. Alexander GE, DeLong MR, Strick PL. Parallel organization of functionally segregated circuits linking basal ganglia and cortex. Annu Rev Neurosci 1986; 9:357–381.
102. Rolls ET. Neurophysiology and cognitive functions of the striatum 1. Revue neurologique (Paris) 1994; 150:648–660.
103. Schultz W, Tremblay L, Hollerman JR. Reward processing in primate orbitofrontal cortex and basal ganglia. Cerebral Cotex 2000; 10:272–284.
104. Kimura M, Yamada H, Matsumoto N. Tonically active neurons in the striatum encode motivational contexts of action. Brain and Development 2003; (25 suppl 1): S20–S23.
105. O'Doherty J, Dayan P, Schultz J, Deichmann R, Friston K, Dolan RJ. Dissociable roles of ventral and dorsal striatum in instrumental conditioning. Science 2004; 304:452–454.
106. Zink CF, Pagnoni G, Martin-Skurski ME, Chappelow JC, Berns GS. Human striatal responses to monetary reward depend on saliency. Neuron 2004; 42:509–517.
107. Pasupathy A, Miller EK. Different time courses of learning-related activity in the prefrontal cortex and striatum. Nature 2005; 433:873–876.

108. Wilke M, Kowatch RA, DelBello MP, Mills NP, Holland SK. Voxel-based morphometry in adolescents with bipolar disorder: First results. Psychiatry Res 2004; 131:57–69.
109. Sanches M, Roberts RL, Sassi RB, et al. Developmental abnormalities in striatum in young bipolar patients: a preliminary study. Bipolar Disord 2005; 7:153–158.
110. Rubinsztein JS, Fletcher PC, Rogers RD, et al. Decision-making in mania: a PET study. Brain 2001; 124:2550–2563.
111. Chang K, Adleman NE, Dienes K, Simeonova DI, Menon V, Reiss A. Anomalous prefrontal-subcortical activation in familial pediatric bipolar disorder: a functional magnetic resonance imaging investigation. Arch Gen Psychiatry 2004; 61:781–792.
112. Dickstein DP, Rich BA, Binstock AB, et al. Comorbid anxiety in phenoytpes of pediatric bipolar disorder. J Child Adolesc Psychopharmacol 2005; 15:534–548.
113. Cecil KM, DelBello MP, Morey R, Strakowski SM. Frontal lobe differences in bipolar disorder as determined by proton MR spectroscopy. Bipolar Disord 2002; 4:357–365.
114. Winsberg ME, Sach N, Tate DL, Adalsteinsson E, Spielman D, Ketter TA. Decreased dorsolateral prefrontal N-acetyl aspartate in bipolar disorder. Biol Psychiatry 2000; 47:475–481.
115. Moore GJ, Bebchuk JM, Hasanat K, et al. Lithium increases N-acetyl-aspartate in the human brain: in vivo evidence in support of bcl-2's neurotrophic effects? Biol Psychiatry 2000; 48:1–8.
116. Manji HK, Moore GJ, Chen G. Lithium up-regulates the cytoprotective protein Bcl-2 in the CNS in vivo: a role for neurotrophic and neuroprotective effects in manic depressive illness. J Clin Psychiatry 2000; 61(suppl 9):82–96.
117. Berridge MJ, Irvine RF. Inositol phosphates and cell signalling. Nature 1989; 341: 197–205.
118. Moore GJ, Bebchuk JM, Parrish JK, et al. Temporal dissociation between lithium-induced changes in frontal lobe myo-inositol and clinical response in manic-depressive illness. Am J Psychiatry 1999; 156:1902–1908.
119. Silverstone PH, Wu RH, O'Donnell R, Ultrich M, Asghar SJ, Hanstock CC. Chronic treatment with both lithium and sodium valproate may normalize phosphoinositol cycle activity in bipolar patients. Hum Psychopharmacol 2002; 17:321–327.
120. Chang K, Adleman NE, Dienes K, Barnea-Goraly N, Reiss A, Ketter T. Decreased N-Acetylaspartate in children with familial bipolar disorder. Biol Psychiatry 2003; 53:1059–1065.
121. Davanzo P, Thomas MA, Yue K, et al. Decreased anterior cingulate myo-inositol/creatine spectroscopy resonance with lithium treatment in children with bipolar disorder. Neuropsychopharmacology 2001; 24:359–369.
122. Davanzo P, Yue K, Thomas MA, et al. Proton magnetic resonance spectroscopy of bipolar disorder versus intermittent explosive disorder in children and adolescents. Am J Psychiatry 2003; 160:1442–1452.
123. Hariri AR, Drabant EM, Munoz KE, et al. A susceptibility gene for affective disorders and the response of the human amygdala. Arch Gen Psychiatry 2005; 62:146–152.
124. Glahn DC, Bearden CE, Niendam TA, Escamilla MA. The feasibility of neuropsychological endophenotypes in the search for genes associated with bipolar affective disorder. Bipolar Disord 2004; 6:171–182.
125. Frangou S, Haldane M, Roddy D, Kumari V. Evidence for deficit in tasks of ventral, but not dorsal, prefrontal executive function as an endophenotypic marker for bipolar disorder. Biol Psychiatry 2005; 58:838–839.
126. Chang KD, Steiner H, Ketter TA. Psychiatry phenomenology of child and adolescent bipolar offspring. J Am Acad Child Adolesc Psychiatry 2000; 39:453–460.
127. Ferrier IN, Chowdhury R, Thompson JM, Watson S, Young AH. Neurocognitive function in unaffected first-degree relatives of patients with bipolar disorder: a preliminary report. Bipolar Disord 2004; 6:319–322.
128. Clark L, Sarna A, Goodwin GM. Impairment of executive function but not memory in first-degree relatives of patients with bipolar I disorder and in euthymic patients with unipolar depression. Am J Psychiatry 2005; 162:1980–1982.
129. Meyer TD, Deckersbach T. No evidence for verbal memory impairment in individuals putatively at risk for bipolar disorder. Compr Psychiatry 2005; 46:472–476.

19 Biological Factors in Bipolar Disorders in Late Life

R. C. Young, J. M. deAsis, and G. S. Alexopoulos
Institute for Geriatric Psychiatry, Weill Medical College of Cornell University, and New York Presbyterian Hospital, White Plains, New York, U.S.A.

HISTORICAL PERSPECTIVE

Current understanding of bipolar disorder (BD) in late life builds on early clinical descriptions by European psychiatrists, and these descriptions emphasized links to brain pathology. Manic signs and symptoms were described in the context of brain lesions and disorders at least as early as the nineteenth century (1). Welt (2) described behavioral "disinhibition" in patients with lesions of the orbital surface of the frontal lobes. This was termed "Witzelsucht" by European psychiatrists (3), and "pseudopsychopathic syndrome" by neurologists (4). These observations set the stage for more recent investigations of brain pathology associated with mania and depression in old age. Early writings mention the association between brain vascular disease and mood disorder in late life (5). Kay et al. (6) and Post (7) proposed that mood disorder with onset in late life reflects in part age-associated brain changes.

In the mid-20th century, investigation of mood disorders in late life was advanced by validation of the distinction between unipolar major depression (UMD) and BD in younger patients (8). The introduction of lithium salts (9) and other mood stabilizers reinforced the utility of the BD category across the age spectrum. It also encouraged both testing distinctions within BD such as atypical or type II patients, and broadening the concept of BD.

Kraepelin discussed the incidence of first episode mania, depression and mixed states across the age span in "manic depressive insanity" (5). In a sample of 903 cases, he observed that mania as first episode of illness was less frequent with increased age, although the incidence tended to increase between age 45 to 50 years. He also observed that, in contrast to depressive first episodes, mixed first episodes declined consistently with age. Yet, he noted that affective episodes could appear first at "old-old" age, for example, 80 years. Kraepelin did not differentiate UMD from BD, however.

Roth and colleagues (10) and other clinicians (7) commented on age and symptom in BD patients. They suggested that although manic syndromes in the elderly were similar qualitatively to those in younger patients, older patients demonstrate attenuation or exaggeration of particular features, and had milder features overall, compared with younger patients.

Those psychiatrists also pointed out that cognitive impairments frequently accompany manic syndromes, particularly in aged manic patients; they described disorientation, and delirium in these patients (5,7,10). Kraepelin observed that dementia was not a necessary outcome of late onset affective episodes, however.

As outlined below, BD in elders provides a potentially rewarding context for studying biological factors mediating and moderating this illness and its treatment. The literature now includes early results of the application of standardized assessment methods and longitudinal characterization, in addition to cross-sectional description. Beyond mood dysregulation, research has recently been focused on comorbidities and behavioral dysfunction in this complex clinical population. While most studies continue to target type I BD, there is new attention being paid to other aspects of the BD spectrum in old age. Investigators have begun to study bipolarity in elders treated in non-psychiatric settings. Research has also begun to address causal factors, vulnerabilities, and pathophysiologies, and the implications of social context and life events in BD elders.

CLINICAL OVERVIEW

Clinical Features and Differential Diagnosis

Reviews of clinical descriptive studies are available (11–13), and this material is summarized in what follows.

Diagnostic criteria for BD, developed in young adult patients, need to be applied in the elderly to test their limits. Exclusion criteria for studies need to be considered carefully, as "casting a wide net" is appropriate to initial studies of this poorly understood population. There has been little systematic investigation of differences in intensity or frequency of particular psychopathologic features in geriatric manic patients, and these reports have suggested extensive overlap with younger patients (14).

The differential diagnosis of mania in late life overlaps that in younger patients and it includes schizophrenic and schizoaffective disorders (15). In older compared with younger patients, mood disorders related to illness or medical treatment are more prevalent, particularly in late onset cases (16). Delirium and dementia both can present manic signs and symptoms (10).

Epidemiology

Elderly patients with manic states and BDs present relatively frequently to psychiatric units serving the aged (17). These patients pose a growing public health challenge. Manic states represent up to one fifth of psychiatric hospitalizations among elders (17). Assessments of community prevalence in elders have been low (18).

While the incidence of BDs is highest in early life, there is a wide range in age at onset (11,19). First episodes of mania can occur in the hundredth decade. A number of studies of geriatric manic inpatients found a median age at onset for first manic episode in the sixth decade (20,21). It appears that various selection factors generate the spectrum of BD patients presenting at geriatric services. One subset of geriatric BD patients has had recurrent early onset BD illness, while another group has had recurrent depressive episodes prior to the late onset of a manic episode (change in polarity), and another has new onset of mood disorder in late life (22). This heterogeneity of antecedent illness course in geriatric BD patients represents an opportunity for clinically and heuristically useful investigation, including characterization using biological measures. Lin-Pl et al. (23) suggest that since age at onset in mixed age patients not only defines subgroups with differing clinical characteristics but also manifests familial aggregation, it may be a characteristic of BD patients that may prove to be a useful phenotype for genetic studies.

Gender differences have received limited study in elders with BD. Two studies have suggested that rates of first hospitalization for mania increase in late life in men but not in women (24,25).

Course

Chronicity, Relapse, Recurrence

Early reports suggested that aged manic patients are vulnerable to prolonged time to recovery or to chronicity (26). Also, limited information regarding patterns of relapse and recurrence in geriatric BD patients suggests that BD elders are at risk for further episodes (13).

Cognitive Dysfunctions and Dementias

Cognitive impairments are described in BD patients when symptomatic and after successful treatment, and these impairments are not explained by comorbid substance abuse (27). Impairments occur in BD patients across the age spectrum but may be more prominent with older age (28,29). Various domains of cognitive function can be impaired. Deficits are described in geriatric manic patients (30,31), and they include executive dysfunction, memory, and speed of processing. BD depressed elders demonstrate memory impairments (29), and although cognition improves with treatment in aged BD patients (32), and while deficits can improve with symptomatic improvement, successfully treated BD elders (31) demonstrate impairments. The relationships between cognitive impairments and other clinical dimensions in elderly BD patients appear complex. Some impairments may reflect the consequences of illness, for example, illness episodes, associated hypothalamic-adrenocortical dysregulation, and somatic treatments. Yet these cognitive impairments may offer a window into the dysfunction of specific brain neuronal circuits in BD elders.

The relationship of BD to dementia is poorly understood. Recent investigation (33) found greater dementia on follow-up in both BD and UMD elders compared with comparison groups with medical illness; this is consistent with an earlier report concerning cognitive performance (34).

Comorbid Substance Abuse

In BD elders there is little data regarding comorbid substance abuse (35). Examination of one administrative database found excess substance abuse in BD elders compared with aged controls, but less substance abuse in elder BD patients compared with younger patients (36).

Mortality

Geriatric BD patients are at risk for high mortality, even compared with geriatric patients with UMD (37). This increased mortality may reflect medical comorbidity and other causes. Suicide rates in mixed-age BD illness are elevated but these rates await study in the elderly.

Causal Factors

Age-associated causal factors may modify the form of illness, that is, clinical features, in early-onset BD. On the causal factors involved in early life illness, such as genetic characteristics, perinatal insults, and other early adverse life events, aging would superimpose other causal factors, for example, effects of illness episodes and treatments or other comorbid conditions.

Age-associated causal factors may also contribute to late onset of BD illness. These may include cardiovascular risk factors and other medical and neurological conditions and their treatments. Late onset illness also may involve familial/genetic mechanisms related to early life BD, and genetic factors related to neurological disease and vascular disease.

Consideration of the *specificity* of relationships between biological and other causal factors and geriatric BD requires comparison with other aged patient groups. In this discussion, we emphasize the comparison of BD with UMD in old age; while the literature concerning causal factors in UMD in late life is more extensive than that in BD, direct comparisons of aged UMD and BD patients are few. Comparisons of findings in chronic psychoses with those in UMD are even more sparse. Despite the possible link between BD in late life to increased risk of dementia, direct comparisons of biological measures between the disorders are not available.

Familial Influences

An important familial component to the etiology of type I BDs in young adults has been established by twin and adoption studies (38). In geriatric patients, late age-at-onset BD is associated with lower rates of familial mood disorder than early onset illness (20); this is analogous to findings in mixed-age BD samples (39). At the same time, age at onset of BD illness aggregates in families (23).

Vascular Disease and Stroke

A vascular component to pathogenesis has been postulated for BD (40), as it has in major depression (41–43). This concept has in part been derived from comorbidity data, and studies of stroke.

Systolic blood pressure in mixed-age manic patients is greater than in controls (44); hypertension can also be noted in geriatric mania (45). Smoking is more prevalent in mixed-age BD patients than in controls (45).

Patients with stroke and other brain lesions can develop mania (46). In a series of eight patients with brain injuries who developed mania (47), all had damage to the right hemisphere. Further, in a series of patients with BD associated with cerebrovascular lesions (48), seven of nine had bilateral hemispheric damage, and one had combined brain stem and bilateral hemispheric damage; none had lesions exclusively located in the left hemisphere; subcortical damage tended to be more frequent than cortical, brainstem, or cerebellar involvement; clinical subtypes and patterns of cycling in these patients resembled those in idiopathic BD. Differences in illness course has been linked to lesion location: unipolar mania was associated with cortical lesions and BD course associated with subcortical lesions in one report (49). In stroke patients, investigators have presented evidence for interaction of causal factors; those with mania had higher rates of familial mood disorder than non-manic patients (50).

Stroke associated with UMD appears to differ in neuroanatomic localization compared with that associated with mania or BD illness. Although there has been disagreement, left anterior lesions are predominant in UMD patients (51).

Other Medical Disorders and Treatments

Krauthammer and Klerman (16) and others have highlighted the association between manic states and a range of medical disorders and treatments. Such "secondary" or "symptomatic" mania is noted more often in late-onset compared with early-onset cases (20).

Manic psychopathology can be detected in patients with dementia. Rates of such co-occurrence may be low in ambulatory care settings (52). The diagnosis of mania in demented patients may involve specific challenges, such as labeling as agitation.

Psychosocial Adversity

A history of precipitating psychosocial events has been noted in some geriatric manic patients (17), and lack of perceived social support is described in BD elders (53). However, there was no systematic comparison to younger patients or to elders with UMD.

LABORATORY MEASURES, VULNERABILITY AND PATHOPHYSIOLOGY IN BIPOLAR ELDERS

Laboratory Measures

Structural neuroimaging has been applied in early studies of BD elders. There has been scant investigation using other types of measures. The investigation of geriatric BDs is has primarily addressed type I disorder, and manic states related to medical disorders. The following discussion focuses on studies that included a sample of BD elders. We present examples of studies of age- or age-at-onset effects in younger BD patients, and mention any direct comparisons with geriatric UMD patients.

Genetic Variation

Studies of candidate genotypes focused on BD elders appear to be lacking. In mixed-age BD patients with wide range of index ages and ages at onset (54), the apolipoprotein (Apo)-E4 allele was linked to patients with both early age at onset and psychotic features. Apo E polymorphisms have been linked to vulnerability to degenerative dementia and to vascular risk factors.

Certain genetic influences associated with BD illness in young adults may be "weaker" in later onset cases of BD (54), and other vulnerabilities such as medical/neurologic comorbidities and other age-associated processes may be necessary to trigger onset of illness in these patients. "Two-hit" pathogenetic models of late onset BD have been discussed (55), which involve a genetic vulnerability to early life BD interacting with other factors, either genetic or non-genetic. Possible second genetic mechanisms proposed included triplet repeat expansion, damage to mitochondrial DNA, and age-specific changes in gene expression.

Neuroimaging

Brain Morphology

There have been few structural neuroimaging studies of BD in older adult (age >50–55 years) or elderly (age >60 years) patients. Table 1 summarizes these studies, which use magnetic resonance imaging (MRI) and computed tomography (CT). The table includes an aggregate of more than 200 patients. Subjects in several studies (56–60) were only on their sixth decade of life. The studies differed methodologically in their focus on global measures, regional volumes, and/or signal hyperintensities (SHs). Of course, younger adults with BD illness also have abnormal brain morphology, including volume abnormalities and SHs (61), and some

TABLE 1 Structural Neuroimaging Studies in Older Adults and Elders with Bipolar Disorders

Author, year	Diag.	N	Age (mean yr)	Age (min. yr)	Imaging	Focus	Assessment	Group differences	Age at onset effect
						Global			
Broadhead and Jacoby, 1990 (14)	BP NC	35 35	72.5 1.3	60	CT	Hemispheres Lateral vent Third vent	Visual and computer ratings	Cortical atrophy: BP > NC VBR: BP = NC Third vent: BP = NC	No
Young, et al. 1999 (112)	BP NC	30 18	71.4 74.4	60	CT	Hemispheres Lateral Vents Third Vent Frontal ventricles	CSW VBR 3VR FHR	CSW scores: BP > NC VBR scores: BP > NC 3VR: BP = NC FHR: BP = NC	CSW pos. associated with age at onset of illness and of mania
Rabins, et al. 2000 (67)	BP UMD SZ NC	14 14 14 21	73.0 69.3 70.1 Matched	60	MRI	Hemispheres	CERAD	Cortical atrophy: BP, UMD > NC Temp sulcus: BP, UMD > NC L Sylvian fissure: BP, UMD > NC	
						Regional			
Beyer (a), 2004 (58)	BP NC	36 29	58.2[a] 61.0[a]	50	MRI	Temporal lobe (hippoc) volume		L hippocampus volume: BP > controls	No
Beyer (b), 2004 (59)	BP NC	36 35	58.8[a] 63.2[a]	50	MRI	Basal ganglia volume Total brain vol Vent vol		R Caudate volume: BP < controls Tot br vol, vent vol: BP = NC	Total brain volume neg. assoc with age at onset
						Signal hyperintensity/ lesions			

McDonald, 1991 (113)	BP	12	68.7[a]	55	MRI	Ventricle size	Global rating	Ventricle size: BP = NC	
	NC	12	68.3[a]			Subcortical SH	SH size and location	Pres of Lesions: BP = NC	
								Large lesion #: BP > NC	
							L versus R	L = R	
							Ant, mid, or post third	Mid third: BP > NC	
						PVH		PVH: BP = NC	
								SH lesions: BP > NC	
Fujikawa, 1995 (60)	BP	30	61.3[a]	NA	MRI	SCI	Area rated:		
	UMD	30	61.3[a]				Perforating	SCI p: BP = UP (LOD)	SCI total:
							Cortical	SCI c: BP = UP (LOD)	BP
							Mixed	SCI m: BP > UP (LOD)	(LOM) > EOA
McDonald, 1999 (56)	BP	32	68.2	60	MRI	PVH	Boyko scale	PVH: BP = NC	No
	NC	40	67.3			SE		SE: BP > NC	
						DWM		DWM: BP > NC	
						SCG 1,2		SCG 1: BP = NC	
								SCG 2: BP > NC	
De Asis, 2006 (65)	BP	40	69.8	60	MRI	Frontal DWM	Boyko scale	DWM: BP > NC	First mania
	NC	15	66.9						DWM
									L N
									R Y
						SCG		SCG: BP = NC	SCG
									L N
									R N

[a]Grp age minimum <60.

Diagnosis: BP, bipolar patients; UMD, unipolar major depression patients; SZ, schizophrenia patients; NC, normal comparisons; LOM, late onset mania; LOD, late onset depression; EOA, early onset affective [(B,UP)10 bipolar and 10 unipolar] patients.

Image: MRI, magnetic resonance imaging; CT, computerized tomography.

Focus: DWM, Deep white matter; SCG, subcortical gray (2 severity measures in McDonald,1999; SCG1, absent/punctate or confluent; SCG2, absent/punctate, rounded with smooth edges or irregular/confluent); PVH, periventricular hyperintensities; SE, subenpendymal; SCI, silent cerebral infarcts: SCIp, subcortical gray and internal capsule, belonging to perforating branch system; SCIc, cerebral cortex and subcortical white matter; SCIm, combination.

Assessment: CERAD, consortium to establish a registry for alzheimer's disease; CSW, cortical sulcal widening; VBR, ventricular brain ration; 3VR, third ventricle ratio; FHR, frontal horn ratio; SH, signal hyperintensities.

Note: N, non significant correlation; NA, not available; Y, significant.

types of morphological abnormalities overlap those in patients with other disorders such as schizophrenia (62,63).

The three studies that included global measures all found greater cortical atrophy in the patients than in aged comparison subjects, and in one the atrophy was associated with late age at onset. One of two studies that examined lateral ventricle–brain ratio (VBR) (64) found higher values in BD patients.

Two studies (58,59) primarily examined regional volumes. One found significant caudate volume decrease on the right in patients, controlling for age and sex. In the other, hippocampal volume on the left was larger in patients compared with controls, and this was associated with lithium exposure. Total brain volume was smaller in patients with later age at onset.

An early study of SHs in older adult and elderly BD patients (57) found increased number of larger SHs with age. Two reports have found greater frontal deep white matter SHs in BD elders compared with controls (56,65), while a difference regarding subcortical gray was found in one (56). Fujikawa (60) found that patients with late onset mania had greater silent cerebral infarcts than geriatric patients with early onset mood disorders. In the deAsis et al. (65) study, right-sided frontal deep white matter SHs were associated with later age at onset of BD illness.

An early study of mixed age BD patients using diffusion tensor imaging found evidence of microstructural disorganization of white matter tracts involving orbitofrontal cortex (66). Such findings are consistent with disconnection of circuits involving subcortical and limbic structures.

Two structural imaging studies (67,60) compared aged BD and UMD patients. The former found cortical atrophy in both BD and UMD aged patients. Fujikawa (60) found that elderly BD patients had greater silent cerebral infarcts than aged UMD patients.

Causal factors implicated in BD in late life may be expressed as abnormal brain morphology. In BD patients, brain morphology abnormalities have been found to be related to familial risk (68). The relationships of SH to familial factors and vascular burden in BD elders has not been examined directly, as they have in UMD (69–71). SHs in normal elders are associated with genetic differences (72).

As with cognitive dysfunctions, some morphological abnormalities in BD elders may be consequences of early life BD illness and its treatment. These factors include effects of repeated episodes of illness, related endocrine dysregulation, comorbid substance abuse, and effects of therapeutic drugs.

Functional Neuroimaging

Aged BD depressed patients examined under resting conditions using single photon emission tomography (73) had lower cerebral blood flow (CBF) in the prefrontal cortices and limbic and paralimbic areas compared with normal controls. Similar differences were noted in geriatric UMD patients in this study, but the patient groups were not directly compared.

Electrophysiological Studies

Studies of electroencephalographic (EEG) measures in geriatric BD appear to be lacking. Age effects on EEG in normals have been described (74).

Neuroendocrine and Neurochemical Studies

With few exceptions, peripheral neuroendocrine function has not been investigated in geriatric BD patients. Lithium antagonizes thyroid axis function in elders (75) as it does in younger patients; the rates of lithium induced-dysfunction may be greater in old age. Change in thyroid axis function occurs in normals with age (76–79). We are unaware of studies of thyroid axis response to lithium in geriatric UMD.

There is anecdotal evidence for gonadal hormone dysregulation in BD in old age.

Late-onset mania has been reported during estrogen replacement (80), consistent with reports of antidepressant effects in young adult UMD. Gonadal hormones apparently have not been directly assessed in late life BD, however.

Catecholaminergic, serotonergic, GABAergic and acetylcholinergic neurotransmitter systems have all been implicated in the pathophysiology of BDs (81). Neurotransmitter and metabolite concentrations in body fluids have not been evaluated in BD elders.

Indirect evidence suggests acetylcholinergic dysregulation in late life BD. Tricyclic antidepressants, which have prominent anticholinergic effects, have been linked to onset of mania in elders (82). Age effects on cholinergic neurons and receptors have been reported in normals (83–85). On the other hand, cholinergic agents can precipitate UMD in elders (86,87).

Signal transduction molecules that are altered by mood stabilizer treatments have been implicated in the pathophysiology of dementia, for example, tau protein (88); this association provides a rationale for preventative intervention trials. Data concerning measures of these molecules in elderly BD patients are not available.

Preliminary investigation of nutrients in elderly psychiatric patients, including BD patients, has suggested associations between serum homocysteine and folate concentrations and abnormal brain morphology on neuroimaging (89). High homocysteine levels were associated with greater SHs on MRI, while low folate levels were associated with low volume of hippocampus and amygdala.

Some morphological abnormalities in BD elders may be consequences of early life mood disorder and its treatment. These factors include effects of repeated episodes of illness and related endocrine dysregulation, comorbid substance abuse, or effects of treatments.

Endophenotypes

The geriatric BD literature does not include study of laboratory measures of biological factors in probands in relationship to presence or absence of familial mood disorders. Endophenotypes (62) have been proposed for early onset BD. The criteria for endophenotypes allow differentiation of abnormalities that are consequences of existing illness from those that reflect causal factors. In young adults, proposed endophenotypes have included cognitive dysfunctions (90) and neuroimaging abnormalities (62,68), both of which are apparently characteristics of remitted BD elders.

Pathophysiology

The pathophysiological abnormalities in geriatric BD are not understood. Models of pathophysiology in BD have been proposed based on emerging concepts derived from clinical findings in patients with coarse brain disease including stroke, traumatic injury, and dementia (1), studies of mixed age BD patients using functional neuroimaging (47), and animal models. While initially focusing on location of lesions of specific brain structures, and differences between right and left-sided lesions, current thinking emphasizes neuronal circuits or networks, that is, frontostriatal and anterior limbic (91,92), that link these structures. Nodal points on these circuits include prefrontal cortex, thalamus, striatum, amygdala, baso-temporal lobe/hippocampus, and cerebellum. Furthermore, functional MRI in BD patients has suggested relative hyperactivity of ventral frontal and limbic regions, for example, orbitofrontal cortex and perigenual cingulate and amygdala, and hypoactivity of dorsal regions including dorsolateral prefrontal cortex and dorsal anterior cingulate. Findings from preliminary structural imaging studies and cognitive investigation in BD elders are consistent with aspects of these models, but such research is at a very early stage.

BIOLOGICAL FACTORS AND THERAPEUTICS

Biological investigation can potentially contribute to management of aged BD patients through characterization of subgroups of patients with poor tolerance for pharmacotherapy or who may be resistant to somatic or psychosocial interventions. This information may help elucidate mechanisms of effects of standard treatments, and help identify patients for whom innovative therapeutic approaches can be tested.

Pharmacokinetic Factors

Age-related changes in physiology, and comorbid conditions and their management, have potential implications for dosing of mood stabilizers and other agents in elderly BD patients. These include hepatic changes, decline in renal clearance, changes in fat/lean body mass ratio, altered binding protein concentrations, and drug–drug and drug–dietary interactions (93). Some of these changes can lead to increased concentration/dose ratio. Unnecessarily high exposure to therapeutic agents can lead to poor tolerability and related treatment failure.

Pharmacodynamics

Side Effects/Tolerability

While it is not clear how much age alone increases risks associated with exposure to defined mood stabilizer exposure, specific comorbidities identify elderly patients at risk for toxicities. Himmelhoch (35) found that poor tolerance of acute lithium treatment in geriatric manic patients was related to neurological disease; patients with dementia or neurologic abnormalities did worse with lithium exposure. Similarly, dementia patients with manic features tolerated moderate valproate doses poorly in a randomized controlled trial (94).

Relationships between tolerated concentrations of mood stabilizers and acute or long-term efficacy are not clear. Identification of clinical or laboratory predictors of response to low dose and low concentration, either intended or necessitated by limited tolerability, could help minimize side effects in such patients. Schaffer and Garvey (95) reported a series of geriatric manic patients treated with 0.4–0.8 mEq/L with favorable result in 14 cases. Other investigators (96) reported that concentrations above 0.8 mEq/L were associated with best anti-manic response.

Geriatric manic states associated with antidepressant treatment were characterized in one study by later age at onset, but not by differing index age (82); the antidepressants were primarily tricyclics. This finding is consistent with findings of another study of tricyclic antidepressant pharmacotherapy (97); it contrasts with a report in young patients (98), and suggests that risk factors for antidepressant-associated mania may differ with age; factors associated with late onset may sensitize to some drug effects.

Laboratory predictors of low adverse outcomes in BD elders are not reported. However, in UMD patients, neuroimaging abnormalities have been associated with side effects of antidepressant agents (99). Genetic variation has also been linked to adverse effects of psychotropics (100,101).

Therapeutic Effects

Age alone may attenuate anti-manic response to mood stabilizer treatment. An early study noted decreased efficacy of naturalistic acute lithium treatment across the age spectrum (102). In another study (103), longer hospitalizations were associated with increased age in a mixed-age manic sample, only 10% of whom were aged ≥60 years.

Neurological disease may modify anti-manic response to mood stabilizers. Dementia may limit benefit from lithium treatment of mania (35). Early findings suggest a relationship between cognitive dysfunction and poor symptomatic outcome of naturalistic inpatient treatment (104).

A retrospective study in young manic patients indicated better outcome of valproate treatment in those with forms of neurological disorder (105). Whether specific neurological dysfunctions predict therapeutic benefit from anticonvulsant treatment remains to be tested in elder BD patients.

Laboratory measures may prove useful in prediction of individual variation in therapeutic response to adequate trials of pharmacotherapy in BD elders. In an early report concerning geriatric BD manic patients, larger VBR was associated with poorer response to naturalistic pharmacotherapy with lithium (106). SHs were associated with poor outcomes in mixed-age patients with BD (107,108). Early reports suggest that poorer outcomes of continuation-maintenance lithium treatment of BD patients are mediated by several genetic polymorphisms (109).

Novel Treatment Strategies

Since some BD elders tolerated conventional treatments poorly or may have modest benefit from them, testing novel treatment approaches is an important aim for research. Early findings regarding biological and clinical factors suggest that particular strategies may be rewarding: for example, cardiovascular interventions such as smoking cessation, blood pressure management, and improved nutrition. Given the overlap between BDs and cognitive impairments in elders, and an association between anticholinergic agents and mania, studies of cholinergic agents may be particularly relevant; cholinesterase inhibition can benefit manic states in young adulthood (110). Another example of innovative treatment is behavioral rehabilitation: given the occurrence of behavioral dysfunction in geriatric BD patients (111), disabilities can be a focus of interventions. Cognitive assessment and laboratory characterization may aid the design of studies to test innovative treatments.

SUMMARY AND PROSPECTS

Age effects on clinical features of BD have received preliminary study, and there is only limited information regarding the laboratory measures and age in BD. Vascular changes and other forms of brain aging may contribute to onset of BDs for the first time late in life. Structural neuroimaging studies have supported this concept. There has been virtually no application of genotyping or peripheral physiological markers to the study of late life BD.

Preliminary evidence suggests that brain change in geriatric BD may predict response to acute pharmacotherapeutic interventions. Evidence regarding predictors of outcomes of long-term pharmacological and psychosocial interventions is entirely lacking.

Ongoing demographic shifts and economic pressures necessitate a more adequate knowledge base in BD elders regarding etiological factors, vulnerabilities, pathophysiologies, and approaches to management. Recent advances regarding genetics and molecular targets of mood stabilizers suggest new research opportunities in late life BD. In BD elders, cognitive and affective neuroscience methods may prove useful as aides in prognostic assessment and treatment. Studies of pathophysiologies of late life BDs may also provide models for the pathophysiology of early life BDs.

ACKNOWLEDGMENTS

Supported by: MH 074511, MH068847, MH067028, HP00157, and MH63638.

REFERENCES

1. Shulman KI. Disinhibition syndromes, secondary mania and bipolar disorder in old age. J Aff Dis 1997; 46:175–182.
2. Welt L. Uber characterveranderungen der menschen infolge von lasionen des stirnhirn. Arch f Klin Med 1888; 339–390.
3. Oppeneheim J. Zur pathologie der grosshirngeschwulste. Arch Psychiatr Nervenkr 1890; 560–587.
4. Blumer D, Benson DF. Personality Changes with Frontal and Temporal Lobe Lesions. New York: Grune and Stratton, 1975.
5. Kraepelin E. Manic-Depressive Insanity and Paranoia. Edinburgh: Livingstone, 1921.
6. Kay DWK, Roth M, Hopkins B. Affective disorders arising in the senium, I: Their association with organic cerebral degeneration. J Ment Sci 1955; 101:302–316.
7. Post F. The Clinical Psychiatry of Late Life. New York, Oxford: Pergamon Press, 1965:79–82.
8. Marneros A, Angst J. Bipolar disorders: roots and evolution. In: Marneros A, Angst J, eds. Bipolar Disorders. Dordrecht: The Netherlands, 2000:1–35.
9. Cade JFJ. Lithium salts in the treatment of psychotic excitement. The Med J Australia 1949; 36:349–352.
10. Slater E, Roth M. Mayer-Gross, Slater, and Roth's Clinical Psychiatry. London: Bailliere, Tindall and Cassell, 1977.
11. Young RC, Klerman GL. Mania in late life: focus on age at onset. Am J Psychiatry 1992; 149:867–876.
12. Shulman KI, Tohen M, Shulman KI, Tohen M, Kutcher SP. Outcome studies of mania in old age. Mood Disorders Across the Life Span. New York: J Wiley & Sons, 1996: 407–410.
13. Dunn KL, Rabins PV, Shulman KI, Tohen M, Kutcher SP. Mania in old age. Mood Disorders Across the Life Span. New York: J Wiley & Sons, 1996:399–406.
14. Broadhead J, Jacoby R. Mania in old age: a first prospective study. Int J Ger Psychiatry 1990; 5:215–222.
15. Young RC, Latoussakis V, de Asis JM, Baran XL, Kalayam B, Meyers BS. Donepezil augmentation in bipolar disorders with cognitive impairment/dementia: NIMH. Abstract, Annual Meeting, New Clinical Drug Evaluation Unit, Phoenix, AZ, U.S.A., 2004.

16. Krauthammer C, Klerman G. Secondary mania. Arch Gen Psychiatry 1978; 35(11): 1333–1339.
17. Yassa R, Nair V, Nastase C. Prevalence of bipolar disorder in a psychogeriatric population. J Affective Disord 1988; 14:197–201.
18. Weissman M, Leaf PJ, Tischler GL, et al. Affective disorders in five United States communities. Psychol Med 1988; 18:141–153.
19. Kennedy N, Boydell J, Kalidindi S, et al. Gender differences in incidence and age at onset of mania and bipolar disorder over a 35-year period in Camberwell, England. Am J Psychiatry 2005; 162(2):257–262.
20. Shulman K, Post F. Bipolar affective disorders in old age. Br J Psychiatry 1980; 136: 26–32.
21. Glasser M, Rabins P. Mania in the elderly. Age and Aging 1984; 13:210–213.
22. Sajatovic M, Gyulai L, Calabrese JR, et al. Maintenance treatment outcomes in older patients with bipolar I disorder. Am J Geriatric Psychiatry 2005; 13:305–311.
23. Lin PI, McIinnis MG, Potash JB, et al. Clinical correlates and familial aggregation of age at onset in bipolar disorder. Am J Psychiatry 2006; 163(2):175–176/PMID:16449464.
24. Spicer CC, Hare EH, Slater E. Neurotic and psychotic forms of depressive illness: Evidence from age incidence in a national sample. Br J Psychiatry 1973; 123: 535–541.
25. Eagles JM, Whalley LJ. Ageing and affective disorders: the age at first onset of affective disorders in Scotland 1966–1978. Br J Psychiatry 1985; 147:180–187.
26. Wertham FI. A group of benign chronic psychoses: prolonged manic excitements with a statisticalstudy of age,duration and frequency in 2,000 manic attacks. Am J Psychiatry 1929; 86:17–78.
27. van Gorp WG, Altshuler L, Theverg DC, Wilkins J, Dixon W. Cognitive impairment in euthymic bipolar patients with and without prior alcohol dependence. Arch Gen Psych 1998; 55:41–46.
28. Savard RJ, Rey C, Post RM. Halstead-Reitan Category Test in bipolar and unipolar affective disorders: relationship to age and phase of illness. J Nerv Ment Dis 1980; 168:297–304.
29. Burt T, Prudic J, Peyser S, Clark J, Sackeim HA. Learning and memory in bipolar and unipolar major depression: effects of aging. Neuropsy, Neuropsychol and Behav Neurol 2000; 13:246–253.
30. Young RC, Murphy CF, Heo M, Alexopoulos GS. Cognitive impairment in bipolar disorder in old age: literature review and findings in manic patients. J Aff Disord 2006:125–131.
31. Gildengers AG, Butters MA, Seligman K, et al. Cognitive functioning in late-life bipolar disorder. Am J Psychiatry 2004; 161:736–738.
32. Wylie ME, Mulsant BH, Pollock B, et al. Age at onset in geriatric bipolar disorder. Am J Ger Psychiatry 1999; 7(1):77–83.
33. Kessing LV, Nilsson FM. Increased risk of developing dementia in patients with major affective disorders compared to patients with other medical illnesses. J Aff Disord 2003; 73:261–269.
34. Dhingra U, Rabins PV. Mania in the elderly: a five-to-seven year follow-up. J Am Ger Soc 1991; 39:581–583.
35. Himmelhoch J, Neil JR, Ray SJ, et al. Age, dementia, dyskinesias, and lithium response. Am J Psychiatry 1980; 137:941–945.
36. Goldstein BI HN, Shulman KI. Comorbidity in bipolar disorder among the elderly: results from an epidemiological community sample. Am J Psychiatry 2006; 163[2(2)]: 319–321.
37. Shulman KI, Tohen M, Satlin A, Mallya G, Kalunian D. Mania compared with unipolar depression in old age. Am J Psychiatry 1992; 142:341–345.
38. McGuffin P, Rjsdijk F, Andrew M, et al. The heritability of bipolar affective disorder and the genetic relationship to unipolar depression. Arch Gen Psychiatry 2003; 60(60):497–502.
39. Moorhead SR, Young AH. Evidence for a late-onset bipolar-I subgroup from 50 years. J Aff Disord 2003; 73(3):271–277.

40. Steffens DC, Krishnan KR. Structural neuroimaging and mood disorders: Recent findings, implications for classification, and future directions. Soc Biol Psychia 1998; 43: 705–712.
41. Alexopoulos GS, Meyers BS, Young RC, Kakuma T, Silbersweig D, Charlson M. Clinically defined vascular depression. Am J Psychiatry 1997; 154:562–565.
42. Alexopoulos GS, Reynolds CF. Evidence-based pharmacotherapy of unipolar depression in late life. In: Bartels S, ed. Evidence Based Treatment in Geriatric Psychiatry. New York: Elsevier Inc., 2005.
43. Krishnan KKR. Arteriosclerotic depression. Med Hypothesis 1995; 44:77–145.
44. Yeates WR, Wallace R. Cerebrovascular risk factors in affective disorder. J Aff Disord 1987; 12:129–134.
45. deAsis JM, Young RC, Alexopoulos GS, Greenwald B, Kakuma T, Ashtari M. Signal hyperintensities in geriatric mania. Abstract, Annual Meeting, American Association for Geriatric Psychiatry, SanDiego, California, U.S.A., 1998.
46. Robinson RG, Boston JD, Starkstein SE, Price TR. Comparison of mania and depression after brain injury: causal factors. Am J Psychiat 1988; 145(2):172–178.
47. Starkstein SE, Mayberg HS, Berthier ML, et al. Mania after brain injury: neuroradiological and metabolic findings. Ann Neurol 1990; 27:652–659.
48. Berthier ML. Poststroke bipolar affective disorder: clinical subtypes, concurrent movement disorders and anatomical correlates. J Neuropsychiatry Clin Neurosci 1996; 8:160–167.
49. Starkstein SE, Federoff P, Berthier ML, Robinson RG. Manic depressive and pure manic states after brain lesions. Biol Psychiatry 1991; 29:773–782.
50. Robinson D, Napoliello MJ. The safety and usefulness of busipirone as an anxiolytic in elderly versus young patients. Clin Ther 1988; 10:740–746.
51. Robinson RG, Kubos KL, Starr LK. Mood disorders in post stroke patients. Importance of location of lesion. Brain 1984; 107:81–93.
52. Lyketsos CG, Corazzini K, Steele C. Mania in Alzheimer's disease. J Neuropsy Clin Neurosci 1995; 7:350–352.
53. Beyer JL, Kuchibhatala M, Cassidy F, Looney C, Krishnan KRR. Social support in older bipolar patients. Bipolar Disord 2002; 5(1):22–27.
54. Bellivier F, Laplanche JL, Schurhoff F, et al. Apolipoprotein E gene polymorphisms in early and late onset bipolar patients. Neurosci Lett 1997; 233:45–48.
55. McMahon FJ, DePaulo JR, Shulman TK. Genetics and age at onset. Mood Disorders Across the Life Span. New York: Wiley-Liss, 1996:35–48.
56. McDonald WM, Tupler LA, Marsteller FA, et al. Hyperintense lesions on magnetic resonance images in bipolar disorder. Biol Psychiatry 1999; 45:965–971.
57. McDonald WM, Krishnan KR, Doraiswamy PM, Blazer DG. Occurrence of subcortical hyperintensities in elderly subjects with mania. Psych Research 1991; 40: 211–220.
58. Beyer J, Kuchibhatla M, Payne M, et al. Hippocampal volume measurement in older adults with bipolar disorder. Am J Geriatr Psychiatry 2005; 12(6):613–620.
59. Beyer J, Kuchibhatla M, Payne M, et al. Caudate volume measurement in older adults with bipolar disorder. Int J Geriatr Psychiatry 2004; 19(2):109–114.
60. Fujikawa T, Yamawaki S, Touhouda Y. Silent cerebral infarctions in patients with late-onset mania. Stroke 1995; 26:946–949.
61. Soares JC, Mann JJ. The anatomy of mood disorders. A review of structural neuroimaging studies. Biol Psych 1997; 41:86–106.
62. McDonald C, Bullmore E, Sham P, et al. Association of genetic risks for schizophrenia and bipolar disorder with specific and generic brain structural endophenotypes. Arch Gen Psychiatry 2004; 61(10):974–984.
63. Hajek T, Carrey N, Alda M. Neuroanatomical abnormalities as risk factors for bipolar disorder. Bipolar Disord 2005; 5:393–403.
64. Young RC, Nambudiri DE, Jain H, Alexopoulos GS. Brain computed tomography in geriatric manic disorder. Biol Psych 1999; 45:1063–1065.
65. de Asis JM, Greenwald BS, Alexopoulos GS, et al. Frontal signal hyperintensities in mania in old age. AM J Geriatr Psychiatry 2006; 14(7):598–604.

66. Beyer JL, Taylor WD, MacFall JR, et al. Cortical white matter microstructural abnormalities in bipolar disorder. Neuropsychoparmacology 2005; 30:2225–2229.
67. Rabins PV, Aylward E, Holroyd S, Pearlson G. MRI findings differentiate between late-onset schizophrenia and late-life mood disorder. Int J Geriatr Psychiatry 2000; 15(10):954–960.
68. Hasler G, Drevets WC, Gould TD, et al. Toward constructing an endophenotype strategy for bipolar disorders. Biol Psychiatry 2006; 60(2):93–105.
69. Ahearn EP, Steffens DC, Cassidy F, et al. Familial leukoencephalopathy in bipolar disorder. Am J Psychiatry 2002; 155:605.
70. Hinckie I, Scott E, Mitchell P, Wilhelm K, Austin MP, Bennett B. Subcortical hyperintensities on magnetic resonance imaging: clinical correlates and prognostic significance in patients with severe depression. Soc Biol Psychiatry 1995; 37:151–160.
71. Greenwald BS, Kramer-Ginsberg E. Krishnan KRR, et al. A controlled study of MRI signal hyperinsities in older depressed patients with and without hypertension. Am J Psychiatry 2001; 49(9):1218–1225.
72. Verpillat P, Alperovitch F, Cambien V, et al. Aldosterone synthase 9CYP11B2) gene polymorphism and cerebal white matter hyperintensities. Am Acad Neurol 2001; 56: 673–675.
73. Ito H, et al. Hypoperfusion in the limbic system and prefrontal cortex in depression: SPECT with anatomic standardization technique. J Nucl Med 1996; 37:410–414.
74. deAsis JM, Yu XL, Young RC. Biological factors in bipolar disorders in late life. In: Soares J, Gershon S, eds. Basic Mechanisms and Therapeutic Implications of Bipolar Disorders. New York: J Wiley, 2000:479–506.
75. Shulman KI, Sykora K, Gill SS, et al. New thyroxine treatment in older adults beginning lithium therapy: implications for clinical practice. Am J Geriatric Psychiatry 2005; 13(4):299–304.
76. Kabadi UM, Rosman PM. Thyroid hormone indices in adult healthy subjects: no influence of aging. J Am Ger Soc 1986; 36:312–316.
77. Sawin CT, Chopra D, Azizi F. The aging thyroid: increased prevalence of elevated serum thyrotropin levels in the elderly. J Am Med Assoc 1979; 242:247–250.
78. van Coevorden A, Laurent E, Decoster C. Decrease basal and stimulated thyrotropin secretion in healthy elderly men. J Clin Endocrin Metab 1989; 69:177–185.
79. Rossmanith WG, Szilagyi A, Scherbaum WA. Episodic thyrotropin (TSH) and prolactin (PRL) secretion during aging and in postmenopausal women. Horm Metab Res 1992; 24:185–190.
80. Young RC, Moline M, Kleyman F. Hormone replacement therapy and late life mania. Am J Geriatric Psychiatry 1997; 5:179–181.
81. Goodwin F, Jamison K. Medical Treatment of Manic Episodes. New York: Oxford University Press, 1990:603–629.
82. Young RC, Jain J, Kiosses D, Meyers BS. Antidepressant-associated mania in late life. Int J Geriatric Psychiatry 2003; 18:421–424.
83. McGeer EG, McGeer PL. Age changes in the human for some enzymes associated with metabolism of catecholamine GABA and acetylcholine. Ada Behav Biol 1975; 16:287.
84. Tune L, Gucker S, Folstein M, Oshida L, Coyle JT. Cerebrospinal fluid acetylcholinesterase activity in senile dementia of the Alzheimer's type. Ann Neurol 1985; 17:46–48.
85. Szilagyi AK, Nemeth A, Martini E, Lendvai B, Venter V. Serum and CSF cholinesterase activity in various kinds of dementia. Eur Arch Psych Neurol 1987; 236:309–311.
86. Wiener PK, Young RC. Late onset psychotic depression associated with carbaryl exposure and low plasma pseudocholinesterase. Am J Psychiatry 1995; 152:646.
87. Friend KD, Young RC. Late-onset major depression with delusions after metoclopramide treatment. J Geriat Psychiatry 1997; 5:79–82.
88. Loy R, Tariot PN. Neuroprotective properties of valproate: potential benefit for AD and tauopathies. J Mol Neurosci 2002; 19(3):303–307.
89. Scott T, Tucker K, Bhadelia A, et al. Homocysteine and B vitamins relate to brain volume and white-matter changes in geriatric patients with psychiatric disorders. Am J Geriatr Psychiatry 2004; 12(6):631–638.

90. Hasler G, Drevets WC, Manji HK, et al. Discovering endophenotypes for major depression. Neuropsychoparmacology 2004; 29:1765–1781.
91. Strakowski S. Bipolar disorder and advances in neuroimaging: a progressive match. CNS Spectr 2006; 114:267–268.
92. Yurgelun-Todd DA, Ross AJ. Functional magnetic resonance imaging studies in bipolar disorder. CNS Spectr 2006; 11(4):287–297.
93. Satlin A, Lipzin B, Salzman C. Diagnosis and treatment of mania. Clinical Geriatr Psychopharmacology. New York: Williams and Wilkins, 1998.
94. Tariot PN, Schneider LS, Mintzer JE, et al. Safety and tolerability of divalproex sodium in the treatment of signs and symptoms of mania in elderly patients with dementia: results of a double-blind, placebo-controlled trial. Current Ther Res 2001; 62(1):51–67.
95. Schaffer CB, Garvey MJ. Use of lithium in acutely manic elderly patients. Clin Gerontologist 1984; 3:58–60.
96. Chen ST, Altshuler LL, Melnyk KA, Erhart SM, Miller E, Mintz J. Efficacy of lithium vs valproate in the treatment of mania in the elderly: a retrospective study. J Clin Psychiatry 1999; 60:181–185.
97. vanScheyen JD, vanKammen DP. Clomipramine-induced mania in unipolar depression. Arch Gen Psychiatry 1979; 36:560–565.
98. Nasrallah A, Lyskowski J, Schader D. TCA-induced mania: differences between switchers and non-switchers. Biol Psych 1982; 17:271–274.
99. Perlis RH, Mischoulon D, Smoller JW, et al. Serotonin transporter polymorphisms and adverse effects with fluoxetine treatment. Biol Psychiatry 2003; 54: 879–883.
100. Mundo E, Walker M, Cate T, Macciardi F, Kennedy JL. The role of serotonin transporter protein gene in antidepressant-associated mania in bipolar disorder. Arch Gen Psychiatry 2001; 58:539–544.
101. Pomara N, Willoughby L, Wesnes K, Greenblatt DJ, Sidtis JJ. Apoliprotein E epsilon 4 allele and lorazepam effects on memory in high-functioning older adults. Arch Gen Psychiatry 2005; 62(2):209–216.
102. van der Velde CD. Effectiveness of lithium carbonate in the treatment of manic-depressive illness. Am J Psychiatry 1970; 123:345–351.
103. Young RC, Falk JR. Age, manic psychopathology and treatment response. Int J Ger Psychia 1989; 4:73–78.
104. Young RC, Murphy CF, DeAsis JM, Apfeldorf WJ, Alexopoulos GS. Executive function and treatment outcome in geriatric mania. Abstract, New Research Program, Annual Meeting, American Psychiatric Association, New Orleans, 2001.
105. Stoll AL, Banov M, Kolbrener M, et al. Neurologic factors predict a favorable valproate response in bipolar and schizoaffective disorders. Clin Psychopharm 1994; 14(5):311–314.
106. Young RC. Striatofrontal impairments and treatment outcome in geriatric mania. Abstract, New Research Program Annual Meeting, APA, 2001.
107. Moore CM, Demopoulos CM, Henry ME, et al. Brain-to-serum lithium ratio and age: an in vivo magnetic resonance spectroscopy study. Am J Psychiatry 2002; 159:1240–1242.
108. Cassidy F, Carroll BJ. Vascular risk factors in late-onset mania. Psychol Med 2002; 32(2):359–362.
109. Kelsoe JR, McKinney R, Gaucher MA, Shekhtman T, Smith G. Sequence variants in the NTRK2 gene predict response to lithium in bipolar disorder: ACNP.
110. Burt T, Sachs G, Demopoulos C. Donepezil in treatment-resistant bipolar disorder. Biol Psychiatry 1999; 45:959–964.
111. Gildengers A, Mulsant BH, Begley AE, et al. A pilot study of standardized treatment in geriatric bipolar disorder. Am J Geriatric Psych 2005; 319–323.
112. Young RC, Nambudiri D, Roe R. VBR in geriatric mania. Society of Biological Psychiatry 1987.
113. McDonald WM, Krishnan KRR, Doraiswamy PM, Blazer DG. Occurrence of subcortical hyperintensities in elderly subjects with mania. Psy Research 1991; 40(4):211–220.

20 Perspectives for New Pharmacological Interventions

Charles L. Bowden
University of Texas Health Science Center at San Antonio, San Antonio, Texas, U.S.A.

ASSESSMENT OF CURRENT STATUS OF THERAPEUTICS

The range of treatments approved for bipolar disorder (BD) has broadened substantially over the past decade, with all treatments other than lithium only having received regulatory approval as of 1995 or later. The expanded range of treatments has had the greatest impact on management of mania both acutely and in maintenance care. Studies of mania have had several advantages over studies of depression or other behavioral facets of BDs. A straightforward paradigm of studying a drug versus placebo in hospitalized manic patients and employment of change in a short, two-item scale has proved a robust model for all currently approved drugs (1,2). Enrollment into such studies has been relatively easy to achieve, as hospitalization is clinically indicated for most such patients, patients are readily assessed while in an inpatient setting, and side effects are relatively well-tolerated for a short period by manic patients.

Fewer studies and fewer treatments are available for depression in BD. Complexities include difficulties in establishing prior manic or hypomanic episodes when cross-sectionally evaluating patients while depressed, higher rates of response on placebo in depression trials, and limited evidence regarding whether depression in bipolar I and II patients can be equated (3–5).

Recent studies provide the first well-designed evidence of maintenance effectiveness of treatments, particularly lithium, lamotrigine, olanzapine, and divalproex, but not yet other bipolar drugs.

Studies of new molecules have been limited in part by current rating scales, which, though adequately sensitive to identify overall reduction from syndromal levels of severity, do not have sufficient numbers of items to provide component analyses that could establish the areas of behavior that might be specific to one drug or a particular combination regimen. Therefore, a putative treatment that might benefit a particular spectrum of bipolar symptomatology, for example, anxiety or irritability, is unlikely to be tested with adequate sensitivity by primary scales. This limitation of both scales and current clinical trial study paradigms has become more evident with data indicating higher comorbidity of BD with other axis 1 disorders than any other axis 1 conditions and studies that establish a relatively consistent group of components, or domains of disturbed behavior in BDs (Table 1) (6–9).

Relatedly, evidence indicates that there is relatively selective efficacy of some mood stabilizers on components of bipolar conditions, rather than across-the-board efficacy (10). Specifically, lithium and divalproex both were markedly superior to placebo in reducing hyperactivity, only divalproex significantly reduced irritability,

TABLE 1 Domains of Disturbed Behavior and Symptoms Identified in Studies of Bipolar Disorder

	Bowden	Suppes	Cassidy	Swann	Altman	Double
Elevated energy	X	X	X	X		X
Psychosis	X	X	X	X	X	X
Irritability	X	X	X	X		X
Affective instability	X	X	X	X		X
Elated	X	X	X	X	X	
Anxious/depressed	X	X		X		X

and neither drug significantly reduced depressive or psychotic symptoms (10). Other studies also support evidence of a pharmacodynamic effect of divalproex on irritability (11–13).

The most recently systematically studied group of drugs, the atypical antipsychotics, appear broadly efficacious in mania. These drugs have a common pharmacodynamic mechanism of blockade of dopamine type 2 (D2) receptors, with lesser effects on serotonergic systems, particularly 5-hydroxy tryptamine type 2 (5HT2) (14).

The behavioral effects of the class of drugs appears to be relatively similar, with most reducing specific manic behaviors, but also nonspecific symptomatology in mania as well, for example, impaired sleep, reduced appetite, and anxiety (15).

A broad range of drugs from other classes, but most with some anti-epileptic properties, have been studied and appear to be ineffective in syndromal benefits in BDs. These include gabapentin, topiramate, levetiracetam, gabatril, verapamil, and zonisamide (4,16,17). However, except for topiramate and gabapentin, the quality of studies of these drugs is marginal.

With these drugs as well as those reviewed earlier, pharmaceutical companies have limited laboratory and clinical investigative tools to test the spectrum of efficacy of new molecules. The exception in approach to characterization of spectrum efficacy is lamotrigine, which was first studied in BD in a large, open trial of bipolar patients in all illness states, with systematic, open rating scales designed to capture evidence of locus of behavioral benefits (3). For the other drugs, research pharmaceutical companies have generally made decisions regarding presumptive area of benefit, and moved directly to studies of that syndrome within BD. However, animal models have been of limited benefit in BD, whether for mania or depression (18).

Specific Targets for New Treatment Development

Some facets of bipolar symptomatology can be understood from descriptive studies. A substantial portion of patients treated with monotherapy regimens for mania or depression continue to have syndromal mania or depression (19–22). Only one randomized study has been conducted in specifically rapid-cycling patients. Whereas lamotrigine was superior to placebo among bipolar II rapid-cycling patients, it was not significantly superior among bipolar I rapid cyclers (23). Maintenance studies indicate that as monotherapy, all drugs tested in adequate designs do not adequately control symptoms while achieving adequate tolerability for more than 30% of patients enrolled (24,25). Recent studies in mania, depression, and maintenance indicate that 15% to 20% higher rates of responding patients result when complementary medications are added to the primary monotherapy treatment. Such studies indicate that the strategy needed to achieve such benefit is generally to add the second drug to a first medication which has not alleviated manic or depressive symptoms despite a trial of reasonable duration and dosage (20,26–28).

Studies that have commenced treatment with the combination regimen have failed to show, or showed less robustly, a difference in favor of the combination regimen over lithium or valproate alone (26). Mixed mania appears to be generally more difficult to treat effectively (29), although divalproex provides acutely better efficacy than lithium (30). However, during maintenance treatment, studies indicate that patients with mixed features at the time of enrollment generally do less well than initially euphoric manic patients, regardless of specific monotherapy treatment, with evidence that such patients are more sensitive to adverse effects and discontinue treatment more often for adverse effects (22).

Taken in the aggregate, these studies indicate that there is need for either new drugs with different mechanisms from those currently in use, and/or for regimens that combine currently available drugs in ways that broaden spectrum of efficacy while providing adequate tolerability. While the need would appear more pressing for mixed mania, rapid cycling, and maintenance care than for mania, all subtypes would appear to have a substantial portion of patients who might have better symptomatic and functional outcomes with more effective regimens and/or new drugs.

Novel Targets, Innovative Approaches

A series of studies indicate that lithium and valproate share substantial overlap of effects on certain neuronal intracellular signaling systems. These include inositol depletion, protein kinase C (PKC) activity, stimulation of the extracellular signal-regulated kinase (ERK) pathway, the Wnt signaling pathway, and increased phosphorylation of glycogen synthase kinase-3 (GSK-3) (31,32). An ERK kinase inhibitor induced effects similar to those of amphetamines, suggesting that the ERK pathway may contribute to antimanic effects of mood stabilizers (33).

In the search for a common mechanism of mood stabilizers, lithium significantly inhibits brain GSK-3 in vivo at concentrations relevant for the treatment of BD (34). Lithium increased the phosphorylation of GSK-3 beta both in cells and in mouse brain after chronic administration, but did not alter the phosphorylation of Akt. GSK-3 is associated with functional roles in neuroprotection and circadian rhythms. Bipolar patients have abnormal circadian rhythm activity in relationship to levels of activity, interests, and sleep; thus this mechanism provides a possible functional pathway by which valproate and lithium impact biological systems that underpin manic behaviors (35).

Lithium and valproate reduce protein kinase C isozymes in rat brain. Both drugs increase DNA binding of the activating protein-1 (AP-1) family of transcription factors for PKC (36). Valproate, but not lithium, is incorporated into neuronal membranes in an active saturable process, and binds at sites of naturally occurring long chain phospholipids (37).

Having overlapping effects does not mean identical pharmacodynamic properties. For example, the effect of lithium on GSK-3 is direct, while that of valproate is indirect, probably through its inhibitory effects on histone deacetylase. Valproate (VPA) blocks histone deacetylase. Inhibition of histone deacetylase is linked to activation of Akt and thereby inhibition of the Akt/GSK-3beta signaling pathway.

Lithium and VPA induce similar changes in the morphology of axons by increasing growth cone size, spreading, and branching (38). VPA increases the DNA binding of AP-1 transcription factor, and the expression of genes regulated by the ERK-AP-1 pathway (39).

Neurokinin (NK)-1-receptor antagonists may affect mood states and substance P (SP) may worsen mood. VPA dose-dependently inhibited SP-induced interleukin (IL-6) synthesis, whereas carbamazepine and lithium showed no inhibitory effect. VPA also downregulated the expression of the substance P receptor (NK-1-receptor) (40).

VPA also activates the peroxisome proliferator-activated receptors (PPAR)-gamma and delta, a mechanism not shared by other antibipolar drugs, and probably contributory to its augmenting insulin responsivity (41,42).

Lamotrigine produces use-dependent inhibition of sodium channels, in a fashion differing from other sodium channel inhibiting anti-epileptic drugs (43–45). Small studies suggest that some drugs that block calcium channels may have some behavioral effects in BDs (46). These overlapping intracellular targets thus provide possible pathways that could be used to identify, and compare in preclinical testing, new molecules with plausible potential for efficacy on some clinical aspects of BDs.

Prospects for New Drug Development

Rating Scale Issues

Rating scales which encompass the spectrum of symptoms of BD should be available and utilized. The benefits of such ratings are that profiles of a specific drug's actions can be identified, and patients selected in early clinical trials who have characteristics are likely to respond to the drug. This strategy stands in contrast to the current approach often used—deciding more on hunch than evidence that a drug will have anti-manic or antidepressant effects, then studying the drug solely or principally in such patients. One of the most persuasive indirect indicators of the limitations of this strategy is that several drugs initially studied in and approved for one disorder are eventually found to have clinically significant benefits in component behaviors of the originally studied syndrome. Examples are gabapentin, shown ineffective in mania (16) but efficacious in social phobia (47); topiramate, also ineffective in mania, but beneficial in at least two conditions associated with impulsivity, binge eating disorder, and alcoholism (48); and specific serotonin receptor inhibitor (SSRIs), all initially studied and approved for depression but more recently shown as efficacious in various anxiety disorders.

Only an integrated scale is likely to provide principle behavioral component factor ratings of use for the purpose of discrimination of the profile of benefit. Individual item analysis can partially suffice for this purpose, but items are inherently less reliable than are factors composed of several statistically related items. Such a scale can still provide manic and depressive subscales which are preferable for certain purposes, for example, primary outcome measures for regulatory trials, but not for new drug development.

Systematic, Open Trials

Open studies in a relatively broad spectrum of behaviorally disturbed individuals can substantially aid in new drug development. Certainly in major depression and BDs, but likely in most other axis I disorders, the range of fundamentally disturbed behaviors is sufficiently broad that determination of the profile of drug action will benefit from studies in a range of patients. The range should differ by severity, and for BD, by syndromal subtype. The development of lamotrigine was speeded and more efficiently targeted by employment of a first large scale open study, rather than a randomized study in a single phase of illness (49,50). An extension of this strategy is that early studies should be considered for application in more than one syndrome, if there is evidence that the component behavioral disturbances on which the drug acts is found in several axis I conditions (31,36,51).

Study of Early Behavioral Effects

As reported previously for major depressive disorder, it is likely that early behavioral improvements, especially if associated with eventually remitted states, will be indicative of the primary profile of a drug's behavioral effects. Conversely, a focus on change at the point that remission has taken place is likely to be less revealing of pharmacodynamic profile, since once recovery is well underway, all aspects of behavior are likely to show secondary, if not primary resolution. Phrased differently, well is well. We have shown major differences in the profile of action of the noradrenergic acting drug desipramine compared with the serotonergic acting drug paroxetine in recent studies by means of these study designs (52,53).

Novel Use of Time to Event Analyses

Expanded application of time to recovered/remitted status has substantial potential to aid in new drug development. Historically, psychotropic drug studies have been short, and have taken global change from baseline to study endpoint as the primary outcome measure, based on a single day's ratings for patients with as little as one day's exposure to the drug, even though trial duration was up to 12 weeks or longer. This chapter is not aimed at all considerations in defining outcome measures and selecting statistical tests, but there are data indicating that larger effect size differences for drug versus placebo comparisons can be obtained when a degree of symptomatic improvement must be maintained for several weeks (54). The time duration can be selected in relationship to length of the study, and should reflect contemporary clinical practices. The statistical techniques supportive of these analyses include mixed model repeated measures, analysis of variance (ANOVA), logistic regression, and Kaplan Meier and log rank survival analyses, with the event required to be sustained. Because early discontinuation is not randomly distributed across treatment arms in studies of BDs, it is generally invalid to treat most, if not all, early discontinuations as right-censored (55).

Selection of Candidate Drugs

A greater emphasis on studies of drugs with novel mechanisms or extensions of mechanisms associated with existing drugs should energize new drug development in bipolar as well as other disorders. Although it is possible that other drugs with D2 receptor blockade properties could prove to have quantitative as opposed to small, incremental differences compared with currently available D2 blocking drugs, the possibility is low, given the overlap of actions of both atypical antipsychotics, and traditional with atypical antipsychotics (56).

It is reasonable to test all new molecules which may have mechanisms related to currently established treatments for BDs carefully. Mechanisms that would bear consideration, but of course are not limited to the above discussed targets, are: GSK-3 inhibition, ERK pathway stimulation, inositol pathway inhibition,

Wnt signaling inhibition, use-dependent sodium channel inhibition, calcium channel inhibition, and histone deacetylase inhibition. For some of these mechanisms, no truly established efficacious drug is currently available. Nevertheless, given small, positive studies with data from more than one drug, were a more potent, or selective calcium channel blocker to be developed, it might have behavioral effects of greater magnitude than drugs studied previously.

Attention to Anxiety

Anxiety states are both linked to BDs through lifetime combinations, and also important components of the behavioral components of BDs. Although it is possible that adequate control of anxiety is provided by drugs such as benzodiazepines, almost no primary analyses of anti-anxiety effects of treatments employed for BD have been conducted. Anxiety disorders show the highest rates of comorbidity with BD of any group of illnesses (6). Further, a lifetime diagnosis of anxiety disorder predicted worse outcomes of bipolar depressed patients (54). Hypercortisolism, which is associated with anxiety and fearfulness, is elevated in bipolar depression and mania, with the highest rates of elevation reported in mixed or dysphoric manic patients (57). Drugs that have some pharmacodynamic impacts on hypothalamic pituitary adrenocortical function, for example, CRH antagonists, would seem plausibly beneficial in aspects of BDs, including anxiety/fearfulness.

The Role of Regulatory Agencies

In the United States, the Food and Drug Administration (FDA) conducts independent analyses of data submitted for new drug indications. The quality of staff in agencies that the author has had opportunities to interact with is consistently high. However, there are often an inadequate number of personnel in relationship to the prompt review of the complex data in multiple categories of documents that must be reviewed, for each drug and each proposed indication. While it is occasionally the case that political considerations influence priorities or policies, in general scientific criteria and reasonable concerns about protection of public health interest drive most actions.

An important point is that regulatory agencies do not set criteria in a vacuum, nor ignore scientific evidence. Further, it is not the FDA or its equivalent that has established criteria that have generally been employed in new drug indication clinical research programs. For example, there is no written statement that submission of a total score on the Montgomery Asberg Depression Rating Scale (MADRS) and report of the proportion of patients who achieve 50% reduction in such score from baseline is required for consideration of an indication for acute depressive episodes in bipolar depression. Rather, regulatory agencies consider proposals from industry scientists, weigh the reasonableness of the proposed study methodology, and provide some, albeit limited, feedback if they recognize issues that seem important but unresolved in the proposed plan of the study. It is the case that precedent plays a major role. In part, this reflects an aim of equipoise within agencies. For reasons of fairness to companies which compete, and some general guidelines that can be understood by all stakeholders in new drug approval processing, agencies are inclined to maintain in effect criteria that were applied in other recent, related applications for a new indication.

However, as scientific evidence changes, considerations for novel endpoints or study designs also change. An important simple example is the recent conduct of studies combining two drugs for treatment of mania in BD. Until the late

1990s, no such studies had been conducted with large samples and with intent of regulatory submission. Many scientists working with mood disorders doubted that the addition of a second agent known to be an effective first agent would provide a statistically significant advantage in a trial. With results that indicated clear additional benefit could be obtained, particularly when the second agent was added to a drug that had not fully controlled manic symptomatology (in effect an enriched, but clinically generalizable strategy) the FDA responded with approval for the combination regimens (20,26). In instances where legitimate major divergent opinion regarding interpretation of evidence-based study designs, the FDA, scientific organizations, and industry have often utilized regularly scheduled or single issue meetings to convene experts from appropriate interest groups to work through the issues for purposes of developing policies.

SUMMARY

This chapter summarizes factors that may facilitate or impede the development of new drug treatments for BDs. What seems most evident as a guide to maximizing opportunities for the much needed addition of new treatments for aspects of BD care is that current paradigms are broadly inadequate. Even where aspects of efficacy are adequate, there are numerous drugs for which tolerability and safety are still problematic. Strategies and solutions for these issues must be multifaceted. The link of drug development strategies to animal and human evidence of targets in central nervous system (CNS) neurons is a given that all persons involved in new drug development recognize. Strategies that build on extension of drugs within a class, although understandable, has had some undesirable consequences. At the least, it impels pharmaceutical companies to overemphasize small differences between compounds. Additionally, it sometimes misleads the lay public in believing that breakthroughs have occurred when differences are actually modest. Further, it deploys limited resources, essentially reducing efforts that might be more productively deployed elsewhere.

A rethinking of strategies for early phase I and II studies is suggested by this author. Reliance on total scores from scales with incomplete spectrum of items to cover the domains of behavioral disturbance in a disorder, certainly one with as much established complexity as BD, can only inadequately establish the spectrum of efficacy of putative new treatments. The approach in terms of more scientifically sophisticated use of rating scales also requires some rethinking about early clinical studies. Research program planners should consider testing in a broader spectrum of patients both within a single syndrome, using current Diagnostic and Statistical Manual-IV (DSM-IV) criteria, but also beyond the bounds of a single syndrome when there are lines of evidence to indicate pharmacodynamic actions that might be clinically important in more than one syndrome.

REFERENCES

1. Tohen M, Goldberg JF, Gonzalez-Pinto, et al. A 12-week, double-blind comparison of olanzapine vs haloperidol in the treatment of acute mania. Arch Gen Psychiat 2003; 60(12):1218–1226.
2. Williams JB, Gibbon M, First MB, et al. The structured clinical interview for DSM-III-R (SCID). II. Multisite test—retest reliability. Arch Gen Psychiatry 1992; 49(8):630–636.

3. Calabrese JR, Suppes T, Bowden CL, et al. A double-blind placebo-controlled prophylaxis study of lamotrigine in rapid cycling bipolar disorder. J Clin Psychiatry 2000; 61:841–850.
4. Bowden CL, Ascher J, Calabrese J, et al. Lamotrigine: Evidence for mood stabilization in bipolar I depression. Presented at the American Psychiatric Association Annual Meeting, New Orleans, LA, May 2001.
5. Bowden CL. Strategies to reduce misdiagnosis of bipolar depression. Psychiat Serv 2001; 52(1):51–55.
6. Kessler RC, Chiu WT, Demler O, Merikangas KR, Walters EE. Prevalence, severity, and comorbidity of 12-month DSM-IV disorders in the National Comorbidity Survey Replication. Arch Gen Psychiat 2005; 62:590–592.
7. Swann AC, Janicak PL, Calabrese JR, et al. Structure of mania: depressive, irritable, and psychotic clusters with distinct course of illness in randomized clinical trial participants. J Affective Dis 2001; 67:123–132.
8. Cassidy F, Forest K, Murry E, Carroll BJ. A factor analysis of the signs and symptoms of mania. Arch Gen Psychiat 1998; 55(1):27–32.
9. Altman EG, Hedeker DR, Janicak PG, Peterson JL, Davis JM. The clinician-administered rating scale for mania (CARS-M): Development, Reliability, and Validity. Society of Biological Psychiatry 1994; 36:124–134.
10. Swann AC, Bowden CL, Calabrese JR, Dilsaver SC, Morris DD. Pattern of response to divalproex, lithium, or placebo in four naturalistic subtypes of mania. Neuropsychopharmacology 2002; 26(4):530–536.
11. Hollander E, Lopez AA, Bienstock CA, et al. A preliminary double-blind, placebo controlled trial of divalproex sodium in borderline personality disorder. J Clin Psychiat 2001; 62(3):199–203.
12. Hollander E, Tracy KA, Swann AC, et al. Divalproex in the treatment of impulsive aggressive: efficacy in cluster B personality disorders. Neuropsychopharmacology 2003; 28(6):1186–1197.
13. Hollander E, Swann AC, Coccaro EF, Jiang P, Smith TB. Impact of trait impulsivity and state aggression on divalproex versus placebo response in borderline personality disorder. Am J Psychiat 2005; 162(3):621–624.
14. Stahl SM. Dopamine system stabilizers, aripiprazole, and the next generation of antipsychotics, part 2: illustrating their mechanism of action. J Clin Psychiat 2001; 62(12):923–924.
15. Calabrese JR, Keck PE Jr, Macfadden W, et al. A randomized, double-blind, placebo-controlled trial of quetiapine in the treatment of bipolar I or II depression. Am J Psychiat 2005; 162(7):1351–1360.
16. Frye MA, Ketter TA, Kimbrell TA, et al. A placebo-controlled study of lamotrigine and gabapentin monotherapy in refractory mood disorders. J Clin Psychopharmacol 2000; 20:607–614.
17. Grunze H, Erfurth A, Marcuse A, et al. Tiagabine appears not to be efficacious in acute mania. J Clin Psychiat 1999; 60:759–762.
18. Vale AL, Ratcliffe F. Effect of lithium administration on rat brain 5-hydroxyindole levels in a possible animal model for mania. Psychopharmacology 1987; 91(3): 352–355.
19. Bowden CL, Brugger AM, Swann AC, et al. Efficacy of divalproex vs lithium and placebo in the treatment of mania. J Am Med Assoc 1994; 271:918–924.
20. Tohen M, Chengappa KNR, Suppes T, et al. Efficacy of olanzapine in combination with valproate or lithium in the treatment of mania in patients partially nonresponsive to valproate or lithium monotherapy. Arch Gen Psychiat 2002; 59:62–69.
21. Tohen M, Vieta E, Calabrese J, et al. Efficacy of olanzapine and olanzapine-fluoxentine combination in the treatment of bipolar I depression. Arch Gen Psychiat 2004; 61(2):1079–1088.
22. Bowden CL, Collins MA, McElroy SL, et al. Relationship of mania symptomatology to maintenance treatment response with divalproex, lithium or placebo. Neuropsychopharmacology 2005; 30(10):1932–1939.
23. Calabrese JR, Shelton MD, Bowden CL, et al. Bipolar rapid cycling: focus on depression as its hallmark. J Clin Psychiat 2001; 62(suppl 14):34–41.

24. Bowden CL, Calabrese JR, McElroy SL, et al. A randomized, placebo-controlled 12-month trial of divalproex and lithium in treatment of outpatients with bipolar I disorder. Arch Gen Psychiat 2000; 57:481–489.
25. Calabrese JR, Bowden CL, Sachs G, et al. A placebo-controlled 18-month trial of lamotrigine and lithium maintenance treatment in recently depressed patients with bipolar I disorder. J Clin Psychiat 2003; 64:1013–1024.
26. Sachs GS, Grossman F, Ghaemi SN, Okamoto A, Bowden CL. Combination of a mood stabilizer with risperidone or haloperidol for treatment of actue mania: A double - blind, placebo-controlled comparison of efficacy and safety. Am J Psychiat 2002; 159:1146–1154.
27. Yatham LN, Goldstein JM, Vieta E, et al. Atypical antipsychotics in bipolar depression: potential mechanisms of action. J Clin Psychiat 2005; 66(5S):40–48.
28. Muller-Oerlinghausen B, Retzow A, Henn FA, Giedke H, Walden J. Valproate as an adjunct to neuroleptic medication for the treatment of acute episodes of mania: a prospective, randomized, double-blind, placebo-controlled, multicenter study. J Clin Psychopharmacol 2000; 20(2):195–203.
29. Keller MB, Lavori PW, Coryell W, et al. Differential outcome of pure manic, mixed/cycling, and pure depressive episodes in patients with bipolar illness. JAMA 1986; 255:3138–3142.
30. Swann AC, Bowden CL, Morris D, et al. Depression during mania: Treatment response to lithium or divalproex. Arch Gen Psychiat 1997; 54:37–42.
31. Bowden CL. Valproate. Bipolar Disorders 2003; 5:189–202.
32. Harwood AJ, Agam G. Search for a common mechanism of mood stabilizers. Biochem Pharmacol 2003; 66(2):179–189.
33. Gould TD, Manji HK. The Wnt signaling pathway in bipolar disorder. Neuroscientist 2002; 8(5):497–511.
34. Gould TD, ChenG, Manji HK. In vivo evidence in the brain for lithium inhibition of glycogen synthase kinase-3. Neuropsychopharmacology 2004; 29(1):32–38.
35. Gray NA, Zhou R, Du J, Moore GJ, Manji HK. The use of mood stabilizers as plasticity enhancers in the treatment of neuropsychiatric disorders. J Clin Psychiat 2003; 64(S)(5):3–17.
36. Singh V, Bowden CL. Concepts surrounding the diagnosis and treatment of mixed states in bipolar disorders. Clinical Approaches in Bipolar Disorders 2005; 4:35–43.
37. Siafaka-Kapadai A, Patiris M, Bowden C, Javors M. Incorporation of [^{3}H]-valproic acid into lipids in GT1–7 neurons. Biochem Pharmacol 1998; 56:207–212.
38. Hall AC, Brennan A, Goold RG, et al. Valproate regulates GSK-3-mediated axonal remodeling and synapsin I clustering in developing neurons. Mol Cell Neurosci 2002; 20(2):257–270.
39. Blaheta RA, Cinatl J Jr. Anti-tumor mechanisms of valproate: a novel role for an old drug. Med Res Rev 2002; 22(5):492–511.
40. Lieb K, Treffurth Y, Hamke M, Akundi RS, von Kleinsorgen M, Fiebich BL. Valproic acid inhibits substance P-induced activation of proten in kinase C epsilon and expression of the substance P receptor. J Neurochem 2003; 86(1):69–76.
41. Horie S, Suga T. Enhancement of peroxisomal beta-oxidation in the liver of rates and mice treatment with valproic acid. Biochem Pharmacol 1985; 34(9): 1357–1362.
42. Lampen A, Carlberg C, Nau H. Peroxisome proliferator-activated receptor delta is a specific sensor for teratogenic valproic acid derivatives. Eur J Pharmacol 2001; 431(1):25–33.
43. Lees G, Leach MJ. Studies on the mechanism of action of the novel anticonvulsant lamotrigine (Lamictal) using primary neuroglial cultures from rat cortex. Brain Research 1993; 612(1–2):190–199.
44. McGeer EG, Zhu SG. Lamotrigine protects against kainate but not ibotenate lesions in rat striatum. Neurosci Lett 1990; 112(2–3):348–351.
45. Leach MJ, Marden CM, Miller AA. Pharmacological studies on lamotrigine, a novel potential antiepileptic drug 2. Neurochemical studies on the mechanism of action. Epilepsia 1986; 27(5):490–497.

46. Post RM, Ketter TA, Pazzaglia PJ, George MS, Marangell L, Denicoff K. New developments in the use of anticonvulsants as mood stabilizers. Neuropsychobiology 1993; 27(3):132–137.
47. Davidson JR. Pharmacotherapy of social phobia. Acta Psychiat Scand Suppl 2003; 417:65–71.
48. Johnson BA, Ait-Daoud N, Bowden CL, et al. Oral topiramate for treatment of alcohol dependence: a randomised controlled trial. Lancet 2003; 9370(361):1677–1685.
49. Bowden CL, Calabrese JR, McElroy SL, et al. The efficacy of lamotrigine in rapid cycling and non-rapid cycling patients with bipolar disorder. Biol Psychiat 1999; 45:953–958.
50. Calabrese JR, Bowden CL, McElroy SL, et al. Spectrum of activity of lamotrigine in treatment-refractory bipolar disorder. Am J Psychiat 1999; 156(7):1019–1023.
51. Porsteinsson AP, Tariot PN, Erb R, et al. Placebo-controlled study of divalproex sodium for agitation in dementia. Am J Geriatr Psychiat 2001; 9(1):58–66.
52. Katz M, Halbreich U, Bowden C, et al. Enhancing the technology of clinical trials and the trials model to evaluate newly developed, targeted antidepressants. Neuropsychopharmacology 2002; 27(3):319.
53. Katz MM, Tekell JL, Bowden CL, et al. Onset and early behavioral effects of pharmacologically different antidepressants and placebo in depression. Neuropsychopharmacology 2004; 29(3):566–579.
54. Bowden CL. Anticonvulsants in bipolar disorder. Aust NZ J Psychiat 2006; 40(5): 386–393.
55. Tohen M, Chengappa KN, Suppes T, et al. Relapse prevention in bipolar I disorder: 18-month comparison of olanzapine plus mood stabiliser vs mood stabiliser alone. Br J Psychiatry 2004; 184(4):337–345.
56. Tamminga CA. Partial dopamine agonists in the treatment of psychosis. J Neural Transm 2002; 109(3):411–420.
57. Swann AC, Stokes PE, Secunda SK, et al. Depressive mania versus agitated depression: biogenic amine and hypthalamic-pituitary adrenocortical function. Biol Psychiat 1994; 35(10):803–813.

21 Physical Comorbidity in Bipolar Disorder

Paul Mackin and Sylvia Ruttledge
School of Neurology, Neurobiology, and Psychiatry, University of Newcastle upon Tyne, Newcastle upon Tyne, U.K.

INTRODUCTION

A number of studies have reported that bipolar disorder (BD) is associated with a mortality rate approximately twice that of the general population (1–5). Suicide is the leading single cause of excess mortality, but natural deaths contribute significantly to reduced life expectancy. The precise magnitude of the problem of physical comorbidity in BD is unclear, and an international project is currently underway that aims to review the worldwide literature as it pertains to physical illness in schizophrenia and mood disorders (6). It is hoped that this project will bring together the various sources of evidence with a view to generating specific suggestions for the improvement of care for people with mental illnesses. Research activity in the field of physical comorbidity in BD has increased considerably over recent years, although output has lagged behind similar research in schizophrenia. It is hoped that this welcome trend will continue, and ultimately lead to evidence-based guidelines for detecting and managing physical illnesses in this population.

This chapter does not aim to provide a comprehensive overview of the problem of physical illness in BD, but rather focuses upon specific disease areas that have attracted the most research interest, namely cardiovascular disease, nutritional and metabolic diseases, and endocrine diseases. All health professionals involved in the management of BD will be acutely aware of the impact of many of the psychotropic drugs on physical health (e.g., weight gain, thyroid dysfunction, etc.). Disentangling the effects of drugs on physical health from the impact of genetic background and lifestyle issues, for example, is far from straightforward. We also summarize the contribution of commonly prescribed psychotropic drugs to physical comorbidity. The chapter concludes with a consideration of the possible reasons for the observed excess of physical illness in people suffering from BD.

CARDIOVASCULAR DISEASE

Coronary heart disease (CHD) and stroke are the principle components of cardiovascular disease (CVD). Approximately 80% of the worldwide burden of CVD occurs in low-income and middle-income countries, but much of our current understanding of the causes and outcome of CVD is derived from developed countries (7). For example, CVD is the leading cause of death in the United States, and stroke ranks third, accounting for nearly 40% of all deaths (8). Almost one million people in the US die of CVD each year, and many of these deaths are preventable (8). CVD is a leading cause of premature, permanent disability, and the economic impact is considerable—around $394 billion in 2005, resulting from health care expenditure, and loss of productivity from death and disability (8).

A recent case-control study in 52 countries has shown that nine easily measured and potentially modifiable risk factors account for over 90% of the risk of an initial acute myocardial infarction, and the effect of these risk factors is consistent in men and women, across different geographic regions and ethnic groups (7). Worldwide, the two most important risk factors are smoking and abnormal lipids (7). Psychosocial factors, abdominal obesity, diabetes, and hypertension are also associated with increased risk of myocardial infarction (7). Modification of these risk factors is of paramount importance in reducing the burden of cardiovascular disease around the world.

There is a burgeoning literature examining the relationship between major depressive disorder and cardiovascular disease, and some of the literature regarding the epidemiology of comorbid coronary artery disease and depression has been reviewed previously (9). The prevalence of depression in cardiac disease is reported to be 17% to 27% (9), and a number of studies have reported that depression may contribute to the progression of existing coronary disease and have a deleterious effect on outcome (10–13). Possible pathophysiological mechanisms have been reviewed previously (13–15). Patients with BD have also been shown to have increased rates of cardiovascular mortality (1,16), although this has been a relatively neglected area of research. Indeed a recent review of medical comorbidities in BD (17) presents no specific data on cardiovascular morbidity in this population, although mean rates are given for comorbid obesity (21%) and Type 2 diabetes (10%), both risk factors for the development of CVD.

Despite the large number of studies examining the relationship between depressive disorder and cardiovascular disease, there are few studies which have investigated the burden of cardiovascular disease in BD. Kilbourne et al. (18) report the prevalence of general medical conditions in a population-based sample of patients diagnosed with BD in the Veterans Administration (VA). In a cross-sectional study of 4310 patients receiving care at VA facilities within the mid-Atlantic region of the United States, general medical conditions were identified from ICD-9 codes recorded on the National Patient Care Database. The total count of general medical conditions between the bipolar cohort and the national VA cohort was compared. In the bipolar cohort the prevalence of hypertension was 34.8%, ischemic heart disease 10.6%, congestive heart failure 3.2%, peripheral vascular disease 2.9%, and stroke 1.7%. It is noteworthy that there was a statistically lower prevalence of cardiovascular diseases in the bipolar cohort compared with the national VA cohort, although the bipolar sample was approximately four to seven years younger than the national cohort with the same condition. Although this is a valuable study that addresses an important gap in the literature, there are several methodological limitations. The data are derived from administrative data sets and diagnoses are not confirmed by formalized procedures. As the recorded conditions are those that have been observed by the provider, the true prevalence of general medical comorbidity may have been underestimated. Conversely, only those patients receiving care (who are more likely to have multiple diagnoses) are included in the database, which may overestimate the true prevalence of medical comorbidity.

The same group has published a similar study, using the same database, which reports general medical comorbidity in older patients with serious mental illness (schizophrenia, schizoaffective disorder, or BD) (19). Of the 8083 patients included in the study, 2446 (30%) had a diagnosis of BD. Overall, older, versus younger, patients were more likely to be diagnosed with general medical comorbidity. The most common comorbid condition among the whole cohort was cardiovascular disease (30.7%); hypertension was 25.5%, congestive heart failure 2.6%, peripheral vascular disease 1.8%, stroke 2.1%, and ischemic heart disease 6.7%. The prevalence rates specific to BD are not given in this study.

Beyer et al. (20) also assessed the presence of general medical conditions in 1379 U.S. outpatients with a diagnosis of bipolar I disorder. Data were extracted from the Duke University Medical Center database. The number of comorbid medical conditions increased as a function of age. Diseases of the circulatory system were present in 13% of patients with BD. Specific cardiovascular diseases are not specified with the exception of "cardiac disease/hypertension," which had a prevalence of 10.7%. This study has similar methodological shortcomings to that of Kilbourne et al. (18). In addition, the lack of a comparison group prevents any evaluation of how medical comorbidity in BD compares with that in the general population.

One study has examined the prevalence of QTc prolongation in a cohort of 65 outpatients from the North East of England receiving antipsychotic drugs (BD = 30.8%, schizophrenia = 30.8%, schizoaffective disorder = 13.8%, other mood disorders = 24.6%) (21). The QTc interval on the electrocardiogram is a measure of ventricular repolarization, and prolongation of the QTc interval is associated with cardiac arrhythmias and sudden death. Only two patients (3%) had prolongation of the QTc interval, and there was a significant correlation between increasing age and QTc interval. The findings of this study should be considered preliminary given the small sample size and the cross-sectional nature of the study, and further investigation of cardiac physiology and its relationship with mood disorders and adverse outcomes is needed.

Strudsholm et al. (22) investigated the risk for pulmonary embolism in patients with BD. Danish national registers were used to examine somatic and psychiatric information on 25,834 patients with BD and 117,815 controls matched for age and sex. Patients with BD had a significantly increased occurrence of pulmonary embolism [increased incidence rate ratio (IRR) = 1.61; 95% CI = 1.38–1.88]. The authors offer several possible explanations for the association between BD and pulmonary embolism, including the effects of restraint (as immobility predisposes to deep vein thrombosis in the lower extremities and pelvis), antipsychotic-induced apathy and consequent immobility, and infectious endocarditis resulting from intravenous drug use. One of the strengths of this study is that all admitted patients were included in the analysis. However, the lack of formalized diagnostic procedures, and the possibility of better detection of other medical conditions (such as pulmonary embolism) in hospitalized patients may have influenced the results of this study.

The paucity of studies of cardiovascular morbidity in patients with BD is of some concern, particularly given that large cohort studies appear to indicate that BD is associated with high levels of medical comorbidity, which includes cardiovascular disease. Well-designed, prospective studies are needed in patients with BDs to investigate in detail how genetic factors, structural/functional/hormonal changes, and psychosocial stress contribute to the development of cardiovascular disease in this population. In addition, studies assessing the effectiveness of psychosocial and behavioral interventions in modifying the risk for cardiovascular disease are urgently needed.

NUTRITIONAL AND METABOLIC DISEASES

Obesity

In recent decades there has been a right-shift in the weight curve for the general population, and the problem of obesity is now often referred to as a pandemic. Obesity poses a particular problem in the management of patients with BD because antipsychotic drugs (23), mood stabilizers such as lithium (24) and valproate (25), and many antidepressants (26,27) are all associated with weight gain. Obesity, in turn, is a component of the metabolic syndrome (MS), and a risk factor for type 2 diabetes mellitus (DM), hypertension, dyslipidemia, ischemic heart disease, and some cancers (28). In addition to pharmacological treatment, risk factors for weight gain and obesity in these patients include comorbid binge-eating disorder, the number of previous depressive episodes, excessive carbohydrate consumption, and low rates of exercise (29).

In a systematic review of 45 studies, patients with BD were at greater risk than the general population for being overweight and obese (30). The prevalence of overweight, obesity, and extreme obesity in BD patients reflects the prevalence locally in the general population—American patients had a higher Body Mass Index (BMI) than European patients (29). A New Zealand study of 89 BD patients and 445 age- and sex-matched controls found that female patients were more often overweight (44% vs. 25%) and obese (20% vs. 13%) than female community control subjects (31). Significantly more patients receiving antipsychotic medications (APs) were obese compared with patients not receiving APs. As a cross-sectional study, it was not possible to tease apart important drug–disease state interactions, that is, whether the higher prevalence of obesity among patients treated with APs was due to the APs, or whether APs may have been associated with greater illness severity, which, in turn, may have been associated with a greater risk of obesity (30). A follow-up lifestyle study showed total fluid intake and intake of sweetened drinks were higher in BD, with total energy intake particularly higher for female patients (31). BD patients reported fewer episodes of low-to-moderate intensity and high intensity physical activity as compared with the "general population." This study was, however, based on self report only.

Drug-induced weight gain is a crucial issue in the management of BD. In a study by Fagiolini et al. (32), most patients received lithium as their primary mood stabilizer. Among the patients treated with lithium, during acute treatment, 14 of the 47 patients (30%) gained at least 5% of their baseline BMI, and during maintenance treatment, 11 of the 45 patients (24%) gained more than 5% of their BMI. The authors speculated that a greater number of prior depressive, but not manic, episodes is associated with an increased likelihood of being overweight or obese at study entry. Further, higher scores on the Hamilton Rating Scale for Depression and negative scores on the Bech-Rafaelsen Mania Scale predicted an increase in BMI during acute treatment. Weight at treatment initiation was inversely related to weight gain during treatment, and the obese group had no significant weight gain during the acute phase of treatment.

Obesity has been associated with a shorter time to bipolar recurrence in the maintenance phase and more depressive recurrences overall (33), but as some commentators have pointed out, this study may have been confounded by a greater number of previous bipolar episodes and higher baseline Hamilton Depression Rating Scale in obese BD patients (28). Again, the direction of causation between obesity and psychiatric outcome is still unclear.

Metabolic Syndrome

Metabolic syndrome (MS) is a constellation of interrelated metabolic risk factors that appear to directly promote the development of atherosclerotic cardiovascular disease (ASCVD). Patients with MS are also at increased risk for developing type 2 diabetes mellitus (DM), which in turn is associated with increased cardiovascular morbidity and mortality. Some atypical APs have been shown to increase the risk of metabolic disturbances. Indeed, patients with BD exhibit risk factors for MS independent of medication use and are predisposed to increased incidence of smoking, poor nutrition, poor health care, and decreased energy expenditure, all of which are risk factors for diabetes, obesity, and dyslipidemia (34).

There are only a few empirical studies of the prevalence of MS in BD patients. In one survey of 103 patients from psychiatric outpatient services in the north of England (32% BD), 12% of patients (overall) had impaired glucose homeostasis disorder (6% impaired fasting glucose and 6% with DM) and 8% met World Health Organization criteria for MS (35). This is likely to be a conservative estimate as measures of blood pressure or urinary albumin excretion were not available.

Diabetes Mellitus

Retrospective chart reviews show up to three times the prevalence of type 2 DM in bipolar 1 disorder compared with national norms (36,37). In another chart review study (38), BMI and psychiatric diagnosis but not medication were associated with new-onset type 2 DM. Chart reviews of inpatients are, however, problematic regarding selection bias and the reliability and validity of the diagnosis of both type 2 DM and BD (28).

The mechanisms of the observed relationship between BD and DM are unclear. It remains to be determined if susceptibility genes for BDs and glucose homeostasis disorders cosegregate. Hypercortisolemia-induced insulin resistance and abdominal adiposity is another mechanism that may account for the observed increased incidence of DM in patients with BD.

There is clear and consistent evidence that commonly prescribed psychotropic medication for the management of BD also contributes to the problem of increased adiposity and disorders of glucose homeostasis in this patient population (39). Currently available studies examining this complex area are usually small and conducted in single sites, are prone to selection and ascertainment bias, are often retrospective, and rarely control for confounding factors. There is a dearth of systematic research regarding predictive risk factors for weight gain and MS, and the impact of psychiatric comorbidity [e.g., binge-eating disorder (30)], course of illness [e.g., number of depressive episodes (33)], and illness–treatment interactions (33,40).

ENDOCRINE DISEASES

Menstrual Abnormalities and Polycystic Ovarian Syndrome

Prospective studies have produced inconsistent findings as to whether there is an association between mood symptoms and the menstrual cycle for women with BD (41,42). Studies show that medications such as selective serotonin reuptake inhibitors, alprazolam, and buspirone may precipitate premenstrual mood disturbances (mania) in vulnerable individuals (43–45).

Although menstrual abnormalities are commonly reported in females prior to diagnosis of BD and initiation of pharmacologic treatment, studies show a correlation between menstrual abnormalities and the use of medications (46). Polycystic ovarian syndrome (PCOS) is a syndrome of ovarian dysfunction and is characterized by hyperandrogenism and menstrual irregularities (47). The associated endocrine profile is elevated testosterone and luteinizing hormone, and low or normal follicle-stimulating hormone. Clinical symptoms include hirsutism, acne, and anovulation. Many women with PCOS also have obesity and insulin resistance (47). Debate surrounds the possible role of anticonvulsants, particularly valproate, in the pathogenesis of PCOS. Much of the available data is limited by the fact that it comes from research in women with epilepsy (48). The relationship between the reproductive endocrine system and BD in women is poorly defined, and whatever data are available are hampered by small patient numbers (49–51). In one of the few "larger" studies, Rasgon et al. (46) in an investigation of the reproductive function and prevalence of PCOS in 80 females with BD found that 65% reported menstrual abnormalities, and 50% reported such abnormalities prior to the diagnosis of BD. Valproic acid was linked to menstrual abnormalities in all but one of these patients. A history of menstrual abnormalities and obesity were predictors for menstrual and hormonal abnormalities (including an increase in the levels of luteinizing hormone, follicle-stimulating hormone, and testosterone over time) following treatment for BD. BMI was significantly positively correlated with free testosterone levels and insulin resistance across all patients, regardless of medication used.

Women with PCOS often develop dyslipidemia, including increased levels of cholesterol, triglycerides, and LDL cholesterol, and decreased levels of HDL cholesterol (52), thus increasing further the risk of cardiovascular disease.

It is therefore important that a thorough history prior to the initiation of medication therapy be taken in order to determine the cause and risks of reproductive endocrine disorders. Patients should be closely monitored for menstrual abnormalities especially when receiving combination therapy with valproic acid to minimize the risk of PCOS (46). Women of reproductive age should be informed of the risks and benefits of their pharmacologic treatment options, and should understand that the impact of managing a chronic mood disorder on reproductive function requires consideration from the outset (53).

Thyroid Abnormalities

Perturbations in the hypothalamic-pituitary-thyroid (HPT) axis, including state-dependent blunting of the serum thyroid stimulating hormone (TSH) response to thyrotropin-releasing hormone (TRH) have been demonstrated in depressed patients (54,55). Thyroid function during manic episodes is less documented but available data show blunted responses of TSH to TRH, similar to those described during depression (56,57) as well as elevations in serum free thyroxine (FT4) index (58) and decreases in serum T3 levels (59). Cassidy et al. (60) compared rates of previously diagnosed thyroid disease in 443 inpatients with BD along sex, race, and manic subtype (mixed vs. pure). Overall, hypothyroidism was more frequent in white patients and females, and increased with age. No differences were noted between patients sampled during mixed or pure manic episodes. The participation of inpatients and its implied severity of illness together with the lack of application of research diagnostic criteria for hypothyroidism limit this study's findings.

Using Danish register data, Thomsen and Kessing (61) examined the risks of hyperthyroidism among three study cohorts of inpatients with a diagnosis of depression, BD, or osteoarthritis. There was a trend towards a significant risk of hyperthyroidism for patients with BD compared with patients with osteoarthritis. The lack of formalized diagnostic procedures in this study, however, could have influenced the results. Moreover, the use of individuals with osteoarthritis as a control group in this study can be regarded as not ideal as the biological association between osteoarthritis and hyperthyroidism is not yet clearly delineated.

Some studies have reported a correlation between rapid-cycling BD and hypothyroidism (62–64). These findings, however, are limited by methodological considerations such as medication status, recruitment bias, broad criteria for defining hypothyroidism, and comparison groups (65). Furthermore, several studies have failed to find an association with hypothyroidism per se and have cast doubt on the direction of causality (66,67).

Psychotropic Drugs

Antipsychotic Drugs

A recently published consensus statement (2004), developed by four North American medical associations, highlighted the problem of antipsychotic-induced metabolic disturbances in patients receiving atypical APs (23). Among the atypical APs, clozapine and olanzipine are associated with significant increases in weight, while ziprasidone and aripiprazole are associated with minimal weight change (68,69). Risperidone and quetiapine appear to be associated with intermediate effects with regard to their propensity to cause weight gain and glucose homeostasis disturbances (23). Crucially, these differences may in turn affect compliance with medication and risk of relapse (40,70).

The consensus panel advised that there be monitoring at baseline and during treatment with APs and suggested that history of obesity, diabetes, dyslipidemia, hypertension, or cardiovascular disease (including family history) be obtained. Further, they advised that height, weight (to calculate BMI), and waist circumference are obtained, together with estimations of blood pressure, fasting blood glucose concentrations, and lipid profiles. These parameters should be regularly monitored following commencement of treatment in order to detect emerging metabolic disease. The consensus statement recommends that if individuals gain more than 5% of their initial weight, consideration should be given to switching to another atypical AP. A "cost/benefit" judgment should be made, of course, to assess the psychiatric status of the patient and the potential benefit of the atypical APs before discontinuing (23).

More recently, results from the first phase of the Clinical Antipsychotic Trials of Intervention Effectiveness (CATIE) trial in patients with schizophrenia revealed that patients taking olanzapine experienced more weight gain and metabolic changes associated with an increased risk of diabetes than patients taking other atypical APs (risperidone, quetiapine, and ziprasidone) (71). Similar large prospective studies examining the impact of antipsychotic treatment on metabolic dysfunction are needed in patients with BD.

The mechanism of AP-associated diabetes is unclear. Altered glucose homeostasis may be mediated through a variety of mechanisms, including increased food intake leading to increased adiposity and insulin resistance, effects on the insulin-signaling pathway, and changes in pancreatic β-cell function causing altered insulin release or altered hepatocyte/myocyte function resulting in impaired insulin sensitivity.

Hyperprolactinemia is a common side effect of some APs (34). Prolactin levels may be elevated in patients treated with either typical or atypical APs. Hyperprolactinemia can be distressing for the individual, and may cause menstrual irregularities, sexual dysfunction, galactorrhea, and gynecomastia. The extent of prolactin elevation appears to be dose related, and increases in prolactin sufficient to suppress the sex steroid axis may result in hypogonadism and perturbations in bone metabolism, including osteoporosis. Data regarding the effects of APs on bone metabolism in relation to patients with BD is limited by small patient numbers, the participation of patients with a schizophrenia diagnosis, limited controls, a lack of prospective studies, and multiple confounders such as hypogonadism, poor diet, and cigarette smoking (72). Hypergonadism may be more likely to occur in women with schizophrenia compared with men taking typical APs: Kinon et al. (73) reported prolactin levels 2.6 times higher in women compared with men. There is no comparable study in BD. There appear to be clear differences between antipsychotic agents with regard to their propensity to cause increases in prolactin release. One study reported hyperprolactinemia in 88% of female patients treated with risperidone (73). Other atypical APs such as olanzapine and quetiapine result in more modest prolactin elevations than equivalent doses of risperidone (74) or haloperidol (75).

Mood Stabilizers

Mood stabilizers are central to the pharmacologic treatment of BD. Some mood stabilizers have been associated with weight gain, including valproic acid, carbamazepine, and lithium. In the case of lithium, both randomized controlled trials and open-label naturalistic outcome studies show significantly more BD patients gaining weight (76,77). Lithium appears to exert insulin-like activity on carbohydrate metabolism in some patients, leading to increased glucose absorption into adipocytes. This effect may stimulate appetite indirectly. Lithium may also have direct appetite-stimulating effects on the hypothalamus. Relieving thirst by consuming high-calorie beverages has been proposed as another weight-gaining mechanism (77). Lamotrigine has been studied in an 18 month randomized controlled maintenance trial in 463 outpatients with bipolar 1 disorder (78). The mean change in body weight at week 76 was 1.2 kg (2.7 lb) for patients receiving placebo, −2.2 kg (−4.9 lb) for lamotrigine, and 4.2 kg (9.3 lb) for lithium. While there was no significant difference in mean weight between patients receiving lamotrigine versus placebo, the differences in weight change were statistically significant between the lamotrigine and lithium groups. The proportion of patients who experienced ≥7% increase in body weight from baseline to final study visit was 7% for the lamotrigine group, 10% for lithium, and 6% for placebo.

While there are few long-term randomized controlled studies of divalproex (valproate) in the treatment of BD, weight gain greater than 5% from baseline was significantly more common in patients receiving divalproex (21%) compared with placebo (7%) (76). As with lithium, similar mechanisms have been proposed for this weight gain, including impaired fatty acid metabolism (79). Patients receiving carbamazepine with major depression, but not with mania, experienced a significant increase in body weight compared with placebo (80). In addition to appetite stimulation, fluid retention and edema have been reported as mechanisms underlying this weight gain (81). Again, given the important role of mood stabilizers in managing BD, the psychiatric status of patients needs to be considered before switching medications possibly causing weight gain, to minimize the risk of destabilizing the patient's psychiatric condition.

Anticonvulsants such as carbamazepine and valproic acid have also been associated with osteopenia (82,83), with the extent of bone loss being related to the duration of treatment. Lithium also has a potential negative impact on bone metabolism given its association with hyperparathyroidism (84,85).

The rate of hypothyroidism in lithium-naïve patients was found to be significantly lower than those treated with lithium (6.3–10.8% vs. 28.0–32.1%) (86). It has been suggested that two categories of bipolar patients are more likely to be at high risk for developing hypothyroidism in the course of lithium treatment. The first category comprises females who have had a longer course of illness predominated by depressive episodes. The second category comprises those with rapid cycling and mixed mania, that is, those who are diagnosed with moderate or severe mania and have experienced a frequent recurrence of mood episodes. In contrast, hypothyroidism is associated with longer illness course in lithium-naïve mania and with more mood episodes in lithium-naïve bipolar depression (86). In a study of adults over 65 years of age (87), lithium users were significantly more likely to be treated with T4 therapy (as a proxy for hypothyroidism) than were valproate users, with hypothyroidism appearing to develop twice as frequently among this age cohort than among a mixed-age population.

Barriers to Care

There are many reasons why people with BD may suffer from increased rates of physical illnesses such as obesity, diabetes, and cardiovascular disease. Poor diet, tobacco smoking, and lack of exercise are all associated with poor physical health, and may play a significant role in increasing physical morbidity in BD (31). In addition, use of medical care often decreases after the onset of a psychiatric disorder (88), and even when patients are engaged in health care services, rates of undiagnosed physical illnesses are often high (35). Other patient characteristics may also contribute to poor detection and diagnosis of physical illness such as impaired ability to verbalize concerns (89,90), poor insight into illness (90), denial of illness (91), or an unwillingness to consult a doctor other than their psychiatrist. When patients are cared for by psychiatrists, primary care physicians, and physicians from other disciplines, there may be a shared assumption that a colleague is taking responsibility for managing a particular medical problem, when in fact the problem is not being attended to at all.

A study by Cradock-O'Leary et al. (92) used the Department of Veterans Affairs database to examine the use of medical services by 17,653 patients who were treated in Southern California and Nevada during the year 2000. Adults of all ages with a diagnosis of BD had an especially high risk of not receiving general medical services. The authors suggest interventions such as improving provider competencies through education and profiling, and organizational interventions such as computerized reminders to prompt mental health professionals to refer to primary care for appropriate screening.

Although there is evidence that BD is associated with increased mortality, some studies suggest that mortality may be reduced in some bipolar patient groups. Attendance at lithium clinics and longer-term use of medication have been shown to be associated with reduced mortality from all causes in people with BD (93–95). These findings are not easy to explain, and may be attributable to a number of factors such as the selection of specific patients to attend such clinics, attendance at these clinics by individuals motivated to adhere to treatment and therefore more likely to be motivated to take a greater interest in physical health issues, or specific effects of medication on longevity.

There are few studies specifically examining the impact of differing models of care on physical well-being and comorbidity in severe mental illness. One randomized trial from the United States valuated an integrated model of primary medical care for a cohort of patients with serious mental disorders, and the authors concluded that on-site, integrated primary care was associated with improved quality and outcomes of medical care (96). There is a growing acknowledgment, backed up by a burgeoning literature on physical comorbidity in severe mental illness, that health professionals involved in the care of people with BD must be mindful of the possibility of coexisting physical illness. There is a need for greater communication and collaboration at the primary/secondary care interface, and for the establishment of clear guidelines outlining responsibilities and protocols for screening and managing physical health and disease in patients with BD.

REFERENCES

1. Tsuang MT, Woolson RF, Fleming JA. Causes of death in schizophrenia and manic-depression. Br J Psychiatry 1980; 136:239–242.
2. Tsuang MT, Woolson RF, Fleming JA. Premature deaths in schizophrenia and affective disorders. An analysis of survival curves and variables affecting the shortened survival. Arch Gen Psychiatry 1980; 37(9):979–983.
3. Weeke A, Vaeth M. Excess mortality of bipolar and unipolar manic-depressive patients. J Affect Disord 1986; 11(3):227–234.
4. Sharma R, Markar HR. Mortality in affective disorder. J Affect Disord 1994; 31(2):91–96.
5. Hoyer EH, Mortensen PB, Olesen AV. Mortality and causes of death in a total national sample of patients with affective disorders admitted for the first time between 1973 and 1993. Br J Psychiatry 2000; 176:76–82.
6. Mackin P. Physical illness in people with mood disorders. European Psychiatry 2006; 21(suppl 1):S37.
7. Yusuf S, Hawken S, Ounpuu S, et al. Effect of potentially modifiable risk factors associated with myocardial infarction in 52 countries (the INTERHEART study): case-control study. Lancet 2004; 364(9438):937–952.
8. Preventing Heart Disease and Stroke. Addressing the Nation's Leading Killers: U.S. Department of Health and Human Services. Centres for Disease Control and Prevention, 2005.
9. Rudisch B, Nemeroff CB. Epidemiology of comorbid coronary artery disease and depression. Biol Psychiatry 2003; 54(3):227–240.
10. Carney RM, Rich MW, Freedland KE, et al. Major depressive disorder predicts cardiac events in patients with coronary artery disease. Psychosom Med 1988; 50(6):627–633.
11. Ladwig KH, Kieser M, Konig J, Breithardt G, Borggrefe M. Affective disorders and survival after acute myocardial infarction. Results from the post-infarction late potential study. Eur Heart J 1991; 12(9):959–964.
12. Frasure-Smith N, Lesperance F. Depression and other psychological risks following myocardial infarction. Arch Gen Psychiatry 2003; 60(6):627–636.
13. Rozanski A, Blumenthal JA, Kaplan J. Impact of psychological factors on the pathogenesis of cardiovascular disease and implications for therapy. Circulation 1999; 99(16):2192–2217.
14. Musselman DL, Tomer A, Manatunga AK, et al. Exaggerated platelet reactivity in major depression. Am J Psychiatry 1996; 153(10):1313–1317.
15. Evans DL, Charney DS, Lewis L, et al. Mood disorders in the medically ill: scientific review and recommendations. Biol Psychiatry 2005; 58(3):175–189.

16. Weeke A, Juel K, Vaeth M. Cardiovascular death and manic-depressive psychosis. J Affect Disord 1987; 13(3):287–292.
17. Dickens C, McGowan L, Percival C, et al. Association between depressive episode before first myocardial infarction and worse cardiac failure following infarction. Psychosomatics 2005; 46(6):523–528.
18. Kilbourne AM, Cornelius JR, Han X, et al. Burden of general medical conditions among individuals with bipolar disorder. Bipolar Disord 2004; 6(5):368–373.
19. Kilbourne AM, Cornelius JR, Han X, et al. General-medical conditions in older patients with serious mental illness. Am J Geriatr Psychiatry 2005; 13(3):250–254.
20. Beyer J, Kuchibhatla M, Gersing K, Krishnan KR. Medical comorbidity in a bipolar outpatient clinical population. Neuropsychopharmacology 2005; 30(2):1–404.
21. Mackin P, Young AH. QTc interval measurement and metabolic parameters in psychiatric patients taking typical or atypical antipsychotic drugs: a preliminary study. J Clin Psychiatry 2005; 66(11):1386–1391.
22. Strudsholm U, Johannessen L, Foldager L, Munk-Jorgensen P. Increased risk for pulmonary embolism in patients with bipolar disorder. Bipolar Disord 2005; 7(1):77–81.
23. American Diabetes Association; American Psychiatric Association; American Association of Clinical Endocrinologists; North American Association for the Study of Obesity. Consensus development conference on antipsychotic drugs and obesity and diabetes. Diabetes Care 2004; 27(2):596–601p.
24. Coxhead N, Silverstone T, Cookson J. Carbamazepine versus lithium in the prophylaxis of bipolar affective disorder. Acta Psychiatr Scand 1992; 85(2):114–118.
25. Macritchie KA, Geddes JR, Scott J, Haslam DR, Goodwin GM. Valproic acid, valproate and divalproex in the maintenance treatment of bipolar disorder. Cochrane Database Syst Rev 2001; 3:CD003196.
26. Sussman N, Ginsberg DL, Bikoff J. Effects of nefazodone on body weight: a pooled analysis of selective serotonin reuptake inhibitor- and imipramine-controlled trials. J Clin Psychiatry 2001; 62(4):256–260.
27. Fava M, Judge R, Hoog SL, Nilsson ME, Koke SC. Fluoxetine versus sertraline and paroxetine in major depressive disorder: changes in weight with long-term treatment. J Clin Psychiatry 2000; 61(11):863–867.
28. Morriss R, Mohammed FA. Metabolism, lifestyle and bipolar affective disorder. J Psychopharmacol 2005; 19(6 suppl):94–101.
29. McElroy SL, Frye MA, Suppes T, et al. Correlates of overweight and obesity in 644 patients with bipolar disorder. J Clin Psychiatry 2002; 63(3):207–213.
30. Keck PE, McElroy SL. Bipolar disorder, obesity, and pharmacotherapy-associated weight gain. J Clin Psychiatry 2003; 64(12):1426–1435.
31. Elmslie JL, Mann JI, Silverstone JT, Williams SM, Romans SE. Determinants of overweight and obesity in patients with bipolar disorder. J Clin Psychiatry 2001; 62(6):486–491; quiz 92–3.
32. Fagiolini A, Frank E, Houck PR, et al. Prevalence of obesity and weight change during treatment in patients with bipolar I disorder. J Clin Psychiatry 2002; 63(6): 528–533.
33. Fagiolini A, Kupfer DJ, Houck PR, Novick DM, Frank E. Obesity as a correlate of outcome in patients with bipolar I disorder. Am J Psychiatry 2003; 160(1):112–117.
34. Masand PS, Culpepper L, Henderson D, et al. Metabolic and endocrine disturbances in psychiatric disorders: a multidisciplinary approach to appropriate atypical antipsychotic utilization. CNS Spectr 2005; 10(10):(suppl14) 1–15.
35. Mackin P, Watkinson H, Young A. The prevalence of obesity, disorders of glucose homeostasis and metabolic syndrome in psychiatric patients taking typical and atypical antipsychotic drugs: a cross-sectional study. Diabetologia 2005; 48(2):215–221.
36. Lilliker SL. Prevalence of diabetes in a manic-depressive population. Compr Psychiatry 1980; 21(4):270–275.
37. Cassidy F, Ahearn E, Carroll BJ. Elevated frequency of diabetes mellitus in hospitalized manic-depressive patients. Am J Psychiatry 1999; 156(9):1417–1420.
38. Regenold WT, Thapar RK, Marano C, Gavirneni S, Kondapavuluru PV. Increased prevalence of type 2 diabetes mellitus among psychiatric inpatients with bipolar I affective

and schizoaffective disorders independent of psychotropic drug use. J Affect Disord 2002; 70(1):19–26.
39. Krishnan KR. Psychiatric and medical comorbidities of bipolar disorder. Psychosom Med 2005; 67(1):1–8.
40. Elmslie JL, Silverstone JT, Mann JI, Williams SM, Romans SE. Prevalence of overweight and obesity in bipolar patients. J Clin Psychiatry 2000; 61(3):179–184.
41. Leibenluft E, Ashman SB, Feldman-Naim S, Yonkers KA. Lack of relationship between menstrual cycle phase and mood in a sample of women with rapid cycling bipolar disorder. Biol Psychiatry 1999; 46(4):577–580.
42. Rasgon N, Bauer M, Glenn T, Elman S, Whybrow PC. Menstrual cycle related mood changes in women with bipolar disorder. Bipolar Disord 2003; 5(1):48–52.
43. Goodman WK, Charney DS. A case of alprazolam, but not lorazepam, inducing manic symptoms. J Clin Psychiatry 1987; 48(3):117–118.
44. Spigset O. Adverse reactions of selective serotonin reuptake inhibitors: reports from a spontaneous reporting system. Drug Saf 1999; 20(3):277–287.
45. Liegghio NE, Yeragani VK. Buspirone-induced hypomania: a case report. J Clin Psychopharmacol 1988; 8(3):226–227.
46. Rasgon NL, Altshuler LL, Fairbanks L, et al. Reproductive function and risk for PCOS in women treated for bipolar disorder. Bipolar Disord 2005; 7(3):246–259.
47. The Rotterdam ESHRE/ASRM–Sponsored PCOS consensus workshop group. Revised 2003 consensus on diagnostic criteria and long-term health risks related to polycystic ovary syndrome (PCOS). Hum Reprod 2004; 19(1):41–47.
48. Vainionpaa LK, Rattya J, Knip M, et al. Valproate-induced hyperandrogenism during pubertal maturation in girls with epilepsy. Ann Neurol 1999; 45(4):444–450.
49. Rasgon NL, Altshuler LL, Gudeman D, et al. Medication status and polycystic ovary syndrome in women with bipolar disorder: a preliminary report. J Clin Psychiatry 2000; 61(3):173–178.
50. O'Donovan C, Kusumakar V, Graves GR, Bird DC. Menstrual abnormalities and polycystic ovary syndrome in women taking valproate for bipolar mood disorder. J Clin Psychiatry 2002; 63(4):322–330.
51. McIntyre RS, Mancini DA, McCann S, Srinivasan J, Kennedy SH. Valproate, bipolar disorder and polycystic ovarian syndrome. Bipolar Disord 2003; 5(1):28–35.
52. Lobo RA, Carmina E. The importance of diagnosing the polycystic ovary syndrome. Ann Intern Med 2000; 132(12):989–993.
53. Freeman MP, Gelenberg AJ. Bipolar disorder in women: reproductive events and treatment considerations. Acta Psychiatr Scand 2005; 112(2):88–96.
54. Bauer M, Whybrow PC. Thyroid hormones and the central nervous system in affective illness: interacts that may have clinical significance. Integr Psych 1988; 6:75–100.
55. Hendrick V, Altshuler L, Whybrow P. Psychoneuroendocrinology of mood disorders. The hypothalamic-pituitary-thyroid axis. Psychiatr Clin North Am 1998; 21(2):277–292.
56. Extein I, Pottash AL, Gold MS, et al. The thyroid-stimulating hormone response to thyrotropin-releasing hormone in mania and bipolar depression. Psychiatry Res 1980; 2(2):199–204.
57. Kiriike N, Izumiya Y, Nishiwaki S, Maeda Y, Nagata T, Kawakita Y. TRH test and DST in schizoaffective mania, mania, and schizophrenia. Biol Psychiatry 1988; 24(4):415–422.
58. Joffe RT, Young LT, Cooke RG, Robb J. The thyroid and mixed affective states. Acta Psychiatr Scand 1994; 90(2):131–132.
59. Kirkegaard C, Bjorum N, Cohn D, Lauridsen UB. Thyrotrophin-releasing hormone (TRH) stimulation test in manic-depressive illness. Arch Gen Psychiatry 1978; 35(8):1017–1021.
60. Cassidy F, Ahearn EP, Carroll BJ. Thyroid function in mixed and pure manic episodes. Bipolar Disord 2002; 4(6):393–397.
61. Thomsen AF, Kessing LV. Increased risk of hyperthyroidism among patients hospitalized with bipolar disorder. Bipolar Disord 2005; 7(4):351–357.
62. Cowdry RW, Wehr TA, Zis AP, Goodwin FK. Thyroid abnormalities associated with rapid-cycling bipolar illness. Arch Gen Psychiatry 1983; 40(4):414–420.

63. Bauer MS, Whybrow PC. Rapid cycling bipolar affective disorder. II. Treatment of refractory rapid cycling with high-dose levothyroxine: a preliminary study. Arch Gen Psychiatry 1990; 47(5):435–440.
64. Bauer MS, Whybrow PC, Winokur A. Rapid cycling bipolar affective disorder. I. Association with grade I hypothyroidism. Arch Gen Psychiatry 1990; 47(5):427–432.
65. Mackin P, Young AH. Rapid cycling bipolar disorder: historical overview and focus on emerging treatments. Bipolar Disord 2004; 6(6):523–529.
66. Joffe RT, Kutcher S, MacDonald C. Thyroid function and bipolar affective disorder. Psychiatry Res 1988; 25(2):117–121.
67. Post RM, Kramlinger KG, Joffe RT, et al. Rapid cycling bipolar affective disorder: lack of relation to hypothyroidism. Psychiatry Res 1997; 72(1):1–7.
68. Allison DB, Mentore JL, Heo M, et al. Antipsychotic-induced weight gain: a comprehensive research synthesis. Am J Psychiatry 1999; 156(11):1686–1696.
69. Casey DE, Haupt DW, Newcomer JW, et al. Antipsychotic-induced weight gain and metabolic abnormalities: implications for increased mortality in patients with schizophrenia. J Clin Psychiatry 2004; 65(suppl 7):4–18; quiz 19–20.
70. McIntyre RS, Konarski JZ. Tolerability profiles of atypical antipsychotics in the treatment of bipolar disorder. J Clin Psychiatry 2005; 66(suppl 3):28–36.
71. Lieberman JA, Stroup TS, McEvoy JP, et al. Effectiveness of antipsychotic drugs in patients with chronic schizophrenia. N Engl J Med 2005; 353(12): 1209–1223.
72. Misra M, Papakostas GI, Klibanski A. Effects of psychiatric disorders and psychotropic medications on prolactin and bone metabolism. J Clin Psychiatry 2004; 65(12):1607–1618; quiz 590, 760–761.
73. Kinon BJ, Gilmore JA, Liu H, Halbreich UM. Prevalence of hyperprolactinemia in schizophrenic patients treated with conventional antipsychotic medications or risperidone. Psychoneuroendocrinology 2003; 28(suppl 2):55–68.
74. Tran PV, Hamilton SH, Kuntz AJ, et al. Double-blind comparison of olanzapine versus risperidone in the treatment of schizophrenia and other psychotic disorders. J Clin Psychopharmacol 1997; 17(5):407–418.
75. Tollefson GD, Beasley CM Jr, Tran PV, et al. Olanzapine versus haloperidol in the treatment of schizophrenia and schizoaffective and schizophreniform disorders: results of an international collaborative trial. Am J Psychiatry 1997; 154(4):457–465.
76. Bowden CL, Calabrese JR, McElroy SL, et al. A randomized, placebo-controlled 12-month trial of divalproex and lithium in treatment of outpatients with bipolar I disorder. Divalproex Maintenance Study Group. Arch Gen Psychiatry 2000; 57(5):481–489.
77. Vendsborg PB, Bech P, Rafaelsen OJ. Lithium treatment and weight gain. Acta Psychiatr Scand 1976; 53(2):139–147.
78. Calabrese JR, Bowden CL, Sachs G, et al. A placebo-controlled 18-month trial of lamotrigine and lithium maintenance treatment in recently depressed patients with bipolar I disorder. J Clin Psychiatry 2003; 64(9):1013–1024.
79. Breum L, Astrup A, Gram L, et al. Metabolic changes during treatment with valproate in humans: implication for untoward weight gain. Metabolism 1992; 41(6):666–670.
80. Joffe RT, Post RM, Uhde TW. Effect of carbamazepine on body weight in affectively ill patients. J Clin Psychiatry 1986; 47(6):313–314.
81. Swann AC. Major system toxicities and side effects of anticonvulsants. J Clin Psychiatry 2001; 62(suppl 14):16–21.
82. Sheth RD, Wesolowski CA, Jacob JC, et al. Effect of carbamazepine and valproate on bone mineral density. J Pediatr 1995; 127(2):256–262.
83. Feldkamp J, Becker A, Witte OW, Scharff D, Scherbaum WA Long-term anticonvulsant therapy leads to low bone mineral density—evidence for direct drug effects of phenytoin and carbamazepine on human osteoblast-like cells. Exp Clin Endocrinol Diabetes 2000; 108(1):37–43.
84. Plenge P, Rafaelsen OJ. Lithium effects on calcium, magnesium and phosphate in man: effects on balance, bone mineral content, faecal and urinary excretion. Acta Psychiatr Scand 1982; 66(5):361–373.
85. Haden ST, Stoll AL, McCormick S, Scott J, Fuleihan Ge-H. Alterations in parathyroid dynamics in lithium-treated subjects. J Clin Endocrinol Metab 1997; 82(9):2844–2848.

86. Zhang ZJ, Qiang L, Kang WH, et al. Differences in hypothyroidism between lithium-free and -treated patients with bipolar disorders. Life Sci 2006; 78(7):771–776.
87. Shulman KI, Sykora K, Gill SS, et al. New thyroxine treatment in older adults beginning lithium therapy: implications for clinical practice. Am J Geriatr Psychiatry 2005; 13(4):299–304.
88. Jeste DV, Gladsjo JA, Lindamer LA, Lacro JP. Medical comorbidity in schizophrenia. Schizophr Bull 1996; 22(3):413–430.
89. Lieberman AA, Coburn AF. The health of the chronically mentally ill: a review of the literature. Community Ment Health J 1986; 22(2):104–116.
90. Massad PM, West AN, Friedman MJ. Relationship between utilization of mental health and medical services in a VA hospital. Am J Psychiatry 1990; 147(4): 465–469.
91. Goldman LS. Medical illness in patients with schizophrenia. J Clin Psychiatry 1999; 60(suppl 21):10–15.
92. Cradock-O'Leary J, Young AS, Yano EM, Wang M, Lee ML. Use of general medical services by VA patients with psychiatric disorders. Psychiatr Serv 2002; 53(7): 874–878.
93. Ahrens B, Grof P, Moller HJ, Muller-Oerlinghausen B, Wolf T. Extended survival of patients on long-term lithium treatment. Can J Psychiatry 1995; 40(5):241–246.
94. Kallner G, Lindelius R, Petterson U, Stockman O, Tham A. Mortality in 497 patients with affective disorders attending a lithium clinic or after having left it. Pharmacopsychiatry 2000; 33(1):8–13.
95. Angst J, Angst F, Gerber-Werder R, Gamma A. Suicide in 406 mood-disorder patients with and without long-term medication: a 40 to 44 years' follow-up. Arch Suicide Res 2005; 9(3):279–300.
96. Druss BG, Rohrbaugh RM, Levinson CM, Rosenheck RA. Integrated medical care for patients with serious psychiatric illness: a randomized trial. Arch Gen Psychiatry 2001; 58(9):861–868.

22 Toward a Pathophysiology of Bipolar Disorders

John F. Neumaier and David L. Dunner
Department of Psychiatry and Behavioral Sciences, University of Washington, Seattle, Washington, U.S.A.

INTRODUCTION

Research findings regarding bipolar disorder (BD) continue to clarify the classification, mechanism of action, and treatment regarding these conditions. However, much research needs to be done in order to elucidate definitive methods of classification and more precise treatment approaches. We stand on the verge of the genomic period and its application to psychiatry and in particular our knowledge about BD. This exciting opportunity will largely allow for more precise classification and treatment of these conditions. However, these research advances will take some time. In this chapter, we review the recent advances in research involving BD and discuss continued gaps in our knowledge regarding these conditions. We also outline some areas for future research.

SUMMARY OF CLINICAL RESEARCH ADVANCES

Two major clinical areas of research should be noted. First of all, there is increasing attention to the concept of bipolar spectrum disorder (BSD), or "soft" BD (1–3). The relationship of patients with depression and mild and often times brief hypomanic periods is a major focus of research interest. Several review articles and research findings point toward a need to loosen the criteria for BD in DSM-V in order to expand the concept of bipolar II and include many individuals who are currently termed "bipolar disorder, not otherwise specified" as having true BD. Data supporting the inclusion of these "soft" bipolar cases as true bipolars are largely validated from family history studies.

Treatment studies of BD have recently pointed to the effects of lithium as an antisuicidal agent (4,5). Although not definitively demonstrated by direct placebo-controlled data, there is mounting evidence from a variety of research methodologies supporting the antisuicide effect of lithium. Since suicide is an unfortunate outcome in many patients with BD, the use of lithium to reduce suicidal behavior would seem to be an important clinical approach, and, hopefully, there will be an increased use of lithium treatment for patents with bipolar conditions. Other treatment studies have pointed toward the use of atypical neuroleptics not only for treatment of mania but more recently for treatment of the acute depressed phase of BD and perhaps for maintenance therapy to prevent recurrences of mania and depression in bipolar patients (6,7). Additional research on anticonvulsants, such as lamotrigine, has also added to our database regarding effective maintenance therapies (8).

Neuroimaging studies have continued to point toward some abnormalities, particularly in brain structure, of bipolar patients (9).

MAIN GAPS IN KNOWLEDGE AND METHODOLOGICAL DIFFICULTIES: BASIC SCIENCES

Despite remarkable advances in the treatment of BD over the last generation, we still have a limited understanding of its pathophysiology. We have no indication of what causes the underlying biochemical changes associated with the symptoms that we can observe and only clues about the mechanisms of action of the treatments that we already use. This is different than many other human conditions in which there is an understanding of at least some aspects of the pathophysiology. For example, myocardial infarctions occur when inadequate oxygen is delivered to the heart muscle, and treatments that change the metabolic supply and demand dynamics may be useful therapeutics. In this example, we also have a clear sense of the differences between causal and therapeutic mechanisms.

The problem of understanding the pathophysiology of BD has been tackled using both human- and animal-based strategies. In humans, genetic associations have been investigated using a variety of strategies and these are discussed elsewhere in this volume. However, genetic strategies are hampered by the possible mismatch between the timing of gene expression and manifestation of the illness. Critical genes may be involved at early developmental stages in the pathogenesis of BD and these may not be easily traced at later stages of adulthood (when symptoms manifest) or in postmortem analyses. While the methodology for analyzing gene expression and protein content in discrete brain regions has leaped forward in recent years, it is possible that even consistent changes detected in brains of affected individuals are not closely linked to the proximate genetic causes of bipolar illness. For example, altered expression of a transcription factor or other regulatory protein during brain maturation may lead to altered neuronal signaling or synaptic organization at a later point that in turn destabilizes mood or impacts decision making or other elements of behaviors associated with mania and depression. The biochemical changes involved in such later events may be critical to the manifestation of symptoms but may not be specific at all to BD. Ironically, time-tested histological methods may give quite sensitive clues implicating altered expression or activity of regulatory factors that guided neuronal development and circuitry wiring decades before the appearance of affective or cognitive symptoms. Some genes may contribute to a general susceptibility to one or more psychiatric disorders, while others may influence a vulnerability to deterioration over time, specific features shared with diseases such as schizophrenia or responsiveness to the therapeutic treatments (10). On the other hand, the investigation of adult brain neurochemistry may yield valuable clues to identify novel proteins or genes that can be targeted for treatment purposes, even if the targets are not closely linked to the original pathophysiology in individuals with BD.

A number of new treatments for BD have been developed in recent years, and it is a natural strategy to try to determine which neurochemical effects of these medications are critical for stabilizing mood. This strategy is certainly likely to lead to the identification of novel treatments, but it may not explain why BD occurs. Even if clues to the pathogenesis of BD are identified based on studies of clinically useful medications, tracing back from effective treatments to biochemical causes of BD will be difficult. For example, there has been continued progress in our understanding of the physiological effects of lithium. Recent evidence has focused on lithium inhibition of inositol monophosphatase leading to depletion of phosphotidyl inositol or inhibition of glycogen synthase kinase-3, a member of the Wnt signaling cascade (11); both of these observations are also reviewed in this volume. Others have suggested that mitochondrial dysfunction lies at the core of BD (12), but it is difficult to separate cause from effect since depressive and manic states may exact different metabolic demands on discrete brain circuits. In addition, the neuroprotective features of lithium have also received considerable attention and may bear on its long-term mechanism of action (13). Some investigators have reasoned that biochemical effects that are common to more than one mood stabilizer are most likely to be essential factors in mood stabilization; however, this presumes that different medications act on the same set of neurons to achieve their therapeutic effects, which is unlikely to be the case. For example, consider that both β-blockers and nitrates can alleviate angina, but they have unrelated molecular mechanisms and sites of action. Furthermore, this strategy may not detect treatments with novel mechanisms of action if different points within a functional neuronal circuit are involved. For example, dihydroxyphenylalanine (L-DOPA) and anticholinergics both reduce the symptoms of Parkinsonism but act at different points in the neural circuit. Similarly, atypical antipsychotics and anticonvulsants are effective in treating BD, yet it is possible that the most easily identifiable neurochemical effects of these medications are far removed from their eventual benefits in bipolar patients. Anticonvulsants can have quite immediate effects on the vulnerability to seizure activity by blocking sodium channels, but the therapeutic effects of these medications develop over weeks to months in BD. Similarly, many investigations have tried to discriminate between the immediate and delayed effects of lithium and other mood stabilizers; each time a new acute or sustained effect of these medications is discovered it raises the hope that a new class of improved medications will become possible. However, acute effects such as channel blockade may or may not be critical to the sustained effects of these drugs (such as changes in gene expression) involved in their ultimate mechanism of action.

We have recently learned from the example of the long and short polymorphisms in the serotonin transporter promoter (5HTTLPR) that genetics alone may not predict the risk for developing depression; Caspi and colleagues have elegantly demonstrated a critical interaction between genetic traits and environmental exposure to stress in modulating the risk for depressive symptoms (14). Similarly, it is well-established that stressful life experiences (15), drugs and alcohol (16), and potentially many other factors influence the onset, severity, and course of BD. Genes and environment are just as likely to interact to produce illness in BD as in major depression. We do not know enough about the mechanistic aspects of how stress influences mood episodes to say whether the stress effect is common or distinct between these disorders. Besides the similarity that stress can exacerbate the course of both unipolar depression and BD, antidepressants can treat the depressive features of both conditions. While antidepressants reliably reduce the risk for relapse and recurrences in unipolar depression, these drugs can aggravate the course of cycling in some bipolar individuals. This important difference suggests that the biological underpinnings of unipolar depression and BD are different, but we still do not know what brain regions and neural circuits account for this. Furthermore, some bipolar individuals seem much more vulnerable to antidepressant-induced cycling than others, and it is possible that this reflects different "subtypes" at least in terms of the vulnerability to changes in monoamine neurotransmission in mania and increased mood cycling.

Animal models have been very helpful in developing new antidepressants but have not led to new, validated treatments for BD. Even in schizophrenia, another disorder with uncertain etiology, animal behavior has been useful in identifying potential neuroleptics that led to a new generation of antipsychotic drugs (although this strategy led to the development of many similar medications, it took the clinical observation that clozapine was unusually effective to open up the conceptual constraints).

MAIN GAPS OF KNOWLEDGE AND METHODOLOGICAL DIFFICULTIES

Clinical Studies

There are four major target areas important in the treatment of bipolar patients: acute mania, acute depression, prevention of mania or hypomania, and prevention of depression. Of these targeted areas, treatment of acute mania is robustly studied with many types of treatments approved and/or shown to be effective. The same cannot be said, however, for acute bipolar depression. This area remains an important area for research since most patients with BD spend the majority of their ill time in the depressed phase. However, only one treatment (the combination of olanzapine and fluoxetine) is officially approved by the Food and Drug Administration (FDA) for bipolar I depression and no treatments are specifically approved for bipolar II depression. Treatment approaches for bipolar depression need to be further researched.

Maintenance studies have suggested that olanzapine and lamotrigine are effective for the prevention of recurrence of mania and depression in bipolar I patients. The difficulties in conducting such studies, however, suggest that further research into prevention of relapse may be limited.

Recently, the vagus nerve stimulator (VNS) has been approved in the United States for patients with treatment-resistant depression. Approximately 10% of the patients in the clinical trails using the VNS were bipolar, suggesting further research of VNS into BD specifically may be of interest. One such study is currently underway in rapid-cycling bipolar patients (17).

We previously have pointed to difficulties in placebo-controlled studies of bipolar I patients because of ethical issues regarding the use of placebo in such patients. We suggest that most patients with BD are not treated with monotherapy and it might be useful for the pharmaceutical industry to change their research designs to gain approval of "add-on" therapies. Knowledge of what works and what does not work for the various phases of BD will be quite important as increasing numbers of such patients are identified.

An important knowledge gap is the anticipated findings from genetic studies that would point to a better system for classification and perhaps a more direct approach toward pathophysiology of BDs and direct treatments of these conditions. Although genetic research in BD is certainly not stalled, it is interesting that major findings continue to be elusive (18). Increasing methodological advances that permit rapid identification of genetic markers for conditions will likely be important in ultimately elucidating the multiple genes thought to be responsible for bipolar conditions.

The clinical interface between neuroimaging techniques and BD also shows a good deal of promise. Various neuroimaging techniques including magnetic resonance imaging (MRI), CT scanning, magnetic resonance spectroscopy (MRS), and others are currently being investigated regarding bipolar patients in order to determine metabolic and structural changes associated with bipolar conditions.

The issue of diagnosis of mania in children is considerably problematic. It is our belief that this controversy will likely be resolved through validation of diagnosis likely from genetic studies. Until then, the issue of the existence and identification of children with BD, particularly mania, remains considerably problematic. Clinical classification studies and treatment studies will not seemingly solve the issue of the correct diagnosis that might be applied to these individuals.

PATHOPHYSIOLOGY OF BIPOLAR DISORDER, INTEGRATION OF AVAILABLE BASIC SCIENCE

Animal models of mental illnesses have always been strongly influenced by the prevailing conceptual frame of the era. Therefore, animal models that have been proposed have evolved from a focus on reflecting the behavioral states to responsiveness to lithium and now to perturbations of the molecular signaling pathways that are presumed to be involved in the disorder. Creating valid disease models is a risky balance of circular reasoning that cannot be easily avoided. BD, along with psychosis, has been particularly difficult to model using animal behavior. Some neurobehavioral disorders, such as seizures, are easier to model than complex behavioral disorders because the manifestations (e.g., epileptiform activity with motor seizure) appear quite similar across species; even depression and anxiety have been modeled using a variety of strategies that have demonstrated good construct and predictive validity (19). This may be because it is easier to interpret these behaviors objectively or because they are similarly present in a range of species. Most attempts to model BD have focused on an attempt to induce a depressive syndrome (these have been well validated) and then a hyperactive syndrome to capture the manic phase. One example is the use of amphetamine, which can induce hyperactivity, sleep disturbance, abnormal sensory gating, etc. in rodents during acute exposure followed by a dysphoric syndrome upon drug discontinuation. Similar symptoms are encountered when humans abuse stimulants, but this psychiatric presentation is very different from BD in many ways. This strategy may also be limited because the amphetamine may only model "downstream" behavioral aspects of mania or depression. Similarly, other hyperdopaminergic states, including the dopamine transporter knockout mouse line, have been considered to model aspects of BD and some of these behavioral effects are reversible by lithium (20). However, this rationale focuses on the mood and activity components (which dopamine may indeed mediate in part) but not the chronobiological aspect of BD. Sleep disturbance is a critical feature of both manic and depressive episodes and sleep deprivation may precipitate mania. There is a strong seasonal component in the cycling of some patients, and the spontaneous development of an acute manic or depressive episode without experimental provocation is an important component of a more valid animal model. Since circadian rhythms usually oscillate predictably in the brain, and lithium may regulate this (21), there is likely to be some link between dysfunction of these reliable biological clocks and the erratic fluctuations that characterize BD (22). It seems likely that this "lesion" lies upstream of dopaminergic function. Thus, animal behavioral models based on mimicking some of the behaviors associated with BD may be useful in developing novel mood-stabilizing treatments, but they are unlikely to capture the essential nature of BD.

Lithium has also been noted to have activity-dependent effects on cyclic AMP levels and other cellular effects (13,23); the idea that a mood stabilizer might have state-dependent effects on signaling mechanisms is appealing and might be a clue to improved animals models of BD. Thus, an improved animal model of bipolar illness would (*i*) show behaviors that reflect both depressive and manic phases in the same animal over time; (*ii*) mimic a range of mood, motivational, cognitive, and activity features of mania and depression; (*iii*) display spontaneous switches between phases as well as precipitated episodes following exposure to sleep deprivation, psychostimulants, antidepressants, etc; and (*iv*) demonstrate reduced behavioral symptoms after chronic administration of a variety of known mood stabilizers. While this is a tall order to fill, it provides caution that any animal models that address only some of these components must be judged carefully so as to avoid overinterpreting the generalized applicability of associated findings.

PATHOPHYSIOLOGY OF BIPOLAR DISORDER, INTEGRATION OF CLINICAL FINDINGS

Studies regarding the mechanism of action of lithium have continued in the hope that elucidation of a definitive mechanism might lead to the development of a distinct pathophysiology for BDs. There are similar studies regarding other treatments of BD, such as atypical antipsychotics and certain anticonvulsants, and it is hoped that knowledge of their mechanism of action might also lead toward the pathophysiology of BD disorder. However, at this point in time, it seems prudent to think of the future as being related to the development of the genetic findings for BD and a pathophysiology being derived from such findings. Treatment studies for BD continue, although, again, most of the attention for new medication development has been focused on major depression and, in the case of BD, on acute mania.

SUMMARY

We stand on the verge of very exciting findings that await the proper research studies. Increasing attention to BD research is likely to have tremendous benefit in terms of our understanding the pathophysiology of these conditions as well as identification of individuals with these conditions and more direct approaches to their treatment.

REFERENCES

1. Dunner DL. Clinical consequences of under-recognized bipolar spectrum disorder. Bipolar Disorders 2003; 5:456–463.
2. Angst J, Cassano G. The mood spectrum: improving the diagnosis of bipolar disorder. Bipolar Disorder 2005; 7(suppl 4):4–12.
3. Akiskal HS. Validating "hard" and "soft" phenotypes within the bipolar spectrum: continuity or discontinuity? J Affect Disord 2003; 73:1–5.
4. Tondo L, Jamison KR, Baldessarini RJ. Effects of lithium maintenance on suicidal behavior in major mood disorders. Ann NY Acad Sci 1997; 836:339–351.
5. Goodwin FK, Fireman B, Simon GE, Hunkeler EM, Lee J, Revicki D. Suicide risk in bipolar disorder during treatment with lithium and divalproex. JAMA 2003; 290:1467–1473.

6. Tohen M, Greil W, Calabrese JR, et al. Olanzapine versus lithium in the maintenance treatment of bipolar disorder; a 12-month, randomized, double-blind, controlled clinical trial. Am J Psychiatry 2005; 162:1281–1290.
7. Calabrese JR, Keck PE Jr, Macfadden W, et al. A randomized, double-blind, placebo-controlled trial of quetiapine in the treatment of bipolar I or II depression. Am J Psychiatry 2005; 162:1351–1360.
8. Goodwin GM, Bowden CL, Calabrese JR, et al. A pooled analysis of 2 placebo-controlled 18-month trials of lamotrigine and lithium maintenance in bipolar I disorder. J Clin Psychiatry 2004; 65:432–441.
9. Lyoo LK, Kim MJ, Stoll AL, et al. Frontal lobe gray matter density decreases in bipolar disorder. Biol Psychiatry 2004; 55:648–651.
10. Murray RM, Sham P, Van Os J, Zanelli J, Cannon M, McDonald C. A developmental model for similarities and dissimilarities between schizophrenia and bipolar disorder. Schizophr Res 2004; 71:405–416.
11. Harwood AJ. Lithium and bipolar mood disorder: the inositol-depletion hypothesis revisited. Mol Psychiatry 2005; 10:117–126.
12. Kato T. Mitochondrial dysfunction in bipolar disorder: from 31P-magnetic resonance spectroscopic findings to their molecular mechanisms. Int Rev Neurobiol 2005; 63:21–40.
13. Bachmann RF, Schloesser RJ, Gould TD, Manji HK. Mood stabilizers target cellular plasticity and resilience cascades: implications for the development of novel therapeutics. Mol Neurobiol 2005; 32:173–202.
14. Caspi A, Sugden K, Moffitt TE, et al. Influence of life stress on depression: moderation by a polymorphism in the 5-HTT gene. Science 2003; 301:386–389.
15. Paykel ES. Life events and affective disorders. Acta Psychiatr Scand Suppl 2003; 418:61–66.
16. Levin FR, Hennessy G. Bipolar disorder and substance abuse. Biol Psychiatry 2004; 56:738–748.
17. Marangell L, Suppes T, Zboyan H, et al. Vagus nerve stimulation for the treatment of rapid cycling disorder. Presented at the 44th Annual Meeting, ACNP, Waikoloa HI, Dec 11–15, 2005. Neuropsychopharmacology 2005; (suppl 1):S180.
18. Mathews CA, Reus VJ. Genetic linkage in bipolar disorder. CNS Spectr 2003; 8: 891–904.
19. Cryan JF, Holmes A. The ascent of mouse: advances in modelling human depression and anxiety. Nat Rev Drug Discov 2004; 4:775–790.
20. Beaulieu JM, Sotnikova TD, Yao WD, et al. Lithium antagonizes dopamine-dependent behaviors mediated by an AKT/glycogen synthase kinase 3 signaling cascade. Proc Natl Acad Sci USA 2004; 101:5099–5104.
21. Abe M, Herzog ED, Block GD. Lithium lengthens the circadian period of individual suprachiasmatic nucleus neurons. Neuroreport 2000; 11:3261–3264.
22. Mansour HA, Monk TH, Nimgaonkar VL. Circadian genes and bipolar disorder. Ann Med 2005; 37:196–205.
23. Beaty O 3rd, Collis MG, Shepherd JT. Action of lithium on the adrenergic nerve ending. J Pharmacol Exp Ther 1981; 218:309–317.

Index

Acetylcholine (ACh), 35
dopamine neurotransmission, modulating, 50
as a stress regulator, 80
Acquired treatment resistance, 312–313
lithium discontinuation-induced refractoriness, 313
types of, 312
Adenylyl cyclases (AC), 111
Adrenergic (monoaminergic)-cholinergic balance hypothesis, 67, 81
Adrenergic receptors, 35
Adrenocorticotropic hormone (ACTH), 91, 145, 191
Advanced sleep phase syndrome (ASPS), 202
Affective disorder patients, spectroscopic studies in, 75–80
acetycholine as stress regulator, 80
cholinomimetic drugs, 77–78
cardiovascular effects of, 77–78
changes in REM sleep in, 75–77
growth hormone supersensitivity, 78
hypothalamic-pituitary-adrenal axis supersensitivity, 78–80
pilocarpine, supersensitive pupillary responses to, 77
Affective psychosis, 213
Afterdischarges (ADs), 299
Aging, bipolar effects of, 365–370
brain morphology, 365–368
functional neuroimaging, 368
genetic variation, 365
laboratory measures, 365
neuroimaging, 365–368
structural neuroimaging studies in, 366–367
vulnerability and pathophysiology in, 365–370
Alphamethylparatyrosine (AMPT), 40
Alprazolam, 391
Alzheimer's disease, 73, 226
Amphetamine, 20, 41
behavioral excitation, 72
hyperactivity, 20
Amygdala, 150, 168
amygdala-kindled seizures, 309
Analysis of variance (ANOVA), 381
Animal studies, kindling/sensitization model in, 300–303
high-licking maternal behavior, 301
low-licking maternal behavior, 301
stress, effects on neurobiology and behavior, 300–301
stressors and substances of abuse, 302–303
Anorexia nervosa, 213
Anticholinergic drugs
mood effects of, 70–71
Anticonvulsants, 328
Antidepressants, 328
monotherapy, 7
neurotrophic and neuroprotective effects, 308
Antihypertensives, 109
Antimanic effects, of physostigmine, 69
Antipsychotic drugs, 393–394
Antipsychotics, atypical, 310, 381
Apoptosis/apoptotic cell death
in human hippocampus, 132–138
bipolars vs. controls, 135
FRET-based quantitative RT-PCR validation, 137
genes associated with, 140
proapoptotic factors, mRNA expression for, 133
schizophrenics vs. control, 134
Arginine vasopressin (AVP), 145
Atherosclerotic cardiovascular disease (ASCVD), 391
Attention deficit hyperactivity disorder (ADHD), 263, 344
Atypical antipsychotics, 310, 381

basal ganglia, 170
Bech Rafaelsen Mania Scale (BR-MAS), 12, 390
Behavioral sensitization, 24
Beta-endorphin secretion, 79
Bioassays, 23
Biological rhythm dysfunction, 189–193
  circadian abnormalities, evidence of, 190–192
  seasonality in bipolar disorder, 192–193
  timing of activity rest, irregularities in, 189–190
Bipolar Affective Disorder Dimension Scale (BADDS), 11
Bipolar disorders (BD), *see also* mood disorder
  animal models, 19–28
    assessment of, 11–12
    bioassays, 23
    endophenotypes, modeling of, 26–27
    problems and possible solutions, 22–27
    psychostimulant-induced models, 19–21
    variability, 25–26
  in bipolar disorder subtypes, 325–337
  bipolar spectrum concept, 9–11
  brain imaging studies in, 161–179
  catecholamine (CA) hypothesis of, 33–55
  cellular endophenotype for, 131–141
  in childhood and adolescence, 343–355
  classification, 1–14
    DSM-IV, 2–4
    ICD-10, 2–4
  cyclothymic disorder, 332–333
  depression, 5–6
  disorder patients, studies of, 96–98, *see also under* serotonergic dysfunction
  genetics of, 233–244
  hypomania, 4–5
  hypothalamic–pituitary–adrenal axis in, 145–154
  infectious etiology hypothesis in, 209–217
  mania, 4–5
  mixed states, 6–7
  modeling, 22
  neurocognitive findings in, 251–267
  physical comorbidity in, 387–396
  psychotic mania, 333–336
  sleep and biological rhythms abnormalities in, 189–203
  stressors in, 275–291
  subtypes, biologic factors in, 325–337
  treatment, pharmacological agents in
    action mechanism, 42–44
[Bipolar disorders (BD)]
  treatments for, 197–201
    mania, 200
    sleep–wake cycle manipulations, 197–200
Bipolar illness, 258
Bipolar remission, sleep in, 196–197
Bipolar spectrum concept, 9–11
  Akiskal's, 10
  boundaries of, 10
  objection, 10
Blood flow studies, 45–46
  and metabolism studies, 171–172
Blood-oxygen level dependent (BOLD) signal, 352
Body mass index (BMI), 390
Borderline personality disorder, 8
Borna disease virus (BDV), 214
Brain derived neurotrophic factor (BDNF), 239–240
  evolutionary conservation of, 239
Brain imaging studies, 161–179, *see also* magnetic resonance imaging
  of catecholamine system in BD, 45–47
    blood flow studies, 45–46
    genetic studies, 46–47
    neurochemical studies, 46
    structural brain imaging studies, 45
  functional neuroimaging, 171–178
  historical background, 161–162
  structural neuroimaging, 162–170
Brain-derived neurotrophic factor (BDNF), 51, 300, 308
Breed true, 241
Brief Psychiatric Rating Scale (BPRS), 335
Brodmann's areas, 164
Bunney-Hamburg Depression Scale, 69
Bupropion, 43
Buspirone, 391

Calcium calmodulin kinase-II (CaMK-II), 300
California verbal learning test (CVLT), 259
  for children (CVLT-C), 347
Cambridge neuropsychological test automated battery (CANTAB), 347
cAMP responding element binding protein (CREB), 113
Carbamazepine, 310, 312, 331, 394–395
  CAMP/PKA signaling cascade and, 115–116
  effect on HPA axis, 153
Cardiovascular disease (CVD), 387–389

Cardiovascular effects
of cholinomimetic drugs, 77–78
Catecholamine (CA) hypothesis of BD, 33–55
abnormalities, 38–40
in norepinephrine and its metabolites, 39–40
in serum, urine, and CSF levels and their tetabolites, 39
brain imaging of, 45–47, *see also under* brain imaging
CA tracts from midbrain to MRS, 37
challenge studies, 40–42
depletion studies, 40
stimulation studies, 41
depression and antidepressants action mechanism, 53
hormonal regulation of, 50
as modulators of MRC, 37–38
mood and, 36–38
mood regulation, 36–37
stress response, 36
neurochemistry and neurophysiology, 34
dopaminergic system, 34
reward mechanisms, 34
neuroendocrine challenge studies of, 41–42
dopamine receptor abnormalities, 41
neuroendocrine receptor abnormalities, 41–42
neurotransmission modulation and implications, 47–52
ACh modulation, 50
5HT modulation, 48
excitatory amino acids modulation, 48–49
GABA modulation, 49
hormonal modulation, 50
by intracellular factors, 50–51
NE modulation of DA neurotransmission, 47–48
neuropeptide modulation, 49–50
noradrenergic system, neurochemistry, 35–36
peripheral tissue, studies in, 43
postmortem studies in, 44–45
as translators of MRC activity, 38
Catechol-o-methyltransferase (COMT), 43, 215
Caucasian BD samples, 240
Cell membrane and signal transduction pathways, 109–123, *see also under* signaling networks
Cellular endophenotype, for bipolar disorder, 131–141
apoptosis in human hippocampus, 132–138
in the hippocampus, 139–141
Cellular machinery
extracellular signal-regulated kinase (ERK) pathway, 119–121
mood stabilizers and PI3K pathway, 118–119
neuroplastic events, 110–122
Central muscarinic regulation of cholinomimetic effects, 80–81
Cerebellum, 170
Cerebrospinal fluid (CSF), 89, 94, 96
Childhood and adolescence, bipolar disorder in, 343–355
broad phenotype diagnostic criteria, 345
cognitive and emotional processing, 346–350
in children, diagnosing, 343–345
narrow phenotype diagnostic criteria, 344
phenotyping system, 344
Choline, in bipolar patients, 75
Cholinergic 2 receptor (CHRM2) gene, 81
Cholinergic-muscarinic dysfunction in mood disorders, 67–82
affective disorder patients, spectroscopic studies in, 75–80
behavioral findings, 68–75
biological findings, 75
central muscarinic regulation of cholinomimetic effects, 80–81
centrally active anticholinergic drugs, 70–71
choline in bipolar patients, 75
cholinomimetic agents, depressive effects of, 72–75
cholinomimetic effects on manic symptoms, 68–69
cholinomimetic-catecholaminergic interactions, 71–72
historical overview and highlights, 67–68
manic symptoms rebound, 69–70
marijuana-physostigmine interactions, 71
muscarinic receptor gene and binding alterations, 81
therapeutic implications, 81–82
Cholinomimetic agents
administration, 69–70
cardiovascular effects of, 77–78
central muscarinic regulation of, 80–81
depressive effects of, 72–75
manic symptoms and, 68–69

Chromosome studies, 235–236
Chronic desipramine binding, 48
Chronic lithium treatment, 175
Chronic psychostimulant treatment, 20
Chronicity, 363
Cingulate hypermetabolism, 171
Cingulated, in bipolar disorder, 166
Circadian abnormalities, evidence of, 190–192
Circadian phase advances, 199–200
Clinical Antipsychotic Trials of Intervention Effectiveness (CATIE), 393
Clinical Global Impressions (CGI), 335
  Clinical Global Impressions scale, Bipolar Version (CGI-BP), 12
Clonidine, 43
Clozapine, in psychotic mania treatment, 335
Cognition, *see also* neurocognitive findings
  attention-emotion interactions and emotion recognition, 348–350
  in childhood and adolescence bipolar disorder, 346–350
  memory, learning, and attention, 347
  response flexibility and motor inhibition, 347–348
  taxonomy of, 252
Cognitive dysfunctions, 363
Cognitive ERP techniques, 228
Combination therapy, 310–311
Comorbid psychiatric illnesses, 263–264
Comorbid substance abuse, 363
Composite International Diagnostic Interview (CIDI), 11
Continuous performance task (CPT), 253, 347
  identical pairs version (CPT-IP), 264
Core haplotype, 242
Coronary heart disease (CHD), 387
Corpus callosum, 167–168
Cortical–subcortical interaction, 55
Corticotropin releasing hormone (CRH), 145, 191
Cortisol-binding globulin levels, 149
Creatine-phosphocreatine (Cr-PCr), 327
Cushing's syndrome, 149, 151
Cyclic AMP (cAMP) signaling cascade and PKA, 113–115
  carbamazepine and, 115–116
  lithium and, 115
Cyclicity in bipolar illness, 193
Cycloid psychosis, 8
Cyclothymia, 1, 3, 8, 326, 332–333
  biology, 332
  epidemiology, 332
  treatment, 333

D-amino acid oxidase (DAO), 239
D-amino acid oxidase activator (DAOA), 239
Declarative memory, 258
Delirium, 362
Dementia, 362–363
Depression, 5–6
  depressed mood, 255
  depressive disorder NOS, 3
  stressors in, 277
Desipramine treatment, 49
Dex/CRH test, 147, 152
Dexamethasone, 147
  metabolism, 152
  suppression test (DST), 146, 288
Diabetes mellitus (DM), 391
Diagnostic analysis of nonverbal accuracy scale (DANVA), 349
Diagnostic and Statistical Manual of Mental Disorders, see under DSM
Diagnostic interview schedule for children (DISC), 346
Disrupted in schizophrenia 1 (DISC-1), 236
Divalproex, 394
  in psychotic mania treatment, 335
Dopaminergic system, 173–174
  neurochemistry and neurophysiology, 34
Dopamines, 199
  ACh modulation of, 50
  agonists, 109
  blockers, 19
  receptor abnormalities, 41
  transporter (DAT), 41
  β-hydroxylase (DBH), 43
Dorsolateral prefrontal cortex (DLPFC), 163, 263, 351
DSM-IV (*Diagnostic and Statistical Manual of Mental Disorders*-IV), 2–4
  classifying mood disorders, 3, *see also under individual entries*
  defining bipolar II disorder, 2
  describing mood disorders, 2, *see also under* mood disorders
DSM-V, 13–14
Dyskinesia, 73

Dysphoric mania, 7
Dysthymia, 1, 3

Elation/grandiosity scores, 69, 71
Elderly, bipolar, 365–370
  brain morphology, 365–368
  functional neuroimaging, 368
  genetic variation, 365
  laboratory measures, 365
  neuroimaging, 365–368
  structural neuroimaging studies in, 366–367
  vulnerability and pathophysiology in, 365–370
Electroencephalographic (EEGs) and event-related potentials (ERPs) in bipolar disorders, 221–229
  clinical EEG abnormalities, 223–224
  cognitive ERP techniques, 228
  conventional ERP abnormalities, 224–225
  future opportunities, 225–228
  lithium effect on, 224
  normal EEG, 221–223
  pharmaco-EEG techniques, 227
  quantitative EEG techniques, 226–227
Endocrine diseases, 391–396
  barriers to care, 395–396
  menstrual abnormalities and polycystic ovarian syndrome, 391–392
  psychotropic drugs, 393–395
  thyroid abnormalities, 392–393
Endophenotypes, 21
  modeling, 26–27
Environmental risk factors, 213–215
Environmental stressors, 275–291
Enzyme studies, 43
Epidemiological Catchment Area (ECA) study, 10, 234
Episode onset, 277–284
  Hammen's interview for, 283
  life events and, relationship between, 279–282
  non–rapid-cycling episode patterns, 285
  stress–episode relationship, 285–287
Episodes, mood, 2
Epstein-Barr virus, 211–212
Euthymia, 258
Evoked response potentials (ERPs), 349
Excitatory amino acids (EAAs), 54
Expressed emotion (EE), 283
Extracellular signal regulated kinase (ERK), 119–121, 379
  antidepressant effects on, 120
  in CNS, 120

5-HT_{2A} receptor, 91
  binding in platelets, 96
5-Hydroxytryptamine (5-HT), 35
5-Hydroxyindole-acetic acid (5-HIAA), 89
[^{18}F]altanserin, 174
[^{18}F]setoperone, 174–175
Fahr's disease, 38
Fenfluramine, 91
Forskolin-induced hypoactivity
  lithium inhibition of, 23
Free thyroxine (FT4) index, 392
Functional neuroimaging, 171–178
  activation studies, 172–173
  in bipolar elders, 368
    electrophysiological studies, 368
    endophenotypes, 369
    neuroendocrine and neurochemical studies, 368–369
    pathophysiology, 369–370
  blood flow and metabolism studies, 171–172
  neurochemical brain imaging, 173–178

GAD65, 131
GAD67, 131
Gamma aminobutyric acid (GABA), 35, 80
  GABA modulation of CA neurotransmission, 49
  GABAergic interneurons, 131
Gene-driven therapeutics, 314
Genes and bipolar disorder, 201–202
Genetics, of bipolar disorder, 46–47, 233–244
  bipolar disorder and schizophrenia, 241–243
  chromosome studies, 235–236
  classical genetic epidemiology, 234–235
  clinical phenotype, 241
  diagnosis and epidemiology, 233–234
  in microbial infection, 209–212
  mode of inheritance, 235
  molecular genetic studies, 236–240
GenMapp algorithms, 133
Glucocorticoids, 150
Glutamate, glutamine, and GABA, 176
Glyceraldehyde-3-phosphate dehydrogenase (G3PDH), 137
Glycogen synthase kinase-3 (GSK-3), 121–122
  glycogen synthase kinase-3 beta (GSK3-β), 202
  inhibition mechanisms, 121
  subtypes, 121

G-proteins, 111–112
abnormalities in bipolar disorder, 112–113
G-protein receptor kinase 3 (GRK3), 240
G-protein-coupled-receptors (GPCRs), 111
lithium and, 112–113
small G proteins, 111
Great Smoky Mountains Study (GSMS), 346
Growth hormone supersensitivity, 78
Guanine nucleotide exchange factors (GEFs), 111
Guanosine diphosphate (GDP), 111
Guanosine triphosphate (GTP), 111
Guanylyl cyclases (GC), 111

[^{3}H]cyanoimipramine, 92
[^{3}H]imipramine, 90, 92, 96
[^{3}H]paroxetine, 90
Haemophilus influenzae, 211
Hamilton Rating Scale for Depression (HAMD), 12, 332, 390
Helicobacter pylori, 211
High-licking maternal behavior, 301
Hippocampal acetylcholine, 80
Hippocampus, 150, 169
Holmes-Rahe Social Readjustment scale, 278
Hormonal modulation of catecholamines, 50
HPA axis abnormalities, 152
Huntington's disease, 38
Hyperandrogenism, 392
Hypercortisolemia, 147, 150–152
Hyperparathyroidism, 395
Hyperprolactinemia, 394
Hypomania, 3–5
antidepressant-induced, 7–8
hypomanic episode, 3, 4
mixed hypomania, 7
Hypothalamic–pituitary–adrenal (HPA) axis, 145–154, 392
abnormalities, 147
activity, 288
carbamazepine, 153
dysregulation, 191
functional integrity, 146
hypercortisolemia, 150–152
lithium, 152
patients with bipolar disorder, 147–149
patients with major depression, 146–147
patients with mood disorders, 149–150
physiology, 145–146
sodium valproate, 153
somatic treatments, 152–153
supersensitivity, 78–80

ICD-10, 2–4, 11
Idazoxan, 43
Illness prognosis
homologous models, 298–299
kindling as a nonhomologous model, 299–300
Impairment, 253–254
Indoleamine hypothesis of depression, 89
Infectious etiology hypothesis, 209–217
environmental risk factors, 213–215
genetic determinants, 210
immunological activity, 210
in microbial infection, 209–212
microbial receptors, 210
specific infectious agents, 214–215
viral etiology, 213–214
viral infections of the CNS and bipolar disorder, 212
Inhibitory syndrome, 73
Inositol depletion, 23, 116
Inositol monophosphatase, 176
Intellectual functioning, 263
impairment, 263
mood state, effect of, 263
Intensive clinical management (ICM), 290
International Classification of Diseases-9 (ICD), 2
Interpersonal and Social Rhythm Therapy (IPSRT), 290
Interview-based life event assessment methods, 278–280
Inventory of Depressive Symptomatology (IDS), 12

Janowsky-Davis Activation-Inhibition Scale, 71, 73–74

Kiddie schedule for affective disorders and schizophrenia for school-age children (K-SADS), 346
Kindling/sensitization model, 285–287, 297–315
animal studies, 300–303,
course of illness and neuroimaging studies, 303–307
age of onset and number of episodes, 307
neurochemistry, 304–307
stressor involvement, 304
historical overview, 297–300
homologous models for affective illness prognosis, 298–299
mood stabilizers, neuroprotective effects of, 313–314

[Kindling/sensitization model]
as a nonhomologous model for illness progression, 299–300
prediction of treatment response, 314
psychopharmacological interventions, 307–314
acquired treatment resistance, 312–313
combination therapy, 310–311
remission, 311–312
Kraepelinian dichotomy, 8, 241

Lamotrigine, 310, 331, 380, 394
Late life bipolar disorders, *see also under* elderly
biological factors, 361–372
pharmacodynamics, 370–371
pharmacokinetic factors, 370
side effects/tolerability, 370–371
therapeutics, 370–371
causal factors, 363–365
familial influences, 364
vascular disease and stroke, 364
clinical overview, 362–365
chronicity, relapse, recurrence, 363
cognitive dysfunctions and dementias, 363
comorbid substance abuse, 363
course, 363
mortality, 363–364
epidemiology, 362–365
historical perspective, 361–362
novel treatment strategies, 371
psychosocial adversity, 365
Laterodorsal tegmental (LDT), 194
Learning and memory 258–260, *see also* verbal learning and memory
Life Events and Difficulties Schedule (LEDS), 277
Limbic-hypothalamo-pituitary-adrenocortical (LHPA), 27
Linkage studies, 236–237, 242
Lithium, 19, 312, 328, 330, 394
as antidepressant, 21–22
antimanic effects, 43
blocking hyperactivity, 20
EEG and, 224
HPA axis, 152
G proteins, 112–113
inhibiting forskolin-induced hypoactivity, 23
lithium discontinuation-induced refractoriness, 313
[Lithium]
lithium-induced myo-inositol reductions, 116
potential neurotrophic and neuroprotective effects, 308
psychotic mania treatment, 335
rapid cycling treatment, 330
type II bipolar disorder treatment, 328
Long-term potentiation (LTP), 300
Longitudinal sleep studies, 196
Low resolution electromagnetic tomographic analysis (LORETA), 227
Low-licking maternal behavior, 301
Lysergic acid diethylamide (LSD), 96

MADRAS depression scale, 70
Magic bullets, 19
Magnetic resonance imaging (MRI), 161
Magnetic resonance spectroscopy (MRS), 175
Magneto-encephalography (MEG), 221
Major depressive disorder (MDD), 3, 90
Malaria parasites, 210
Mammalian circadian rhythms, 191
Mania, 4–5, *see also* hypomania
antidepressant-induced, 7–8
cholinomimetic effects on, 68–69
"classic" mania, 4
dysphoric mania, 7
manic disorder, 3
Manic Intensity Scale scores, 69, 71
manic syndrome, 4
medication for, 200–201
serotonergic function in, 94–96
cerebrospinal fluid, 94
challenge studies, 95
imaging studies, 95
medication studies, 95–96
platelet studies, 94–95
stressors, 277
treating and preventing, 200
MARCKS (myristoylated alanine rich C kinase substrate), 117–118
Marijuana-physostigmine interactions, 71
Matrix reasoning, 261
Medial temporal structures, 168–169
Memory trace, 299
Menstrual abnormalities, 391–392
Mesocorticolimbic (MCL) system, 34, 39
Methamphetamine, 41
Methodological difficulties, 402–405
Microbial infection, genetic factors, 209–212

Mineralocorticoid receptors, 150
Mini International Neuropsychiatric Interview (MINI), 11
Mismatch negativity (MMN), 223
Mixed depression, 9
Mixed disorder, 3
Mixed states, 6–7
Molecular genetic studies, 236–240
  brain derived neurotrophic factor, 239–240
  D-amino acid oxidase activator (G72)/G30 locus, 239
  functional candidates, 238–239
  gene studies, 237–240
  linkage studies, 236–237
Monoamine oxidase (MAO), 43
Monoaminergic neurotransmitter systems, 109
Montgomery Åsberg Depression Rating Scale (MADRS), 12, 70, 382
Mood disorders,
  brief history, 1–2
  catecholamine and, 36–38
  cholinergic-muscarinic dysfunction in, 67–82, episodes, 2
  genetic studies, 234
  serotonergic dysfunction in, 89–98
  specifiers, 2
Mood stabilizers, 394–395
  neuroprotective effects, 313–314
  PI3K pathway and, 118–119
Mood state, effect of, 254
Mood-regulating circuit (MRC), 36
  catecholamines
    tracts from midbrain to MRS, 37
    as modulators, 37–38
    as translators of MRC activity, 38
Mortality, 363–364
Muscarinic receptor gene, 81
Mycobacteria tuberculosis, 211
Myo-inositol (mI), 176, 353

N-acetylaspartate (NAA), 175
N-methyl-D-aspartate (NMDA), 48, 225
Neocortex, 263
Neuregulin 1, 242
Neurochemical brain imaging, 173–178
  PET and SPECT neuroreceptor studies, 173–178
    biochemical findings, 175–178
    dopaminergic system, 173–174
    serotonergic system, 174–175
    in unipolar disorder, 177–178
Neurocognitive findings in bipolar disorder, 251–267
  attention/vigilance, 253–254
  clinical course, 254, 255
  impairment, 253–257
  intellectual functioning, 263
  learning and memory, 258–260
  mood state, 254, 255
  neuropsychological findings, 266
    attention, 266
    intellectual functioning, 266
    reasoning and problem solving, 266
    verbal learning and memory, 266
    visual learning and memory, 266
    working memory, 266
  reasoning and problem solving, 260–263
  speed of processing, 254–255, 266
  working memory, 255–258
Neuroendocrine challenge studies
  of CA receptor function in BD, 41–42, *see also under* catecholamines
  depletion studies and, 91
Neuroendocrine receptor abnormalities, 41–42
Neuroimaging
  in bipolar elders, 365
  neuroimaging, structural, *see under* structural imaging
  in pediatric bipolar disorder, 350, *see also under* pediatric bipolar disorder
Neurokinin (NK)-1-receptor antagonists, 380
Neuropeptide modulation of CA neurotransmission, 49–50
Neuropeptide Y (NPY), 49
Neurotrophins, 239
  neurotrophin-3 (NT-3), 51
NIMH-Life Chart Methodology (NIMH-LCM), 12
Non–rapid-cycling episode patterns, 285
Noradrenergic (NA) system, neurochemistry, 35–36
Norepinephrine
  catecholamine abnormalities, 39–40
  in reward mechanisms, 35–36
Nutritional and metabolic diseases, 390–391
  atherosclerotic cardiovascular disease (ASCVD), 391
  diabetes mellitus (DM), 390–391
  metabolic syndrome (MS), 391

Obesity, 390
Olanzapine, 163, 394
for psychotic mania, 335–336
for rapid cycling, 331–332
Opioid, 35

Paired Associates Learning task, 260
Parachlorophenylalanine (PCPA), 91
Parkinson's disease, 50, 70, 199
Paroxetine, 264
Pattern recognition test, 257
Pediatric bipolar disorder
functional neuroimaging in, 352–354
neurochemistry, 353–354
neuroimaging, 350
prevalence, 346
structural neuroimaging in, 350–352
Pedunculopontine tegmental (PPT), 194
Peripheral cholinergic supersensitivity, 68
PET and SPECT neuroreceptor studies, 173–178
Pharmaco-EEG techniques, 227
Pharmacological agents, in BD treatment, 42–44
dopamine system, medication effects on, 42
enzyme studies, 43
neuroendocrine system, medication effects on, 42–43
in peripheral tissue, 43
receptor studies, 43–44
signal transduction mechanisms, 44
Pharmacological interventions 377–383
current status of therapeutics, 377–383
new drug development, 380–383
anxiety, 382
candidate drugs, 381–382
rating scale issues, 380
regulatory agencies, 382–383
study of early behavioral effects, 381
systematic open trials, 380–381
use of time to event analyses, 381
targets for new treatment development, 378–379
Phenotype diagnostic criteria, 344
Phosphatidylinositol 4,5-bisphosphate [PI(4,5)P_2], 116
Phosphatidylinositol-3-kinase (PI3K) signaling pathway, 118
Phosphodiesterases (PDE), 111
Phosphoinositide 3-kinases (PI3Ks), 111
Phosphoinositide signaling cascades and protein kinase C (PKC), 116–118
Phosphoinositide-dependent kinase 1 (PDK1), 118
Phospholipase C (PLC), 111, 116
Phospholipases A2 (PLA2), 111
Phosphomonoester (PME) level, 176
Physical comorbidity in bipolar disorder, 387–396
cardiovascular disease, 387–389
endocrine diseases, 391–396
nutritional and metabolic diseases, 390–391
Physostigmine, 69, 78–79
anergic effects, 79
Pilocarpine, 68
supersensitive pupillary responses, 77
Platelet studies, 90–91, 94–95
Polycystic ovarian syndrome (PCOS), 391–392
Polyphosphate-1-phosphatase, 176
Porsolt forced swim test, 22
Positron emission tomography (PET), 93, 171, 221
Post-traumatic stress disorder, 26
Postmortem studies in bipolar disorder, 44–45
Post's kindling model, 286
Prefrontal cortex, 163–167
cingulate in bipolar disorder, 166
subgenual prefrontal cortex, 165
Probenecid blocks, 94
Profile of Mood States Scale, 74
Proliferator-activated receptors (PPAR), 380
Protein kinase C (PKC), 379
phosphoinositide signaling cascades and, 116–118
Proton echoplanar spectroscopic imaging (PEPSI), 49
Psychostimulants, 109
psychostimulant-induced models, 19–21
lithium effects, 19
Psychotic illness, bipolar disorder and, 52
Psychotic mania, 326, 333–336
biology, 334–335
epidemiology, 333
treatment, 335–336
clozapine, 335
divalproex, 335
lithium, 335
olanzapine, 335–336
quetiapine, 336
risperidone, 336

Psychotropic drugs, 393–395
  antipsychotic drugs, 393–394
  mood stabilizers, 394–395
Psychotropic medications, effects of, 264–265

Quetiapine, 393, 394
  in psychotic mania treatment, 336

Rapid cycling, 9, 14, 326, 328–332
  biology, 330
  epidemiology, 329–330
  prevalence of, 329
  treatment, 330–332
    carbamazepine, 331
    comparisons of treatments, 332
    lamotrigine, 331
    lithium, 330
    olanzapine, 331–332
    valproate, 331
Rapid eye movement (REM), 67, 75–77, 193
Rating scales, 380
Reasoning and problem solving, 260–263
  clinical course, impact of, 262–263
  impairment, evidence for, 262
  Matrix Reasoning, 261
  mood state, effect of, 262
  Raven's Progressive Matrices, 261
  Wisconsin Card Sorting Test (WCST), 261
Receptor studies, 43–44
Recurrence or relapse, 363
Remission, achieving and maintaining, 311–312
Research diagnostic criteria (RDC), 346
Reward mechanisms
  dopamine, 34
  norepinephrine, 35–36
Rey-Osterrieth complex figure test (ROCFT), 257, 260
Rhythm stabilization, 200
Risperidone, 393–394
  in psychotic mania treatment, 336

Schedules for Clinical Assessment in Neuropsychiatry (SCAN), 11
Schistosoma mansoni, 210
Schizoaffective disorder, 3, 8, 52, 241
Schizophrenia/schizophrenics, 131, 213, 226
  bipolar disorder and, 241–243
    family studies, 241
[Schizophrenia/schizophrenics
  bipolar disorder and]
    gene studies, 242–243
    linkage studies, 242
    twin studies, 241–242
  core haplotype, 242
  Disrupted in Schizophrenia 1, 242
  Neuregulin 1, 242
Scopolamine, 70, 81
Seasonal affective disorder (SAD), 193
Seasonality in bipolar disorder, 192–193
Selective serotonin reuptake inhibitors (SSRIs), 42, 198, 391
Sensitization
  kindling and behavioral, 285–287
    episode sensitization, 285
    stressor sensitization, 285
Serotonergic dysfunction in mood disorders, 89–98
  cerebrospinal fluid studies (CFS), 96
  challenge studies, 97
  genetic studies, 97–98
  historical view, 89
  in major depression, 90–91
    cerebrospinal fluid (CSF), 89
    neuroendocrine challenge and depletion studies, 91
      platelet studies, 90–91
      postmortem studies, 91
  platelet studies, 96–97
  postmortem studies, 97
  serotonergic function in mania, 94–96
  serotonin and 5-HIAA, 91–94
    5-HT_{2A} receptor, 92–93
    HT_{1A} receptor, 92
    imaging studies, 93–94
    serotonin transporter, 92
Serotonergic system, 174–175
Serotoninergic axons, 152
Sertraline, 264
Severe mood and behavioral dysregulation (SMD) diagnostic criteria, 345
  exclusion criteria, 345
  inclusion criteria, 345
Signal transduction mechanisms/ pathways, 44
  cell membrane, 109–123
Signaling networks, 110–122
  carbamazepine and CAMP/PKA signaling cascade, 115–116

[Signaling networks]
cyclic AMP (cAMP) signaling cascade and PKA, 113–115
extracellular signal-regulated kinase (ERK) pathway, 119–121
glycogen synthase kinase-3 (GSK-3), 121–122
G-proteins, 111–112
mood stabilizers and PI3K pathway, 118–119
phosphatidylinositol-3-kinase (PI3K) signaling pathway, 118
phosphoinositide signaling cascades, 116–118
Simultaneous and delayed matching to sample task, 257
Single manic disorder, 3
Sleep abnormalities, 193–197
biological rhythms abnormalities and, 189–203
in bipolar depression, 194–195
in bipolar mania, 195
genes and bipolar disorder, 201–202
longitudinal studies, 196
sleep and circadian rhythms, 200–201
biological rhythm dysfunction, 189–193
sleep deprivation, 197–198
mechanisms, 198–199
sleep in bipolar remission, 196–197
sleep regulation, 194
Slow processing speed, 255
Slow wave sleep (SWS), 193
Social rhythm disruption (SRD), 289–290
Sodium valproate, 176
effect on HPA axis, 153
Somatic treatments, effects of, 152–153
Somatostatin, 49
Span of apprehension task (SPAN), 253
Spatial recognition test, 257
specific serotonin receptor inhibitor (SSRIs), 380
Specifiers, mood, 2
longitudinal course specifiers, 2
Speed of processing, 254–255
Stanley Foundation Bipolar Network, 6
Stress and episode sensitization, 298–299
Stressful life events, 190
Stressors in bipolar disorder, 275–291
depression and mania, 277
episode onset, 277–284
HPA axis dysfunction and bipolar episodes, 288–289
methodological issues, 276–277
[Stressors in bipolar disorder]
Post's hypotheses, 286
recovery from an episode, 284
social rhythm disruption and bipolar episodes, 289–290
stress–episode relationship, 285–287
Stroop color and word tests (SCWT), 253
Structural neuroimaging, 162–170
cerebellum, 170
corpus callosum, 167–168
medial temporal structures, 168–169
in pediatric bipolar disorder, 350–352
prefrontal cortex, 163–167
thalamus, 169–170
basal ganglia, 170
total brain volume, 162–163
white matter, 167–168
Subtypes of bipolar disorder, 325–337
bipolar I, 326
bipolar II, 326-328
cyclothymic disorder, 326
psychotic mania, 326
rapid cycling, 326
biology of subtypes, 327
epidemiology of subtypes, 326
treatment, 328
Supersensitive pupillary responses to pilocarpine, 77
Suprachiasmatic nucleus (SCN), 191
Systematic Treatment Enhancement Program (STEP), 329

Temporal-hippocampal system, 258
Thalamus, 169–170
3-methoxy-4-hydroxyphenylglycol (MHPG), 39, 72
Thyroid abnormalities, 392–393
Thyrotropin-releasing hormone (TRH), 392
Toxoplasma gondii, 215
Treatment-discontinuation-related refractoriness, 312
Tryptophan hydroxylase (TPH), 91, 97
Twin studies, 241–242

Unipolar depression, 90, 234
vs. bipolar depression and, 2, 51
Unipolar major depression (UMD), 361
Unipolar–bipolar dichotomy, 8

Vagus nerve stimulator (VNS), 404
Valproate, 310, 312, 331
neurotrophic and neuroprotective effects, 308

Valproic acid, 394–395
Vascular disease and stroke, 364
Velocardiofacial syndrome (VCFS), 43, 236
Venlafaxine, 120
Ventral prefrontal cortex (VPFC), 347
Ventral tegmental area (VTA), 34, 309
Verbal declarative memory impairments, 258
Verbal learning and memory
  clinical course, impact of, 259–260
  impairment, evidence for, 258–260
  mood state, effect of, 259–260
Vermis area V3, 170
Viral etiology in bipolar disorder, 213–214
Viral infections of the CNS and bipolar disorder, 212
Visual backward masking, 253
Visual spatial memory tasks, 257
  pattern recognition test, 257
  spatial recognition test, 257
    pattern recognition memory task, 257
    Rey-Osterreith complex figure test, 257
    simultaneous and delayed matching to sample task, 257
[Visual spatial memory tasks
  spatial recognition test]
    visual-spatial paired associate learning task, 257
Voxel-based morphometry (VBM), 163, 352

Wechsler Adult Intelligence Scale version III (WAIS-III), 255
White matter, 167–168
Wisconsin Card Sorting Test (WCST), 261, 263
Working memory, 255–258
  clinical course, impact of, 257–258
  conceptualization, 256
  impairment, 256–257
  mood state, effect of, 257
  tripartite organization of, 256

X-linkage, 235

Young Mania Rating Scale (YMRS), 12, 332, 335

Ziprasidone, 393